Conference on Challenges for Development in Mining Science and Mining Industry 2018

IOP Conference Series: Earth and Environmental Science
Volume 262

Novosibirsk, Russia
1 - 5 October 2018

Editor:

Mikhail Kurlenya

ISBN: 978-1-5108-8857-9
ISSN: 1755-1307

Printed from e-media with permission by:

Curran Associates, Inc.
57 Morehouse Lane
Red Hook, NY 12571

Some format issues inherent in the e-media version may also appear in this print version.

Institute of Physics
Dirac House, Temple Back
Bristol BS1 6BE UK

Phone: 44 1 17 929 7481
Fax: 44 1 17 920 0979

techtracking@iop.org

Additional copies of this publication are available from:

Curran Associates, Inc.
57 Morehouse Lane
Red Hook, NY 12571 USA
Phone: 845-758-0400
Fax: 845-758-2633
Email: curran@proceedings.com
Web: www.proceedings.com

TABLE OF CONTENTS

IOP Conf. Series: Earth and Environmental Science **262** (2019) 011001 doi:10.1088/1755-1315/262/1/011001

2018 Conference Proceedings

Challenges for Development in Mining Science and Mining Industry

October 1–5, 2018

Novosibirsk, Russia

It is a pleasure for us to have this opportunity and offer our foreign readers the selected papers presented at the 2018 Conference on Challenges for Development in Mining Science and Mining Industry.

The Conference has been hosted by the Institute of Mining of the Siberian Branch of the Russian Academy of Sciences under auspices of the Presidium of the Siberian Branch RAS and the Scientific Council on Mining of the Russian Academy of Sciences.

The Conference provided a platform for the discussion of basic and applied aspects of mining sciences, with emphasis laid on state-of-the-art, main achievements and development trends in the mining industry: improved control in R&D projects of integrated mineral mining based on the principles of partnership; systems of diagnostics and monitoring of geomechanical behavior of rock in geotechnologies; umbrella programs of environmental protection in the course of mining.

Mikhail Kurlenya

Academician RAS

Editor-in-Chief

IOP Conf. Series: Earth and Environmental Science **262** (2019) 011002 doi:10.1088/1755-1315/262/1/011002

Peer review statement

All papers published in this volume of *IOP Conference Series: Earth and Environmental Science* have been peer reviewed through processes administered by the proceedings Editors. Reviews were conducted by expert referees to the professional and scientific standards expected of a proceedings journal published by IOP Publishing.

IOP Conf. Series: Earth and Environmental Science **262** (2019) 012001 doi:10.1088/1755-1315/262/1/012001

Lever-type percussion mechanisms for removal of adhered deposits from holding reservoirs

ES Abdraimov, BB Bakirov* and MI Shadiev

Abdraimov Research Center for Problems of Machine Engineering, National Academy of Sciences, Bishkek, Kyrgyz Republic

E-mail:*Bakirov.57@list.ru

Abstract. The authors discuss designs of Abdraimov's lever-type percussion mechanism with the longest piston rod for removal adhered deposits from ash-and-slag pipelines and hoppers. The kinematic and kinestatic test results are presented for three layouts of Abdraimov's mechanism with the longest base. The basic parameters of the layouts for high-frequency percussion plants for cleaning holding reservoirs and surfaces are determined and recommended.

1. Introduction

Modern ecological setting dictates mining, oil processing, energy generating and other industries to introduce and apply advanced technologies ensuring no-waste removal of coal and oil contamination (subject to low cost cleaning).

When thermal station burn low-quality solid fuel (with low combustion heat and high ash content), much ash left requires disposal. Currently most power plants remove ash and slag by hydraulic method and store in ash-disposal areas. One tenth of as is only used economically—in manufacturing of building materials, in construction and agriculture.

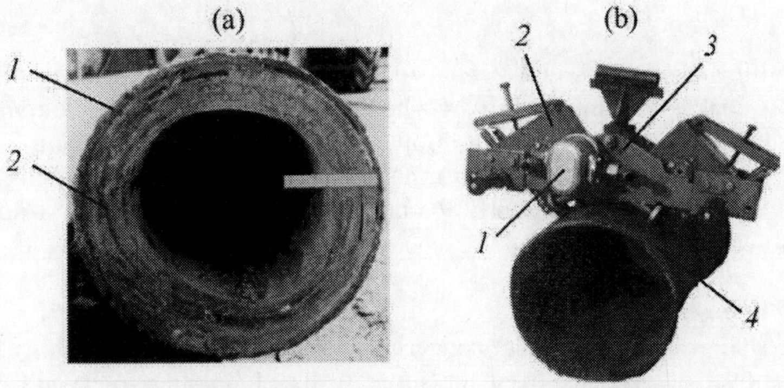

Figure 1. Ash-and-slag pipeline and portable machine MP-1: (a) ash-and-slag line (*1*—steel pipe; *2*—hard ash and slag layer); (b) portable machine MP-2 for ash-and-slag deposit removal (*1*—fluid-power motor KHSH-32; *2*—vibropercussion mechanisms; *3*—frame; *4*—ash-and-slag pipe).

Wide application of hydraulic ash-and-slag removal is explained by its advantages: continuous removal of huge amounts of ash-and-slag to great distances (to 30 km), full mechanization of haulage and dumping, comparatively simple and reliable equipment. However, the method has disadvantages as well. The most essential drawbacks are high water consumption (10–30 m^3 of water per 1 t of ash and slag) and contamination of natural water bodies with process water with high content of toxic substances after hydraulic ash-and-slag removal. Inside ash-and-slag pipelines, hard ash layer accrues, which worsens capacity of the lines, and it is required to undertake preventive treatment, i.e. remove hard deposits on the inner surfaces of equipment and surface utility lines included in the system of collection and discharge of combustion products [1, 2].

During operation of ash-and-slag pipelines, deposit accumulation results in formation of hard ash layer *2* up to 80–120 mm thick (Figure 1). This diminishes useful diameter and worsens passability of the pipes.

One of the problem of heat power plants is elimination of coal bridging in hoppers as dynamic choking-up always takes place above hopper inlets during flow of granular fuels (Figure 2) [3, 4]. During free outflow, choking is continuously broken under the pressure of overlying layers. Above smaller size inlets, choking is more stable and can terminate fuel outflow from hopper.

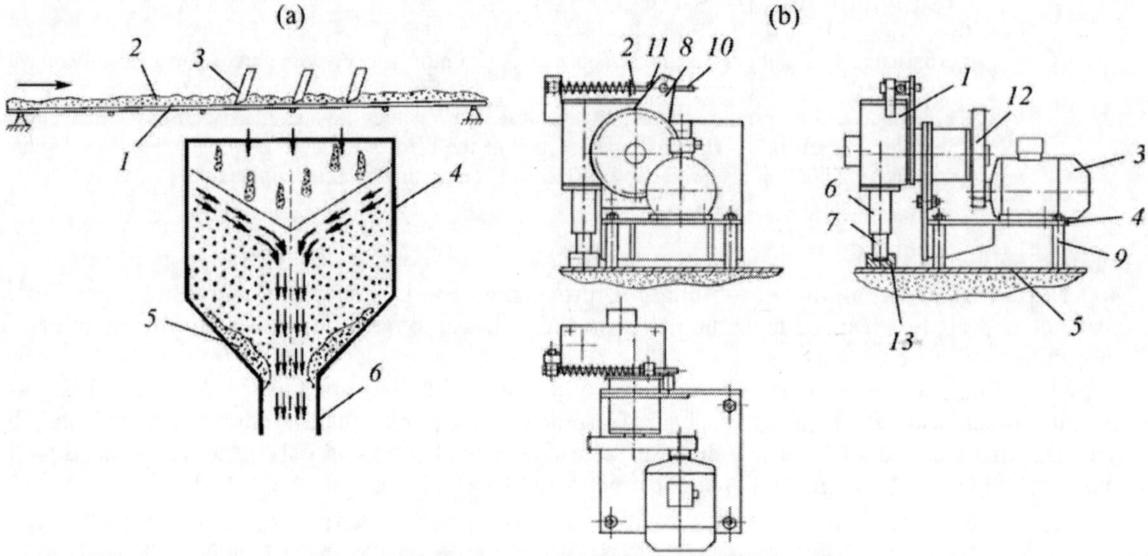

Figure 2. Filling of hoppers and percussion machine design: (a) filling of hopper (*1*—conveyor; *2*—coal; *3*—filling facility; *4*—hopper; *5*—choking-up; *6*—outlet); (b) MB-10 machine to break chokes in receiving hopper (*1*—percussion mechanism; *2*—hold-down; *3*—electric motor; *4*—assembly platform; *5*—hopper walls; *6*—guiding sleeve; *7*—waveguide; *8*—overhanging support; *9*—bolts; *10*—guiding bolt; *11*—adjusting nuts; *12*—reduction unit; *13*—shoe).

2. Technology for deposits removal

Based on the afore-said, it is of the current concern to develop efficient technology for high-quality and high-productive removal of deposits from the inner walls of holding reservoirs using the lever-type percussion mechanisms proposed by Abdraimov [5].

Configurations of Abdraimov's lever-type percussion mechanism depend on the size of the largest component: with the longest piston rod (l_2 max) (Figure 3a), with the longest yoke (l_3 max) (Figure 3b) and with the longest base (l_0 max) (Figure 3c), and the lengths of the elements should obey the one of the conditions below:

$$l_0 - l_1 = l_2 - l_3, \quad (1)$$

$$l_0 - l_1 = l_3 - l_2, \quad (2)$$

$$l_0 + l_1 = l_2 + l_3, \quad (3)$$

where l_1, l_2 and l_3 are the lengths of the crank arm, piston rod and the yoke, respectively; l_0 is the distance between the crank arm and yoke supports.

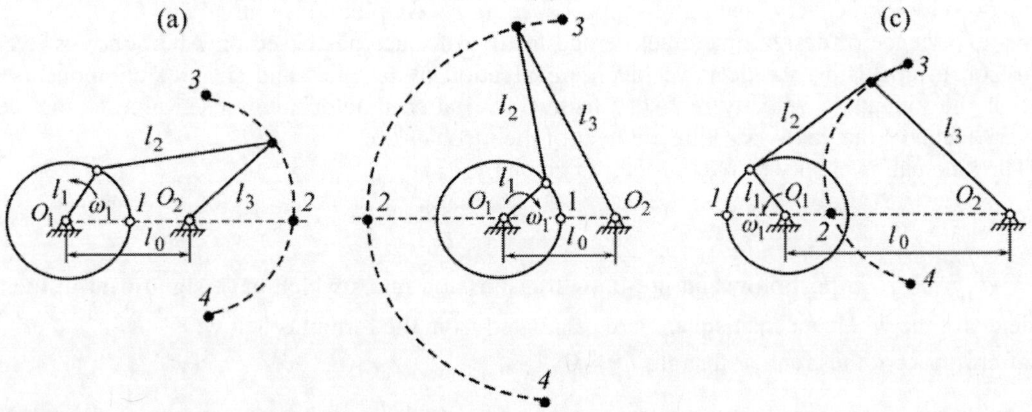

Figure 3. Kinematic scheme of crank arm-and-yoke variable structure percussion mechanism by Abdraimov.

In the course of research and development, the scientific team of the Institute of Machine Science within the Kyrgyz Republic Academy of Sciences has designed different percussion machines and facilities based on Abdraimov's lever mechanisms, such as: sleeve-free automatic press; jacks with manual, electric and hydraulic drives; manual jackhammers with different drives; compacting machines; drilling rifs; ash-and-slag removal device to clean ash-and-slag pipelines, feed hoppers of heat power plant boliers; hammer attachments with blow energy to 2.5 kJ; shock pulse generators of high capacity for deep soil compaction for construction of various structures, and to reinforce beds for railway and motor roads, as well as for vibration generation in soil (to stimulate oil and gas recovery) with unit blow energy up to 50 kJ.

Abdraimov's lever-type percussion mechanisms with variable structure feature high efficiency exceeding the known analogs, which implies reliable operation and wide application area. Their operation within percussion machines differs from conventional (hydraulic and pneumatic) mechanisms by simplicity of manufacture and maintenance in activities which go without bulky hydraulic stations and compressors.

According to the analysis of tested schemes, most of the listed percussion machines and facilities are designed sing one of the variable structure mechanisms, namely, the longest piston rod l_2 (Figure 3a) [6]. At request of the Bishkeke Heat Power Plant, the Laboratory of Theory of Mechanisms and Machines at the Institute of Machine Science has designed portable percussion plant MP-1 for cleaning ash-and-slag pipelines [7].

The portable plant MP-1 is intended to remove ash and slag deposits from pipes by vibropercussion. The plant has two percussion mechanisms actuated by hydraulic drive. The vibropercussion mechanisms are mounted on the common base and make impact on the pipe using special tools—waveguides. General view of this structure is shown in Figure 1b. Hydraulic drive of the percussion machine MP-1 can be powered by mobile hydraulic plant or hydraulics of a road construction machine (tractor, hydraulic crane, shovel, etc.) available.

Moreover, the Laboratory has designed the percussion machine MVB-10 to remove chocking in hoppers (Figure 2b). This machine is composed of a variable-structure percussion mechanism, single gearing, electric drive 4A80V4U3, hold-down and an assembly platform. The main output parameters of the vibropercussion machines for removal of adhered deposits are frequency *f*, blow energy *A* and percussion capacity *N*.

Aiming at improving capacity of these percussion machines at preset output characteristics, Abdraimov's lever-type percussion mechanisms with the longest base were subjected to kinematic and force testing. Based on the test results, the curves of the transmission ratios of the mechanism, kinematic velocity recovery factor and support spacing r were plotted (Figure 4 and 5) [8, 9].

As experience of design, manufacture and testing of machines based on Abdraimov's lever-type percussion mechanisms, the decisive factor in selection of schemes and sizes of components is the value of the kinematic velocity recovery factor. The latter should have close value to the velocity recovery factor of the yoke—striking element of the mechanism.

The kinematic velocity recovery factor is given by:

$$R_{kin} = \frac{U_{31+}}{U_{31-}}, \tag{4}$$

where U_{31+}, U_{31-}—post-blow and pre-blow transmission ratios which have significant influence on efficiency of the whole mechanism and are calculated from the formulas below:

after impact at the crank arm angle $\varphi_1 = 0$:

$$U_{31+} = U_{31}(0) = \frac{(\lambda_0 \lambda_2 / \lambda_3)^{1/2} - 1}{1 + \lambda_0}, \tag{5}$$

before impact at $\varphi_1 = 2\pi$:

$$U_{31-} = U_{31}(2\pi) = -\frac{(\lambda_0 \lambda_2 / \lambda_3)^{1/2} + 1}{1 + \lambda_0}, \tag{6}$$

where $\lambda_1 = l_1 / l_1 = 1$, $\lambda_2 = l_2 / l_1$, $\lambda_3 = l_3 / l_1$, $\lambda_0 = l_0 / l_1$ are the relative lengths of the elements.

These parameters govern performance of percussion machines and facilities. Prior to selecting a layout and size ratio of percussion mechanism elements, it is necessary to analyze performance of the machine using the calculated values of blow energy, factor R_{kin} and blow frequency.

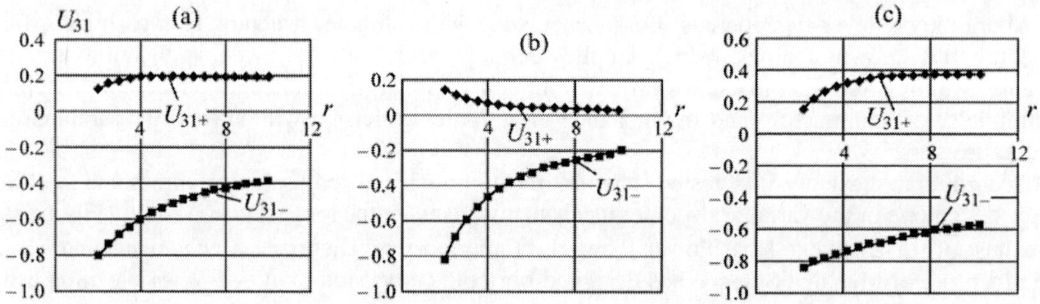

Figure 4. Curves $U_{31}(r)$ for different schemes of the variable-structure percussion mechanism with the longest base: (a) $l_1 < l_2 = l_3 < l_0$; (b) $l_1 < l_2 < l_3 < l_0$; (c) $l_1 < l_3 < l_2 < l_0$.

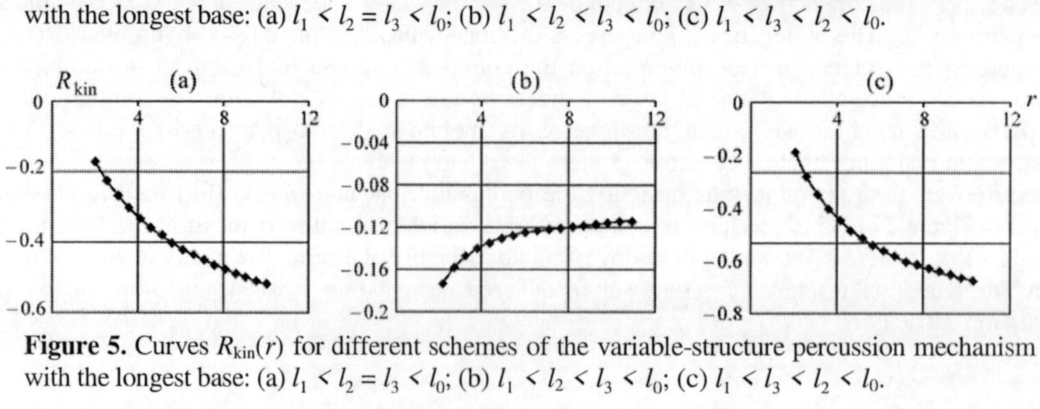

Figure 5. Curves $R_{\text{kin}}(r)$ for different schemes of the variable-structure percussion mechanism with the longest base: (a) $l_1 < l_2 = l_3 < l_0$; (b) $l_1 < l_2 < l_3 < l_0$; (c) $l_1 < l_3 < l_2 < l_0$.

3. Conclusions

Based on the research results, the recommendations have been made on selecting parameters of vibromachins for cleaning holding reservoirs:

1. It is advisable to accept the value of the kinematic velocity recovery factor in the range of $R_{kin} = 0.3–0.5$.

2. The collision velocity $[V_{col}]$ shol not exceed 10–12 m/s in order to undamaged the reservoir surface; for this reason, the blow energy A_{blow} should be low.

Finally, the research findings enable: selecting layouts and length ratios of the percussion mechanism components to be included in the structure of percussion machines meant for cleaning pipes, hoppers and other holding reservoirs from difficult adhered deposits; recommending the layout with the length ratio $l_1 < l_3 < l_2 < l_0$ and parameter $r = 5$, which is able to increase shock power and, thus, efficiency of percussion machine.

References

[1] Gavrilov EI 1987 *Fuel and Transport Facilities and Ash and Slag Removal at Thermal Power Plants: Textbook for High Schools* Moscow: Energoatomizdat (in Russian)

[2] Ryzhkin VYa 1987 *Thermal Power Plants: Textbook for High Schools* VCh Girschfeld (Ed) Moscow: Energoizdat (in Russian)

[3] Abdraimov S and Dzhumataev MS 1994 Variable structure mechanisms and application in engineering *Int. Conf. Proc.* Almaty pp 145–148 (in Russian)

[4] Karimov AA, Teshebaev A and Dykanaliev KM 2003 Machines for eliminating delays in the fuel supply and ash removal systems of CHP *Proceedings of II International Conference: Problems of Mechanics of Modern Machines* Ulan-Ude: VSTU pp 5–8 (in Russian)

[5] Abdraimov S and Dzhumataev MS 1993 Swivel-Lever Mechanisms of Variable Structure, Bishkek: Ilim (in Russian)

[6] Bakirov B, Abdraimov ES and Shadiev MI 2017 Development of the mechanisms of variable structure in Kyrgyzstan International *Scientific and Practical Conference Proceedings: Fundamentals of Mechanics* Novokuznetsk: MashinoStroenie No 2 pp 6–9 (in Russian)

[7] Abdraimov ES, Fokin YuA, Karimov AA and Erkebaev M 2001Creation of a specialized plant for cleaning coal bunkers of CHP *Proceedings of the Conference dedicated to the I Congress of Engineers of Kyrgyzstan and the 10th anniversary of the formation of the IA KR*, Bishkek: Tekhnologiya, pp 50–54 (in Russian)

[8] Bakirov B and Shadiev MI 2017 Analysis of the influence exerted by geometric parameters in various layouts of Abdraimov's mechanisms with the longest base on the energy characteristics of percussion machines *Vestnik KRSU* No 5 (17) pp 13–14

[9] Bakirov B, Abdraimov ES and Shadiev MI 2018 Background for creation of high-frequency machines based on variable-structure lever-typer percussion mechanism *Journal of Advanced Researchin Technical Science* (North Charleston,USA) No 8 pp 40–44

Dependence of stress state of monolithic lining on weight of process equipment in underground mines

SV Antsiferov*, AS Sammal and VG Dvoryankin

Tula State University, Tula, Russia

E-mail: *antsser@mail.ru

Abstract. The analytical method is developed to design monolithic lining for tunnels as function of weight of process equipment in underground mines. The method is implemented as a computer program which makes it possible to find and adjust regularities of stress state formation in mine support and adjacent rock mass.

1. Introduction

Mineral mining, processing and storage use underground structures and special process equipment of considerable weight. Mine planning, design, construction and operation requires assessment of stress of support structures in order to determine their load-bearing capacity. Different methods used in stress state assessment can be divided into experimental (in-situ or lab-scale) and computational. According to [1], support design should take into account both external effects and internal loads generated, for instance, by heavy process equipment, haulage machines or piled materials.

Widely applicable computer engineering allows using dedicated software packages implementing numerical methods, for example, the finite element method, and conventional engineering analysis based on hypotheses and assumptions of structural theory [2]. Despite some essential shortages, such methods are widely applied in practical design.

For some years, the Tula State University has been engaged in research aimed to improve the theory of modeling interaction between underground structures and surrounding rock mass as elements of a unified system exposed to deformation, as well as to develop new strict analytical methods of mine support design (tunnel lining). These methods rest upon an integrated scientific and methodological basis—continuum mechanics, in particular, elasticity theory [3], geomechanics and mechanics of underground structures [2, 4].

Developed at the Tula State University, the design methods for monolithic, multi-layered (hybrid), shotcrete plus rock bolting support types (lining) for deep and shallow underground excavations (tunnels) subject to static loading—rock weight, groundwater pressure, tectonic forces and seismic effect of earthquakes [2, 5–7]—are recommended for practical application in various purpose mine design by effective normative legal documentation both in Russia and abroad.

The lining design method for a deep tunnel of arbitrary cross section under load generated by weight of heading and haulage machines, hoisting facilities, jacks and expandable mechanism thrust against lining is proposed in [8]. It should be emphasized that down to recent time, there is yet not a strict mathematical method for design of mine support constructed in situ immediately nearby ground surface and subjected to local loads due to weight of heavy mining and transportation machines operating in mines.

IOP Conf. Series: Earth and Environmental Science **262** (2019) 012002 doi:10.1088/1755-1315/262/1/012002

The science-based methods to calculate such loads are in high demand in design and construction of special-purpose underground storages or repositories for safe keeping of radioactive waste and spent nuclear fuel. The pressure produced by lead radioactive waste containers on the floor of underground structures can reach considerable values and exert appreciable influence on stresses ad strains in support (lining). Irrespectively of the type of rocks and repository, it should be provided that the containers are removed from the storage, and this can be accompanied by essential loading exerted by weight of the containers and transport equipment on lining.

2. Problem statement

In connection with this, based on the mathematical modeling of interaction between circular cross-section tunnel constructed in subsurface and rock mass as elements of a unified system exposed to deformation under internal local load, the relevant problem of elasticity and creep was formulated and its rigorous solution was obtained. The analytical model of the problem is depicted in Figure 1.

Rock mass is modeled by a semi-infinite, uniform, isotropic, linearly deformable ponderable medium S_0 with a straight-line boundary L. Material of the medium S_0 is characterized y the bulk density γ, deformation modulus E_0 and Poisson's ratio ν_0. The circular hole with the boundary L_0 and radius R_0, located in the medium S_0 at a distance H from L and supported by rings S_1 with the inner boundary L_1 and radius R_1, models the tunnel lining. The ring S_1 is made of a material with the deformation characteristics E_1 and ν_1.

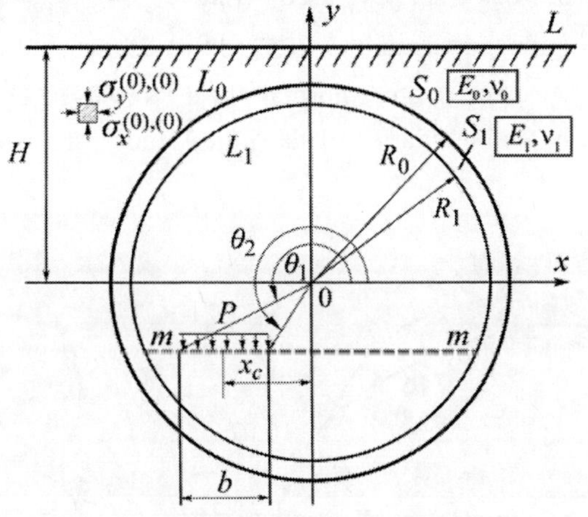

Figure 1. Analytical model.

3. Mathematical modeling results

The problem schematically shown in Figure 1 is solved as a sum of solutions of two plane elasticity problems for a ponderable medium free from the load on the boundary L_1 (problem on rock mass weight) and for an imponderous medium with its boundary L_1 subjected to the local distributed load (problem on weight of an object located in tunnel).

The weight of rocks in the domain S_0 is modeled as a field of unequal initial stresses, while the composite stress state in S_0 is presented as a sum of the initial and unknown additional stresses conditioned by the reinforced hole in the medium. The weight of the tunnel lining as against rock mass weight is neglected.

The boundary conditions in the formulated plane elasticity problem govern the absence of the load at the straight line boundary L of the medium S_0, fulfillment of continuity conditions by vectors of total stresses and displacement at the line L_0 of the interface between the domains S_0 and S_1 having different deformation characteristics, and the absence of external load at L_1 in the problem on rock mass weight, or the presence of the vertical load P on the part $[\theta_1, \theta_2]$ of the inner boundary L_1 in the problem on weight of equipment inside the tunnel [9]. In the reduced formulation of the plane elasticity problem, distributed load along the long axis of the tunnel is assumed as infinite; the influence of the finite value of this load on the stress state of the lining can be taken into account using adjusting empirical coefficients.

The boundary-value problems of the theory of complex variable function, consistent with the problems on rock mass weight and locally distributed internal load, are solved using the mathematical apparatus provided by the theory of complex variable functions—complex potentials introduced by Kolosov–Muskhelishvili [3], analytical extension of complex potentials, properties of complex Fourier and Laurent series [10, 6].

The computational technique implemented as a software package for personal computers, makes it possible to determine stresses and forces (bending moment and longitudinal forces) in the tunnel lining, as well as stress state of surrounding rock mass. The computation results allow finding and adjusting regularities of stress state formed in the monolithic lining of the circular tunnel under local vertical load on the inner surface of the lining.

Below in this article, the authors offer the results obtained in the computer modeling of stress state of the monolithic concrete lining under weight of rocks and heavy process equipment situated inside the tunnel at the following input data: occurrence depth of the tunnel is $H = 10.4$ m, outer and inner radii of the lining are $R_0 = 5.2$ and $R_1 = 4.7$ m, respectively (thickness of the concrete lining is $\Delta = 0.5$ m); deformation characteristics of rocks are $E_0 = 20$ MPa, $\nu_0 = 0.35$; concrete—$E_1 = 36000$ MPa, $\nu_1 = 0.2$; the load application range is from $\theta_1 = 205°$ to $\theta_2 = 260°$ (Figure 1).

Figure 2 shows contour lines of relative normal shear stresses σ_θ / P at the points of the outer (Figure 2a) and inner (Figure 2b) boundaries of the lining under the action of the locally distributed inner load.

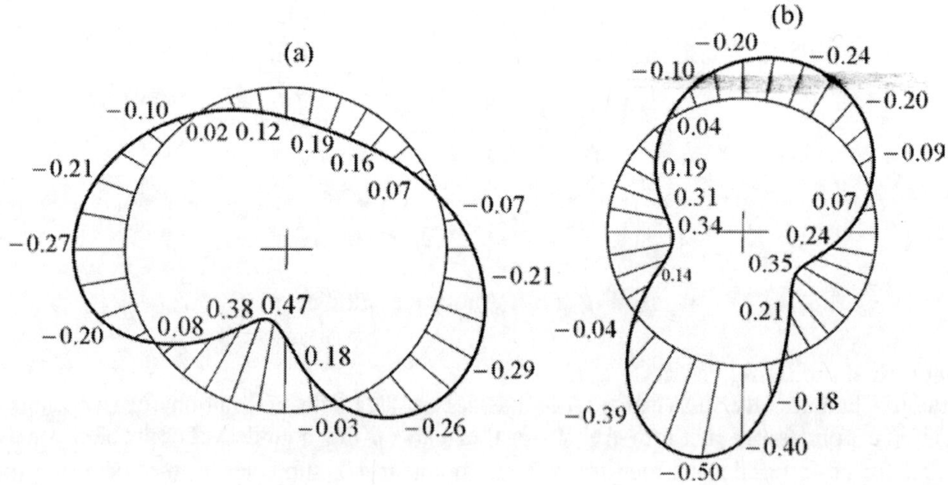

Figure 2. Contour lines of relative stresses σ_θ / P at the points of (a) outer and (b) inner boundaries of the lining under action of internal locally distributed load.

It follows from Figure 2 that the internal locally distributed load at the points of the outer and inner boundaries of the tunnel lining induces both compressive (negative) and tensile (positive) stresses. The maximum values of the tensile stresses at the outer boundary and compressive stresses at the inner boundary appear in the radial sections in the middle of the load application segments at the level $m - m$ ($\theta \approx 240°$), see Figure 1.

Figure 3 presents contour line of the normal shear stresses σ_θ at the boundary points under the action of rock weight.

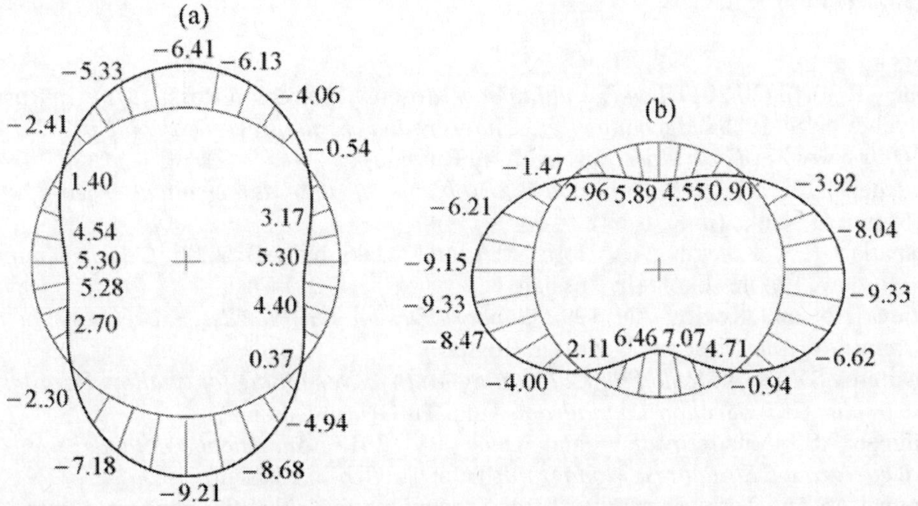

Figure 3. Contour lines of stress σ_θ, MPa, at the points of the lining boundary due to weight of rocks: (a) outer boundary; (b) inner boundary.

It is evident in Figure 3 that the maximum compressive stresses under rock weight are not higher than 9.21 and 9.23 MPa at the outer and inner boundaries, respectively. The maximum tensile stresses are 5.30 and 7.07 Mpa at the outer and inner boundaries of the lining in radial sections at θ = 270°, which conforms with the lower point of the vertical diameter.

Figure 4 depicts the dependence of the tensile and compressive stresses σ_θ^{max} / P on the deformation modulus E_1 of concrete in case of various thickness of the lining. The input data are: H = 10.4 m, R_0 = 5.2 m; E_0 = 20 MPa, $\nu_0 = 0.35$; the load application range is from θ_1 = 205° to θ_2 = 260°, which agrees with the width b = 1.6 m at the level m–m in Figure 1. The range of the deformation modulus corresponds to the concrete grades B15 to B60; Poisson's ratio of concrete is $\nu_1 = 0.2$.

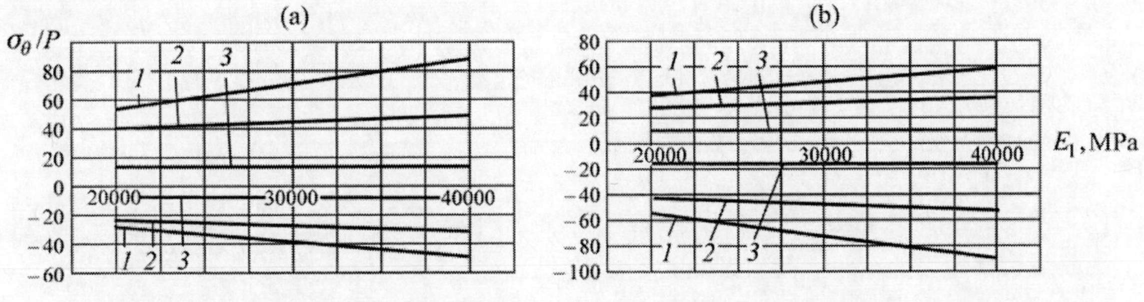

Figure 4. Dependence of σ_θ^{max} / P on E_1 at the (a) outer and (b) inner boundaries of the lining. Lines *1*, *2* and *3* stand for the lining thickness Δ = 0.2 m, Δ = 0.5 m and Δ = 1.0 m, respectively.

It follows from the dependences in Figure 4 that maximum stresses in the lining grow gradually with the concrete modulus of deformation by the linear law. The most intense growth of the stresses is observed in the thinner lining—lines *1* and *2* in Figure 4. The increase in the deformation modulus of concrete at the lining thickness Δ = 1.0 m almost has no effect on the value of the stresses (lines *3*).

The authors believe that the developed rigorous analytical method implemented as a computer program is a convenient tool for the support design in various purpose underground excavations, including lining of traffic tunnels.

References

[1] Code CR 20.13330.2011 *Loads and Impacts* Moscow:Minregion Rossii 2011 (in Russian)
[2] Bulychev NS 1989 *Mechanics of Underground Structures in Problems and Examples: University Textbook* Moscow: Nedra (in Russian)
[3] Muskhelishvili NI 1966 *Some Basic Problems of the Mathematical Theory of Elasticity* Moscow: Nauka (in Russian)
[4] Kasparian EV, Kozyrev AA, Iofis MA and Makarov AB 2006 *Geomechanics. Tutorial* Moscow: Vyssh. shkola (in Russian)
[5] Fotieva NN and Kozlov AN 1992 *Support Design for Parallel Excavations in Seismically Active Regions* Moscow: Nedra (in Russian)
[6] Antsiferov SV 2014 *Method for Calculating Multi-Layer Lining for Shallow Parallel Tunnels of Circular Cross-Section: Monograph* Tula: TulGU (in Russian)
[7] Sammal AS, Antsiferov SV and Deev PV 2013 *Analytical Methods for Calculating Underground Structures: Monograph* Tula: TulGU (in Russian)
[8] Sammal AS and Tormysheva OA 2014 A technique for estimating the stress state of lining for transport tunnels under the action of internal local loads *Transport. Stroit.* No 7 pp 16–18
[9] Antsiferov SV, Dvoryankin VG and Fotieva NN 2017 Lining design to sustain weight of equipment in tunnels *Izv. TulGU. Nauki o Zemle* No 4 pp 263–273
[10] Fotieva NN and Antsiferova LN 1997 Calculation of multilayered lining of shallow tunnels *Mechanics of Underground Structures: Collection of Scientific Papers* Tula: TulGU pp 9–25 (in Russian)

IOP Conf. Series: Earth and Environmental Science **262** (2019) 012003 doi:10.1088/1755-1315/262/1/012003

Perspectives to process old gold-bearing tailings at Kumtor gold ore concentrator, Kyrgyz Republic

GB Asanaliev[1]*, BT Tolobekova[1] and GA Nagoeva[2]

[1]Institute of Geomechanics and Subsoil Development, Academy of Sciences of the Kyrgyz Republic, Bishkek, Kyrgyz Republic
[2]Institute of Mining and Mining Technologies, Bishkek, Kyrgyz Republic

E-mail: *gany_asan@mail.ru

Abstract. Research findings on old tailings at Kumtor gold deposit are presented The data on the chemical, phase, grain-size analysis of flotation tailings and cyanidation cakes are reported. Heap leaching process is described. Economic advantages of promising processing of mining wastes by the integrated conventional and innovative technologies in view to support metal production from accumulated mining and concentrator wastes in Kyrgyz Republic are demonstrated.

1. Introduction

High-altitude gold deposit Kumtor explored by Soviet geologists in 1989 in Kyrgyz Republic is a unique gold deposit which initial proved gold reserves amounted to 731 t of gold. The development of this-kind mineral deposit was repeatedly postponed because of mining complexity under permafrost conditions. However in 1992 the government of Kyrgyz Republic entered into the contract with Kyrgyzaltyn state corporation and Camcco Corporation, Canada. The contract granted exclusive rights to Comeco Corp. to develop Kumtor deposit. Within the terms of the contract Kumtor Gold Co. was founded. The company promoters were Kyrgyzaltyn Corp. on behalf of Kyrgyz Republic government and Cameco Corporation on behalf of Canada with 1/3 : 2/3 capital interest, respectively.

In 1997 the commercial exploitation of the deposit was launched. The average production rate amounted to 17 t of gold /year (550 thousand oz.), and up to 2 t of silver/year (65 thousand oz.) In 2003 the Kumtor restructuring agreement signed between the government of Kyrgyz Republic and Cameco Corporation involved rescission of the Contract dated by 1992 and foundation of a new joint stock company, registered as Centerra Gold Inc. in Toronto, Canada. Above 120 Mt of gold-bearing ore of average grade within 4 g per 1 t of ore (480 t of gold) was processed at the gold ore concentrator for 21 year operation of the gold mine. More than 360 t of marketable gold were produced. The low requirement for the run-of-mine ore grade, inefficient techniques for ore processing, the valuable content in concentrator tailings, as a rule, appreciably exceeded the admissible level (Table 1). Preliminarily calculated metallurgical balance of Kumtor tailing dumps indicates the content of more than 100 t of gold, 160 t of silver, more than 4000 t of copper.

The poor prospected vein gold reserves, scaling down of geological prospecting activity, deficit of promising areas in the open acreage give grounds to re-estimate the rating of non-traditional mineral resources and to assess commercial value of gold-bearing wastes at Kyrgyzstan gold mines. The advance in mineral processing enables to consider such mineral mining and processing wastes as large man-made (techno-genic) mineral deposits. In foreign countries the gold-bearing waste materials are involved into re-processing. Thus in Republic of South Africa the annual gold recovery rate from waste materials amounts to 3–5 t [4]. In Kyrgyz Republic Makmal, Tereksai and other operating and idle mines afford the promising potential for additional gold recovery from wastes. Specific legislation concerning the utilization of techno-genic objects, lack of information on actual ore loss in the course of exploitation of currently-idle mines, poor geological and technical documentation make impossible the prompt involvement of "ready-made" techno-genic deposits into production. Investment attractiveness of techno-genic mineral foundations is enhanced by their location in developed industrial regions in the vicinity of operating ore preparation plants, provided that dumps are on the daylight surface, as in this case the material mass is disintegrated and low expenditure are demanded to involve waste material into production.

Table 1. Chemical composition of the pulp discharged from Kumtor gold ore concentrator.

Year	2008			2009			2010			2011		
Element	Au	Ag	Cu	Au	Ag	Cu	Au	Ag	Cu	Au	Ag	Cu
January	0.37	0.73	20.30	0.29	1.13	26.32	0.93	1.21	26.08	0.93	1.69	42.03
February	0.36	0.97	22.23	0.24	1.11	29.99	0.84	1.27	21.36	0.60	1.19	34.62
March	0.58	1.40	17.98	0.47	0.97	35.03	0.63	1.06	18.57	0.61	1.15	33.90
April	0.42	1.02	17.84	0.55	1.10	32.17	0.41	1.03	18.88	0.52	1.08	36.41
May	0.59	1.41	31.70	0.40	1.01	23.52	0.45	1.02	17.98	0.58	1.07	32.93
June	0.58	1.10	30.44	0.49	0.94	25.70	0.44	1.12	21.91	0.61	1.13	43.74
July	0.46	1.13	24.41	0.78	1.51	28.39	0.28	1.08	21.56	0.71	1.27	47.47
August	0.52	0.96	40.48	0.67	1.36	46.77	0.25	0.90	23.23	0.84	1.81	31.55
September	0.75	1.23	60.82	0.38	1.03	41.19	0.26	0.87	25.91	0.82	1.88	24.02
October	0.76	1.09	44.15	0.32	2.07	26.78	0.30	1.09	26.40	0.82	1.46	27.77
November	0.74	1.22	25.70	0.94	1.51	37.58	0.78	1.27	28.60	0.86	1.33	29.11
December	0.56	1.13	21.19	1.19	1.79	41.58	1.85	2.78	51.51	0.87	1.42	39.52

Table 2. Chemical composition of Kumtor tailing specimens.

Parameter	Tailings	Abundance in host rock
Ag, mg/l	410	0.04–0.3
Al, %	7.83	8.32
As, mg/l	8.9	1–13
Sb, mg/l	4.1	1.5
Ba, mg/l	42	425
Ca, %	**5.86**	4.15
Cd, mg/l	0.15	0.03–0/3
Co, mg/l	24.1	25
Cr, mg/l	80	100
Cu, mg/l	28	4–87
Fe, %	**6.28**	5.63
K, %	**4.3**	2.09
Mg, %	1.45	2.33
Mn, %	1.8	< 0.15
Mo, mg/l	5.79	1.5
Na, %	1.16	2.36
Hg, mg/l	0.023	0.09
Ni, mg/l	56.6	2–130
Pb, mg/l	30.2	6–20
Zn, mg/l	72	16–105

In 2009 Alex Stuart laboratory, Kumtor Gold, performed chemical analysis of tailings of gold ore concentrator (Table 2). The pulp was sampled at different points of the tailing beach, the dressing capacity of the dump material was studied. Geological specific features of techno-genic foundations depend on origin and composition of gold-bearing material subjected to first techno-genic and later to hypergenic effects (low temperature, pressure, oxidation, etc.). At present the attempts are made to substantiate the process for gold recovery from old tailings of beach-type dumps by drainage of a tailing massif and division it into non-productive horizons, including poor-grade tailings in the top section of the techno-genic mas and productive horizons at the base of techno-genic mass with higher gold content. Thus, hypergenic transformation of tailings under influence of acid aggressive sulfate-calcic medium at the techno-genic water-bearing formation leads to redistribution of gold from a disseminated state to horizons specified by enriched precious metal content of 1–10 g/t, applicable for commercial processing. The establishment of boundaries of a hypergenically modified productive horizons, removal of tailings from non-productive horizons and further processing of modified tailings at productive horizons are efficiently realized by applying the heap leaching process [1]. It is reasonable to emphasize that at present there is a number of scientifically substantiated innovative structures for a leaching module in the permafrost zone. The advantages of heap leaching process run as follows: low total and specific capital expenditures, low maintenance costs, lack of energy- and material-intensive operations, high labour performance, few personal.

2. Increasing the efficiency of gold mining

The present research work aims at improving efficiency of Kyrgyzstan gold deposit development at the expense of integrated conventional and innovative processes for metal production from accumulated mining wastes and ore processing at the final stages of mineral deposit exploitation (2, 3, 5).

The mineralogical analysis is executed by counting in monomineral gold fractions. Gold is lost in flotation tailings because gold remains associated with nonmetal minerals. Non-floatable gold in flotation tailings occurs as intergrowths with nonmetal minerals mostly with carbonates (Table 3, 4).

Table 3. Granulometric characteristic of flotation tailings, cakes of cyanidation of flotation tailings and concentrates.

Ssize class, μm	Yield of granulometric classes, %		
	Flotation tailings	Cyanidation cakes	
		Flotation tailings	concentrate
+ 150	21.47	23.72	0.14
53–150	27.72	29.88	2.65
25–53	15.03	14.38	20.57
5–25	20.16	13.25	47.20
−5	15.62	18.77	29.44

Table 4. Gold content in mineral different-density fractions, g/cm^3.

Fraction size, μm	Flotation tailings		Cakes of cyanidation of flotation tailings		Cakes of cyanidation of concentrate	
	2.68–3.10	< 2.68	2.68–3.10	< 2.68	2.68–3.10	< 2.68
+ 150	0.65	0.50	0.49	0.51	—	—
53–150	0.51	0.41	0.24	0.42	—	—
25–53	0.50	0.37	0.15	0.17	0.67	0.96
5–25	0.28	0.17	0.11	0.17	0.48	0.84
−5	—	—	—	—	—	1.42

The gold loss in flotation tailings is attributed by poor opening of finely disseminated gold and calaverite in host rocks, while gold is lost in cakes of cyanidation of the concentrate because of insufficient opening of gold-bearing minerals among pyrite and their incomplete dissolution (Figure 1). The loss of gold and calaverite inclusions in pyrite is 40.3 %; the ore specimen ground to 77 % of 25 μm, gold and calaverite grains of 7 μm in size remain confined within pyrite (Figure 2). Grinding down to less than 10 μm guarantees opening of all the gold inclusions within 3 μm size range. Grinding down to 10 μm the expectant gold content in cyanidation cakes can be about 1.5 g/t, assuming that gold loss in adsorption of gold by slimes is eliminated. The experimental data on are close to gold content (1.4 g/t), obtained in the product of intensive cyanidation. In this case the low gold leaching (0.84 g/t) is explained by finely dispersed gold content in the pyrite structure and gold-bearing mineral inclusions of less than 3 μm in pyrite.

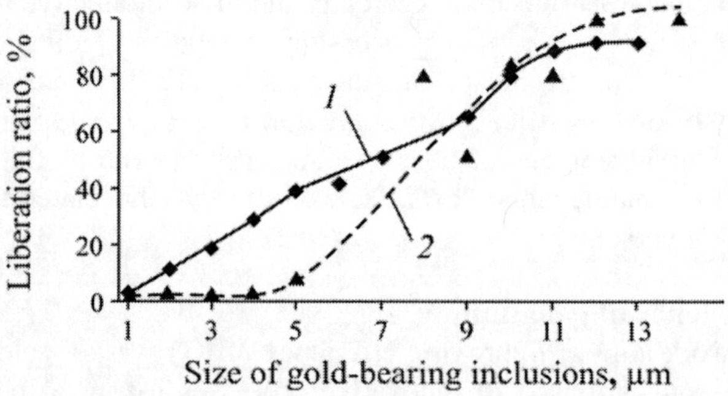

Figure 1. Relationship of opening degree of gold (*1*) and calaverite (*2*) in pyrite versus their grain size.

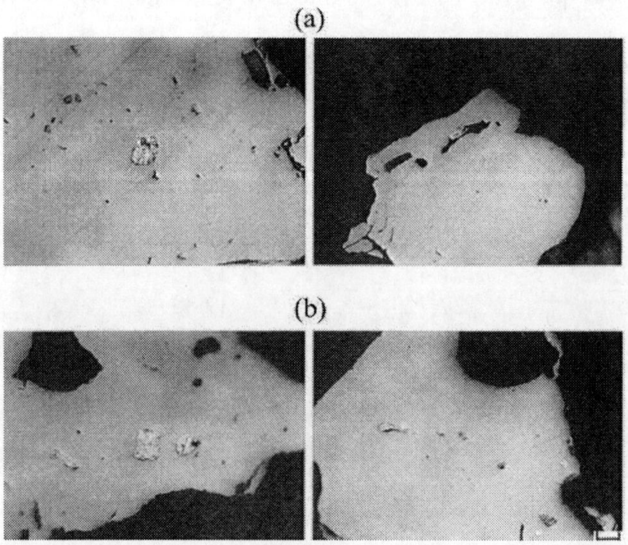

Figure 2. Inclusions of (a) virgin gold and (b) calaverite in pyrite.

The investigation into phase composition of gold in tailings of the gold ore concentrator revealed the following reasons for gold loss.

• Gold loss in flotation tailings:

—insufficient opening of gold and calaverite due to their fine dissemination in host rocks is responsible for 65 % of total loss values; the average 12 μm inclusions contain gold mainly associated with carbonates (37%);

—loss in slime size class of 5 μm (15%);
—unfloatable free gold and calaverite (10 %).
• Gold loss in cakes of cyanidation of the concentrate:
—gold-bearing minerals of 1–7 μm in size, confined within pyrite (40.3 gold loss);
—insoluble gold and calaverite, free and in intergrowths (28%);
—finely dispersed gold and calverite of less than 1 μm in size (18%);
—adsorption of gold by carbon substance under applied process (15%).
• Gold loss in cakes of cyanidation of flotation tailings:
—insufficient liberation of virgin gold and calaverite from gangue (65%);
—due to adsorption by carbon substance (24.6%).

3. Conclusions

It is scientifically justified that the processing of techno-genic gold-bearing mineral materials can provide the production cost of 1 ounce of gold 1.5–2 times lower as compared to that in conventional technologies, moreover, the acting laws afford appreciable preferences (The Tax Code of Kyrgyz Republic, article 300. point 3), the zero tax on mineral resource utilization is granted.

It is verified by computations that innovative techniques for additional recovery of precious metals from technogenic waste materials enable to expend operation life of active mines by 15–20 years, along with concurrent solution to economic, mineral resources, ecological and social problems of the region and the state as a whole.

References

[1] Dementev VE, Druzhina GJ and Gudkov SS 2004 *Heap Leaching of Gold and Silver* Irkutsk: Irgiredmet (in Russian)
[2] Shemetov PA, Sutenkov VN and Naimova RS 2009 *Prospects of Mining and Processing Waste Management at Gold Deposit* pp 2–4 (in Russian)
[3] Logachev AV 2009 Scientifically-Methodological Production Framework for Final Mining Stages *Cand. Tech. Sci. Dissertation* Novocherkassk (in Russian)
[4] Seluk DA and Seluk IN 2012 *Method of Extraction of Gold from the Stale Tails of Aggradational Tailings* Moscow pp 3–6
[5] Gurin KK 2013 Research and development of method for gold recovery from concentration plant waste *Cand. Tech. Sci. Dissertation* Moscow (in Russian)

IOP Conf. Series: Earth and Environmental Science **262** (2019) 012004 doi:10.1088/1755-1315/262/1/012004

Locating harmonic microseismic sources using phases of signals and spectral transformations

AV Azarov[1]* and AS Serdyukov[1,2]

[1]Chinakal Institute of Mining, Siberian Branch, Russian Academy of Sciences, Novosibirsk, Russia
[2]Trofimuk Institute of Petroleum Geology and Geophysics, Siberian Branch, Russian Academy of Sciences, Novosibirsk, Russia

E-mail: *antonazv@mail.ru

Abstract. The paper presents the algorithm of locating harmonic microseismic signals using phases of the signals and spectral transformations. Different spectral transformations are considered and their properties required for the set problem solving are identified. The determination of the harmonic signals' location is exemplified using synthetic data.

1. Introduction

Usually geodynamic processes in a geological medium are the sources of seismic emission bearing information on the processes. One of the methods to study the seismicity is microseismic monitoring [1]. The important task of microsesimic monitoring is locating seismic sources.

The way to locate the seismic sources depends on the signal/noise ratio. When the signal/noise is sufficiently high, it is possible to pick arrival times of waves of separate events in seismograms. In this case, the seismic sources can be located using the kinematic methods [2–4]. Such methods are applicable in downhole monitoring, i.e. when receivers are located in observation borehole [5, 6].

When the receivers are located on the surface (surface monitoring), the signal/noise ratio is usually low, and it is impossible to pick wave arrival times. For this reason, hypocenters of seismic events are located using semblance methods based upon stacking waves [7, 8]. Such methods include emission tomography [9]. This technique is based on determination of semblance measure for each point of space. To calculate semblance measure, signals from all receivers are summed with time delays that conform with travel times of wave from the chosen point of space. The semblance function can be calculated both in time and frequency domains [10, 11].

The studies [12–14] suggest locating the seismic sources using the time-reverse modeling. The method of time-reverse modeling works at low signal/noise ratios, which makes it applicable in surface monitoring. The method consists in propagating of wave field recorded by all receivers backward to space in the known velocity model of the medium. Seismic fields of the receivers interfere in space, and the energy focuses in the coordinates of the seismic source. A similar method is suggested in [12] where image resolution is improved by semblance-weighted deconvolution.

In [15] the problem of determination of microseismic event parameters is considered using the methods of statistical theory of random process parameters. This approach provides the opportunity to locate the seismic sources in case of high level of noise [16].

Another method of locating signals is called beamforming [17]: it is widely used in sonars, radars seismicity (seismic signal processing), radio astronomy and other areas. Beamforming acts as a spatial filter which amplifies events from the wanted direction and suppresses events from the other directions [18]. There are two basic implementations of beamforming. The first implementation is based on calculation of phase shifts on receivers, compensation of seismic traces and their summation. It is similar to the above-mentioned calculation of semblance function. The other implementation is computation of cross-correlation matrix constructed using the receiver data [19]. This procedure also requires information on phases of receivers from the test point of space, which makes the algorithm commensurable with the seismic emission tomography by computational complexity. In the present-day seismics the cross-correlation beamforming is often used to study surface waves [20, 21].

This paper deals with locating harmonic sources [22]. Such sources can be observed, for instance in microseismic monitoring of rock ruptures filled with fluid [23, 24]. The algorithm proposed for the set problem solution is based on the phase information of signal and on using spectral transformations. The proposed approach differs from the above listed techniques and is convenient for locating microseismic sources by monitoring systems composed of groups of receivers arranged in lines.

2. Method of locating harmonic sources

Let us discuss implementation of the proposed approach. There are harmonic sources in a medium as well as a group of receivers arranged in line at equal distances. It is required to locate a source using this system.

The algorithm starts with a time–frequency representation of signal on each receiver. To solve this problem we will use spectral transform [25]. The classical approach is application of the Short-Time Fourier Transform (*SFFT*):

$$S_n(\tau, f) = \int_{-\infty}^{+\infty} s_n(t) w(t-\tau) e^{-i2\pi f t} dt, \tag{1}$$

where $s_n(t)$—signal on a *j*-th receiver; *f*—signal frequency; $w(t)$—window function with a support in finite interval named the window width. A deficiency of the approach is finding the window width ensuring the best time–frequency resolution. This problem is suggested to be solved using other spectral transformations such as Stockwell transform (*ST*) or synchrosqueezing transform (*SST*). Let as discuss Stockwell transform [26] which is a special case of continuous wavelet transform (*CWT*):

$$S_n(t, f) = \int_{-\infty}^{+\infty} s_n(t) \frac{|f|}{\sqrt{2\pi}} e^{-\frac{(\tau-t)^2 f^2}{2}} e^{-i2\pi f t} dt. \tag{2}$$

The key advantage of (2) is automatic selection of window width depending on signal frequency. As frequency increases, the window width decreases, and, vice versa, with lower frequency, the window is wider, which finally leads to the high time–frequency resolution. *ST* is suitable for the analysis of both pulsed and long in time signals. On the other hand, in case of harmonic signals, the best time–frequency resolution can be ensured by *SST* [27]:

$$s(t) = \sum_{j=1}^{J} A_j(t) \cos(\theta_j(t)), \tag{3}$$

$s(t)$ is a signal; $A_j(t)$ is the signal amplitude; $k_j = \frac{1}{2\pi}\frac{d}{dt}\theta_j(t)$ is frequency of a *j*-th component of expansion; *J* is a number of the expansion components. Thus, implementation of *SST* consists in finding A_j and θ_j for expansion of (3) at each moment of time. When using *SST* as with *ST*, there is no need to select an optimal size for the window as this is done automatically. This property makes *ST* and *SST* preferable over the classical *STFT*. It is worthy of mentioning that all described transforms preserve signal phases, which is important for implementation of the proposed approach.

The next step is finding frequency of the source, plotting vector of phases on this frequency and applying additional spectral transformation of the phase vector to find the wave arrival direction. Let us analyze this step in more detail. Let a harmonic source emits at frequency f_0 at a time t_0. Then, having time–frequency representation of signal $S_m(t,f)$ on each receiver *m*, we can plot vector of phases:

$$g = \left(S_1(t_0,f_0)/A_1 \quad S_2(t_0,f_0)/A_2 \;\ldots\; S_N(t_0,f_0)/A_N\right), \tag{4}$$

where *N* is a number of receivers; $A_n = \left|S_n(t,f)\right|$, i.e. each value in the vector is normalized and we, for this reason, take into account only phase characteristics of the signal. Then, the vector *g* is subjected to additional spectral transformation:

$$G(n,k) = F\left(g(n)\right) \quad n = 1 \,..\, N, \tag{5}$$

where *F* is some discrete spectral transform (*SFFT*, *ST* or *SST*); *G(n,k)* is a space–frequency spectrogram; *m* is a number of component; *k* is a spatial frequency (wave number). In the resultant *G(n,k)*, by the spectrum peaks, the spatial frequency *k* of the incident wave is found. After that, the apparent velocity ($V_{app} = f/k$) and arrival direction ($cos(\alpha) = V_{real}/V_{app}$) of wave are determined, where V_{real} is a wave velocity in the medium. In a homogeneous medium direction of arrival leads to the source. When a set of beams emergent from receivers at the found angles is plotted, the source location is found as their intersection. In a heterogeneous medium, the beams are constructed using information on the velocity structure of the medium in order to take into account refraction effects. The described algorithm can be presented as follows:

1. Time–frequency representation of signals on all receivers;
2. Finding frequency f_0 of the source;
3. Selection of time t_0 and plotting vector of phases **g**
4. Space–frequency representation on phase information on receivers;
5. Finding spectrum peaks in the spatial spectrogram in order to determine the spatial frequency *k* and calculation of incidence angle of plane-wave signal.

This section describes the algorithm of locating a source by one-dimensional linear observation system. However, usually the surface microseismic monitoring uses three-dimensional systems In this case the algorithm is employed if it is possible to distinguish groups of receivers arranged in line within the three-dimensional monitoring system.

3. Locating a source: Illustration

Let us consider a two-dimensional uniform linearly elastic medium (Figure 1a). Let a harmonic point source be in the medium at a point (1100, 500). Receivers are set at a spacing of 2 m along the straight line *y*=0. P-wave velocity in the medium is 2500 m/s.

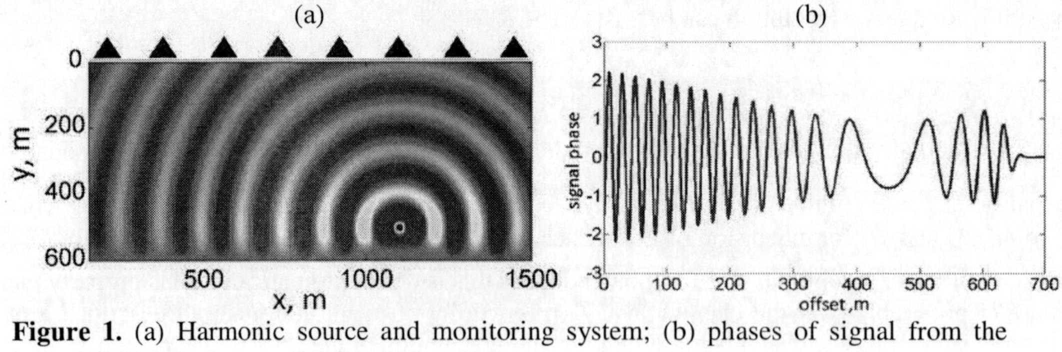

Figure 1. (a) Harmonic source and monitoring system; (b) phases of signal from the harmonic source at a certain time moment.

The synthetic seismograms are calculated using approaches proposed in [28, 29]. Then, we take *ST* of signals on each receiver and plot the vector of phases (4). By taking an argument (phase) of each element of the resultant vector we obtain a spatial signal. Figure 1b depicts such signal in the considered example at the source frequency of 20 Hz. Later on, the vector of phases is subjected to additional spectral transformation (spectral transform of spatial signal). Figure 2 shows illustrations of various transforms of spatial signal for a source of 100 Hz. It is shown that the best frequency–space resolution is achieved using *SST* (Figure 2c). Under the same condition *ST* fails to detect low frequencies of the spatial signal. For this reason *ST* brings considerable errors of the source location. Numerical experiments show that with a suitable selected window width, the method *STFT* (Figure 2a) provides the accuracy of spatial signal frequencies' calculations as *SST*. At the same time *SST* is preferable as it enables more accurate distinguishing of signals from sources located at different spatial points owing to high space–frequency resolution.

Analysis of the obtained spectrograms reveals that, when the incidence angle of wave changes sign, spectrum peaks move from one part of the spectrum to another (from top to bottom and vice versa). This property has been used to recover the wave incidence angles. Relocation of the spectrum peaks in the spectrogram can be explained by the switch of the signal to the opposite phase. Thus, analysis of spatial spectrograms can be useful in solution of problems connected with locating of sources with different focus mechanisms [30].

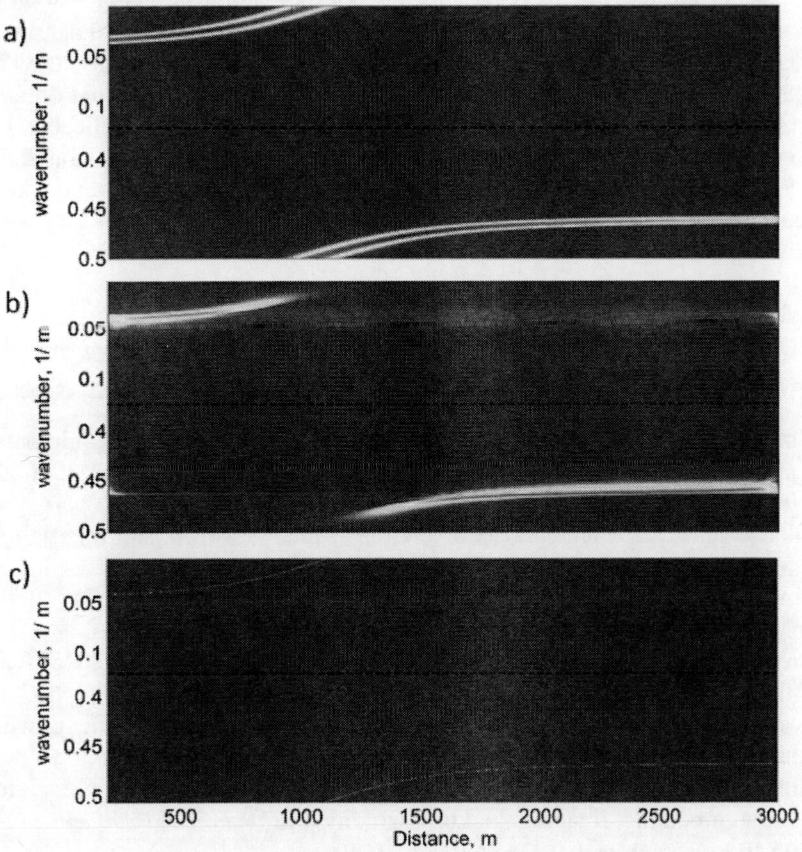

Figure 2. Spectrograms of spatial signal: (a) *STFT*; (b) *ST*; (c) *SST*.

The spatial frequency (wave number) of incident wave is found by detection of spectrum peaks in spectrogram (Figure 2). Thus, we have a wave number for each receiver. Then, the direction of wave arrival is calculated. In the homogeneous medium under consideration we locate a source by plotting a set of direct beams emitted from receivers at the found angles. The intersection point of the beams is the location of the source. Figure 3 shows examples of locating of harmonic sources with the frequencies of 10 and 50 Hz. Evidently, the proposed approach allows locating harmonic sources.

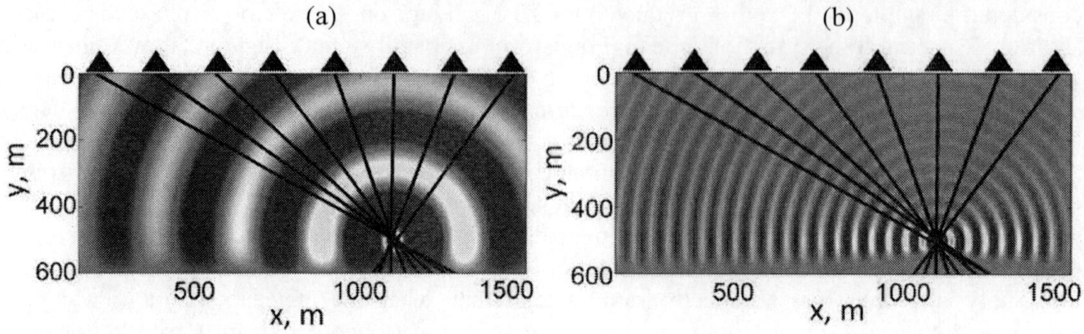

Figure 3. Locating a harmonic source using the proposed approach: (a) source frequency is 10 Hz; (b) source frequency is 50 Hz.

4. Conclusions

The paper has proposed the way to locate harmonic microseismic sources using spectral transformations and phase information of the signal. It has been shown that for the time–frequency representation of a signal, transforms *ST* and *SST* are preferable as against the classical *STFT*. The basic advantages consist in elimination of selecting the width of window and obtaining the best time–frequency resolution. Furthermore, it has been found that *ST* is inapplicable to space–frequency representation of signal. Using synthetic data, the author have demonstrated workability of the proposed algorithm. Based on the obtained results, it is possible to draw a conclusion that the use of various spectral transforms can improve efficiency of microseismic monitoring algorithms and assist in solving arising problems.

Acknowledgements

The research was supported by Russian Science Foundation, Project No 18-77-00043.

References

[1] Serdyukov SV, Azarov AV, Dergach PA and Duchkov AA 2015 Equipment for microseismic monitoring of geodynamic processes in underground hard mineral mining *Journal of Mining Science* Vol 51 No 3 pp 634–640

[2] Grechka V and Yaskevich S 2013 Inversion of microseismic data for triclinic velocity models *Geophysical Prospecting* Vol No 6 pp 1159–1170

[3] Warpinski N 2009 Microseismic monitoring: Inside and out *Journal of Petroleum Technology* Vol No 11 pp 80–85

[4] Kurlenya MV, Serdyukov AS, Duchkov AA and Serdyukov SV 2015 Wave tomography of methane pockets in coal bed *Journal of Mining Science* Vol No 4 pp 617–622

[5] Maxwell SC et al 2010 Petroleum reservoir characterization using downhole microseismic monitoring *Geophysics* Vol No 5 pp 75A129–75A137

[6] Serdyukov AS and Duchkov AA 2015 Hybrid kinematic-dynamic approach to seismic wave equation modeling, imaging, and tomography *Mathematical Problems in Engineering* 543540 http://doi.org/10.1155/2015/543540

[7] Chambers K, Kendall JM, Brandsberg-Dahl S Rueda J 2010 Testing the ability of surface arrays to monitor microseismic activity *Geophysical Prospecting* Vol 58 No 5 pp 821–830

[8] Gharti HN, Oye V, Roth M, Kühn D 2010 Automated microearthquake location using envelope stacking and robust global optimization *Geophysics* Vol 75 No 4 pp MA27–MA46

[9] Hogoev EA, Kolesnikov YI 2011 Application of emission tomography for study of geodynamically active zones *Tekhnologii Seismorazvedki* Vol 1 pp 59–65

[10] Chebotareva I a, Kushnir AF and Rozhkov MV 2008 Elimination of high-amplitude noise during passive monitoring of hydrocarbon deposits by the emission tomography method. *Izvestiya, Physics of the Solid Earth* Vol 44 pp 1002–1017
[11] Haldorsen JBU, Brooks NJ and Milenkovic M 2013 Locating microseismic sources using migration-based deconvolution *Geophysics* Vol No 5 pp KS73–KS84.
[12] Artman B, Podladtchikov I and Witten B 2010 Source location using time-reverse imaging *eophysical Prospecting* Vol 58 No 5 pp 861–873
[13] Zhu T 2014 Time-reverse modeling of acoustic wave propagation in attenuating media *Geophysical Journal International* Vol 197 No 1 pp 483–494.
[14] O'Brien GS et al. 2011 Time reverse location of seismic long-period events recorded on Mt Etna *Geophysical Journal International* Vol 184 No 1 pp 452–462
[15] Kushnir A, Varypaev A, Dricker I, Rozhkov M and Rozhkov N 2014 Passive surface microseismic monitoring as a statistical problem: location of weak microseismic signals in the presence of strongly correlated noise *Geophysical Prospecting* 2014 Vol 62 No 4 pp 819–833
[16] Kurlenya MV, Serdyukov AS, Serdyukov SV and Cheverda V A 2010 Seismic approach to location of methane accumulation zones in a coal seam *Journal of Mining Science* Vol 46 No 6 pp 621–629
[17] Johnson DH and Dudgeon DE 1993 Array signal processing: concepts and techniques *Englewood Cliffs: PTR Prentice Hall* pp 1–523
[18] Van Veen BD and Buckley KM 1988 Beamforming: A versatile approach to spatial filtering *IEEE assp magazine* Vol No 2 pp 4–24
[19] Ruigrok E, Gibbons S and Wapenaar K 2017 Cross-correlation beamforming *Journal of Seismology* pp 1–14
[20] Behr Y et al 2013 Source directionality of ambient seismic noise inferred from three-component beamforming *Journal of Geophysical Research: Solid Earth* Vol 118 No 1 pp 240–248
[21] Lehujeur M et al 2017 Vertical seismic profiling using double-beamforming processing of nonuniform anthropogenic seismic noise: The case study of Rittershoffen, Upper Rhine Graben, France *Geophysics* Vol 82 No 6 pp B209–B217
[22] Skazka VV, Serdyukov SV, Erokhin GN and Serdyukov AS 2013 Near-field range of the direct-impact seismic source *Journal of Mining Science* Vol 49 No 1 pp 60–67
[23] Azarov AV and Serdyukov SV 2018 Seismicity of rock rupture filled with fluid *IOP Conference Series: Earth and Environmental Science* Vol 134 No 1 pp 012002
[24] Azarov AV and Serdyukov AS 2018 Excitation of seismic waves at the motion of fluid in the rock massive crack *IOP Conf. Ser.: Earth and Environmental Science* Vol 134 No 1 pp 012003.
[25] Tary JB, Herrera RH, Han J and M. van der Baan 2014, Spectral estimation—What is new? What is next?, *Rev. Geophys.* 52 pp 723–749
[26] Stockwell R, Mansinha L and Lowe R 1996 Localization of the complex spectrum: The S transform *IEEE Transactions on Signal Processing* 44 (4) pp 998–1001
[27] Daubechies I, Lu J and Wu H T 2011 Synchrosqueezed wavelet transforms: An empirical mode decomposition-like tool *Appl. Comput. Harmonic Analysis* Vol 30 No 2 pp 243–261
[28] Azarov AV, Serdyukov AS and Nikitin AA 2015 Numerical modeling of microseismic noise in the block-hierarchical heterogeneous environments *GIAB* No 9 pp 328–335
[29] Kurlenya MV, Serdyukov AS, Azarov AV and Nikitin AA 2015 Numerical modeling of wavefields of microseismic events in underground mining *Journal of Mining Science* Vol 51 No 4 pp 689–695
[30] Azarov AV, Serdyukov AS and Yablokov AV 2016 Methods of the focal mechanism determination of microseismic events based on modeling full wave fields in horizontally stratified media *GIAB* No 10 pp 131–143

Synthesis of elements for waver gears of tunneling assemblies

AV Baris* and VP Gileta

Novosibirsk State Technical University, Novosibirsk, Russia

E-mail:*pmx08@mail.ru

Abstract. Wave gears have high load capacity and small weight and dimensions parameters, therefore they are increasingly used in mining aggregates. The method of synthesis of elements of the wave transfer, which is based on the method of vector contours, is given in the work. The method takes into account the design features of the finger supports and the achievement of continuity of contact of the fingers with the generator, the rigid wheel and the holes of the guide disk. The proposed technique allows creating wave transmissions with a range of transmission ratios of 6 – 80. The adequacy of the synthesis technique is proved by the operability of waveguide samples, whose flexible wheels are made in the form of a finger chain. A new wave transmission is used in the developed microtunneling equipment.

1. Introduction

Wave gears, although recently invented, enjoy increasingly wider application. They ensure transmission ratios from 80 to 300, possess high loading capability, have efficiency to 0.9 and make it possible to reduced weight and size of drives by 1.5–2.5 times as compared with the regular toothed gearings [1–5]. Due to these advantages, wave gears can be used in power reducers in mining machines. For example, wave gear is included in cutter drive gear box of heading machine SM-130 [6]. A wave gear with spacer links is used in transmission of the tunneling machine—geokhod [7].

Design of a wave gear is the same since invention. A wave gear is composed of three main elements: flexible wheel rigidly connected with the output shaft; rigid wheel fixed in the body and a wave generator in the form of a cum with the input shaft. The application area of the wave gears is limited by transmission ratios higher than 80 [3–5]. The underlying condition is the design of the flexible wheel which as a thin-wall envelope with toothed periphery. At the transmission rations lower than 80, flexible wheels can break down due to high bending stresses.

In order to ensure transmission ratios of wave gears from 6 to 80, it is proposed to use a finger chain in the capacity of the flexible wheel [8]. In this case, deformation of the flexible wheel takes place owing relative turn of the fingers. Working ability of a wave gear is governed by quality shaping of teeth on the flexible and rigid wheels [4, 5, 8].

2. Design features of teethed gears

Having the most accurate description of teeth geometry on flexible and rigid wheels is very important for syntheses of force wave teethed gears. The conventional synthesis of wave gears assumes that the flexible wheel deformations are insignificant, and engagement of flexible and toothed is considered as plane for this reason. In the developed chain wave gear, fingers represent the teeth of the flexible wheel; fixed on the backer plate, fingers move not in parallel to the gear axis but carry out rotational motions along conical trajectories. The inclination of the fingers to the gear axis varies continuously and is 2–5°. The faces in the new gear are synthesized based on the spatial motion of the fingers, which ensures that the fingers have linear contact with the faces of the rigid wheel teeth, holes of the guide disk and generator.

In the wave gear structure in Figure 1, the flexible element is a chain of fingers coupled by plates. The fingers are mounted on the guide backer plate. Each plate couples two fingers and forms a chain hinge. The rigid wheel has internal teeth and is fixed in the body. As against the conventional flexible wheel, the finger chain lacks torsional stiffness; for this reason, the new gear is equipped with the guide disk with special-shape holes arranged on the output shaft. The guide disk is manufactured as a complete unit with the backer plate at which, using third-class spherical kinematic couples, the ends of the fingers are clamped. On the other side, the fingers interact with the cam generator on the input shaft. The number of the fingers exceeds the number of the rigid wheel teeth and equals the number of the guide disk holes.

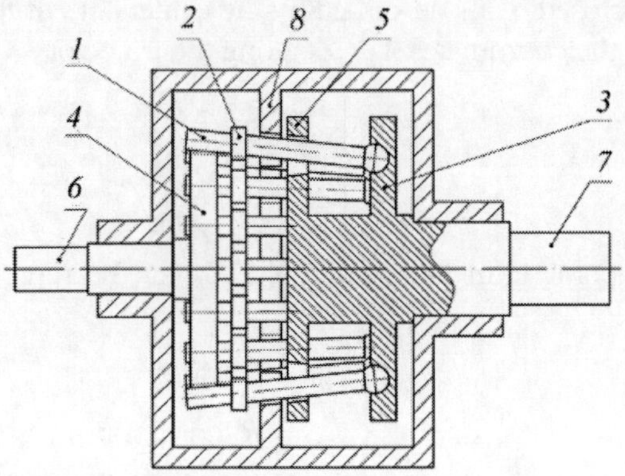

Figure 1. Wave gear: *1*—finger; *2*—plate; *3*—backer plate; *4*—generator; *5*—guide disk; *6*—input shaft; *7*—output shaft; *8*—rigid wheel.

The transmission ratio is given by the expression [3–5 with number z_g of the flexible wheel teeth is replaced by the number n_g of fingers in the chain; "minus" points at rotation of the input and output shafts in different directions:

$$i = -\frac{n_g}{K_z U}, \tag{1}$$

where K_z is the divisible factor equal to difference between the numbers of rigid wheel teeth and fingers in the same deformation wave; U —is the number of deformation waves.

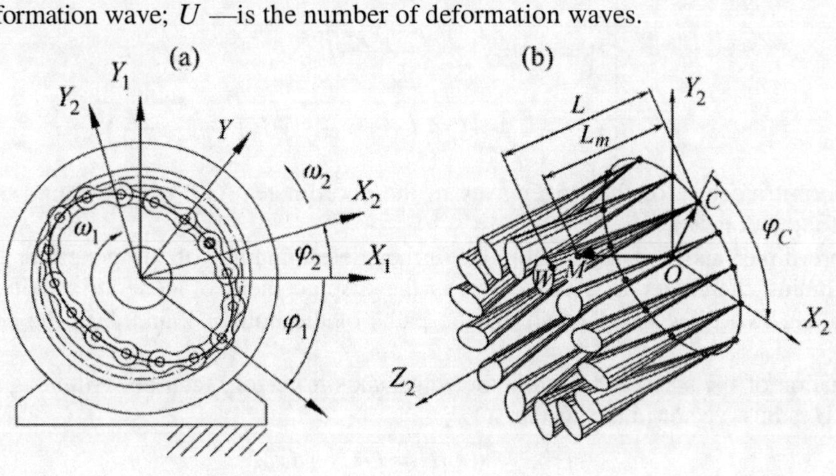

Figure 2. Sketch of determining trajectories of finger axes: (a) calculation coordinate systems; (b) motion of finger axes in coordinate system connected with guide disk.

3. Finger motion modeling

The study of the law of finger motion is based on the coordinate method applied in the theory of toothing [9]. Let there be introduced three coordinates *XYZ*, $X_1Y_1Z_1$ and $X_2Y_2Z_2$ connected, respectively, with generator, rigid wheel and guide disk (Figure 2a).

The axes *Z*, Z_1 and Z_2 coincide with the axis of the chain wave gear. The planes *XY*, X_1Y_1 and X_2Y_2 coincide among themselves and with the plane in which the finger-and-backer plate clamp points *C* being the centers of spherical kinematic pairs. The spacing of the neighbor points *C* is equal to the chain pitch p_C.

Let us consider motion of fingers in the coordinates $X_2Y_2Z_2$ as rotational motions of solids relative to immobile points *C* (Figure 2b). The axes of the fingers move along the conical trajectories with the tips at the points *C* along the closed ovals described by the centers *W* of the chain hinges [8].

Let the point *W* move in the coordinates $X_2Y_2Z_2$ along the trajectory:

$$\begin{cases} x_{2W} = f_{x2}(t), \\ y_{2W} = f_{y2}(t), \\ z_{2W} = f_{z2}(t). \end{cases} \tag{2}$$

The coordinates of the point *C* in the system $X_2Y_2Z_2$ can be represented by the coordinates of radius–vector $\overline{OC}$:

$$\begin{cases} x_{2C} = R_c \sin\phi_C, \\ y_{2C} = R_c \cos\phi_C, \\ z_{2C} = 0. \end{cases} \tag{3}$$

The coordinates of the point *M* in the axis of the finger can be defined by the vector:

$$\overline{OM} = \overline{OC} + \overline{CM}. \tag{4}$$

Here, the vector $\overline{CM}$ coincides with the vector $\overline{CW}$.

Denote the vector length $\left|\overline{CM}\right| = L_m$ and present the conical surface as the geometrical places of the points *M*. According to (4):

$$\begin{cases} x_{2M} = x_{2C} + \dfrac{x_{2W} - x_{2C}}{L} L_m, \\ y_{2M} = y_{2C} + \dfrac{y_{2W} - y_{2C}}{L} L_m, \\ z_{2M} = \dfrac{L_m}{L}\sqrt{L^2 - \left(x_{2W} - x_{2C}\right)^2 - \left(y_{2W} - y_{2C}\right)^2}. \end{cases} \tag{5}$$

The motion trajectory of the finger axes in the coordinates *XYZ* can be found by transform of coordinates for the point *M*.

It is more direct and convenient for the further determination of the generator face to use the method of finding the finger axis trajectory as the displacement of generatrix coincident with the finger axis along two directrixes which are the paths of the points *W* and *C* in the coordinates *XYZ* (Fig. 3a).

The equation of the surface, which is the finger axis trajectory, can be written as the coordinates of the point *M* radius–vector in the system *XYZ*

$$\overline{OM'} = \overline{OC'} + \overline{CM'}. \tag{6}$$

The equation of the generator face is found similarly to the finger axis trajectory in the system $X_2Y_2Z_2$.

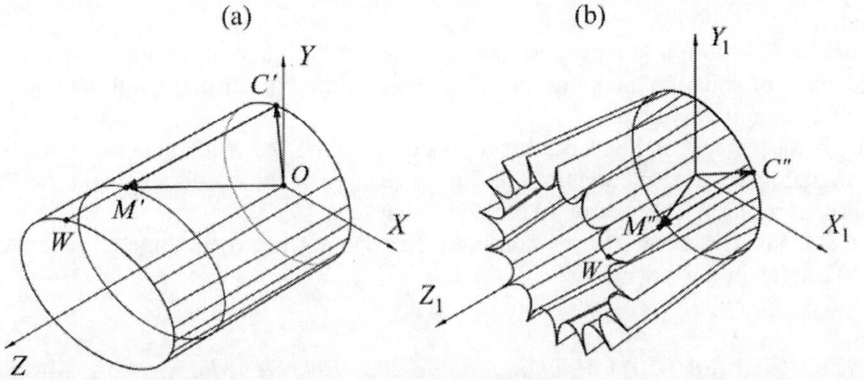

Figure 3. Sketch of finding trajectory of finger axes: (a) in coordinate system connected with generator; (b) in coordinate system connected with rigid wheel.

The coordinates of the radius–vector $\overline{OM}'' = \overline{OC}'' + \overline{CM}''$ in the system $X_1Y_1Z_1$ (Fig. 2b) will be:

$$\begin{cases} x_{1M} = x_{1C} + \dfrac{x_{1W} - x_{1C}}{L} L_m, \\ y_{1M} = y_{1C} + \dfrac{y_{1W} - y_{1C}}{L} L_m, \\ z_{1M} = \dfrac{L_m}{L} \sqrt{L^2 - (x_{1W} - x_{1C})^2 - (y_{1W} - y_{1C})^2}. \end{cases} \quad (7)$$

Motion of fingers in the coordinate systems XYZ and $X_1Y_1Z_1$ can be characterized as compound motion of solids. In this case, relative motion of fingers in the system $X_2Y_2Z_2$ is their rotation relative to the points *C*. Translation motion is the rotation of the points *C* around the gear axis. This is the case of summation of rotations around unparallel axes.

The synthesis of work faces should ensure conditions when lateral cylindrical surfaces of the flexible chain fingers is in continuous contact with generator, rigid wheel teeth and guide disk holes. The work faces of the generator, rigid wheel teeth and guide disk holes will be the enveloping surfaces of the interfaced fingers of the chain [8]. Since the faces of the fingers are cylindrical, the problem reduces to determining equally distant surfaces to the trajectories of the finger axes. This method can be assumed as the analytical synthesis of engagements based on the classical apparatus of differential geometry. This method also enables synthesizing work faces of the generator cam.

The synthesis method is implemented in engineering of prototype wave gears with the transmission ratios 6, 8 and 10 (Fig. 4). The working ability of these wave gears proved efficiency of the procedure. The wave gear with the flexible wheel represented by the finger chain was used in the microtunneling equipment [10].

Figure 4. Prototype wave gear with transmission ratio 8.

4. Conclusions

Representation of the motion of fingers in the guide disk relative to immobile points takes into account structural features of spherical supports of the fingers on the backer plate and the one-way connection between the fingers and the plates.

Motion of the fingers in the coordinate systems connected with generator and rigid wheels is compound, including relative motions of the fingers relative to the rigid wheel and translational rotation of spherical support of the fingers around the wave gear axis.

The developed procedure ensures continuity of contact between fingers, generator, rigid wheel and guide disk holes in the wave gear.

References

[1] Frolov KV et al (Eds) 2011 *Machine Engineering. Encyclopedia* Moscow: Mashinostroenie Vol 4 (in Russian)
[2] Kubo A 2914 The future of mechanical power gears *Theory and Practice of Toothed Gears: Int. Symp. Proc.* Izhevsk: IzhGTU pp 50–56 (In Russian)
[3] Frolov KV et al (Eds) 1995 *Machine Engineering. Encyclopedia* Moscow: Mashinostroenie Vol 4 (in Russian)
[4] Ivanov MN 1981 *Harmonic Drives* Moscow: Vyssh. Shkola (in Russian)
[5] Ginzburg EG 1969 *Toothed Wave Gears* Moscow: Mashinostroenie (in Russian)
[6] Kovalev VA, Khoreshok AA and Gerike BL 2015 Diagnostics of the condition of selective-action heading machines *Ugol* No 9 pp 42–46
[7] Aksenov VV, Efremenkov AB, Timofeev VYu, Blashchuk MYu and Beglyakov VYu 2010 Review of transmissions *GIAB* OV No 3 pp 55–66
[8] Gileta VP and Baris AV 2014 Chain wave gear *Izv. Vuzov. Stroit.* No 5 pp 84–90
[9] Litvin FL 1960 *Theory of Gearing* Moscow: Nauka (in Russian)
[10] Gileta VP, Baris AV and Vanag YuV 2017 Layout and variables of air-powered drive of small-size pneumatic winch *IOP Conference Series: Earth and Environmental Science* Vol 53 Art 012023

IOP Conf. Series: Earth and Environmental Science **262** (2019) 012006 doi:10.1088/1755-1315/262/1/012006

Upgrading of deviation measurement system for more exact determination of sub-horizontal hole profile

VD Baryshnikov*, VG Kachalsky and DV Baryshnikov

Chinakal Institute of Mining, Siberian Branch Russian, Academy of Sciences, Novosibirsk, Russia

E-mail: *vbar@misd.ru

Abstract. Tests of a developed system for measuring deviation of drill holes reveal its shortcomings. Regarding subhorizontal holes, the key problem is measurement of inclinations and deviations from vertical line at a huge number of points (up to 200) by the measuring device moving inside a casing string to 100 m long. The authors analyze influence of different-type data channels on reliability, accuracy and stability of the data recording in the system based on analog and digital signal converters. Applicability of liquid and solid-state integral deviation sensors for pointing of the down-the-hole measuring device is discussed. The paper proposes to use digital microprocessor methods for measuring and converting signals with software support at the upper and lower levels of the deviation determination system.

1. Introduction

Safety of mining can be provided using various methods, including integrated geomechanical assessment of properties and behavior of rock mass. Protection from water inrushes in mines under water bodies on the ground surface, or under aquifers involves creation of safety pillars, and the only reliable tool to maintain mining safety and prevent risk of flooding is instrumental monitoring of deformation of such pillars.

One of the main characteristics of movement in undermined rock strata is vertical displacements (subsidence).

In top-downward cut-and-fill, the cemented backfill is a support for the protective pillar. Safety of the pillar essentially depends on sagging of the underlying backfill. Sag of a check layer in the backfill is determined based on the difference between final and initial profile of a measurement hole drilled in the backfill. These profiles are determined by deviation measurements. Applicability of directional survey in backfill control is discussed in [1, 2]. The deviation survey consists is sequential measurement of inclinations at equal spacing along a hole [3]. Reconstruction of the hole profile uses available standards and guidelines [4, 5] by averaging the measured inclinations:

$$z_n = z_0 + x\sum_{i=1}^{n} \sin\left(\frac{\alpha_i + \alpha_{i-1}}{2}\right),$$

where z_0 is the vertical mark of the hole mouth according to survey data; x is spacing; of measurements; $n = l_{\text{hole}} / x$ is the number of measurements; α_i is the inclination of casing at a point i. In a general case, a hole profile can be represented by a smooth curve.

Deviation devices have different channels to receive and measure signals proportional to inclinations of the device. Currently, international manufacturers offer a wide selection of inclination sensors differing in principles of inclination measurement and intended for solving various problems connected with human activities [6].

As a rule, deviation devices use sensors in the form of spherical gyroscopes (accelerators with magnetic rotor suspensions), one-axis gyros, magnetoresistive inclination sensors, angular accelerometers, liquid-type pickups, etc. Inclination sensors made on the basis of liquid-type pickups are less sensitive to vibrations as against solid-state accelerometers, and are immune to magnetic masses nearby (drill tubes, casings, reinforcement, etc.) as compared to inclination meters with magnetic elements.

2. Development of downhole measurement system

For measuring inclinations at preset points, a system comprised of a downhole deviation device, signal converter and a personal computer is designed. The downhole deviation device consists of a sealed housing with two liquid sensors of inclination in vertical and horizontal planes. The tests of the device have revealed a number of its blind sides such as:

1. Insufficient measurement reliability due to considerable noise in outputs of deviation devices manufactured in Russia. The upgraded system is equipped with the deviation devices produced by Seiko, Germany (Figure 1). German devices have housing of protection class IPG65, which ensures reliable operation in wide temperature range from -40 to $+85$°C, at humidity up to 100%, inclination range of ±30° for the main sensor and ±45° for the ancillary sensor, with resolution of 0.005° and linearity of ±0.06° for readings within the whole range of measurement [5].

2. In case of elongation of data channel, application of deviation devices with weak analog output results increases capacitive load and diminishes signal amplitude, which degrades dynamic characteristics of the measurement circuit and, as a consequence, adds stabilization time of the deviation device readings. In the upgraded system, wire interface connection with PC (RS-232) is replaced by Bluetooth, which ensures operator's comfort. The upgraded deviation device has larger spacing (distance between lower supports) from 25 to 50 cm. The outer diameter of the devices is diminished from 70 to 55 cm.

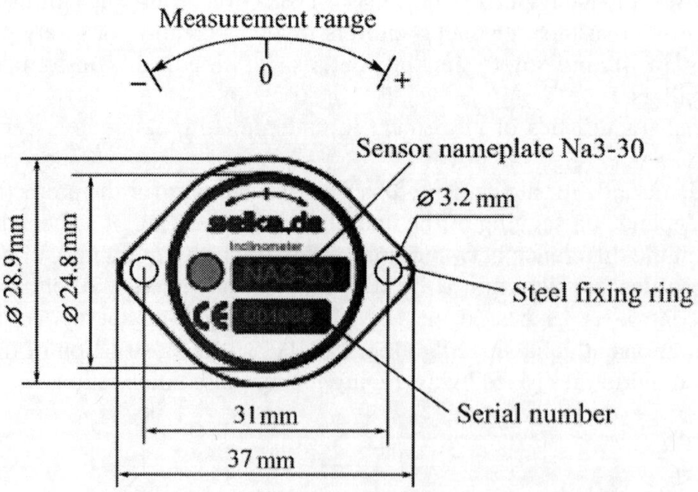

Figure 1. Seika deviation device and dimensions.

The flow chart of the upgraded downhole tool in Figure 2 comprises of two deviation devices, p-c board with AD converter chips, sensor reference processor and chip of connection with RS-485 channel.

Signal conversion device is arranged in an individual housing and connected to tablet computer via Bluetooth.

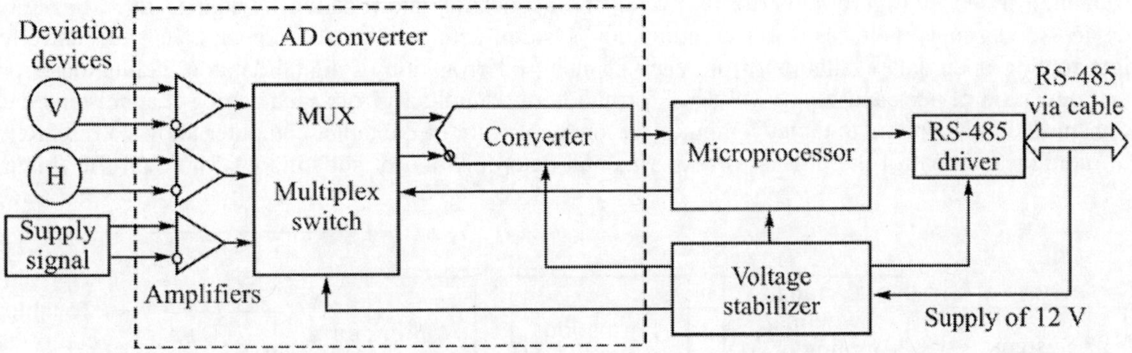

Figure 2. Flow chart of signal processing from deviation devices (V—vertical deviation device; H—horizontal deviation device).

Receiving and processing of signals from deviation devices, as well as their conversion to standard protocol RS-232 for wireless transfer via Bluetooth is implemented by a signal conversion device equipped with (Figure 3):

- RS-232–RS-485 interface converter (module ADAM-4520, Advantech);
- Bluetooth wireless communication modul (BT-0240 RS232 Adapter [7];
- Li–Ion accumulator battery, capacity 12 A·h.

The interface converter ADAM-4520 and wireless adaptor Bluetooth provide wireless communication in standard Bluetooth V2.0. The accumulator battery maintains operation of the downhole tool and recording station for not less than 24 h.

The engineering problem to be solved is improvement of efficiency and accuracy of the downhole tool—two-coordinate deviation system—by means of more accurate orientation of the tool and recording of inclinations at control points in the hole relative to horizontal and vertical planes.

The electronic devices in the flow charts in Figure 3 are placed in a sealed housing of the downhole tool. The communication and electric connection are implemented by a coupler with four-wire cable channel RS-485 and supply.

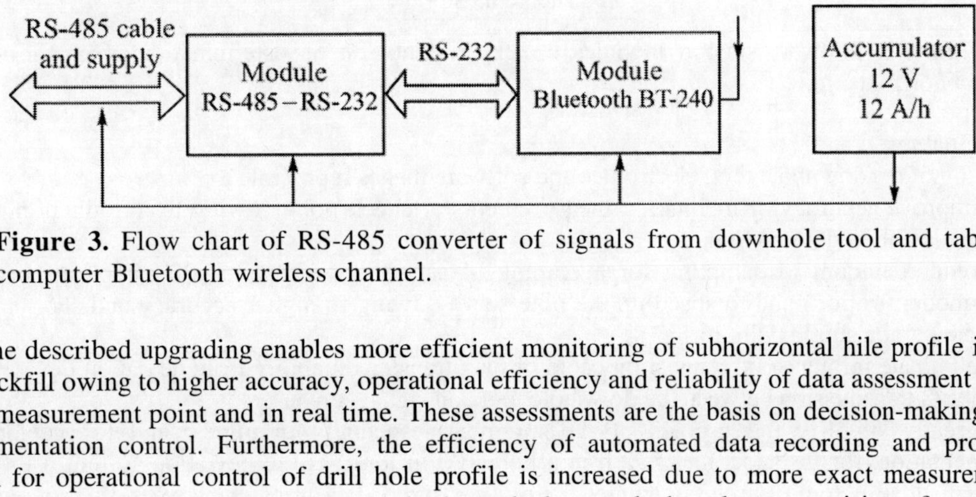

Figure 3. Flow chart of RS-485 converter of signals from downhole tool and tablet computer Bluetooth wireless channel.

The described upgrading enables more efficient monitoring of subhorizontal hile profile in rocks and backfill owing to higher accuracy, operational efficiency and reliability of data assessment directly at the measurement point and in real time. These assessments are the basis on decision-making on the experimentation control. Furthermore, the efficiency of automated data recording and processing system for operational control of drill hole profile is increased due to more exact measurement of inclinations in the check hole relative to horizontal plane and, thus, better precision of stress state determination in rocks.

Program support is greatly amended. The new version is composed of two parts with regard to the block structure of the software system [8]. First, software of the microprocessor, the flow chart of which is demonstrated in Figure 4, performs cyclic and sequential interrogation of signals from deviation devices and supply voltages via a commutation system, actuation of AD conversion, expectation of interruption upon data availability from each channel, interrogation of digital data on inclinations from two deviation devices and supply voltages, formation of data blocks from each analog channels and data transmission on demand to tablet computer. Second, software of the tablet computer allows a researcher to monitor and control the process of experimental determination of subhorizontal hole profile (Figure 5).

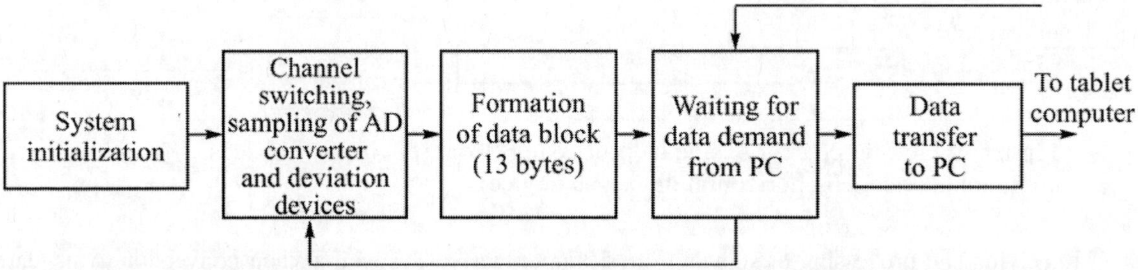

Figure 4. Flow chart of software support for microprocessor system of downhole tool.

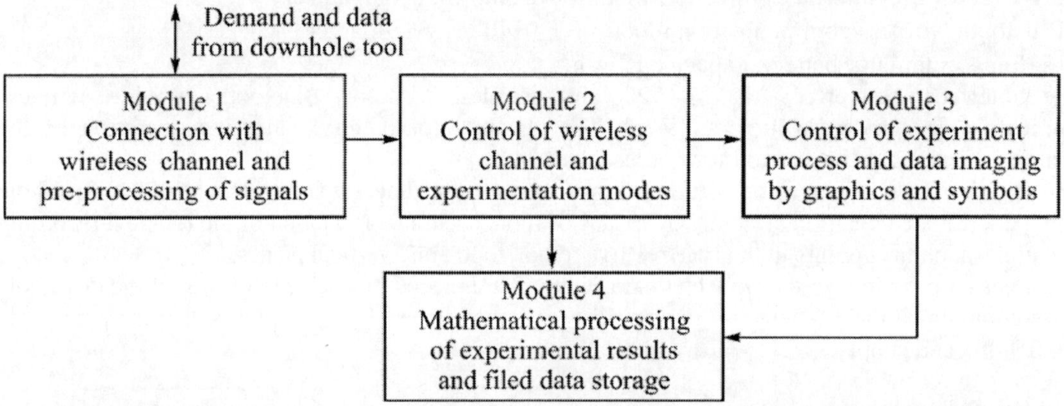

Figure 5. Software support modules for experimentation on determination of profile of subhorizontal holes.

3. Conclusions

Upgrading of the downhole tool, electronics and software makes it possible to:

—improve accuracy of inclination measurements at check points owing to introduction of high-precision deviation devices Seiko;

—reduce amount of sampling for averaging (4 samplings as against 20 samplings before the system modernization) and to speed up the hole survey owing to higher accuracy and shorter time of measurement at a single point to 1–2 s;

—eliminate influence of connection cable of data transfer reliability from deviation devices due to discrete data transmission between the downhole tool and signal converter;

—greatly simplify the process of measurement, especially in mines, and to accelerate the experimentation, for instance, to 1.5–2 h in a hole 100 m long, as the deviation system uses a tablet computer and Bluetooth wireless channel for the measured data transfer between the computer and signal converter;

—implement on –the-fly control over the experiment;

—use express-analysis and compare the earlier and current measurement of the check hole profile, promptly detect jumps in the profile and locate zones of possible dog-legs;

—real-time assess experimental results on-site and to compare them with the earlier obtained data.

References

[1] Baryshnikov VD, Baryshnikov DV and Kachalsky VG 2012 Displacement control in undermined backfill *Nonlinear Geomechanics and Geodynamics in Deep Level Ming; II Sino–Russian Conference Proceeings* Novosibirsk 2012*ing: International Conference Proceedings* Novosibirsk: Nauka pp 385–388

[2] Baryshnikov VD, Baryshnikov DV and Kachalsky VG 2010 Displacement control in undermined backfill *Proceedings of IV International Conferece on Geomechanics* Bulgaria pp 127–135

[3] http://doidpo.rusoil.net/pluginfile.php/14563/mod_resource/content/2BD.pdf Basic principles of inclino-metry

[4] *RD 153-39.0-072-01 Guidlelines on Geophysical Survey and Cabled Instumentation Operatioon in Oil and Gas Wells* Approved by the Ministry of Energy of Russian Federation, Order No. 134 Dated 7 May 2001 (in Russian)

[5] *Register of EAGO Standards. Systems for Certification of Geophysical Products* Available at: http://eago.1gb.ru/en/uploads/files/Cert_EAGO_Standards.pdf www.eago.ru

[6] https://ru.wikipedia.org/wiki/Inklinometer

[7] http://www.seika.de/english/pdf_e/NA_e.pdf

[8] Leontiev AV and Petrov VE 1997 Block-modular principle of construction of geomechanical measuring and computing systems *Journal of Mining Science* No 1 pp 110–119

IOP Conf. Series: Earth and Environmental Science **262** (2019) 012007 doi:10.1088/1755-1315/262/1/012007

Stress state of rock mass during stoping under pit bottom in the presence of slide rocks on the pit bottom

VD Baryshnikov* and LN Gakhova

Chinakal Institute of Mining, Siberian Branch Russian, Academy of Sciences, Novosibirsk, Russia

E-mail: *vbar@misd.ru

Abstract. The article studies influence of slide rocks on the bottom of Internatsionalny open pit mine on stress state of rock mass during stoping under the pit bottom. The stress distribution during extraction of ore reserves from the initial layer under the pit bottom is determined. The post-limiting deformation zones in ore body during formation of a crown pillar are delineated using the Mohr–Coulomb criterion. The authors give a predictive estimate of alteration in stress state of the crown pillar with the advance of stoping.

1. Introduction

Internatsionalnaya kimberlite pipe represented by the subvertical oval ore body with the transverse cross-section 60×104 m is developed by ALROSA the top-down cut-and-fill method with layers 4–5 high and stopes 5–6 m wide. For safe extraction of ore reserves under the open pit mine bottom in complex hydro-geomechanical conditions, it is planned to arrange a protective crown pillar 35 m thick.

Safe mining of kimberlite resources calls for persistent improvement of standards and procedures based on integrated researches [1]. The most effective approach to the assessment of crown behavior, alongside with rock mechanics [2–5], instrumental and visual techniques, are the the mathematical modeling methods which allow description of features and regular patterns of stress state in the crown in case of the alteration of geotechnical situation in rock mass under the pit bottom and hydrological situation within the open pit void [6, 7].

This study aims to analyze and predict geomechanical situation under the open pit mine bottom in such cases as wet slide rocks on the pit bottom and advance of stoping under the bottom.

2. Stress state estimation

Stress state under the open pit bottom was estimated based on numerical calculation data obtained using the method of boundary integral equations and elastic model adapted to the open pit mine conditions [6–8]. The problems were solved for the real geometry of the open pit. The depth of Internatsionalny mine is 315 m (bottom level is + 85 m). Currently, it is filled with slide rocks up to the level of + 120 m Water level in the pit is + 115 m.

The natural field of stresses for the mine is assumed as $\sigma_y^0 = -\gamma H$, $\sigma_x^0 = -\lambda\gamma H$, where σ_x^0, σ_y^0 are the horizontal and vertical stresses; $\lambda = 1$ is the lateral earth pressure coefficient [9]; γ is the bulk density of rocks, MN/m^3; H is the mining depth, m. The calculations included the average modulus of deformation E = 7.5 GPa and Poisson's ratio $\nu = 0.25$ [10–12]. The interpretation of the calculated results used such values of stresses σ_s which permitted employment of the Mohr–Coulomb criterion [5, 8]:

$$\sigma_s = \frac{\sigma_1 - \sigma_3}{2\cos\varphi} + \frac{\sigma_1 + \sigma_3}{2}\mathrm{tg}\varphi,$$

where $\sigma_1 > \sigma_2 > \sigma_3$ are the principal stresses; φ is the internal friction angle.

From the visual observations and numerical calculations for a pilot extraction block (PEB 2), it has been found that failure of rock mass in the open pit bottom takes place in the zones where $\sigma_s > 4$ MPa [6–8]. Thus, $\sigma_s^{\text{к}} = 4$ MP is assumed as the critical value.

After completion of open pit mining, zones of higher horizontal stresses and vertical stress relaxation appear in the pit bottom rocks. It is planned to form a crown pillar by cutting and filling in the initial layer at a distance of 35 m from the pit bottom (Figure 1). After extraction of the first-phase stopes in the initial cut layer, the compressive stresses σ_x at the top of the crown grow up to $-16 \div -18$ MPa (Figure 1a).

Relaxation of vertical stresses is observed under the pit bottom and in the roof of the first-phase stopes (Figure 1b). In the pillars between these stopes, σ_y reaches – 12 MPa. Completion of mining in the initial layer results in the increase in σ_x to $-20 \div -22$ MPa (Figure 1c) and total relaxation of σ_y along the height of the crown (Figure 1d).

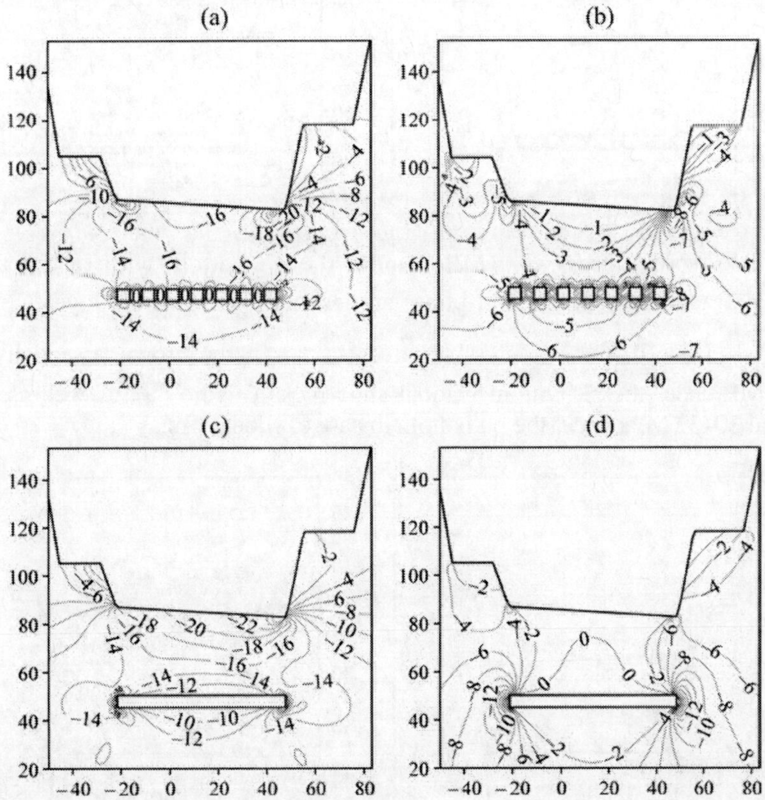

Figure 1. Horizontal stress σ_x and vertical stress σ_y (MPa) in rock mass under the open pit bottom after (a), (b) extraction of the first-phase stopes in the initial layer and (c), (d) upon completion of mining in the layer.

Thus, the horizontal stresses σ_x in the crown are the determinant, and the further analysis of the stress state uses the calculated data on σ_x.

When the open pit is filled with slid rocks up tot the level of + 120 m (Figure 2) and when the fill gets wet to the level of 115 m, additional load on the crown is [5]:

$$P_z = \zeta \gamma_1 H_1 + (1-\zeta)\gamma_2 H_2,$$

where $\zeta = 0.7$ is the degree of fragmentation in slide rocks; $\gamma_1 = 0.022$ is the bulk density of rocks, MN/m^3; $\gamma_2 = 0.01$ is the bulk density of water, MN/m^3; H_1 is the thickness of slide rocks; H_2 is the thickness of wet rocks.

Additional load on the crown induces an increase in the compressive stresses σ_x under the pit bottom, especially in the concentration zones during the first-phase stoping (Figure 2a), and in σ_x by $\approx 10\%$ after completion of stoping in the first layer. In the meanwhile, stress distribution in the roof of the initial cut-and-fill layer is insignificant (Figures 1c and 2b).

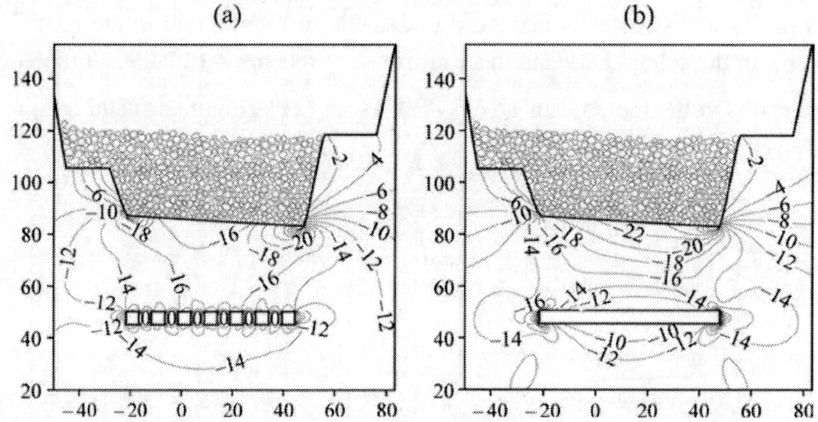

Figure 2. Horizontal stress σ_x (MPa) under the pit bottom with regard to additional load after extraction of (a) the first-phase stopes and (b) the whole layer.

As stoping is advanced, the compressive stresses σ_x in the crown essentially grow across its thickness: to 28 MP under the pit bottom without slide rocks (Figure 3a), as well as to 26 MPa in the stoping roof and to 30–32 MPa under the pit bottom in case of slide rocks.

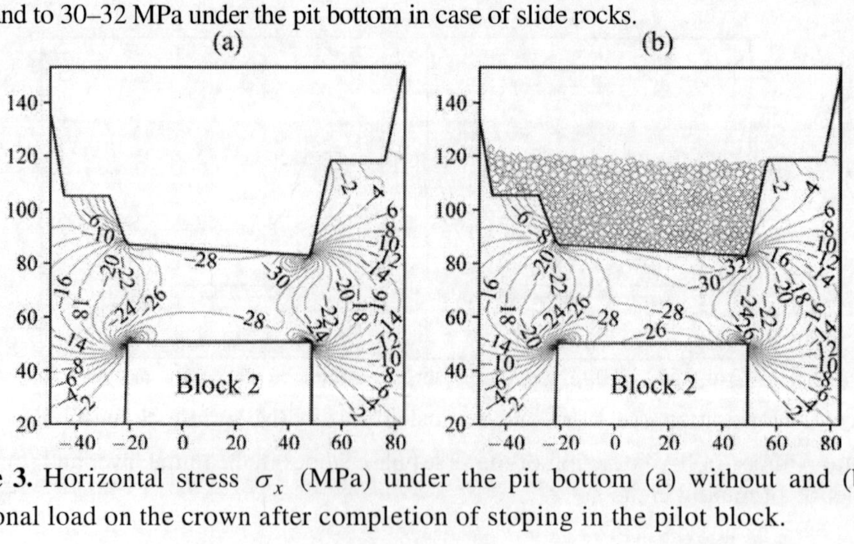

Figure 3. Horizontal stress σ_x (MPa) under the pit bottom (a) without and (b) with additional load on the crown after completion of stoping in the pilot block.

3. Inelastic strain zones behavior

Behavior of inelastic strain zones with taking into account additional load in the open pit allows drawing some conclusions (Figure 4). During the first-phase stoping in the initial layer, the inelastic strain zones are formed under the pit bottom, at the corner areas of the pit bottom and pit walls, as well as near the stopes at the ends of the layer (Figure 4a). Compete extraction of ore reserves from the initial layer changes essentially the deformation pattern in the crown and nearby it: the inelastic strains embrace considerable portion of the crown; in the roof of the layer at the interface of enclosing rock mass and kimberlites, the inelastic strain zones expand to 2–2.5 m depthward (Figure 4b).

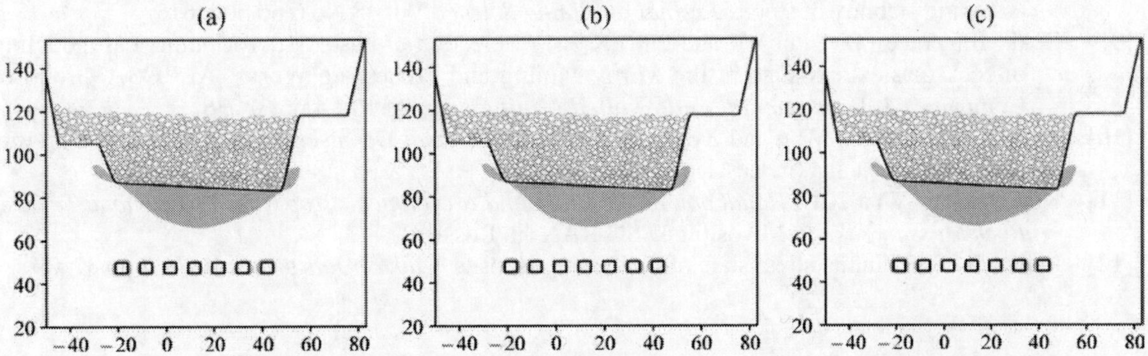

Figure 4. Formation of inelastic strain zones during stoping advance in the initial layer under open pit mine bottom: (a) the first-phases stopes are extracted; (b) initial layer stoping is completed; (c) mining in block 2 is completed.

After total extraction of block 2, the inelastic strain zones spreads across and along the thickness of the crown. The inelastic strain zones develop at a distance of 200 m from the pit bottom and inward enclosing rock mass (Figure 4c).

In this manner, the pre-estimate of stresses and strains induced in rock mass under the open pit bottom by slide rocks on the bottom and by stoping proves the essential increase in the horizontal stresses which cause inelastic strains in the ore crown pillar and in enclosing rock mass at the open pit bottom. In particular, at the present boundary conditions in the accepted geomechanical model, the post-limiting deformation domains at the final stage of stoping in block 2 encompass a greater portion of the crown and the lower part of rock mass at the pit bottom and pit walls up to the level of 105–110 m.

Acknowledgements

This study has been supported by ALROSA.

References

[1] Trubetskoy KN, Kaplunov DR and Rylnikova MV 2011 Problems and prospects of geotechnological innovation in the sphere of integrated kimberlite resources management *Problems and Ways of Efficient Diamond Mining: International Conference Proceedings* Novosibirsk: Nauka (in Russian)

[2] Mathews KN, Hoek E, Wyllie DC and Stewart SBV 1980 Prediction of stable excavation spans for mining at depths below 1,000 meters in hard rock *Golder Associates Report to Canada Centre for Mining and Energy Technology (CAANMET)*, Department of Energy and Resources Ottawa Canada

[3] Laubscher DN 1990 A geomechanics classification system for the rating of rock mass in mini design *J. S. Afr. Min. Metall.* No 10 pp 257–273

[4] Barton N 1993 Application of Q-system and index tests to estimate shear strength and deformability of rock masses *Workshop on Norwegian Method of Tunneling* New Delhi pp 66–84

[5] Bulychev NS 1989 *Mechanics of Underground Structures* Moscow: Nedra (in Russian)
[6] Baryshnikov VD and Gakhova LN 2017 Stress-strain state of adjacent rock mass under slice mining of steeply dipping ore bodies *J. Fundamental and Applied Mining Science* Vol 4 pp 32–36
[7] Baryshnikov VD and Gakhova LN 2015 Selection of efficient flow sheets for initial cutting in top-down mining in Internatsionalny mine *Journal of Mining Science* Vol 51 No 6 pp 1165–1172
[8] Kurlenya MV, Baryshnikov VD and Gakhova LN 2012 Experimental and analytical method for assessing stability of stopes *Journal of Mining Science* Vol 48 No 6 pp 609–615
[9] Bokiy IB, Zoteev OV, Pul VV and Pul EK 2017 Selection of basic data for numerical modeling of rock mass stress state in the Mirny Mining and Processing Works, ALROSA Group of Companies *J. Fundamental and Applied Mining Science* Vol 4 pp 32–36
[10] Kazeev AA, Izakson VYu and Zvonarev NK 1995 *Thermal Geomechanics of Diamond Deposits* Novosibirsk: Nauka (in Russian)
[11] Konovalenko VYa 2012 *Handbook on Physical and Mechanical Properties of Diamond Deposit Rock Mass in Yakutia* Novosibirsk: SO RAN (in Russian)
[12] Iudin MM 2007 Initial stress state of kimberlite deposits *Nauka i Oobrazovanie* No 3 pp 37–39

Experimental assessment of stresses in enclosing rock mass of Aikhal mine

VD Baryshnikov*, AP Khmelinin and DV Baryshnikov

Chinakal Institute of Mining, Siberian Branch, Russian Academy of Sciences, Novosibirsk, Russia

E-mail: *v-baryshnikov@yandex.ru

Abstract. Experimental assessment of stresses has been carried out in the vicinity of temporary roadway outside the influence zone of stoping in Aikhal mine. The analysis of the research results shows that the values of stresses experimentally found in adjacent rock mass at the depth of mining are local and cannot be used for correct determination of stress state of an intact rock mass.

1. Introduction

Natural stress state of rock mass is a certain factor of power in assessment of mining conditions, stability of underground excavations and support design. Using stress distribution hypothesizes by Dinnik or Heim as boundary conditions in numerical modeling of rock mass stress state is not always correct and requires instrumental confirmation. The only one reliable method of estimating natural stresses in tectonically active areas is experimental research.

Currently Aikhal mine, ALROSA, extract mineral reserves under the open pit bottom at a depth of 300–400 meters in rhomb-shaped stopes arranged in staggered order, in descending cut-and-fill with cemented backfill. Optimized location and support of temporary roadways in rock mass enclosing complex morphology ore body requires detail justification [1]. The experimental assessment of stresses, given lack of information on initial stress state, is a live issue in Aikhal kimberlite pipe mining.

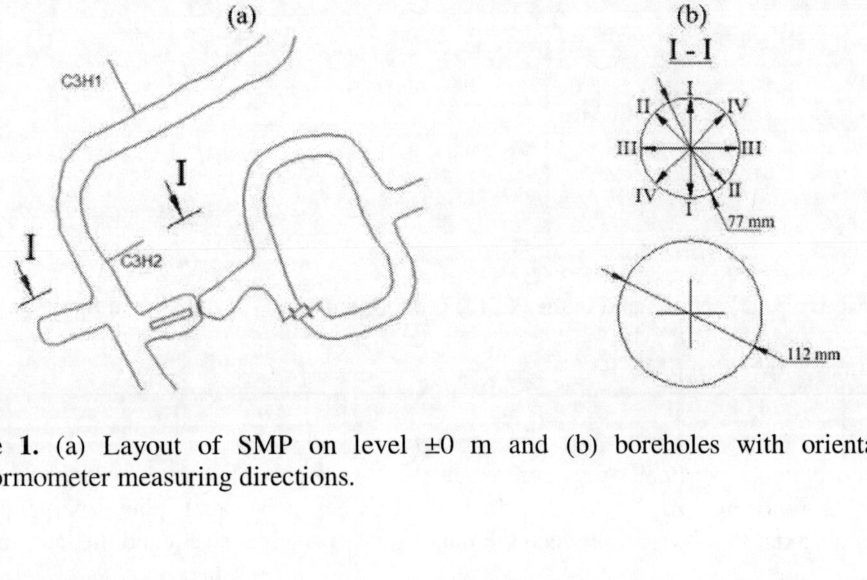

Figure 1. (a) Layout of SMP on level ±0 m and (b) boreholes with orientation of deformometer measuring directions.

2. Research methods and approaches

In 2017 the first cycle of stress state assessment was carried out using parallel borehole method. This method [2] and technical devices for its implementation [3] were developed at the Chinakal Institute of Mining SB RAS. The designed automated measurement system was successfully trialed in actual stress–strain analysis in salt rock mass in Mir mine [4].

Measuring and perturbing boreholes were drilled by Husqvarna drilling machine with diamond boring bit Ø77 mm and Ø112 mm. This small size machine drills holes to 2.5 m long. The installation depth of the deformometer was set based on the data of video filming on the borehole wall condition using a video probe, in order to avoid damage or cracking in the hole walls in the measurement interval.

3. Results and discussion

The plot of radial displacements of walls in the measuring borehole under step-by-step loading of walls in the perturbing borehole subjected to uniform pressure at SMP 1 is presented in Figure 2. The experimental data analysis confirms that the dependence of radial displacements on the loading pressure is linear and there are no residual displacements after unloading of the perturbing borehole. The afore-said allows the elasticity theory apparatus to be used to estimate deformation properties and stresses in rock mass.

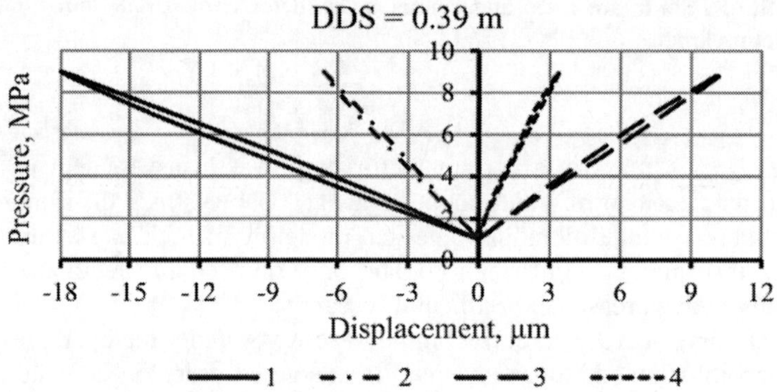

Figure 2. Radial displacements of walls in the measuring borehole under loading and unloading of the perturbing borehole.

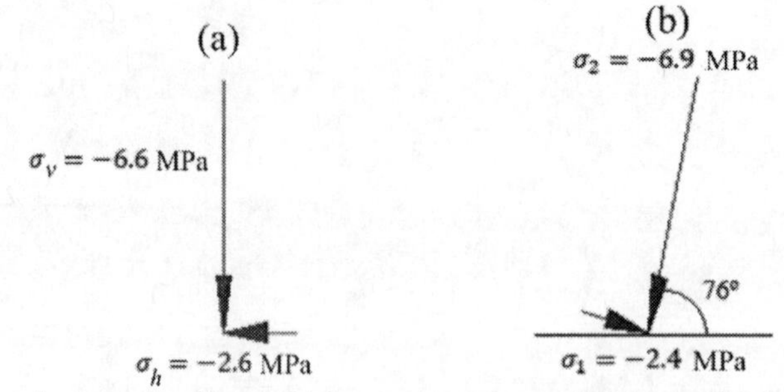

Figure 3. Directions and values of (a) components and (b) quasi-principal stresses.

Table 1. Stresses (MPa) at SMS 1.

DDS, m	σ_v	σ_h	τ_{vh}	σ_1	σ_2	$\Gamma\sigma_2$, deg
0.39	–6.6	–2.6	1.1	–2.4	–6.9	76

Comments: σ_v—vertical stress component; σ_h— horizontal stress component; τ_{vh}—shearing stress component; σ_1, σ_2—quasi-principal stresses in vertical plane perpendicular to the borehole axis; $\Gamma\sigma_2$—angle between the major quasi-principal stress and the horizon.

The resultant stresses determined at the depth of the deformometer setting up (DDS) of 0.39 m from the excavation walls at SMP 1 are presented in Table 1 and Figure 3.

In accordance with the calculations, the vertical component of stress on level ±0 m under overlying rock weight is $\gamma H = 10$ MPa. The concentration coefficient of the vertical component at DDS = 0.39 m from the numerical calculations of stress distribution around the excavation was $K = 1.62$. Taking that coefficient into consideration, the theoretical value of the vertical stress component is $\sigma_v^t = 10\ \text{MPa} \times 1.62 = 16.2\ \text{MPa}$ and it sufficiently (by 2.5 times) exceeds the experimental value $\sigma_v = 6.6$ MPa. This means that rocks nearby the excavation wall are unloaded. So we cannot correctly interpret experimental data and determine stress state of intact rock mass.

After that the enclosing rock mass stresses were measured at SMS 2 at DDS of 0.6 m, 1.1 m, 1.77 m and 2.44 m. The curves of the measuring borehole wall displacements in radial direction under loading–unloading of the perturbing borehole are plotted in Figure 4.

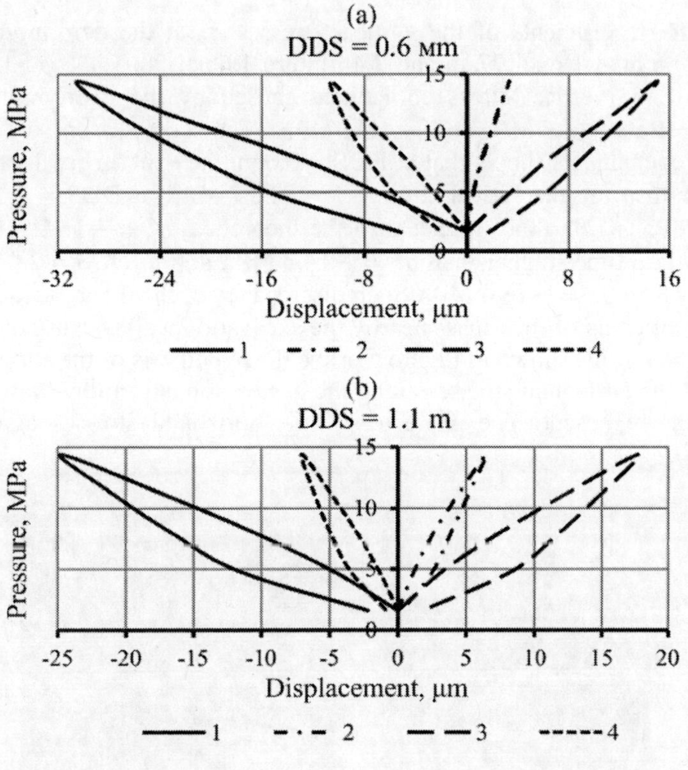

Figure 4. Radial displacement of walls in the measuring borehole under loading–unloading of the perturbing borehole at (a) DDS =0.6 m and (b) DDS =1.1 m.

The values of stresses (MPa) determined in the experiment are given in Table 2 and Figure 5.

Table 2. Stresses (MPa) at SMS 2.

DDS, m	σ_v	σ_h	τ_{vh}	σ_1	σ_2	$\Gamma\sigma_2$, deg
0.6	−12.8	−3.4	2.6	−2.7	−13.5	76
1.1	−16.8	−7.2	1.3	−7.0	−16.9	82
1.77	−17.1	−6.0	-2.3	−5.5	−17.5	79
2.44	−12.4	−3.4	-4.0	−1.8	−13.9	69

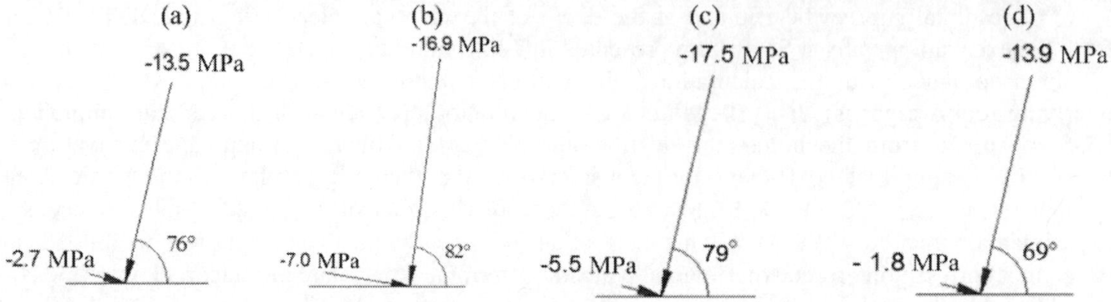

Figure 5. Directions and values of quasi-principal stresses at SMS 2: (a) DDS = 0.6 m; (b) DDS = 1.1 m; (c) DDS = 1.77 m; (d) DDS = 2.44 m.

The concentration coefficients of the vertical stresses σ_v at the deformometer depth from the excavation wall of 0.6 m, 1.1 m, 1.77 m and 2.44 m are 1.66, 1.58, 1.43 and 1.34 respectively. The theoretical values of the vertical stress component caused by the overlying rock weight σ_v^t are -16.7 MPa, -15.9 MPa, -14.3 MPa and -13.4 MPa. The comparison of the theoretical and experimental stress components shows that at DDS = 0.6 m, the experimental value $\sigma_v = -12.8$ MPa is 1.3 times smaller than the theoretical stress $\sigma_v^t = -16.7$ MPa; at DDS = 1.1 m the experimental value $\sigma_v = -16.8$ MPa is 1.05 times higher than the theoretical $\sigma_v^t = -15.9$ MPa; at DDS = 1.77 m $\sigma_v = -17.6$ MPa is 1.23 times higher than $\sigma_v^t = -14.3$ MPa; and at DDS = 2.44 m $\sigma_v = -12.4$ MPa is 1.08 times smaller than $\sigma_v^t = -13.4$ MPa. Such discrepancy of the theoretical and experimental data is caused by partial softening of rock mass nearby the excavation wall because of drilling and blasting. Damage of the borehole walls shown in Figure 6 prove the hypothesis of the softened rock mass.

The values of the horizontal stresses along the excavation axis allow drawing a conclusion that the tectonic stresses are absent. The character of the horizontal stresses agrees with the Dinnik hypothesis best at all.

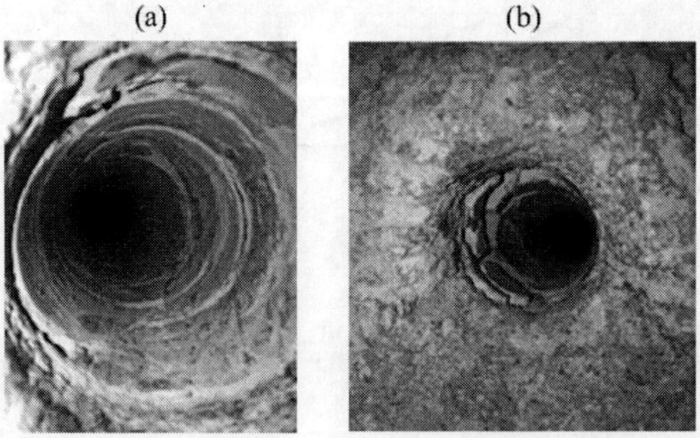

Figure 6. Damage of borehole walls at (a) SMS 1 and (b) SMS 2.

4. Conclusions

The pre-estimation of stress state in intact rock mass proves the absence of the tectonic stresses, and the lateral earth pressure coefficient is 0.5.

The research results show that stresses measured nearby the excavation wall characterize only stress state at the measurement point and not allow stress state assessment in the intact rock mass correctly.

Stress measurements are to be conducted at the distance of softened or partly damaged walls of excavations, outside its influence zone, in order to get correct assessment of natural stress state parameters of rock mass. Such approach requires using relevant methods of measurements and providing the measurements by drilling.

References

[1] Genzel GN, Voropaev BP, Yakushenko MV, Zelberg AS, Kramskov NP and Lobanov VV 2002 Deactivation of Mir open pit mine and flooding protection of underground mine in transition period *Proceedings of Scientific Conference: Mirny-2001* Moscow: Ruda Metally (in Russian)

[2] Kurlenya MV, Baryshnikov VD, Popov SN et al 1981 Method to determine stress and strains in rocks *Otkrytiya Izobreteniya* No 40

[3] Baryshnikov VD and Kachalsky VG 2010 Automation instrumentation to measure rock mass stresses in parallel-drilled holes *Journal of Mining Science* Vol 46 No 3 pp 338–342

[4] Baryshnikov VD and Baryshnikov DV 2017 Experimental research data on stress state in salt rock mass around an underground excavation *J. Fundament. Appl. Min. Sci.* Vol 4 No 2 pp 32–36

[5] Zhirnov AA, Abdrakhmanov SU, Shaposhnik YuN and Konurin AI 2018 Rock mass stability estimation and selection of mine support design at Orlov complex ore deposit *Gornyi Zhurnal* No 3 pp 51–57 DOI: 10.17580/gzh.2018.03.08

[6] Konovalenko VYa 2012 Reference *Book on Physical Properties of Rocks at Diamond Deposits in Yakutia* Novosibirsk (in Russian)

IOP Conf. Series: Earth and Environmental Science **262** (2019) 012009 doi:10.1088/1755-1315/262/1/012009

Process for assessment of stress-strain state of a rock mass and its defects using data on shear measurements at outcrop surface

AI Chanyshev[1,2]*, IM Abdulin[1] and OE Belousova[1]

[1]Chinakal Institute of Mining, Siberian Branch, Russian Academy of Sciences, Novosibirsk, Russia
[2]Novosibirsk State University of Economics and Management, Novosibirsk, Russia

E-mail: *a.i.chanyshev@gmail.com

Abstract. It is proposed to assess the stress-strain state of a rock mass and its defects by measuring displacements of mine working contour. Algorithms and software to solve the elastic and elastic-plastic problems are worked out for a case when a working contour shifts to failure domain. To compute of the working contour displacements, when all the shears except creep-strain occurred, the researchers set forth the process for continuous stress release in the contour of the preset geometry at continuous monitoring and recording of shear benchmark data at its boundary. The process is demonstrated on shear measurements on a circular hole in a rubber plate under tension load.

1. Introduction

It is the common knowledge that elastic and plastic deformations run almost instantly unlike creep deformations. Actually, when mine workings are driven in a rock mass, elastic and plastic deformations, including displacements corresponding to these deformations happen instantaneously at contours of mine workings. That is the reason why there is no sense to speak of displacements of benchmarks at working contours, provided that they do not relate to creep strain in a rock mass. Such statement of the "industrial" experiment with driving mine workings lead to loss of important information on shears at contours of mine workings. This information can be useful in search for stress values in "infinity" and defect imperfections of a rock mass with identification of hard inclusions and closed cavities [1–5]. The present paper proposes the process for determination of actually occurred displacements at mining contours using stress, applied in infinity.

2. Mathematical model and solution

The new-proposed approach is justified. For the sake of simplicity we consider a plane strain version. Assume, there is coordinate system xOy and deformation along axis z is plane. Let plane xOy is constrained with stress $\sigma_x^\infty = -p$, $\sigma_y^\infty = -p$, ($\sigma_z^\infty = -2\nu p$). The study case is presented in Figure 1a.

For plain strain the elasticity relations are:

$$\sigma_r - \sigma_\theta = 2\mu(\varepsilon_r - \varepsilon_\theta),\ \sigma_r + \sigma_\theta = 2k(\varepsilon_r + \varepsilon_\theta),$$
$$\varepsilon_r = \frac{du}{dr},\ \ \varepsilon_\theta = \frac{u}{r},\ \ 2\mu = E/(1+\nu),\ \ 2k = 2\mu/(1-2\nu). \tag{1}$$

Here, radial displacements u are counted at a certain preset point O. Under stress compression of the plane $\sigma_x^\infty = -p$, $\sigma_y^\infty = -p$ we obtain stresses in polar coodinates $\sigma_r = \sigma_\theta = -p$.

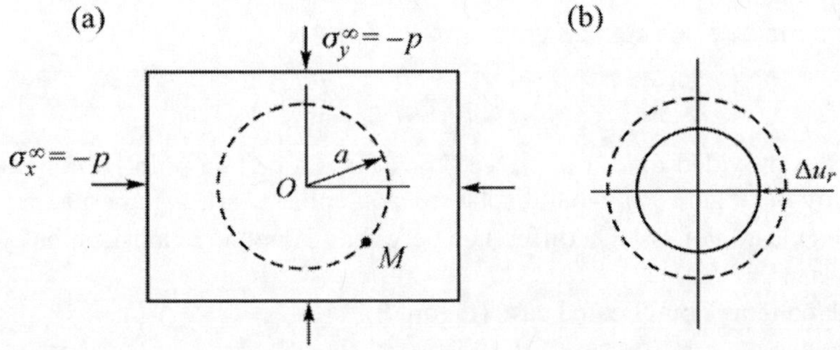

Figure 1. Scheme of mine working driving in a rock mass: (a) in the initial hydrostatic stress field; (b) at the complete stress release at the contour concurrently to driving of a working.

From (1) it follows that $\dfrac{du}{dr} + \dfrac{u}{r} = -\dfrac{p}{k}$ and $\dfrac{du}{dr} - \dfrac{u}{r} = 0$. Solving this system of equations we find:

$$u = -\frac{p}{2k} r. \tag{2}$$

Thus, under stress load $\sigma_x^\infty = -p$, $\sigma_y^\infty = -p$ displacements of kind (2) take place.

Next, in the plane under consideration we drive a working of radius a, this means that stress σ_r at its contour should be equal to zero. To gain this effect and to compensate compressive stress $\sigma_r = -p$ we apply tension stress $\sigma_r = +p$, vanishing in the infinity of the contour of the working of radius a. We continue to keep to assumption that the material deforms under the elasticity law (1).

Solving differential equations of the problem we obtain that additional stress $\tilde{\sigma}_r$ should be distributed under the law:

$$\tilde{\sigma}_r = C_1 + \frac{C_2}{r^2}. \tag{3}$$

At $r \to \infty$ $\sigma_r = 0$, so $C_1 = 0$ in (3). Further, at $r = a$ $\tilde{\sigma}_r$ should be positive and equal to p. As a result, formula for $\tilde{\sigma}_r$ is:

$$\tilde{\sigma}_r = \frac{pa^2}{r^2}. \tag{4}$$

Thereto, $\tilde{\sigma}_\theta = -\dfrac{pa^2}{r^2}$. Then $\tilde{\varepsilon}_r + \tilde{\varepsilon}_\theta = 0$, wherefrom we find $\tilde{u} = \dfrac{C_3}{r}$. Substituting this term to the first relation (1) allows definition $C_3 = -\dfrac{pa^2}{2\mu}$, then displacement additional to (2) is:

$$\tilde{u} = -\frac{pa^2}{2\mu r}, \tag{5}$$

It is shown in Figure 1b.

It is obvious that tension shear (5) directed inside of the mine working correspond to tension stresses (4). Overlapping of two stress states gives the resultant stress sum:

$$\sigma_r = -p\left(1-\frac{a^2}{r^2}\right),\quad \sigma_\theta = -p\left(1+\frac{a^2}{r^2}\right). \tag{6}$$

Displacements u can be also summed up:

$$u = -\frac{p}{2k}r - \frac{pa^2}{2\mu r}. \tag{7}$$

A question arises, provided that in a stressed rock mass a certain point at contour of a would-be working is marked by M (Figure 1a), what displacements: either (5) or (7) are obtained after the stress state is released. It is clear, at $\nu = 1/2$ difference between (5) and (7) vanishes, but you should mind that it happens only at $\nu = 1/2$.

Now let consider a more complicated case (Figure 2).

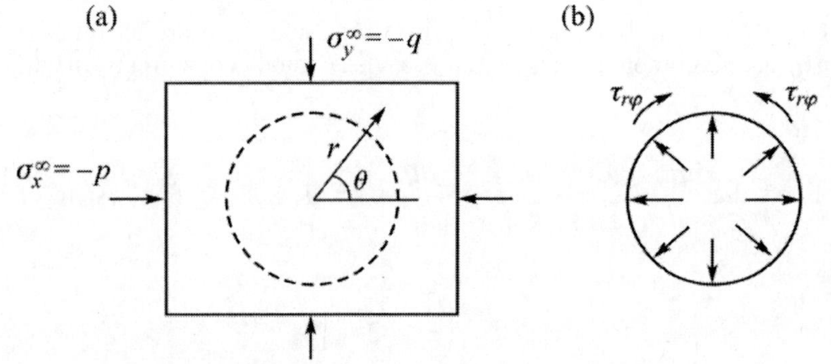

Figure 2. (a) Rock mass under two loads *p* and *q* and (b) formation of tangential stresses at contour of a would-be working.

Let a rock mass be under two loads $\sigma_x^\infty = -p$, $\sigma_y^\infty = -q$, $(p, q > 0)$. Introducing polar coordinates *r*, θ, at plane *xOy* for stresses $\sigma_r, \sigma_\theta, \tau_{r\theta}$, we obtain:

$$\sigma_r = -p\cos^2\theta - q\sin^2\theta,\quad \sigma_\theta = -p\sin^2\theta - q\cos^2\theta,$$
$$\tau_{r\theta} = \frac{\sigma_x^\infty - \sigma_y^\infty}{2}\sin 2\theta = -\frac{p-q}{2}\sin 2\theta. \tag{8}$$

In homogeneous stress and strain field $\sigma_x^\infty = -p$, $\sigma_y^\infty = -q$ correspond to displacements u_x, u_y or displacements u_r, u_θ are:

$$\begin{cases} u_r = -\dfrac{(1-2\nu)}{4\mu}(p+q)r + \dfrac{1}{2\mu}(q-p)r\cos 2\theta, \\ u_\theta = -\dfrac{p-q}{4\mu} r\sin 2\theta. \end{cases} \tag{9}$$

It follows from (8) that compressive stresses σ_r, positive and negative $\tau_{r\theta}$ act in the contour of a would-be working with equation $r = a$. To drive a working with boundary conditions $\sigma_r = 0$, $\tau_{r\theta} = 0$ requires to apply opposite-sign stresses to σ_r, $\tau_{r\theta}$ at the boundary $r = a$:

$$\sigma_r = p\cos^2\theta + q\sin^2\theta,\quad \tau_{r\theta} = \frac{p-q}{2}\sin 2\theta.$$

To define stresses, displacements in a rock mass with a working $r = a$ Kolosov–Muskhelishvili formula is applicable [6]:

$$\sigma_r - i\tau_{r\theta} = \Phi(z) + \overline{\Phi(z)} - e^{2i\theta}[\overline{z}\Phi'(z) + \Psi(z)]. \tag{10}$$

Considering (8) and $e^{2i\theta} = z / \overline{z} = z^2 / a^2$, $z\overline{z} = a^2$, we obtain:

$$\frac{p+q}{2} + \frac{p-q}{2}\frac{z^2}{a^2} = \Phi(z) + \overline{\Phi(z)} - \frac{z^2}{a^2}\left[\frac{a^2}{z}\Phi'(z) + \Psi(z)\right]. \tag{11}$$

Solution (11) is found as:

$$\Phi(z) = \frac{B_2}{z^2}, \quad \Psi(z) = \frac{A_2}{z^2} + \frac{A_4}{z^4}. \tag{12}$$

Substituting (12) into (11) and comparing coefficients at equal powers z we find the system of three equations for determination of three unknowns B_2, A_2, A_4. Their values are equal to:

$$A_2 = -\frac{p+q}{2}a^2, \quad B_2 = \frac{p-q}{2}a^2, \quad A_4 = \frac{3}{2}(p-q)a^4.$$

Further, using (10) we find stresses:

$$\sigma_r = \frac{p+q}{2}\left(\frac{a}{r}\right)^2 + 2(q-p)\cos 2\theta\left(\frac{a}{r}\right)^2 - \frac{3}{2}(q-p)\cos 2\theta\left(\frac{a}{r}\right)^4,$$

$$\tau_{r\theta} = \left[(p-q)\left(\frac{a}{r}\right)^2 - \frac{3}{2}(p-q)\left(\frac{a}{r}\right)^4\right]\sin 2\theta, \tag{13}$$

$$\sigma_r + \sigma_\theta = 4\,\mathrm{Re}\,\Phi(z).$$

We should note that stresses (13) at $z \to \infty$ tend to zero.

To determine displacements u_r, u_θ the second group of Kolosov–Muskhelishvili's formula is available:

$$2\mu(u_r + iu_\theta) = e^{-i\theta}[\aleph\varphi(z) - z\overline{\varphi'(z)} - \overline{\psi(z)}], \tag{14}$$

where $\varphi(z) = \int \Phi(z)dz = -\frac{p-q}{2}\frac{a^2}{z}$, $\psi(z) = \frac{p+q}{2}\frac{a^2}{z^2} - \frac{1}{2}(p-q)\frac{a^4}{z^3}$, $\aleph = 3 - 4\nu$. After identification of imaginary and real components in (14) we have:

$$\frac{2\mu u_r}{a} = -\frac{p+q}{2}\left(\frac{a}{r}\right) - \frac{p-q}{2}\left(\frac{a}{r}\right)\left[\aleph + 1 - \left(\frac{a}{r}\right)^2\right]\cos 2\theta,$$

$$\frac{2\mu u_\theta}{a} = \frac{p-q}{2}\left(\frac{a}{r}\right)\sin 2\theta\left[\aleph - 1 + \left(\frac{a}{r}\right)^3\right]. \tag{15}$$

At $r = a$ from (15) it follows:

$$\frac{2\mu u_r}{a} = -\frac{p+q}{2} - \frac{p-q}{2}\aleph\cos 2\theta, \quad \frac{2\mu u_\theta}{a} = \frac{p-q}{2}\aleph\sin 2\theta. \tag{16}$$

If in (16) values θ are assumed equal to $\theta = 0$, $\theta = \pi$ we obtain in total:

$$2\mu u_r\big|_{0,\pi} = -\frac{p+q}{2} - \frac{p-q}{2}\aleph, \quad u_\theta\big|_{0,\pi} = 0. \tag{17}$$

If in (16) values θ are assumed equal to $\theta = \frac{\pi}{2}$, $\theta = -\frac{\pi}{2}$ we have:

$$2\mu u_r\big|_{\pi/2} = -\frac{p+q}{2} + \frac{p-q}{2}\aleph. \tag{18}$$

From (17), (18) we derive formulas to find σ_x, σ_y values in "infinity":

$$p+q = -\frac{2\mu}{a}(u_r\big|_0 + u_r\big|_{\pi/2}),\ \ p-q = -\frac{2\mu}{\aleph a}(u_r\big|_{\pi/2} - u_r\big|_0). \tag{19}$$

These formulas are analogous to formulas derived in the stress-release method [7].

3. Laboratory test results

It should be outlined that the considered scheme to form a mine working is extremely primitive because of disregard for rock mass failure inside the contour $r = a$. Moreover, the unloading and secondary loading can happen because of both tension loads and shears within/out of the contour. In view of this fact after a mine working is formed, $r = a$ can appear at its contour concurrently as displacements of both kinds (2) and (5); along with (9) and (15).

Two problems are stated in the present paper. The first problem deals with experimental research aimed at development of a process for computing displacements at working contour. The second concerns the formula to determine a load in "infinity" based on values of displacements at working boundary and to establish defects of a rock mass.

To solve the first problem we implement the following experiments. We take a sheet of an elastic material, let it be rubber. We make an attempt to load it by two ways, shown in Figure 3. The first scheme in Figure 3a implies that a circle of 20 cm in diameter with center coinciding with the center of the rubber sheet is plotted on the rubber sheet of 1×1 m dimensions and 1.5 mm in thickness. 12 points are evenly marked at circle boundary like in a clock.

Next, interior of the circle is cut out; the rubber sheet with a circular hole is placed vertically with fixation of its upper edge and is subjected to a tension test, viz., different loads from 10 to 20 kGs are applied to the bottom edge of the sheet. A digital camera is used to identify displacements u_r and u_θ at 12 points of the circle.

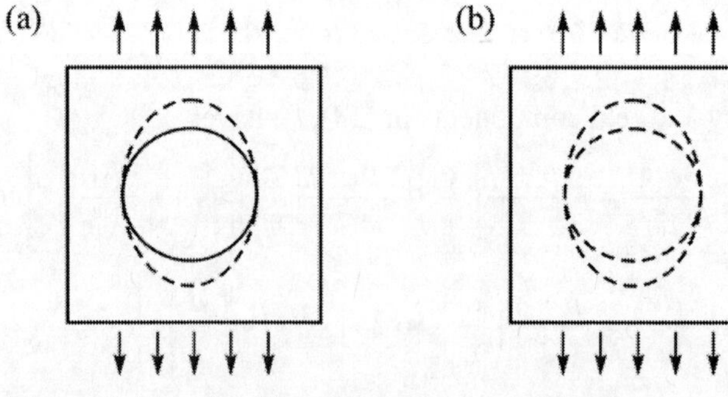

Figure 3. Two loading schemes for a rubber-like material.

The second loading scheme in Figure 3b implies the following. The initial rubber sheet is positioned vertically, the preset tension load is applied at the lower edge (the top edge of the sheet is fixed). The test sheet is in a tension state; cliché of the circle of the same radius as in Figure 3a is placed on the sheet surface and delineated. 12 points are marked at the boundary of the new-made circle. The circle is cut under tensile state of the test sheet. A digital camera is used to record displacements u_r, u_θ at the circle boundary.

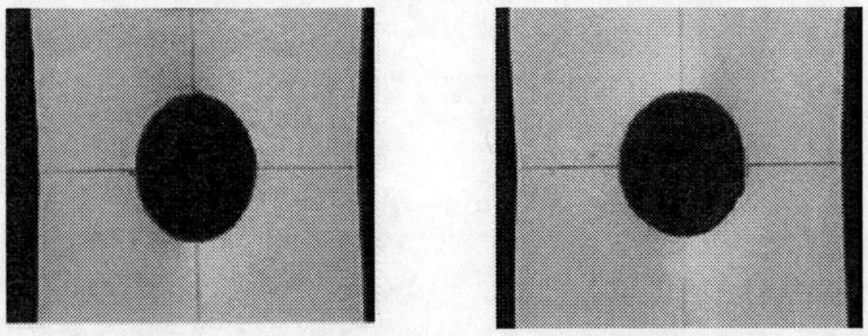

Figure 4. Tension of material sheets with a hole by applying two loading schemes.

Next, displacements obtained by two methods are compared. Loading results shown in Figure 4 demonstrate that tension of a material sheet with a hole leads to greater displacements u_r in direction of tension load action as compared to tension of a material, free of an initial hole. Difference is not significant, but it is available.

The proposed process to determine displacements is deposition of an initial contour of a mine working on an even surface in a rock mass, its unloading follows; and displacements of hole contour are recorded under unloading.

4. Conclusions

Displacements of a rock mass are determined for cases without mine workings and considering formation of mine workings on the basis of elasticity relationships.

The experiments are undertaken on rubber-like materials in view to estimate displacements in a rock mass loaded by two methods.

Acknowledgements
This study was supported by the Russian Foundation for Basic Research, Project No 18-05-00757a.

References

[1] Schwab AA 1989 Ill-posed static problems in the theory of elasticity *Izv. AN SSSR. Mekh. Tverd. Tela* No 6 pp 24–27
[2] Chanyshev AI 2001 To the problem of deformable medium failure Part I Basic equations *Journal of Mining Science* Vol 37 No 3 pp 273–288
[3] Chanyshev AI and Abdulin IM 2008 Characteristics and the relations on them at the stage of post-limit deformation in rocks *Journal of Mining Science* Vol 44 No 5 pp 451–463
[4] Kabanikhin S 2009 *Inverse and Ill-Posed Problems* Novosibirsk: SO RAN (in Russian)
[5] Alekseev AS 1967 Inverse dynamic seismic problems *Some Methods and Algorithms for Interpreting Geophysical Data* Moscow (in Russian)
[6] Muskhelishvili NI 1949 *Some Basic Problems of the Mathematical Theory of Elasticity* Moscow (in Russian)
[7] Baklashov IV 2004 *Geomechanics: Textbook for High Schools. Fundamentals of Geomechanics* Moscow: MGGU Vol 1 pp 208 (in Russian)

Critical sizes of cone-shaped opening in open mining

AI Chanyshev[1, 2]*, OE Belousova[1], LL Efimenko[2] and OA Lukyashko[1]

[1]Chinakal Institute of Mining, Siberian Branch Russian, Academy of Sciences, Novosibirsk, Russia

[2]Novosibirsk State University of Economics and Management, Novosibirsk, Russia

E-mail: *a.i.chanyshev@gmail.com

Abstract. In case of unstable cone-shaped pitwalls and dumps, the overlying rock weight is of importance. Different weight of overlying rocks generates different vertical pressures per horizontal sections of pits and dumps. In case that the type of a failure surface is known for the present slop of pitwall or dump, it is possible to detect a critical section of the maximum pressure. The location of the critical section is governed by the height and the bottom radius of a pit or a dump. This study relates the destruction zone location with the height and bottom radius of open pits (dumps). The results of the theoretical and experimental studies are reported.

1. Introduction

This study focuses on failure of rocks mass around a cone-shaped excavation. The aim is to interrelate such parameters as: pitwall slope, bottom radius, pitwall height and slope of possible failure zone. The research is based on the provision that rock mass failure takes place under overlying rock weight, thus, there is some pitwall height at which pressure generated by the overlying rock weight is maximum (derivative of the pressure-to-depth function goes to zero!). This pressure value is assumed as a critical pressure initiating failure in lower-lying rocks. The connection of possible failure and cone-shaped excavation is found thereof. Similar approaches are described in [1–16].

2. Cone-shaped excavation and critical section

Figure 1 shows a cone-shaped excavation *LMKB* with the bottom radius a, height $QA = H$ and slope α of the pitwall *KB*. The angle β stands for the slope of the sliding surface in failure. This angle can be free from the pitwall slope α (which conforms with the initial anisotropy of the medium), or can be related with α and parameters of the medium such as internal friction angle [17, 18]. However, in any case, the angle β at the preset values of α and the medium parameters is constant: $\beta = \beta(\alpha)$. We use this circumstance in determination of critical parameters of the cone-shaped excavation in Figure 1.

Let us calculate the volume of the ring with the trapezoidal section *EBCD* in Figure 1. To this end, we first find the volume of the cylinder with the bottom radius *FD* and height *DC* (Figure 1). Notice that $FD = FN + ND$, where $FN = a$, $ND = h\,\mathrm{ctg}\,\beta$, $DC = H - h$; then:

$$V_{\text{cylinder}} = \pi[a + h\,\mathrm{ctg}\beta](H - h).$$

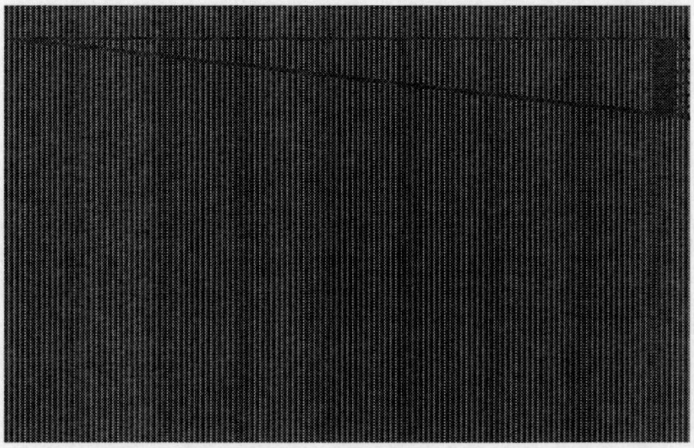

Figure 1. Schematic representation of the cone-shaped excavation.

Now, we find the volume of a conical collar with the bottom and top radii *FE* and *AB*, and the height *DC*. This volume is equal to the difference of volumes of two cones with the triangular sections *OAB* and *OFE*. The volume of the cone with the radius *AB* and height *OA* is given by:

$$V_1 = \frac{1}{3}\pi(AB)^2[H + a\,\mathrm{tg}\alpha] = \frac{1}{3}\pi(a + H\,\mathrm{ctg}\alpha)^2[H + a\,\mathrm{tg}\alpha].$$

The volume of the cone with the radius *FE* and height *OE* is:

$$V_2 = \frac{1}{3}\pi(FE)^2[h + a\,\mathrm{tg}\alpha] = \frac{1}{3}\pi(a + h\,\mathrm{ctg}\alpha)^2[h + a\,\mathrm{tg}\alpha].$$

There from, we find the volume of the unknown ring:

$$V_{\text{ring}} = \pi(a + h\,\mathrm{ctg}\beta)^2[H - h] - \frac{1}{3}\pi(a + H\,\mathrm{ctg}\alpha)^2[H + a\,\mathrm{tg}\alpha] + \frac{1}{3}\pi(a + h\,\mathrm{ctg}\alpha)^2[h + a\,\mathrm{tg}\alpha]. \quad (1)$$

Determine the area *S* of the ring with the external radius *FD* and internal radius *EF* as the difference of two areas:

$$S = \pi(FD)^2 - \pi(EF)^2 = \pi[(a + h\,\mathrm{ctg}\beta)^2 + (a + h\,\mathrm{ctg}\alpha)] = \pi h(\mathrm{ctg}\beta - \mathrm{ctg}\alpha)[2a + h(\mathrm{ctg}\beta + \mathrm{ctg}\alpha)]. \quad (2)$$

Multiply (1) by ρg and obtain the weight of the ring with the section *EBCD*. Divide the weight by the area *S* and have the pressure generated by the ring on the lower lying rocks. Introduce the denotations:

$$\frac{a}{H} = \tilde{a}, \quad \frac{h}{H} = x. \quad (3)$$

Then the pressure *p* generated by the ring weight on the surface *ED* in Figure 1 is:

$$\frac{p}{\rho g H} = \frac{V_{\text{cylinder}}}{H^3} \cdot \frac{H^2}{S} =$$

$$= \frac{3(\tilde{a} + x\mathrm{ctg}\beta)(1 - x) - (\tilde{a} + \mathrm{ctg}\alpha)(1 + \tilde{a}\mathrm{tg}\alpha) - (\tilde{a} + \mathrm{ctg}\alpha)^2(1 + \tilde{a}\mathrm{tg}\alpha) + (\tilde{a} + x\mathrm{ctg}\alpha)^2(x + \tilde{a}\mathrm{tg}\alpha)}{3x(\mathrm{ctg}\beta - \mathrm{ctg}\alpha)[2\tilde{a} + x(\mathrm{ctg}\beta + \mathrm{ctg}\alpha)]}. \quad (4)$$

At the fixed values of a, H, α, β, the relation (4) represents the function of the relative pressure $p/\rho g H$ on *x*. This means that at some height *h* in a cone-shaped excavation (Figure 1), the pressure can be maximum. For calculating this pressure, we find the first derivate with respect to $x = h/H$ in (4) and put it to zero. As a result, we arrive at the fourth-order equation for *x*:

$$(\text{ctg}\beta+\text{ctg}\alpha)(\text{ctg}^2\alpha-3\text{ctg}^2\beta)x^4+4\tilde{a}(\text{ctg}^2\alpha-3\text{ctg}^2\beta)x^3+$$
$$+6\tilde{a}(\tilde{a}\text{ctg}\alpha-2\tilde{a}\text{ctg}\beta-\text{ctg}\beta\text{ctg}\alpha)x^2+2\text{ctg}\alpha(\text{ctg}\beta+\text{ctg}\alpha)(\text{ctg}\alpha+3\tilde{a})x+2\tilde{a}\text{ctg}\alpha(\text{ctg}\alpha+3\tilde{a})=0. \quad (5)$$

Solution of (5) yields x at the input α, β, $\tilde{a}$. Let us analyze its roots.

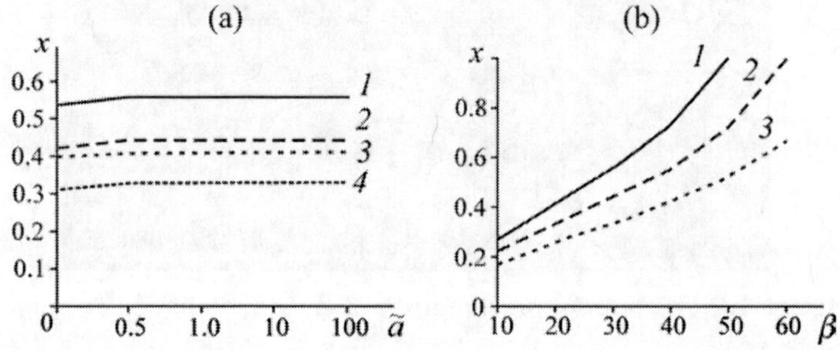

Figure 2. Curves of x versus (a) $\tilde{a}$ at different α, β: *1*—α = 60°, β = 30°; *2*—α = 50°, β = 30°; *3*—α = 50°, β = 20°; *4*—α = 60°, β = 20°; and (b) β at different α: *1*—60°, *2*—50°; *3*—70°; $\tilde{a}=1$.

Figure 2 shows the variation in x (from 0 to 1) as function of $\tilde{a}$ at different values of α and β (Fig. 2a) and as function of β at different α and $\tilde{a}=1$ (Fig. 2a). The relationship between (in the range of 0–1) and the angle α at different values of β and $\tilde{a}=1$ is illustrated in Fig. 3a. With the known x, it is possible to calculate the relative $p/\rho gH$ from (4); the results are demonstrated in Fig. 3b.

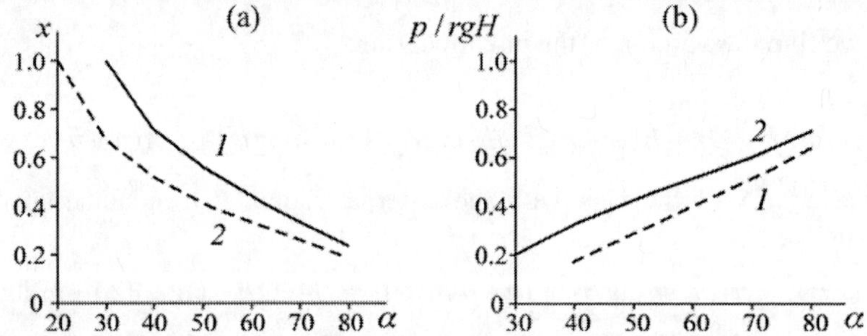

Figure 3. Relationships of (a) x and α at $\tilde{a}=1$; (b) p / ρgH and α at: *1*—β = 30°; *2*—β = 20°.

3. Critical pressure

As regards the critical pressure on the surface *ED* in Figure 1, let us examine the face *KD* at the slope β relative to the horizontal axis r (Orz is a cylindrical coordinate system). it is assumed that in failure, the shearing stress on *ED* is insignificant, and the principal pressure is exerted by the vertical stress σ_z which creates shearing at the face *KD*. In order to find the shearing stress, we set a normal $\vec{n}$ and a tangent $\vec{t}$ to *KD*:

$$\vec{n}=(-\sin\beta,\cos\beta), \quad \vec{t}=(\cos\beta,\sin\beta).$$

Considering that the stress tensor T_σ in the axial symmetry is given by:

$$T_\sigma=\begin{pmatrix}\sigma_r & \tau_{rz} & 0\\ \tau_z & \sigma_z & 0\\ 0 & 0 & \sigma_\varphi\end{pmatrix},$$

we have that the stress vector at the face with the normal $\vec{n}$ is:

$$\vec{p}_n = (-\sigma_r \sin\beta + \tau_{rz}\cos\beta)\vec{e}_r + (-\tau_{rz}\sin\beta + \sigma_z\cos\beta)\vec{e}_z,$$

and the shearing stress is:

$$\tau_n = \tau_{zr}\cos 2\beta + \frac{\sigma_z - \sigma_r}{2}\sin 2\beta, \quad (6)$$

where $\vec{e}_z$, $\vec{e}_r$ are the orts of the coordinate system *Orz*.

Destruction of rocks under the surface *ED* affected by a press-tool requires that the modulus of the shearing stress τ_n equals or exceeds the shearing strength of rocks:

$$|\tau_n| \geq \tau_s.$$

Assuming that $\tau_{rz} = 0$ and σ_r is smaller than σ_z on the surface *ED*, we have in a zero approximation that $(|\sigma_z|\sin 2\beta)/2 \geq \tau_s$ or:

$$|\sigma_z| \geq \frac{2\tau_s}{\sin 2\beta}. \quad (7)$$

The angle β and shearing strength are fond experimentally. Equating of the pressures (4) and (7) produces the critical height of an open pit and the wanted relation between the open pit параметрами α, β, a and H at the moment of instability.

In the framework of this study, test series on failure of a cone-shaped excavation in a granular medium have been performed.

4. Conclusions

1. The authors have located the critical horizontal section in the cone-shaped excavation, at which pressure du to overlying rock weight is maximum.
2. The relation between the parameters of the cone-shaped excavation at the moment of instability is obtained.

Acknowledgements

The study was supported by the Russian Foundation for Basic Research, Project No 18-05-00757a.

References

[1] Frank U2014 Multi-perspective enterprise modeling: Foundational concepts, prospects and future research challenges *Software and Systems Modeling* 13(3) pp 941–962

[2] Brown C 2012Autonomous vehicle technology in mining *Engineering and Mining Journal* 213(1) pp 30–32

[3] Hahn S, Pastor S and Thompson R 2015 Development of mine haul road surfacing condition monitoring through digital image processing *Engineering and Mining Journal* 67(9) pp 34–45

[4] Trubetskoy KN, Rylnikova MV, Vladimirov DYa and Pytalev IA 2017 Provisions and prospects for introduction of robotic geotechnologies in open pit mining *Gornyi Zhurnal* Issue 11 pp 60–64

[5] Abroskin AS 2015 Use of modern systems of automation of open cast mining *Bulletin of the Tomsk Polytechnic University, Geo Assets Engineering* 326(12) pp 122–130

[6] Rylnikova MV, Yun AB and Terentieva IV 2015 Prospects and development strategy of Jezkazgan deposit *Gornyi Zhurnal* Issue 5 pp 44–49

[7] Kozyrev AA, Semenova IE, Rybin VV and Avetisyan IM 2015 Stress redistribution in deep open pit mine zhelezny at kovdor iron ore deposit Journal of Mining Science Vol 51 No 4 pp 659–665

[8] Kozyrev AA, Rybin V and Konstantinov K 2015 Assessment result on geomechanical state of near-wall rock mass in open-pits of the Kola Region by integrated instrumental methods

I*nternational Multidisciplinary Scientific GeoConference Surveying Geology and Mining Ecology Management—SGEM* Vol 3 Issue 1 pp 111–118

[9] Stacy TR 2007 Slope stability in high stress and hard rock conditions *Proceedings of the 2007 International Symposium on Rock Slope Stability in Open-Pit Mining and Civil Engineering* Perth Australia pp 187–200

[10] Makarov PV and Bakeev Russian Academy of Sciences 2015 Simulation of spallation life of metals in relation to operating stresses in the nanosecond loading time range *AIP Conference Proceedings: International Conference on Advanced Materials with Hierarchical Structure for New Technologies and Reliable Structures* Vol 1683 No 020134

[11] Sheshenin SV, Artamonova NB, Frolova YV and Ladygin VM 2015 Defining the elastic properties and the tensor of the pore-pressure transfer in rocks using the averaging method, *Moscow University Geology Bulletin* Vol 70 No 4 pp 354–361

[12] Frolova YV 2010 Patterns of transformations in the compositions and properties of Icelandic hyaloclastites during lithogenesis *Moscow University Geology Bulletin* Vol 65 No 2 pp 104–114

[13] Frolova YV 2015 *Laboratory Methods for the Study of Hard Rocks* Moscow: KDU (in Russian)

[14] Zhabko AV 2013 Calculation theory of stability offoundations and slopes *Proceedings XV International ISM Congress* Aachen Germany pp 85–97

[15] Melnikov NN and Kozyrev AA 2015 Changes in the geodynamic regime of the geological environment during arge-scale mining operations in deep quarries *GIAB* Special Issue 56 pp 7–23

[16] Kasparyan EV and Kozyrev AA 2015 Geomechanical problems in open mining *GIAB* Special Issue 56 pp 134–143

[17] Podyminogin GM and Chanyshev AI Estimate of maximum permissible height of pit wall based on a rigid-plastic model *Journal of Mining Science* Vol 51 No 3 pp 448–455

[18] Podyminogin GM and Chanyshev AI 2015 Determination of ultimate pitwall parameters in axisymmetric rigid-plastic model of rocks *Journal of Mining Science* Vol 51 No 4 pp 679–688

IOP Conf. Series: Earth and Environmental Science **262** (2019) 012011 doi:10.1088/1755-1315/262/1/012011

Vibratory–percussive disassembly in recovery of electric-centrifugal oil well pumps

VV Chervov

Chinakal Institute of Mining, Siberian Branch Russian, Academy of Sciences, Novosibirsk, Russia

E-mail: vchervov@yandex.ru

Abstract. Repair of borehole centrifugal pumps involves complete disassembly and removal of steel parts from a long steel casing made of thick-walled pipes. In this paper, the technology of removal of steel parts from the pipe using a pneumatic hammer with impactor portion 400 kg in weight at the compressed air flow rate not more than 3 cubic meters per minute is discussed.

1. Introduction

It is expedient to operate high-rate wells using centrifugal pumps [1–4]. Given medium to high drainage of formation fluid (100–500 m^3/day and more), centrifugal pumps offer the higher economic efficiency and the easiest maintenance. A centrifugal pump is put into a well beneath the fluid level in oil well tubing and actuated by electric motor arranged below and supplied via a special cable. The motor arranged directly at the pump allows transmission of great power. In a casing string with a diameter of 168 mm at a pump head of 1000 m, the useful power of the centrifugal well pump exceeds 100 kW

Centrifugal well pumps with a head of 1915 m ensure fluid withdrawal to 125 m^3/day from wells with internal casing diameter of 130 mm; at the head of 550 m—900 m^3/day from will with internal casing diameter of 148.3 mm.

Withdrawal of highly corrosive fluids uses centrifugal well pumps with basic parts made of materials resistant to corrosion.

Overhaul life of centrifugal well pumps at medium and high drainage is longer than that of sucker-rod pumps and makes on average 260–320 days, for instance, at reservoirs in Bashkiria and Tatarstan [5, 6].

2. Features of centrifugal well pumps operation

A centrifugal well pump installation (Figure 1) consists of a hydraulically protected compensator of submersible electric motor; sealed protector; multi-section pump; flat and round cables fastened to the tubing by steel belts; back valve facilitating priming after the pump stop; bleed valve to drain fluid from the tubing in order to simplify pump removal; as well as equipment of the well mouth, operator station, transformer and various ancillary facilities. The electric pump unit is placed down the well on the oilwell tubing. The submersible motor is supplied via the round cable running from the ground surface down the tubing to the pump unit and passing into the flat cable. The flat cable is used with intent to reduce total diameter of the electric pump unit in order to ensure easy and damage-free up and down motion of the pump in a well. Round cables can be totally replaced by flat cables of the proper cross-section [7, 8].

IOP Conf. Series: Earth and Environmental Science **262** (2019) 012011 doi:10.1088/1755-1315/262/1/012011

Centrifugal pump installations operate in wells with different internal diameter of casing strings, from 121.7 to 148.3 mm.

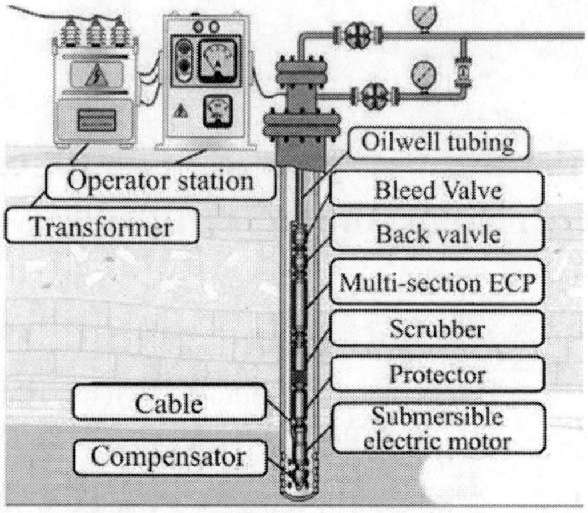

Figure 1. Centrifugal well pimp installation.

The diameter constraint necessitates the increase in length of the electric pumping units at the motor capacity up to 250 kV. The range of diameters for the pumping units is 116–142.5 mm, the length can be more than 25 m (Figure 2).

Figure 2. Multi-section electric centrifugal pumps.

The basic components of an electric centrifugal pump (ECP) are the impeller and diffuser (Figure 3). The impeller increases potential and kinetic energy of fluid flow by speeding it up in the blade system of the impeller and by build-up of pressure.

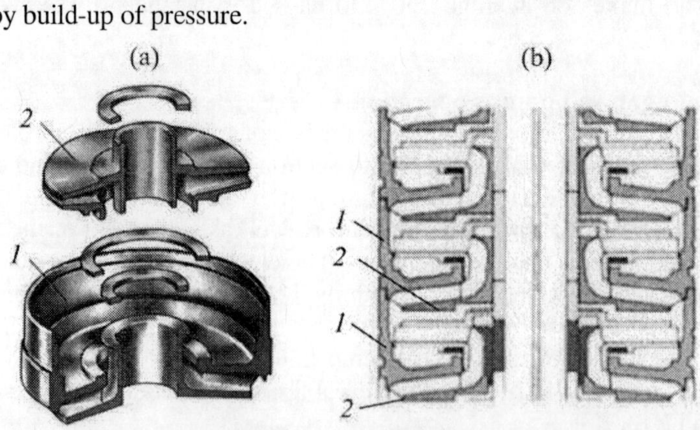

Figure 3. (a) Stage and (b) design of ECP: *1*—diffuser; *2*—impeller.

IOP Conf. Series: Earth and Environmental Science **262** (2019) 012011 doi:10.1088/1755-1315/262/1/012011

Small dimensions of centrifugal pumps dictate that diffusers are always designed as guide vanes. The key function of the diffuser is fluid withdrawal from the impeller, speeding down the fluid flow rate and simultaneous conversion of the kinetic energy to the potential energy (pressure build-up), transition of the fluid flow to the next impeller or to the discharge connection. Designs of impellers and guide vanes, as well as the pump performance depend on the planned feed and head of a stage.

Sequentially connected on the common shaft, the ECP stages are arranged in the casing (thick-wall tube). Usually, ECP casings are manufactured with one of three outer diameters: 92.5, 102 and 103.5 mm.

Repair or recovery of ECP requires that impeller and guide vanes are removed from the casing. After long-term operation of a pump, removal of the parts from the casing, specifically, guide vanes arranged inside the thick-walled tube with minimum clearances, is greatly difficult the clearances are clogged with sand. In this case, sand particles should be milled to finer fractions in order that these parts are taken from the casing. Actually, it is anything but simple to apply each guide vanes with the force sufficient for sand milling and the part removal from the thick-walled tube.

Another problem of the same technological difficulty is removal compacted soil from the pipe driven with its open end to the full length of a pass. The method used to clean the pipe makes use of simultaneous counteraction of percussion and static pressure on the plug [9]. This method ensures removal of soil plugs even from tubes of small diameters of 40–100 mm [10].

Application of high forces to the first guide vane (end most in the line) results in its failure and damage of the inner cylindrical surface of the casing, which is discarded as a consequence. Therefore, in repair maintenance, an ECP is dismantled by hitting on the casing (tube) end, or by throwing the pump to a hard surface from a height more than 1 m. The casing is imparted with acceleration after hitting or deceleration after throwing, while inner parts hold immobile relative to the casing due to inertia or move toward its outlet. As a result, internal stresses arise between the inner surface of the casing and the seating surface of the guide vane. Under the action of these stresses, internal contact bonds are broken as san particles are milled, and the parts are removed from the casing tube.

To throw ECP casing from height, the devices should have height more than the length of the casing. Furthermore, a mechanism is required to turn the casing into vertical position, including grip and hoist engine. In throwing, the impact frequency is much lower than one impact per second. Efficiency of impeller and guide vane removal is low at high expenditures.

It is simpler and more efficiently to remove impellers and guide vanes from pump casings using pneumatic hammers with variable impact force at minimum frequency [11], which can be sourced by compressors of minimum air flow rate, at level of 3 m^3/min.

3. A pneumatic hammer application

For driving a pipe with a diameter of 100 m in soil a length to 40 m, it is recommended to use Typhoon-40 (T-40) pneumatic hammer with an impactor portion weight of 40 kg and unit blow energy of 400 J (Figure 4). The tests of T-40 in horizontal operation show that the hammer impact frequency is within the range of 2.5–5.8 Hz (air flow rate 3.6–6.5 m^3/min); the lower limit of stable operation in debugging of T-40 is 1.8–2.1 Hz (air flow rate 2.6–3 m^3/min).

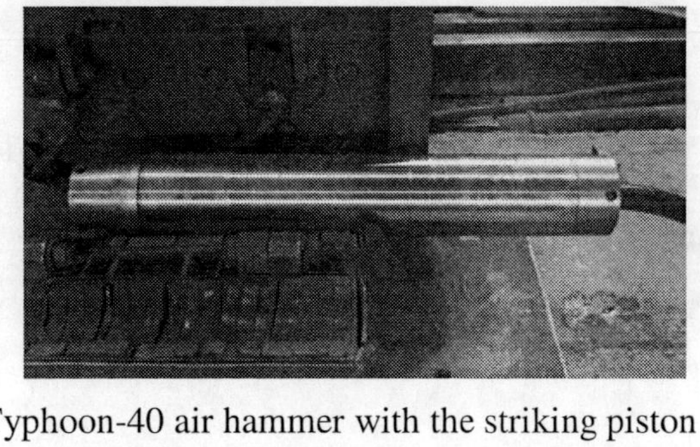

Figure 4. Typhoon-40 air hammer with the striking piston 40 kg in weight.

Typhoon-40 air hammer was manufactured and delivered under a contract to Pyt-Yakh town in Khanty Mansi Automnomous Okrug—Yugra, 100 km away from Surgut. Inside Severnaya industrial zone, in an engineering building, a special test bench was constructed for implementing the technology of ECP disassembly using T-40 air hammer (Figure 5).

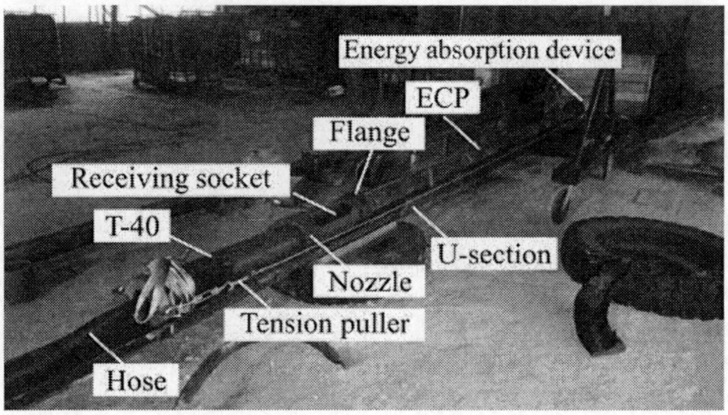

Figure 5. ECP disassembly test bench.

The tests bench for disassembly of ECP and removal of impeller and guide vanes from the pump casing consists of a guide U-section put on supports (tyres) of different height to obtain a reverse slope. At on end of the U-section, an energy absorption device is arranged on a frame connected with the U-section. ECP casing is laid in the U-section and thrust on the energy absorption device made as a piston chamber 210 mm in diameter filled with wood chips [12]. At the other end of ECP casing, an adapter flange with a nested receiving socket with discharge opening is arranged. T-40 air hammer is connected with the receiving socket via a nozzle with the conical seating surfaces using a tension puller containing steel-wire ropes and turnbuckles. Compressed air at a pressure of 0.6 MPa is fed from an electric compressor at a rate of 3 m^3/min to the air hammer along a hose with a diameter of 32 mm and 20 m long.

The procedure of ECP disassembly and removal of impeller and guide vanes includes stages below:

1. Placement of ECP in the U-section, installation of the flange, receiving socket, nozzle, air hammer and tension puller;
2. Compressed air feed to the air hammer;
3. Breakage of bonds holding guide vanes in the ECP casing under impact impulses imparted to the casing and displacement of parts in the line of the slope (backward) by gravity into the receiving socket;
4. Removal of impeller and guide vanes from the socket through the discharge opening.

Average duration of disassembly cycle for one ECP is 20–25 min.

4. Conclusions

1. Application of pneumatic hammer to disassembly of electric centrifugal pumps at an industrial scale has totally proved security and high efficiency of removal of the pump parts from its steel casing under the action of impact impulses.
2. Variable impact force of the air hammer with the piston weight of 40 kg enables using compressed air source as a low-power electric compressor with a capacity not more than 3 m^3/min without reduction in the process efficiency and at the same energy of unit blow.

Acknowledgements

The study was carried out in the framework of Basic Research Program No. AAAA-A17-117122090003-2.

References

[1] Abdulin FS 1983 *Oil and Gas Production* Moscow: Nedra (in Russian)
[2] Bukhalenka EI 1990 *Oil Equipment* Moscow : Nedra (in Russian) l
[3] Makhmudov SA 1987 *Installation, Operation and Repair of Downhole Pumping Units* Moscow: Nedra (in Russian)
[4] Molchanov AG 1985 *Oilfield Machinery* Moscow: Nedra
[5] Mikhailov KF 1995 *Handbook of Oil Field Mechanics* Moscow: Gostekhizdat (in Russian)
[6] Mishchenko RI 1984 *Oilfield Machinery and Mechanisms* Moscow: Gostehizdat (in Russian)
[7] Muravyov VM 1989 Oil and Gas Wells Operation Moscow: Nedra (in Russian)
[8] Ovchinnikov VA 1993 Oil Equipment Vol II Moscow: VNIIneftemashin (in Russian)
[9] Smolyanitsky BN, Chervov VV, Trubitsyn VV, Tishchenko IV and Veber IE 1999 RF Patent No 2130997 Method of cleaning the pipe from the ground core and the device for its implementation *Byull. Izobret.* No 15
[10] Chervov VV and Kondratenko AS 2006 Pneumatic hammer Typhoon-70 and a new method of cleaning tubes from the soil plugs *Mekhanizats. Stroit.* No 8 pp 8–12
[11] Chervov VV 2016 Minimum impact frequency of pneumatic hammer with elastic valve in air distribution system *J. Fundament. Appl. Min. Sci.* Vol 3 No 2 pp 217–221
[12] Chervov VV and Smolentsev AS 2007 Test stand for pneumatic hammers *Journal of Mining Science* Vol 43 No 6 pp 618–624

Design parameters of the device for adjusting a trajectory of pneumatic puncher motion in soil

BB Danilov*, BN Smolyanitsky and DO Cheshchin

Chinakal Institute of Mining, Siberian Branch, Russian Academy of Sciences, Novosibirsk, Russia

E-mail: *bbdanilov@mail.ru

Abstract. The calculation data are presented on forces required to deflect the rear end of the pneumatic puncher body at a wanted angle, given that a trajectory of its motion in soil should be controlled and adjusted. These data are taken as the basis to determine forces acting on elements of the device designed to change a trajectory of the controllable pneumatic puncher. The researchers determined the pressure value in the elastic chamber of the device for adjusting a puncher trajectory, under which it is feasible to deflect the tail section, responsible for control of a puncher motion track. Design dimensions of the load-bearing elements of the device under consideration are computed.

1. Introduction

The pneumatic puncher is considered the ease-in-operation and highly efficient punching facility for hole-making in a consolidated soil mass [1–5]. This process allows making holes of diameter up to 300 mm, including pioneer holes intended for further expansion [6]. However the scope of puncher application is restricted with hole-making length, as an increase in hole length is inevitably related to less probability for the puncher to reach precisely a final design point under ever-increasing influence of random deflecting factors [7]. Technological capabilities of these machines can be expanded by providing their motion along a preset trajectory, may be a curvilinear one, when a new-made hole reaches a design-prescribed underground area [8, 9].

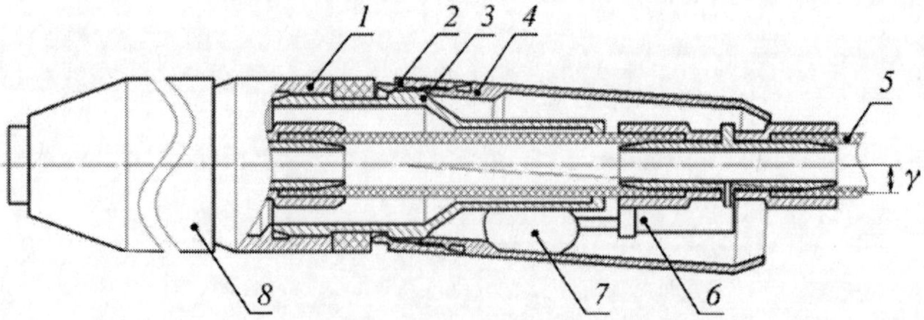

Figure 1. Scheme of the device for control of a puncher trajectory (with deflection of the rear section): *1*—screw of a pneumatic puncher; *2*—screw; *3*—sleeve pipe; *4*—deflecting section; *5*—air-intake hose (sleeve); *6*—controller of the elastic chamber; *7*—elastic chamber; *8*—pneumatic puncher body.

2. Main characteristics of air puncher

The pilot design of a controllable air puncher with deflecting tail section is developed with the purpose to investigate experimentally the control of pneumatic puncher motion in a soil mass (Figure 1) [10, 11]. Pressure supply to elastic chamber *7* causes variations in geometric shape and creates the force capable to deflect a pneumatic puncher to a correct direction.

Computation of the specifications of the deflecting device to control a pneumatic puncher rests on relationships describing interaction of its components with a soil mass [12] and experimental study evidence [13]. Commercial pneumatic puncher IP 4605 is selected as the basic study object. Its geometric and energy parameters are reported below.

Body radius r, m	0.0475
Body length l_2, m	1.51
Gross mass of the machine m, kg	55
Impactor mass m_y, kg	22
Impact frequency v, Hz	5.4
Energy of a single impact E_i, J	100
Working compressed air pressure p_c, Atm	6
External radius of air-supply line r_0, mm	18

Sandy soil, widely available in the Novosibirsk Region, is selected as a soil mass model. Its specifications are:

Density ρ, kg/m^3	1700
Porosity e	0.67
Density in terms of impactor DorNII [14] C, impact	4–5
Soil–metal friction factor μ	0.7

With a view to design a deflecting mechanism it is necessary to determine a length of deflectable tail section of the body, deflection angle, pressure in the elastic chamber in order to generate a required deflecting force, as well as radial dimensions of basic load-bearing elements of the structure with account for restrictions dependent on a hole diameter.

The calculation scheme of interaction between a deflecting section of the air puncher and soil is shown in Figure 2.

The input data are a type and physical and mechanical properties of soil, a hole diameter, m; maximum radius of puncher motion trajectory in soil (displacement of the body Δy per 1 meter of hole length), m; deflection angle γ of the tail section; nominal compressed air pressure p_c, Pa.

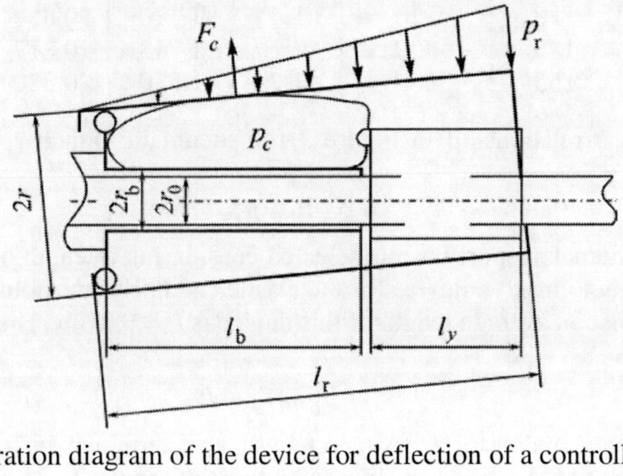

Figure 2. Operation diagram of the device for deflection of a controllable air puncher.

Of prime importance are the parameters, affecting Δy and calculated; they are the length of deflecting rear section l_r, length of support bush l_b and external radius of support bush r_b.

3. Determination of air puncher deflection

It is established based on the data on stabilization of pneumatic puncher motion that an average deflection of a puncher per 1 meter of hole length is equal to $\Delta y = 16$ mm/m [15]. According to the evidence reported in [14], such deviation can be reached at different combinations of length and deflection angles of the rear section of the air puncher; these combinations can be computed from the relationship plotted in Figure 3.

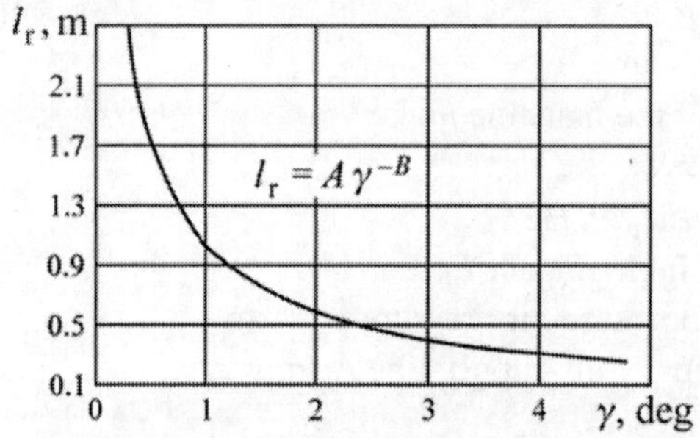

Figure 3. Relationship of a length of deflectable tail section versus of its deflection angle.

Values of coefficients *A* and *B* for the most popular air puncher models are cited in the table. They correspond to sandy soil of density ρ = 1700 kg/m^3 and ρ = 1800 kg/m^3, specific for Novosibirsk specimens [16].

Table 1. Coefficients *A* and *B*.

Pneumatic puncher model	*D*, mm	*E*,J	*L*, m	ρ = 1700 kg/m^3		ρ = 1800 kg/m^3	
				A	*B*	*A*	*B*
IP4610	55	15	1.2	0.505	0.873	0.509	0.876
MP-55	55	20	1.15	0.552	0.874	0.559	0.875
PP66K	67	65	1.24	0.954	0.829	0.954	0.833
PP66D	67	80	1.46	1.218	0.818	1.217	0.819
SO144	71	65	1.35	1.030	0.820	1.015	0.827
MP-80	80	70	1.21	0.803	0.853	0.805	0.856
IP4605	96	100	1.5	0.996	0.835	0.980	0.840
IP4603	130	220	1.4	1.110	0.850	1.095	0.854
MP-145	145	400	1.83	1.743	0.817	1.713	0.823
SO134	155	500	1.7	1.570	0.820	1.534	0.832

As per the eable, the relationship in Figure 3 for pneumatic puncher IP 4605 is described by empirical formula:

$$l_r = 0.996\gamma^{-0.835}. \tag{1}$$

The length of the internal support bush is selected considering a length of deflectable rear section with account for necessity to insert a device for the elastic chamber control inside of it (Figure 1). It is found in sketch design that an average length of this device is $l_y = 0.07$ m. Then a length of the internal bush is assumed equal to:

$$l_b = l_r - l_y. \tag{2}$$

The maximum external radius of the support bush is determined in terms of provision of the required cross-section area of the exhaust channel according to formula:

$$\max r_b = \sqrt{(r-t)^2 - 4.5(r_0 - t_0)^2}\,, \tag{3}$$

where r is external radius of the deflected section, m; t is thickness of a wall of the deflected section, m; r_0 is external radius of inlet pot, m; t_0 is thickness of inlet pot wall, m. Then at t = 0.004 m, t_0 = 0.003 m; the maximum radius r_b = 0.03 m.

Pressure in the elastic chamber, required for punching the deflectable rear section of the air puncher body into a soil mass, is determined depending on deflection angle from expression:

$$p_c = \frac{0.776\,(l_r)^{2.34}\, p_{\text{str}}^{0.33}}{(l_b - l_{cd})^2\, r_b}\,(E r\,\text{tg}(\gamma))^{0.67}, \tag{4}$$

where E is general modulus of soil mass strain, Pa; p_{str} is structural strength of the soil mass, Pa; [17, 18]; l_{cd} is length of control device, m; γ is deflection angle of the rear section, deg.

Substituting length of the deflectable rear section (1) into (4) it is possible to calculate the pressure in the elastic chamber which is high enough to provide deflection of the rear section by angle γ. This relationship is plotted in Figure 4.

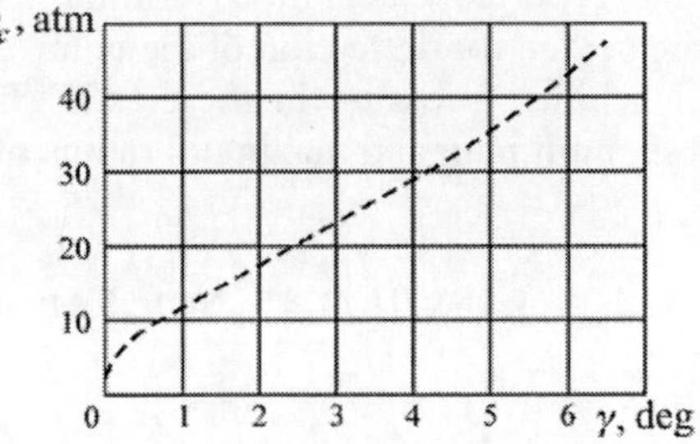

Figure 4. Pressure p_c required to turn the rear section of the body by angle γ.

Evaluation of pressure in the elastic chamber reveals that working pressure of the air used as an energy carrier in an impact device operation is not sufficient to generate the force, necessary for deformation of hole wall, subjected to deflection by the rear section of the body; a higher pressure is demanded. Considering insufficient volume of the elastic chamber, high-pressure compressors of small capacity, such as Kraftmann ARCTURUS with maximum pressure of 3.5 MPa can be employed as accessory facilities [19].

Pressure of 3.3 MPa in the elastic chamber can induce the maximum deflection of the rear section by 4.6°. Then the length of deflectable rear section, identified from the curve in Figure 3 or from formula (1), is equal to $l_b = 0.9964.6^{-0.835} = 0.279$ m.

To lower specific amount of metal per a structure, it is feasible to manufacture the support bush of less radius, provided that the bearing surface for the elastic chamber is sufficient (Figure 5).

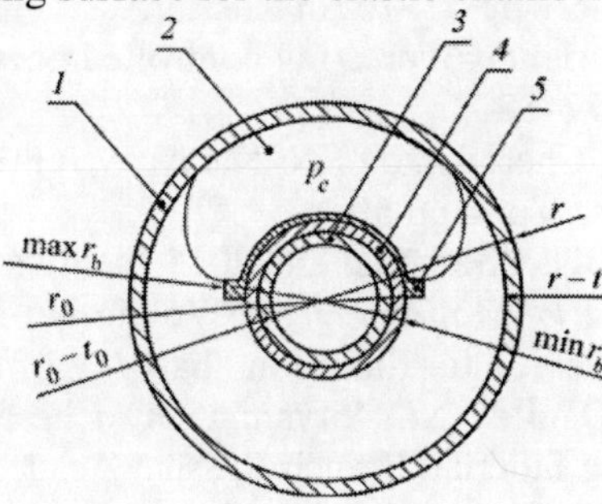

Figure 5. Geometric parameters of the cross-section of the device for deflection of the air puncher: *1*—deflectable rear section; *2*—elastic chamber; *3*—air intake hose; *4*—support bush; *5*—bearing surface

Considering formula (2) and values l_3, strength condition of the support bush is used to determine its minimum external radius for pressure p_c:

$$\min r_b = \sqrt{\frac{0.636}{\sigma_{ys}}[l_b^3 p_c + \sqrt{2.468\sigma_{ys}^2 r_0{}^2 + l_b^4 p_c^2}]}, \quad (5)$$

where σ_{ys} is yield stress of bush material, Pa.

For steel 35X at $\sigma_\text{т} = 750$ MPa [20] the minimum radius of the support bush is $\min r_\text{в} = 0.022$ m. The rational length of the deflectable rear section of the device body, which provides the structure strength and a satisfactory value of rotary motion of the rear section to compensate random deviations of the air puncher within a soil mass is $l_b = 0.279$ m under 3.3 MPa pressure in the rear chamber. Then from formula (2) the support bush length l_b = 0.209 m in the range of its external radius $r_b = 0.022$–0.030 m.

4. Conclusions

The established relationships and reported calculation formulas enable to determine basic design parameters and dimensions of the device for deflection of the controllable air puncher, manufacture of the basis of serial production machines: pressure in the elastic chamber, length of the deflectable section, length of the support bush, minimum and maximum radius of the support bush.

Acknowledgements

The work is fulfilled within research project VIII.74.3.1, State Registration No 01201353208.

References

[1] Gurkov KS, Klimashko VV, Kostylev AD et al 1990 Pneumatic hammers Institute of Mining Siberian Branch Russian Academy of Sciences Novosibirsk (in Russian)
[2] Rybakov AP 2005 Basics of trenchless technologies Moscow Press Bjuro No 1 (in Russian)
[3] Yao Ningpinga, Zhang Jie, Jin Xinga, Huang Hanjinga 2014 Status and development of directional drilling technology in coal mine *Procedia Engineering* Vol 73 pp. 289–298
[4] Shadrina A, Saruev L, Vasenin S 2016 The technology improvement and development of the new design-engineering principles of pilot bore directional drilling *IOP Conf. Series: Earth and Environmental Science* 43 (2016) 012068. doi:10.1088/1755-1315/43/1/012068
[5] Ratskevich GI, Kozlov VA, Kostylev AD 1978 Application of pneumatic machines percussion device for underground structures *Building equipment* No 5 pp 8–10 (in Russian)
[6] Sudnishnikov BV, Kostylev AD, Tupitsin KK 1970 Pneumatic punchers in construction and mining *Journal of Mining Science* No 2 pp 44–49
[7] Kostylev AD, Chepurnoy NP 1969 Investigation of the accuracy of drilling wells with pneumatic punchers with a different cylindrical body part, Pneumohydraulic power pulse systems Novosibirsk Issue 2 pp 62–70 (in Russian)
[8] Kostylev AD 1996 The experience of creating controlled pneumatic punchers *Journal of Mining Science* Vol 32 No 6 pp 77–82
[9] Kostylev AD, Tupitsyn KK, Cherednikov EN, Karavaev AT 1985 Controlled pneumatic hammer *Journal of Mining Science* Vol 4 pp 59–63
[10] Danilov BB, Smolyanitsky BN, Cheshchin DO 2015 Justification of basic diagrams of horizontal drilling deflectors *Journal of Mining Science* Vol 51 No 3 pp 553–561
[11] Pat. 156648 RF 2015 The Device for direction changing of motion of pneumo-punch machine BN Smolyanitskiy BB Danilov ND Siryamin DO Cheshchin the applicant and the patent holder Institute of Mining Siberian Branch Russian Academy of Sciences No 2015115184/03; publ 22.04.2015, Issued 2015, *Bu .No 31.* (in Russian)
[12] Danilov BB, Smolyanitskiy BN, Chanyshev AI, Cheshchin DO 2017 Determination of forces to change the trajectory of the pneumatic punch in the ground *Journal of Mining Science* Vol

53 No 4 pp 69–79

[13] Danilov BB, Cheshchin DO 2016 Determination of the location of the center of rotation of the body of the pneumatic punch in the ground under the influence of a deflecting force *Interexpo GEO-Siberia 2016 XII International Scientific Conference Subsoil Use. Mining. Directions and technologies of prospecting, exploration and development of mineral deposits. Geoecology* Novosibirsk SSUGT Issue 3 pp 45–48 (in Russian)

[14] GOST 5180-84 1993 Soils Methods for laboratory determination of physical characteristics Moscow Izd-vo standartov (in Russian)

[15] Kostylev AD Chepurnoy NP 1969 Some results of the study straightness punching holes in the ground hammers *Journal of Mining Science* No 5 pp 41–44 (in Russian)

[16] SNiP 2.05.02-85 2013 Car roads Moscow Izd-vo standartov (in Russian)

[17] Osipova MA, Sviridov VL 2013 Structural strength as a criterion for assessing the deformability of loess soils *Polzunovskiy vestnik* Vol 4–1 pp 26–28 (in Russian)

[18] Osipova ON, Dyba VP, Galashev YuV 2016 Depth study of compressible thickness of the base in full-scale experiment *Procedia Engineering* Vol 150 pp 2245–2249

[19] Model range of compressors Kraftmann Arcturus, URL: http://www.kraftmann-compressor.ru/compressor-kraftmann-arcturus.html [Model line of compressors Kraftmann Arcturus, URL: http://www.kraftmann-compressor.ru/compressor-kraftmann-arcturus.html (reference data 05.04.2018)]

[20] Anuriev VI 2006 *Directory of the Constructor-Machine Builder* Moscow (in Russian)

IOP Conf. Series: Earth and Environmental Science **262** (2019) 012013 doi:10.1088/1755-1315/262/1/012013

Substantiation of basic diagram and determination of energy and design parameters of the volumetric-type hydrohammer for vibroimpact punching of holes in soil

BB Danilov* and AA Rechkin

Chinakal Institute of Mining, Siberian Branch, Russian Academy of Sciences, Novosibirsk, Russia

E-mail: *bbdanilov@mail.ru

Abstract. The feasibility to use the ring elastic valve in volumetric-type hydrohammers is substantiated. The relationships of impact power versus the ratio of working area, diameter and length of interchamber throttle are determined by applying a simulation model of hydrohammer mechanism. Rational parameters of the study mass-size dimensional-type hydraulic hammers are selected.

1. Introduction

The use of a hydraulic drive mechanism preserves its significance and actuality in design of present-day impact mining machines due to such advantages of the hydraulic drive as appreciable increase in power, higher performance efficiency as compared to pneumatic impact machines, consequently, higher rock-breaking efficiency [1, 2]. The volumetric-type impact machines are distinguished for the best efficiency factor among hydraulic impact machinery. The volumetric-type impact machines are machines where motion and impact action of the anvil-piston is forced under pressure of a fluid supplied into working chambers of the cylinder [3]. A pneumatic hydraulic accumulator is usually used in most this-type machines with the purpose to save energy during an idle stroke of the anvil and to give off energy in its working stroke.

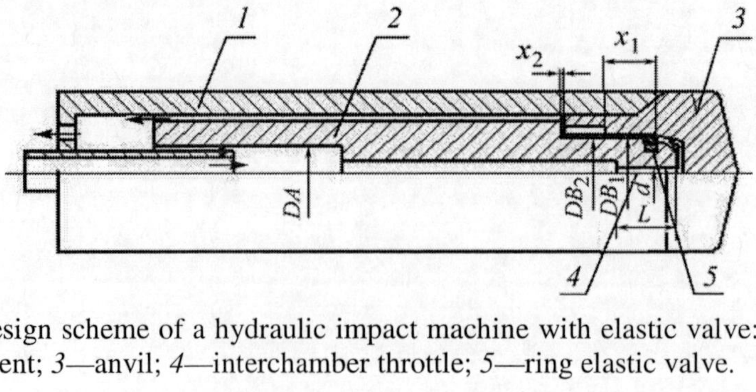

Figure 1. Design scheme of a hydraulic impact machine with elastic valve: *1*—body; *2*—striking element; *3*—anvil; *4*—interchamber throttle; *5*—ring elastic valve.

The design complexity and necessity to manufacture their details with high specific precision are imperfections of hydraulic volumetric-type machines. A way to simplify their design is to use elastic shut-off-and-regulating elements intended to distribute flows of the working fluid [4, 5]. At the Institute of Mining SB RAS the researchers developed pneumatic impact machines Typhoon where the ring elastic valve is employed as an elastic shut-off-and-regulating element to lock the back-stroke chamber [6]. The long-term successful application of these machines in construction and mining industries, design simplicity and operational reliability provide all the grounds to consider applicability of this design (Figure 1) in hydraulic impact volumetric-type machines.

2. Simulation model development

The simulation model of impact mechanism operation is developed in SimulationX software to complete a design and energy parameters of a hydraulic impact machine with a ring elastic valve. The calculation scheme of the machine is shown in Figure 2.

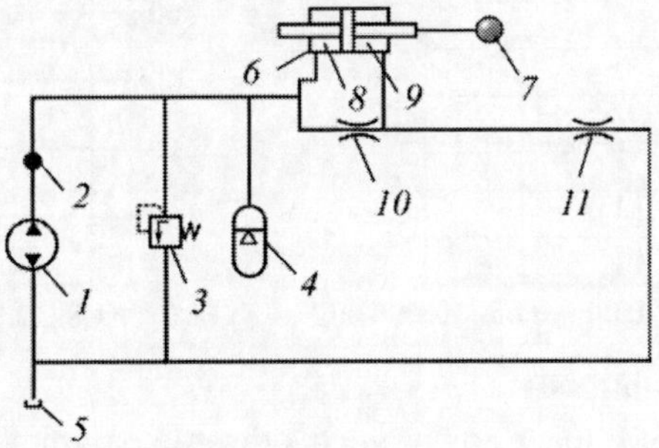

Figure 2. Calculation scheme of a hydraulic impact machine with a ring elastic valve: *1*—pump; *2*—fluid volume; *3*—safety valve; *4*—pneumatic hydraulic accumulator; *5*—overflow, *6*—two-stock hydrocylinder with variable area of the back-stroke chamber, *7*—mass of the striking element; *8*—direct-stroke chamber; *9*—back-stroke chamber; *10*—interchamber throttle; *11*—locking throttle.

To determine initial parameters, the sketch design of the striking device is executed, then computation experiments were undertaken using earlier design parameters: mass of the striking element m = 4 kg, diameter of the ring-shaped elastic valve prior to extension (a direct stroke) DB_1 = 0.025 m, post-extension diameter of the ring elastic valve (a back stroke) DB_2 = 0.028 m, coordinate of valve locking (or distance from the striking element to anvil at valve locking) x_2 = 0.001 m, working stroke of the striking element x_1 = 0.05 m. The work of the ring elastic valve is modeled by locking throttle *11* (Figure 2), when the valve in its direct stroke reaches coordinate of the back-stroke chamber locking, thereto piston area is increasing from the side of this chamber after the impact. The valve is considered operating instantly, quite discretely. Diameter of the direct stroke chamber DA, diameter d and length L of interchamber throttle were measured in view to select optimal energy parameters of the mechanism. The pre-impact velocity of the striking element, impact frequency, impact energy, and impact power were considered as basic estimation criteria.

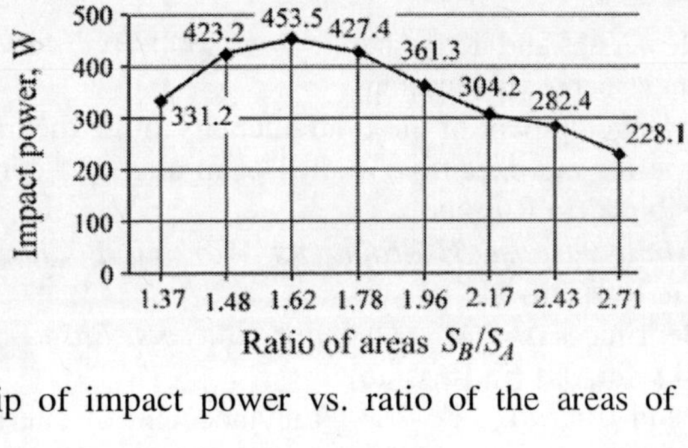

Figure 3. Relationship of impact power vs. ratio of the areas of front S_B and rear S_A chambers.

Figure 3 demonstrates the resultant relationship of impact power vs. ratio between areas of the front and rear chambers. In this case ratio of 1.78 is assumed the most optimal one, as at ratio of 1.62, providing a greater value of impact power, we fall outside the critical value of pre-impact velocity which it is limited to ≈ 4 m/s in reasonable terms of impact strength and long-term operation life.

At the selected ratio of working areas the geometric parameters of interchamber throttle were varied (Figure 4). The resultant relationship of impact power versus throttle length at fixed diameter of $1.8 \cdot 10^{-3}$ m indicates the actually proportional growth of impact power with increase in throttle length (Figure 4a). There are grounds to speak of value of 0.05 m as the optimal one, because higher values lead to excess of the above established limit of the pre-impact velocity.

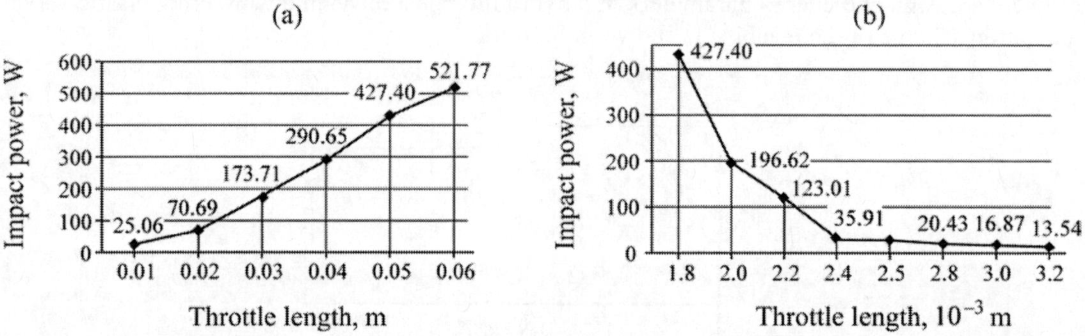

Figure 4. Relationship of impact power versus throttle parameters: (a) length; (b) diameter.

In final computations the diameter of interchamber throttle varied at fixed length and fixed area ratio, established at previous stages. The relationship in Figure 4b indicates hyperbolic growth of impact power with increase in throttle diameter, however, the researchers failed to obtain a solution for throttle diameter less than $1.8 \cdot 10^{-3}$ m.

3. Conclusions

The simulation model constructed in the present research justifies the feasibility to apply the design scheme with the ring elastic valve in hydraulic volumetric-type hammers and enables to determine optimal parameters using the criteria of the pre-impact velocity of the striking element, impact frequency, impact energy, impact power, etc.

Numerical experiments on the model made it possible to establish relationships of impact power versus some design parameters, to find rational ratios of working areas of direct- and back-stroke chambers and geometric parameters of the interchamber throttle to provide the maximum impact power of the selected type–size mechanism to design the hydraulic hammer for the vibration-impact hole-making process.

References

[1] Arkhipenko AP, Fedulov AI 1991 *Hydraulic Hammers* Novosibirsk: IGD SO AN SSSR (in Russian)

[2] Ushakov LS, Kotylev YuE and Kravchenko VA 2000 *Hydraulic Machines of Impact Action* Moscow: Mashinostroenie (in Russian)

[3] Gorodilov LV 2010 Development of the fundamentals of the theory of hydraulic impact systems of volumetric type for the executive bodies of mining and construction machinery *Dr Tech Sci Thesis* Novosibirsk (in Russian)

[4] Galdin NS 2005 *Multi-Purpose Hydroimpact Working Bodies of Road-Building Machines* Omsk: SibADI (in Russian)

[5] Galdin NS and Bedrina EA 2003 *Dynamic Buckets Based on Hydraulic Hammers for Excavators* Omsk: SibADI (in Russian)

[6] Smolyanitskiy BN and Chervov VV 2014 Enhancement of energy carrier performance of air hammers in underground construction *Journal of Mining Science* Vol 50 No 5 pp 918–928

IOP Conf. Series: Earth and Environmental Science **262** (2019) 012014 doi:10.1088/1755-1315/262/1/012014

Support design for permanent mine roadways in tectonically active areas

PV Deev*, MA Petruhin and AA Tzukanov

Tula State University, Tula, Russia

E-mail: *dodysya@yandex.ru

Abstract. The authors propose an analytical method of support design for arbitrary cross-section mine roadways subjected to the action of tectonic forces and seismic loads of earthquakes. As distinguished from the existing methods, the new technique allows taking into account influence of a close interface of different rocks. The seismic analysis includes the action of the waves reflected from the interface and the waves refracted through it. The method is based on an analytical solution of 2D elastic problem on stress state of an elastic ring supporting an arbitrary hole in one of two half-infinite media with the common boundary. The solution of the problem has been obtained using the theory of analytical functions of complex variables, Cauchy type integrals, conformal mapping and complex series. The examples of the support design are given.

1. Introduction

Many minerals occur in the zones of active neotectnics. Nonuniform rate movements of tectonic blocks induce additional stresses in rocks, which should be taken into account in mine planning and design.

At the present time, parameters of a tectonic stress field can be found to a sufficient accuracy using geomechanical and geophysical methods of stress measurement [1, 2]. Estimates of stress state of underground structures are obtained from analytical solutions of problems in continuum mechanics [3, 4] and by numerical modeling [5]. The analytical methods of solving elastic problems are advantageous over the numerical techniques though restrained by the limited number of the available analytical solutions.

The method proposed in [6] for the support design for noncircular mine roadways under the action of tectonic forces is based on the analytical solution of 2D elasticity problem on stress state of a ring supporting an arbitrary shape hole in an infinite uniform medium under conditions of initial stress state. The problem is solved using the theory of analytical functions of complex variable. The approach put forward in [7] enables taking into account technological nonuniformity of rocks in the support design under the action of tectonic forces. The technological nonuniformity is understood by the authors as one or a number of zones having different characteristics from the surrounding rock mass around a roadway. Such zones appear as a result of rock failure under drilling and blasting, or when injection reinforcement is applied. A feature of the elasticity problem set and solved in [7] is modeling of the nonuniform rock mass zone and support as variable thickness rings, which allows considering nonuniform domains of arbitrary configuration. Inclusion of many domains–layers in the analytical models makes it possible to image smooth variation in deformability of rocks with distance from the roadway boundary.

IOP Conf. Series: Earth and Environmental Science **262** (2019) 012014 doi:10.1088/1755-1315/262/1/012014

2. A new method of support design for a mine roadway

This study offers the method of support design for a mine roadway located in the tectonically active rock mass near an interface of rocks of different deformability. The analytical model of the support is depicted in Figure 1.

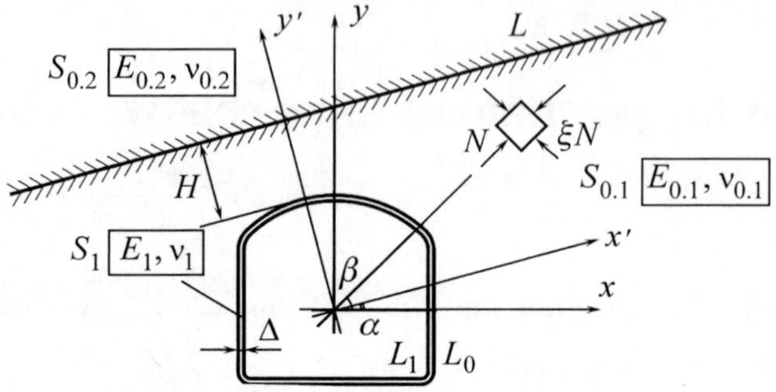

Figure 1. Analytical model of support for roadway located near interface in tectonically active rock mass.

In the figure, the deformable media $S_{0,1}$ and $S_{0,2}$, deformability of which is characterized by the deformation moduli $E_{0,1}$, $E_{0,2}$ and Poisson's ratios $\nu_{0,1}$, $\nu_{0,2}$, simulate rock mass composed of two types of rocks. At the distance H from the interface L occurring at the α relative to horizontal line, there are a hole supported by the ring S_1 made of a material with the deformability characteristics E_1 and ν_1. The medium $S_{0,1}$ experiences uniform initial field of the principal stresses N and ξN such the the principal stress N is oriented at the angle β relative to horizontal line. At the lines L and L_0, the vectors of stresses and displacements are continuous; the internal boundary L_1 of the ring is free from the external forces.

Using the theory of analytical functions of complex variables, in particular, complex potentials proposed by Kolosv–Muskhelishviliy, Cauchy type integrals, conformal mapping and complex series, the set problem was solved as iteration of step-by-step refinements of influence exerted by the interface of two media on the stress state of the ring simulating support of a mine roadway.

Aiming to take into account the support–rock interaction, an adjusting factor α^* is introduced in the calculations; this factor can be found using the procedure described in [8].

The calculations are performed for a solid concrete lining in a roadway having cross-section illustrated in Figure 2.

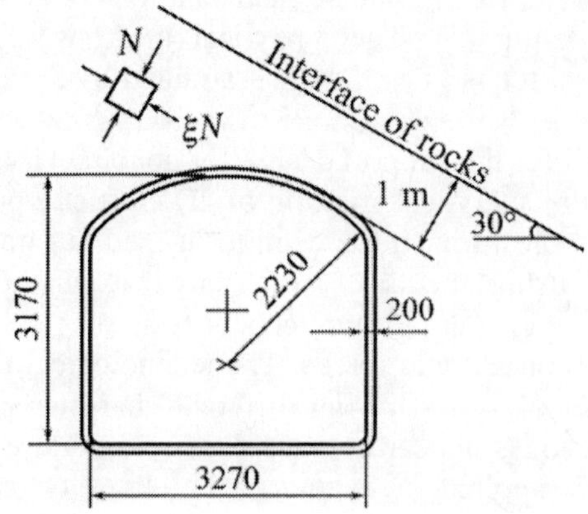

Figure 2. Cross-section of roadway under analysis.

The deformation characteristics of the support and surrounding rocks are: E_1 = 20 000 MPa, $\nu_1 = 0.2$, $E_{0,1}$ = 50 000 MPa, $\nu_{0,1} = 0.25$, $E_{0,2}$ = 1000 MPa, $\nu_{0,2} = 0.35$. The initial stress field in rocks includes the major compressive stress oriented in perpendicular to the interface ($\beta = 60°$), and the ratio of the principal stresses is $\xi = 0.5$.

Figure 3 depicts the calculated results for the support for the described roadway as the distribution diagrams of normal shear stresses at the internal and external boundaries of the lining—$\sigma_\theta^{(in)}$ and $\sigma_\theta^{(ex)}$, respectively. The values of the stresses are given as fractions of of the product α*N. The dashed lines show the calculation without regard for the influence of the rock interface.

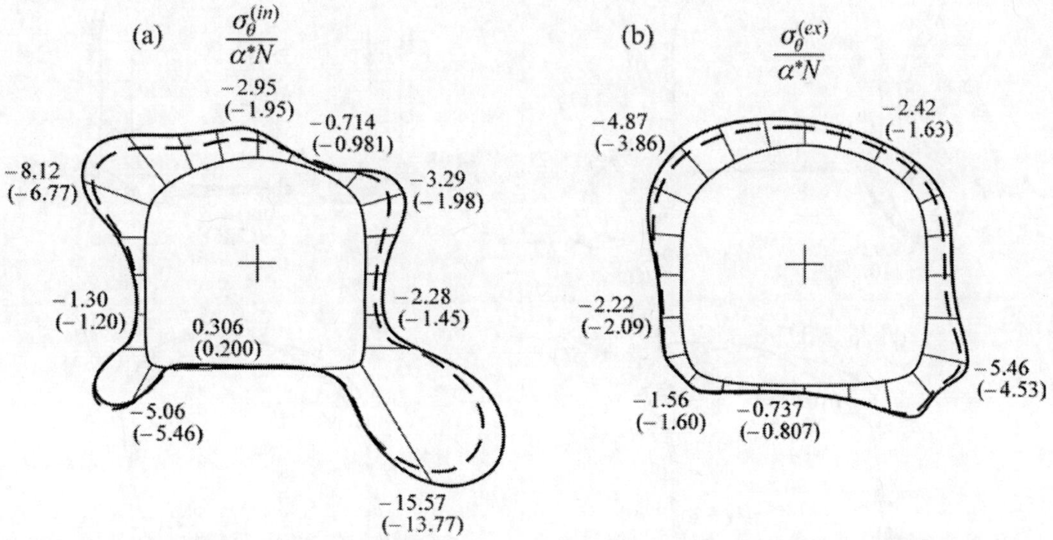

Figure 3. Normal shear stresses at (a) internal and (b) external boundaries of the support cross-section under the action of tectonic forces.

It is seen in the figure that the influence of the interface shows itself as a noticeable increase in the normal shear stresses in the support. The highest increment in the stresses is observed in the roof and at the juncture of the right-hand wall with the floor.

The developed method can be used in support designs under seismic loads of earthquakes. In this case, 2D elasticity problem in Fig. 1 assumes that initial stresses are absent in the medium $S_{0,1}$ which is though subjected to infinite compression along the interface by unequal stresses determined in conformity with the currently customary approach [6, 9] with regard for reflection and refraction of seismic waves at the interface [10]. This problem is solved using the methods applied to to above problem solution.

The calculated results on the support design under the action of vertical seismic waves of tension–compression are shear are presented below. The distribution diagrams of normal shear stresses at the internal $\sigma_\theta^{(in)}$ and external $\sigma_\theta^{(ex)}$ boundaries along the cross-section of the lining are plotted as fractions of the normal P and shear S stresses at the front of P- and S-waves. As before, the dashed lines illustrate calculations regardless of the influence of the interface.

The stresses P and S can be evaluated using the formulas [9]:

$$P = \frac{1}{2\pi} A K_1 \gamma c_1 T_0, \quad S = \frac{1}{2\pi} A K_1 \gamma c_2 T_0,$$

where A is coefficient equal to the earthquake magnitude; K_1 is a coefficient of permissible damage (for underground structures, the product AK_1 is assumed as 0.025, 0.05 and 0.1 in case of seismic activity at magnitudes 7, 8 and 9, respectively); γ is the bulk density of rocks; T_0 is the prevailing PPV period; c_1 and c_2 are the velocities of P- and S-waves, respectively.

According to the distribution diagrams in Fig. 4, the interface near the roadway decrease stresses cause by the tension–compression wave and increase stresses induced by the shear wave.

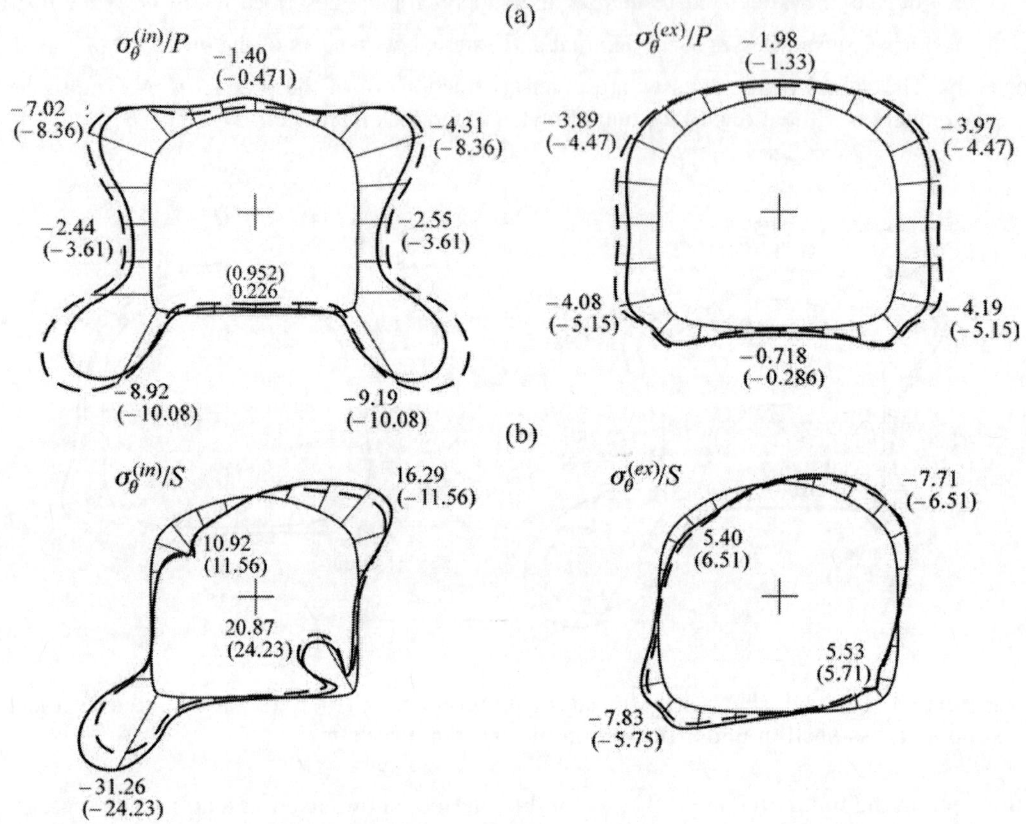

Figure 4. Normal shear stresses in lining in roadway under the action of seismic waves: (a) tension–compression and (b) shear.

It is worthy of mentioning that stress state of rocks near the interface essentially depends on the direction of seismic waves [9], which is impossible to determine before an earthquake. Therefore, in order to get a holistic insight into the seismic load exerted by an earthquake on a roadway located near an interface, it is required to analyze different directions of seismic waves in case of various locations of the roadway relative to the interface. Such research is laborious and complex in terms of interpretation. The present study's authors think that it is wiser to plot envelopes of maximum tensile and compressive stresses arising in support of mine roadways in case of different directions of seismic waves. First proposed in [6], this approach, after little modification, can be applied to the case analyzed in this study.

3. Conclusions

1. The method has been proposed for the support design for underground excavations located near the interface of rocks of different deformability under the action of tectonic forces and seismic loads of earthquakes. The new analytical solutions are obtained for 2D elasticity problems on stress state of a ring supporting an arbitrary configuration hole near a straight interface of two elastic media when the interface is subjected to infinite stresses modeling seismic loads, or one of the media experiences initial stress field.

2. The presence of the interface results in the decrease in the seismic stresses in the support in case of the vertical tension–compression wave and increases the stresses in the conditions of the vertical shear wave.
3. If a thick layer of weaker rocks occurs near the roadway, stresses in the support grow.

Acknowledgements
This study has been supported by the President of the Russian Federation, Grant No. MD-1280.2017.5.

References

[1] Kozyrev AA 1983 Diagnosis and identification of tectonic stress fields in seismically active regions in search of earthquake precursors *Mechanics of Rocks in Underground Construction and Mineral Mining at Great Depths* Leningrad (in Russian)
[2] Rybin VV and Savchenko SN 2017 Reconstruction of effective stresses within the Kovdor group of deposits on the basis of full-scale measurements *Proc. Fersman Scientific Session of the Kola Science Center of the Russian Academy of Sciences* Apatity (In Russian)
[3] Fotieva NN, Bulychev NS, Deev PV and Vladova VV 2008 Design of support of multiple non-circular workings in tectonic areas *Archives of Mining Science. Quarterly. Contribution of Strata Mechanics to Mining Technology and Work Safety* Cracow Vol 53 Issue 3
[4] Muir Wood AM 1975 The circular tunnels in elastic ground *Géotechnique* Vol 25 Issue 1 pp 115–127
[5] Halim IS and Chen WP 2001 The use of numerical method in the design of large underground tunnel and chamber in rock *DC Rocks* American Rock Mechanics Association
[6] Fotieva NN 1980 Support Designs for Underground Structures Seismically Active Regions Moscow: Nedra (in Russian)
[7] Fotieva NN, Bulychev NS and Sammal AS 1997 Designing tunnel linings in tectonic regions with the rock technological heterogeneity to be taken into account *IX International Conference on Computer Methods and Advances in Geomechanics* Wuhan: Balkema pp 1315–1320
[8] Bulychev NS and Fotieva NN 1999 Once more about the modern theory of underground structures design *Proceedings of the World Tunnel Congress 1999* Oslo pp 327–333
[9] RF Construction Code SP 268.1325800.2016. Transport Facilities in Seismic Regions. Design Regulations (in Russian)
[10] Deev PV and Petruhin MA 2017 Seismic analysis of mine support *J. Fundamet. Appl. Min. Sci.* Vol 4 No 2 pp 37–41 (in Russian)

Hybrid opencast/underground process to mine Kuzbass coal deposits

VA Fedorin*, VYa Shakhmatov, BA Anferov and LV Kuznetsova

Federal Research Center of Coal and Coal Chemistry, Siberian Branch, Russian Academy of Sciences, Kemerovo, Russia

E-mail: fva@icc.kemsc.ru

Abstract. Modular technological structures of mine sites adapted to hybrid coal mining technologies in Kuzbass, protected by the Russian Federation patents are proposed. These intangible assets are of particular value to business community focused on research and commercialization of the results of intellectual activity. Therefore, the intellectual property can be considered as a promising credit instrument in the scientific concept of "intellectual underground mine of coal opencast". High efficiency of the hybrid method for development of coal seams exceeds the average profit in Kuzbass by 1.5 times. In the long term, the unmanned Highwall geotechnology is considered as an intermediate stage of the hybrid mining method, Highwall mining technology is a fully autonomous, mobile, high-performance and economic coal mining system that allows fully mechanized underground mining at an opencast.

1. Introduction

The long-term scientific-and-technological advance in the fuel and energy complex implies the high priority of coal mining technologies in terms of "intellectual opencast" and "intellectual mine" concept permitting 15–20% reduction in coal production costs thanks to more efficient management of the intellectual property [1, 2]. So far, the present-day coal industry is distinguished for the poor scientific-and-technological potential and low innovation activities.

In Kuzbass the coal mining industry is based on new scientific approaches and innovative principles in construction of production facilities and structure-and-network theory to open gently sloping coal seam deposits. In 2017 38 fully mechanized mine workings actively operated with the coal production rate of approximately 80 Mt, viz., 2 Mt per a working in average. The modular geotechnological process for opening and preliminary development of coal seams was employed. Many-years experience serves the basis in development of technological solutions for autonomous modular mine sections at operating opencasts where the combined opencast-underground mining processes are used to extract flat-lying seams within the concept "intellectual mine at coal opencast". The technological solution rests on the elaborate mining-technological procedure for opening and preliminary development of mine sections with adaptation to coal mining process at an opencast, including a high-performance stoping face with the use of the integrated unmanned deep seam mining (IDSM) process and a conventional complex geotechnology on extraction of coal seams of 3–5 and more meters in thickness.

At Institute of Coal, SB RAN, the modular mining-technological structures at mine sections adapted to hybrid coal mining technologies are developed and protected by Russian Federation patents

[3–5]. These intangible assets are intellectual property of specific value for business activities, oriented to research and commercialization of intellectual products [6].

The research-technological grounds for the integrated development of coal deposits should be aimed at development and modernization of complex geotechnologies with respective intellectual property oriented to an innovation component of Patent portfolio, which cost can manifold exceed the cost of any other property owned by a coal mining company. In this connection the intellectual property can be considered as a perspective credit instrument in the integrated subsoil management [2].

2. Integrated deep seam mining

The authors' research work is based on research and methodological findings of research scientists of IPKON RAS [7], IM SB RAS [8], and NTTs-NIIOGR on the integrated geotechnological development of coal deposits. The present research work resulted in a novel scientific concept on the complex process of integrated in-depth exploitation of coal deposits: one of three proposed applications is protected by Russian Federation Patent:

- the first one: open coal mining up to design depth according to the economic design criterion, the use of available underground transport infrastructure at the final operation stage;
- the second (open-underground bench): a single high bench known as Highwall without castigation of coal pit walls with unmanned component in integrated deep mining of coal seams and handling of broken coal through communication systems available at the opencast;
- the third: underground mining by modular geotechnological structure at mine sections with application available production infrastructure of a coal openpit.

In perspective long-term projects Highwall unmanned geotechnology, namely, an integrated deep seam mining (IDSM) can be considered as a component of hybrid openpit/underground mining process. IDSM is a completely autonomous, mobile, highly efficient and economic coal mining system allowing the complete mechanized underground seam mining at a coal opencast.

In the coal mining industry the complex geotechnologies comprise the information on integration of open and underground coal mining techniques in space and time, regularities in behavior of coal pit–underground mine system in a rock mass, as well as technical, economic, ecological and management interrelations.

The technological complex of the integrated coal mining process is considered as a combination of functionally interrelated facilities for open and underground mining operations to fulfill the design technological processes under restricted mining-geological conditions to realize the geotechnological potential.

Practical application of the integrated mining process in Kuzbass is dated back to 2000-s as a result of scientific-methodical substantiation of the complex physicotechnical geotechnology at Institute of Coal SB RAS and technico-economic justification of investment ventures at Sibirginsky coal open mine (Novosibirsk), Mokhovsky openpit (Baikaimsky Mine). In 2017 coal production by the integrated mining process reached 59424 thou t or 25 % of the total coal production at Kuzbass mines (Figure 1). 57 466 thou t fall on the underground mining and 124 560 thou t fall on open mining.

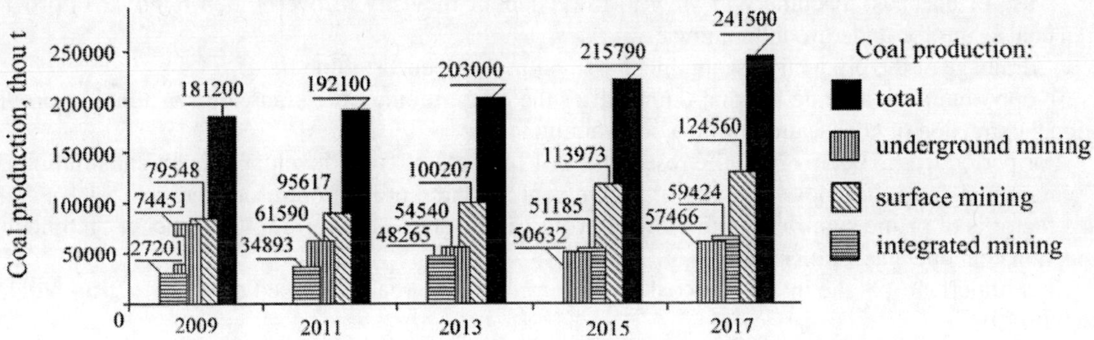

Figure 1. Annual coal production by different mining processes.

For the period from 2009 to 2017 coal production by the integrated mining process was increased more than twofold from 27 to 59 Mt/year in Kuzbass.

The coal production by the integrated mining process in Kuzbass was analyzed with singling out coal production by open and underground mining techniques separately, using the modular geotechnological structure at mine sections at coal opencasts with the use of available production infrastructure (Figure 2).

It is established that coal production at modular mine sections of coal open mines increased approximately 2.5 times for 8 years from 11.4 to 27.4 Mt. (The coal production share by open mining in the integrated mining process increased twofold from 15.8 to 32 Mt).

The application scope of schemes for opening and development of coal open-mine fields by applying the modular integrated mining-technological structure embraces practically all the range of conditions at many-years operating, new-designed and perspective open mines on flat-lying coal seams in Kuzbass. It is of specific importance for mining at underground areas of the first order specified with the critical overburden ratio; as the operation life of a coal mine can be notably extended with no appreciable investments and properly-minded subsoil management. It is reasonable to use the integrated mining technology to drain openpit fields at coal deposits with complex hydrogeological conditions; it can be efficient at coal opencast sections where integration of mining and drainage operations is imperative. In 2017 12 coal mining entities applied the integrated mining technology and yielded approximately one forth of coal production in Kuzbass.

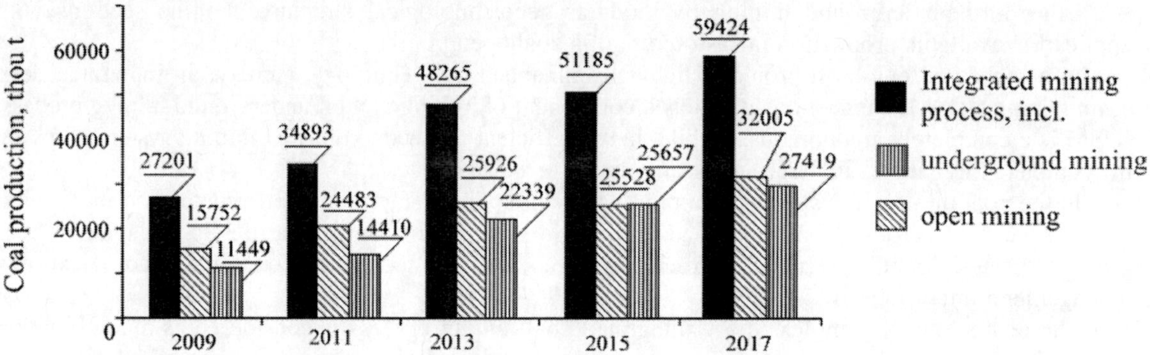

Figure 2. Structure of coal production by the integrated mining process in Kuzbass, thousand tons per year.

Performance of the integrated mining process is higher thanks to a number of factors:

—combined use of available production infrastructure, power lines, auto- and railway lines, substations, moto-transport fleet to handle coal, etc.;

—handling of opencast coal at conveyors of underground section of the mine, thus optimizing the transportation performance;

—use of opencast machinery in view to lower labour intensity in overburden removal in order to open coal seams for underground mining;

—drainage of the production seam thickness within the opencast mine field;

—opportunity to arrange internal dumps after the open mining operations are cancelled to provide rational extraction of coal seams within a coal mining lease.

For period from 2006 to 2017 the researchers of Institute of Coal developed technical solutions on the integrated geotechnology for comprehensive coal resource utilization, protected by 15 RF Patents This factor is of prime significance in view to protect innovations concerning the concept "Intellectual mine at a coal opencast" in Russia.

It is important that the integrated coal seam mining can be realized based on a production block (a module) [9].

At the Institute of Coal SB RAS under development is the methodical approach to the consecutive formation of mining technological models and their optimization based on the all-round analytical data. In 2017 the principal parameters of the integrated coal mining geotechnology: production cost, price, profit, profitability were analyzed by using the methodology for evaluation and selection of effective technological schemes of coal seam mining under actual mining and geological conditions at Kuzbass operating collieries. Economic and financial parameters are reported for 12 Kuzbass mining entities with total coal production of 59 424 thou t of coking coal brands in the Table. The cost of coal production by the integrated mining technique is averaged and equals 2286 Rub/t, coal price being 3513 Rub/t [10]. Profitability of marketed products estimated as a ratio of balance profit to current expenditures amounts to 45%, profitability in terms of net profit is 39%.

Table 1. Economic and financial parameters of coal production by the integrated mining technique in 2017.

Parameter	Value
Marketable coal production by open/underground mining technique, total, thou t:	59424
Open mining	35005
Underground mining	27419
Expenses on marketable product production, MRub	135843
Production cost of 1t of coal, Rub	2286
Price of 1 t of coal, Rub	3513
Marketable product, MRub	208755
Production profits, MRub	72912
Property tax, etc., MRub	11819
Pre-tax profit, MRub	61093
Profit tax, MRub	15273
Net profit, MRub	45820
Net profit, Rub/t	771
Profits margin, %	45

3. Conclusions

At Institute of Coal SB RAS the modular mining-technological structures of mine sections, adapted to the integrated coal mining technique in Kuzbass are developed and protected by Russian Federation Patents. These intangible assets are of specific value for business oriented to research work and commercialization of intellectual products. Intellectual property can be considered as a perspective credit instrument in the scope of scientific concept "intellectual mine at a coal opencast". High performance of the integrated technique to mine coal seams is established, the profitability exceeds 1.5 times the average profit value in Kuzbass.

References

[1] Forecast of scientific and technological development of the fuel and energy sector of Russia for the period up to 2035 (Approved by the Ministry of Energy of the Russian Federation on 14.10.2016) 106 pp. [Electronic resource]. URL: TEK document-66647.pdf

[2] Lebed VA, Misiti Jacopo and Rozhkova OV 2017 Intellectual property as a promising credit instrument in the sphere of integrated development of subsurface resources seams *Naukoemk. Tekhnol. Razrab. Ispolz. Mineral. Resurs.* No 3 pp 116–119

[3] Fedorin VA, Shakhmatov VYa, Mikhailov AYu 2018 Combined method of development of

coal seams of Kuzbass based on the synthesis of open and underground mining processes *Vestn Nauch. Tsentra VostNII po Prom. Ekol. Bezop.* No 1 pp 32–40
[4] Mikhailov AY, Varfolomeyev EL 2015 Development of an open-underground coal mining method in the Tersi geological and economic region *GIAB* No 10 pp 51–58
[5] Tatarinova OA, Varfolomeev EL 2017 Definition and classification of open-underground (combined) method of development of coal deposits of Kuzbass *Vest. KuzGTU* No 6 pp 25–30
[6] Anferov BA et al 2015 *State and Prospects of Development of Public–Private Partnership Projects in the Context of Integrated Development of Subsurface Resources* AE Kontorovich, SM Nikitenko and EV Goosen (Eds) Kemerovo: Sib. Izd. Gruppa (in Russian)
[7] Kaplunov DR and Rylnikova MV 2012 Combined Mining of Ore Deposits: Tutorial Moscow: Gornaya kniga (in Russian)
[8] Ordin AA and Klishin VI 2009 *Optimization of Technological Parameters of Mines Based on Lagged Models* Novosibirsk: Nauka (in Russian)
[9] Chernikh NG 2017 Method of complex open-underground mining of coal seams *Naukoemk. Tekhnol. Razrab. Ispolz. Mineral. Resurs.* No 3 pp 157–161
[10] *Coal Industry of Kuzbass: Main Indicators of Work in 2017: Monthly Collection* Kemerovo: IVTC No 12 81 pp

Modeling geomechanical processes in underground longwall and shortwall mining of thick coal seams with elements of robotization

VN Fryanov and LD Pavlova*

Siberian State Industrial University, Novokuznetsk, Russia

E-mail: *ld_pavlova@mail.ru

Abstract. A variant of the technological scheme for thick layer mining with top coal caving and pre-destruction of coal by means of advanced parallel chambers is proposed. Based on the results of numerical modeling, the optimum dimensions are justified for chambers, interchamber coal pillars as well as crown pillar between the chambers and immediate roof rocks. It has been proved that after the advanced chambers are made, the strength roof coal is reduced by a factor of 2 to 2.5, and therefore the conditions are created for uniform coal caving without dilution. A variant of the technology of unmanned coal mining in the chambers is proposed.

1. Introduction

The problem of safe and efficient thick coal mining in the conditions of spontaneous combustion hazard and dynamic phenomena in Kuzbass has been of increasing concern for the late 50 years [1, 2]. The rate of development of technology and equipment for mining thick coal seams lags far behind the technology for medium-thick coal mining.

The fully mechanized thick coal mining technology using system KTU-3M has a number of variants with slicing, hydraulic mining and coal transport [2–7]. Recently, the fully mechanized top coal caving technology is efficiently implemented [2, 6, 7]. At the same time, due to uncontrolled pre-fracturing and destruction of top coal, the coal loss reach 0.9–4.2 t/m^2 of mined-out seam, or 18% in the total extraction panel [8].

In order to develop methods and means for pre-fracturing and destruction of top coal, it is necessary to determine laws of stress state formation in top coal. The complexity of the problem is constituted by the requirement to preserve stability of top coal near the face for elimination of uncontrolled coal caving in the operation zone of a cutter–loader and to create conditions for dynamic coal breakage in the zone of coal loading on conveyor.

The major constraint of pushing the limits of application of the longwall top coal caving technology is the absence of procedures and guidelines for evaluation of geomechanical parameters ensuring safe and efficient coal mining in difficult conditions.

2. The objective and program of the research

This study based on the revealed stress distribution patterns, shapes and sizes of pre-fracture zones in top coal proposes the method of preliminary weakening by means of advanced extraction of top coal by rooms and pillars using robotic equipment.

The objective of the research is to validate parameters of rooms and pillars in the upper layer to ensure stability of top coal at the face of the lower layer and destruction of coal in the zone of loading on conveyor.

The research program included stage-wise implementation of:

—analysis of stress state of rocks in the traditional LTCC technology without advanced room and pillar in the upper layer;

—finding of regular patters of stress distribution in rock mass during partial extraction of coal in the upper layer by advanced rooms in case of varied spatial arrangement of rooms along the coal seam height, width of rooms and rib pillars;

—justification of recommendations for selecting rational parameters of rooms and pillars in the pre-set geological and geotechnical conditions.

3. Research methods

The numerical modeling assumed a geometric layout vertically along line *I–I* (Figure 1). The plain strain problem for classical elasticity–plasticity has the following boundary conditions: the top of the model is free from loading, the horizontal displacements of the side boundaries and the vertical displacements of the bottom boundary are zero.

The finite element-based modeling results are estimated using the in situ measurement data. The ratio of residual and initial strengths of coal is determined using the values of principal stresses found from the Mohr–Coulomb failure envelope.

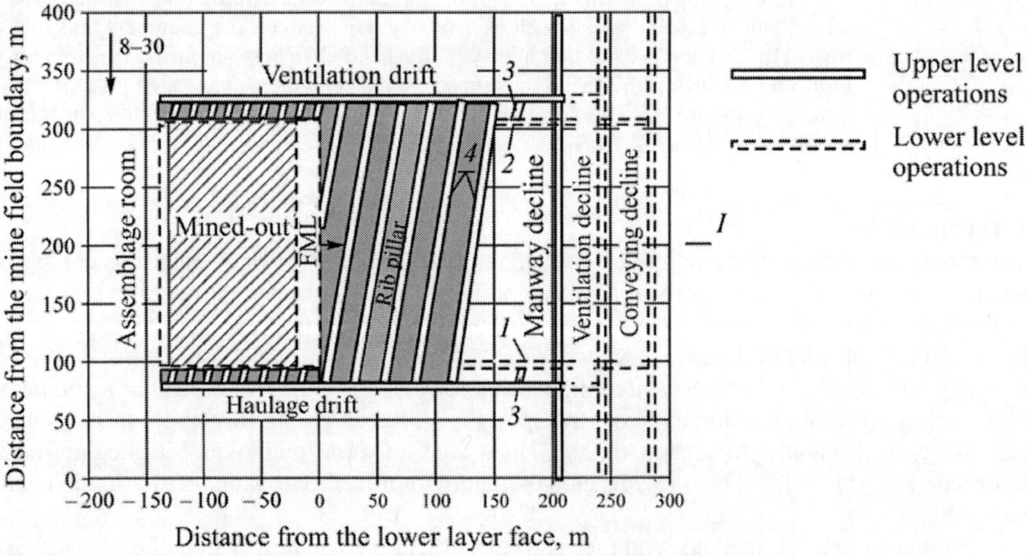

Figure 1. Process layout of thick coal seam mining with shotwall pre-fracture of upper layer and cutting of lower layer and loading of coal on armored face conveyor: *1*—conveyor drift in the lower layer; *2*—ventilation drift in the lower lower; *3*—cross connections; *4*—rooms; FML—fully mechanized longwalling in the lower layer.

4. Results and discussion

The analysis of the influence exerted by rooms in partial extraction of coal in the upper layer involved a few variants of spatial arrangement of the rooms and rib pillars relative the coal seam thickness. Figure 2 depicts the results of vertical stress distribution near the face of the lower layer before cutting rooms in the upper layer. The rational variant of room arrangement in the undercut layer in Figure 3 shows that in the coal seam 9.4 m thick, the crown pillar with a thickness of 0.9 m is left at the seam roof, and the pillar at the floor of the rooms is 3.7 m thick.

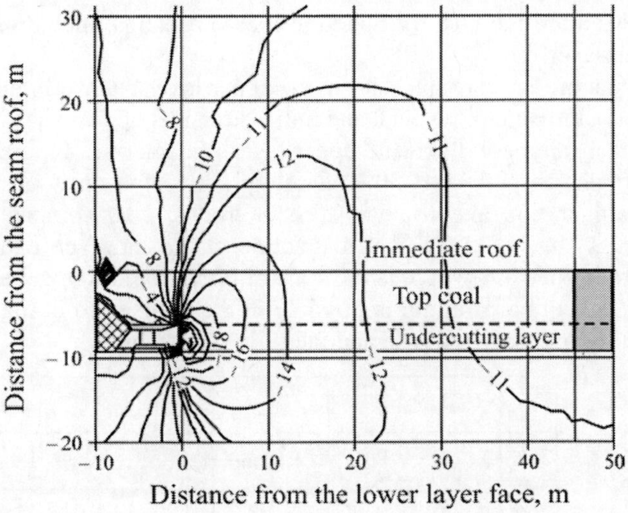

Figure 2. Contour lines of vertical stresses in rocks near the faces in the LTCC technology.

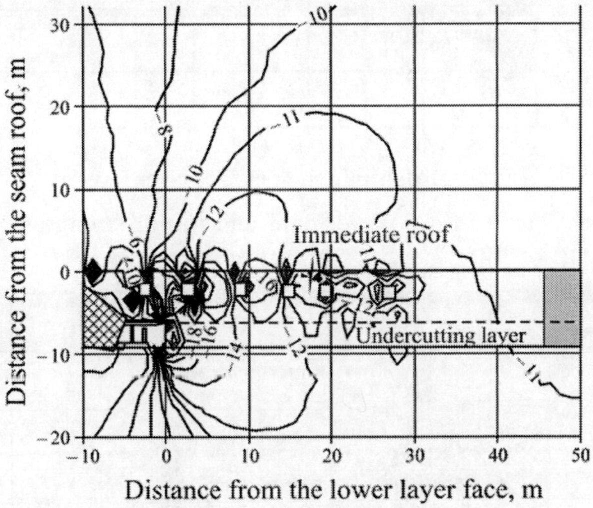

Figure 3. Contour lines of vertical stresses in rocks near the faces in the LTCC technology after advanced room and pillar in the upper layer.

The comparison of the vertical stress distributions in rocks before and after advanced room and pillar in the upper layer in Figures 2 and 3, respectively, reveals some regular patterns as follows:

—the maximum vertical stress concentration factor is observed near the fully mechanized longwall face (Figure 2) and equals 2.2; it reduces by 10% approximately after advance room and pillar (Figure 2);

—during advanced shortwall mining in the upper layer, two zones of abutment pressure arise: ahead of the fully mechanized longwall face in the lower layer and ahead of the room closest to intact rocks;

—rib pillars are the stress raisers: the vertical stress concentration factors in the pillars are by 1.2–1.3 times higher than before the preliminary cutting in the upper layer;

—the influence of advanced rooms on the vertical stresses at the face in the lower layer is much less than the pressure of overhanging undercut roof rocks, i.e., shortwalls unalter the geomechanical situation on the whole and are the tool for handling local problems, including intensification of coal pre-fracture in the upper layer.

In the analysis of coal pre-fracture intensity in the upper layer before (Figure 4) and after advanced room and pillar (Figure 5), the ratio of residual and initial strengths of coal was determined.

In the traditional coal mining without the upper layer pre-fracture (Figure 4), the strength of coal above the roof support decreases by 40–50%. Directly above the units of roof support, under the influence of their bearing ration and triaxial stress of rocks, coal compacts and large blocks are formed. The latter fact is confirmed by coal distraction at the interface of immediate roof and the seam. The most dangerous zone at the face is the zone of operation of the cutter–loader as reduction in coal strength to 30–40% of initial strength is possible in the course of cutting in the lower lower; for this reason, the probability of loose coal rash and face slip is very high.

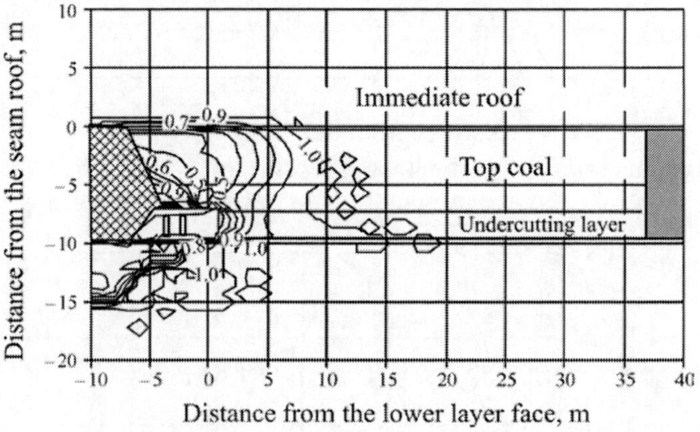

Figure 4. Contour lines of ratio of residual and initial strengths of coal in the upper layer.

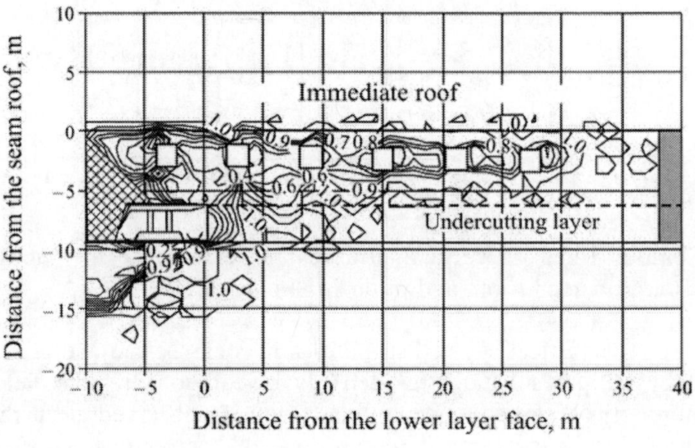

Figure 5. Contour lines of ratio of residual and initial strengths of coal after advanced room and pillar in the upper layer.

In case of shortwalling in the upper layer (Figure 5), strength of coal in the layer decreases by 40–50%. Ahead of the face in the lower layer, the ratio of residual and initial strengths of coal in Figures 4 and 5 is almost the same, i.e., the room and pillar mining in the upper layer has almost no influence on the strength of coal in the undercutting layer.

Immediate roof rocks are nondestructed; thus, rock falls and coal dilution are improbable. The advanced room and pillar in the upper layer reduces impact of secondary roof caving behind the face [9].

IOP Conf. Series: Earth and Environmental Science **262** (2019) 012016 doi:10.1088/1755-1315/262/1/012016

At the same time, in the vicinity of rooms and in the rib pillars in the upper layer, the strength of coal is reduced by 2–2.5 times, i.e., top coal caving will be uniform in the whole height of the upper layer.

For the technology of pre-fracturing of coal in upper layer by room and pillar, it is proposed to use robotic equipment with remote control by operators accommodated in safe places [10–12].

The unmanned coal cutting technology is to comply with the requirements below:

—the rooms and rib pillars should be unsupported not to violate the LTTC conditions;

—the cross section area of the rooms should be within 1–2 m^2 as larger areas can result in formation of roof arches and rock failure in the immediate roof;

—electric energy is prohibited in the advanced room and pillar as high rate methane drainage in this case can result in explosion of air and methane mixture.

For the room and pillar method implementation, it is suggested to use augering.

Conclusions

Based on the numerical modeling result and expert estimation, it is found that stability of immediate roof rocks requires a coal crown 1.0–1.5 thick to be left above the rooms of advanced pre-fracturing. In order to preserve top coal stability above the face in the operation zone of cutter–loader, it is recommended that the thickness of coal layer between the floor of rooms in the upper layer and the upper boundary of the undercutting layer is not less than 2 m.

The applicability and efficiency of top coal pre-fracturing without reduction of load-bearing capacity of immediate roof and coal above and beneath the advanced rooms has been proved.

References

[1] Strelnikov DA, Kozhevin VG and Gorbachev TF 1959 *Coal Mining in Kuzbass* Moscow: Ugletekhizdat (in Russian)

[2] Klishin VI, Shundulidi IA, Ermakov AYu and Solovyov AS 2013 *Top Coal Caving Technologies for Thick Gently Dipping Seams* Novosibirs: Nauka (in Russian)

[3] Krylov VF Zapreev SI, Berdyugin BA and Skrylyov SG 1973 *Development of Thick Gently Dipping Beds* Moscow: Nedra (in Russian)

[4] Muchnik VS, Holland EB and Marcus MN 1986 Underground Hydraulic Coal Mining Moscow: Nedra (in Russian)

[5] Gromov Yu, Bychkov YuN and Kruglikov VI 1985 *Ground Control in Thick Gently Dipping Coal Mining* Moscow: Nedra (in Russian)

[6] Sencus VV and Ermakov AYu 2016 Testing results of top coal caving technology in thick gently dipping seams *Nauk. Tekhn. Razrab. Ispolz. Min. Resurs.* No 2 pp 97–98

[7] Klishin VI 2013 Susbatntiation of top coal caving technologies for thick gently and steeply dipping seams *GIAB* No 6 pp 36–47

[8] Kalinin SI, Novoseltsev SA, Galimardanov RKh, Renev AA, Filimonov KA, Timoshenko AM and Fedorovich AP 2011 Longwall Top Coal Caving in Thick Seam Kemerovo: KuzGTU (in Russian)

[9] Reuter M, Krach M, Kissling W and Weksler Ju 2017 Geomechanics of the face in secondary roof fractures *J. Fundament Appl. Min. Sci.* Vol 4 No 2 pp 140–144

[10] Fryanov VN, Pavlova LD and Temlyantsev MV 2017 Theoretical approaches to creation of robotic coal mines based on the synthesis of simulation technologies *IOP Conference Series: Earth and Environmental Science* Vol 84 pp 1–8 (012001). Available at: http://iopscience.iop.org/issue/1755-1315/84/1

[11] Pavlova LD and Fryanov VN 2018 Geomechanical evaluation of deep-level robotic coal mining by the results of numerical modeling *Gornyi Zhurnal* No 2 pp 48–52

[12] Pavlova LD, Fryanov VN and Kornev ES 2011 Geomechanical forecasting in robotic coal mining *Geodynamics and Stress State of the Earth;s Bowels: All-Russian Conf. Proc.* Novosibirsk: IGD SO RAN pp 163–169 (in Russian)

Evaluation of stress state in rock mass surrounding underground structures of waterworks

LN Gakhova

Chinakal Institute of Mining, Siberian Branch Russian, Academy of Sciences, Novosibirsk, Russia

E-mail: gahoval@mail.ru

Abstract. Using the data of the numerical modeling of stress state in rock mass surrounding deep-level structures of waterworks, the author finds regular patterns and specific features of stress distribution under stage-wise underground construction. It is shown in the paper that as cross-section of an underground structure is enlarged, inelastic strain zones arise nearby the structures, and parameters of these zones are determined.

1. Introduction

When waterworks structures are arranged under the ground, the governing factor of stability for such structures is the effective stress state, and the role of the latter considerably grows with increasing depth. As a rule, stability of underground structures is evaluated using the estimates of stress state in surrounding rock mass. The classifications based on the analysis of mechanical properties of rocks may be inefficient in this case [1, 2]. Numerical modeling of stresses in integration with the full-scale observation of deformation processes in the vicinity of underground structures provides reliable estimation and prediction of stress–strain state during both construction and operation of underground waterworks [3, 4].

The sequence of formation of a mined-out void in surrounding rock mass for placement of waterworks structures is another determinant of their stability and a factor influencing stress state of rocks.

2. Numerical studies of the stress state and the results

This paper presents the analysis of changes in stress state of rock mass in the course of stage-wise construction of rock-fill embankment at Rogun Hydropower Plant, namely, powerhouse hall and transformer chamber, on the left-hand side of the Vakhsh River at a depth of 350 m. it is assumed that the transformer chamber (200 m long, 20 m wide and 40 m high) is completely formed, and cross-section of the powerhouse hall (length of 220 m, width of 22 m and maximum height of 78 m) is made in three stages: the first stage is formation of arch and cutting of 1/3 of the cross section; the second stage is cutting of 2/3 of the cross section; the third stage is full opening of the cross section.

The powerhouse hall and transformer chamber are arranged in hard rock mass composed of intercalating sandstone and siltstone. According to the data in [6], the rock layers are inclined at an angle of 70–75° relative to horizon. Sandstone and siltstone have uniaxial compression strength σ_c ranging from 100 to 200 and from 60 to 80 MPa, respectively [7].

The model of the linearly deformable medium assumes that the stratified rock mass is quasi-isotropic; the interfaces of the layers are free from sliding but in rigid cohesion. The calculations are performed using boundary integral equations [5]. The ratio of dimensions of the powerhouse hall and transformer chamber allows formulating plane stress problem.

For the numerical calculation of stresses in underground waterworks structures of hydroelectric power plants, in compliance with the research data on elastic, deformation and strength characteristics of enclosing rock mass [7], it is recommended to use: Poisson's ratio $\nu = 0.26$, elasticity modulus $E = 3.5–4.5 \cdot 10^4$ MPa, cohesion $C = 0.41–0.7$ MPa, internal friction angle $\varphi = 35°$. In situ stress state is assumed to be as follows [6]: $\sigma_x^0 = -35$ MPa, $\sigma_y^0 = -26$ MPa.

The calculated results are analyzed using numerical values of the stress tensor components (σ_x, σ_y, τ_{xy}) and shearing stresses σ_s, which allows delineating potential zones of post-limit deformation based on comparison of the stresses σ_s and cohesion of rock mass using the Mohr–Coulomb criterion [3, 4]:

$$\sigma_s = \frac{\sigma_1 - \sigma_3}{2\cos\varphi} + \frac{\sigma_1 + \sigma_3}{2}\,\mathrm{tg}\varphi,$$

where $\sigma_1 > \sigma_2 > \sigma_3$ are the principal stresses; φ is the internal friction angle.

The distribution of the shearing stress in the roof, sidewalls and floor of the powerhouse hall void in Figure 1 agrees with the results of the transversely isotropic model of rock mass [6].

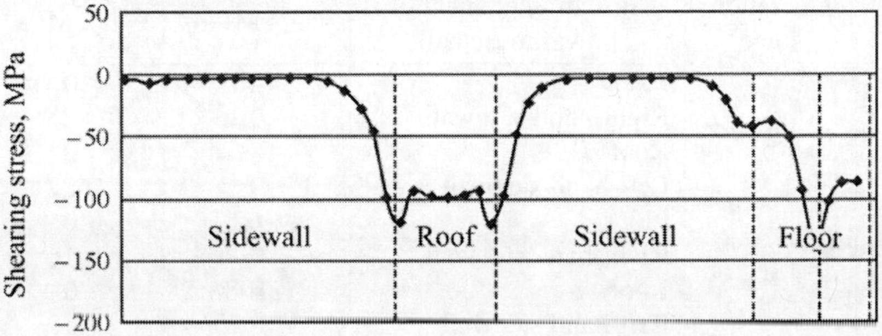

Figure 1. Distribution of shearing stress (MPa) along the boundary of powerhouse hall.

Behavior of stresses in rock mass surrounding the powerhouse hall and transformer chamber during their construction is described in Figures 2, 3 and in the table below.

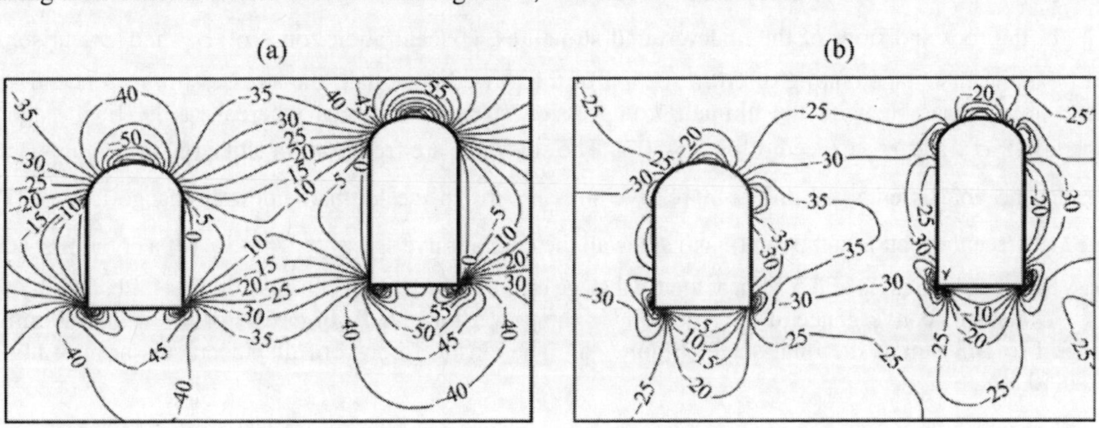

Figure 2. (a) Horizontal and (b) vertical stresses (MPa) in surrounding rock mass after the first-stage cutting of cross section of powerhouse hall.

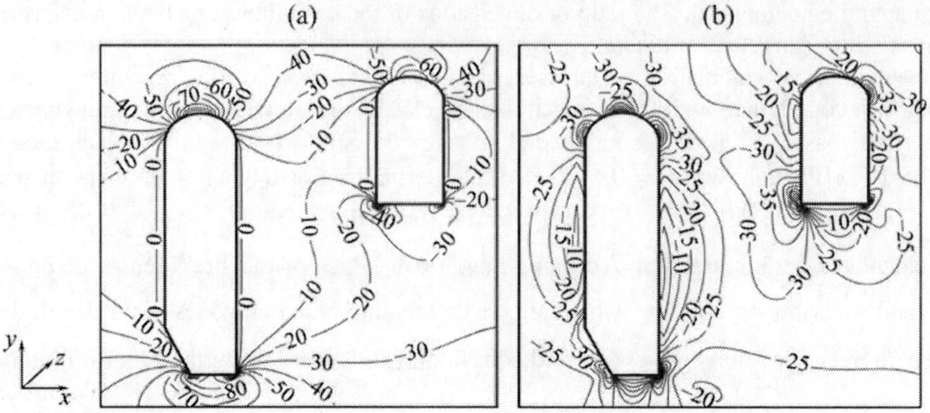

Figure 3. (a) Horizontal and (b) vertical stresses (MPa) in surrounding rock mass after the powerhouse hall void is cut out to the full cross section.

Table 1. Maximum and minimum stresses at the boundary of powerhouse hall during stage-wise cutting of its full cross section.

Operation stage	Structural element of powerhouse hall	σ_x, MPa	σ_y, MPa
I	Floor	−40	0
	Right-hand sidewall	0	−25
	Roof	−8	0
	Left-hand sidewall	0	−27
II	Floor	−62	0
	Right-hand sidewall	0	−16
	Roof	−108	0
	Left-hand sidewall	0	−12
III	Floor	−105	0
	Right-hand sidewall	0	3.5
	Roof	−126	0
	Left-hand sidewall	0	−5

In the roof and floor of the underground structures, concentration zones of σ_x and σ_y arise. As early as the first-stage cutting of cross section of the powerhouse hall, the stresses in the concentration zone are comparable with the ultimate compression strength of rocks whereas at the final stage of operation, σ_x and σ_y exceed σ_c by 30–50%. The sidewalls are free from in situ stresses (Figures 2 and 3). As the void is enlarged, the compressive stresses σ_y in the left-had sidewall change from −27 to −5 MPa (see the table). In the right-hand sidewall the compressive stress $\sigma_y = -25$ MPa at the first stage pass into tensile stresses of 3.5 MPa at the final stage of the cross section cutting (Figures 1, 2b, table).

The elastic convergence of the sidewalls in the powerhouse hall (Figure 4) grows from 280 mm at stage I to 543 mm at the final stage (Figure 3a). The elastic floor–roof displacements are not higher than 50 mm.

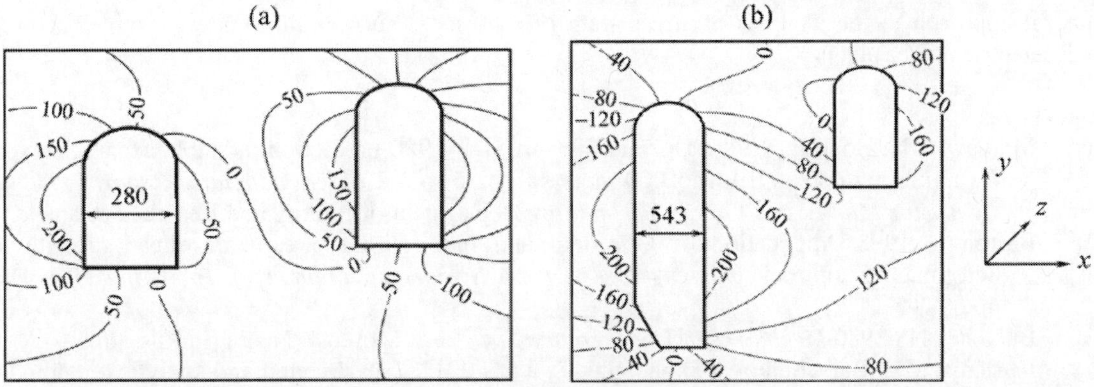

Figure 4. Elastic displacements (mm) in horizontal direction, u_x, after completion of (a) stage I and (b) stage III of cross-section cutting in the powerhouse hall.

Development of the inelastic strain zone in surrounding rock mass by stages of cutting the powerhouse hall until full cross section opening is depicted in Figure 5.

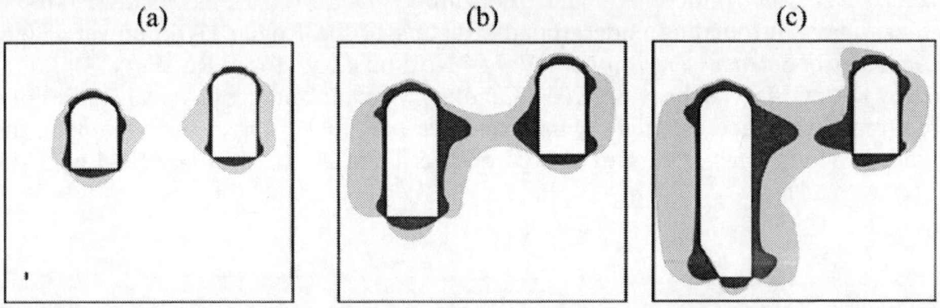

Figure 5. Zones of inelastic strains in damaged (grey) and weakly damaged (dark) rocks by stages of full cross section cutting of powerhouse hall: (a) stage I; (b) stage II; (c) stage III.

At stage I of the powerhouse hall cross section cutting, the inelastic strain zones extend to 1–2.5 m in hard rocks and to 5–6 m in damaged rocks. When the full cross section is opened (Figure 5c), the zones of inelastic strains in sidewalls extend to 1–3 m in hard rocks. In the pillar between the powerhouse hall and transformer chamber, the inelastic strain zones almost merge (Figure 5c). in the roof of the powerhouse void during its stage-wise cutting, σ_s alter insignificantly.

3. Conclusions

The study has revealed the stress state behavior in rock mass during construction of underground waterworks structures at Rogun Hydropower Plant. In the powerhouse hall void cut to the full cross section:

—the sidewall rock mass is completely free from the action of in situ stresses;

—in the concentration zones, σ_x and σ_y exceed the ultimate compression strength of rocks by 30–50%;

—the inelastic strain zones embrace upperlying rock mass from the side of the transformer chamber;

—the convergence of sidewalls in the powerhouse hall reaches 543 mm.

The comparison of the calculations using the boundary integral equations and the data of the transversely isotropic model proves good agreement of the results and applicability of the boundary integral approach to the analysis of stress state of rock mass surrounding underground structure of hydroelectric power plants.

References

[1] Mathews KN, Hoek E, Wyllie DC and Stewart SBV 1980 Prediction of stable excavation spans for mining at depths below 1,000 meters in hard rock *Golder Associates Report to Canada Centre for Mining and Energy Technology* Department of Energy and Resources Canada

[2] Barton N 1993 Application of Q-system and index tests to estimate shear strength and deformability of rock masses *Workshop on Norwegian Method of Tunneling* New Delhi pp 66–84

[3] Bulychev NS 1980 *Mechanics of Underground Structures* Moscow: Nedra (in Russian)

[4] Kurlenya MV, Baryshnikov VD and Gakhova LN 2012 Experimental and analytical method for assessing stability of slopes *Journal of Mining Science* Vol 48 No 4 pp 609–615

[5] Gakhova LN 2013 Solution of two-dimensional and three-dimensional elastic problems of a bedded rock mass using boundary intergal equations *Goeydnmaics and Stress State of the Earth's Interior: XX Conference Proceedings* Novosibirsk: IGD SO RAN pp 185–190 (in Russian)

[6] Nizomov DN, Khodzhiboev AA and Khodzhiboev OA 2011 Transversally isotropic model of rocks mass surrounding underground structurs of the Rogun Hydropower Plant *Tajikistan Academy of Sciences Transactions* Vol 54 No 5 pp 420–426 (in Russian)

[7] Nazirov DA and Davlatshoev SK 2017 Labroatory-scvale and geophysical studies of the elastic, defromational and strength characteristics of rock mass surroundingh tunnels and underground rooms of the Rogun Hydropower Plant *Stroit. Arkhitekt.*No 4 pp 34–38

Mining company as an effective element in coal supply to the regions of the Far North

VL Gavrilov* and VI Fedorov

Chersky Institute of Mining of the North, Siberian Branch, Russian Academy of Sciences, Yakutsk, Russia

E-mail: *gvlugorsk@mail.ru

Abstract. The companies belonging to chains of mining, transportation and consumption of coal to the hard-to-reach regions use their own, often contradictory criteria for the assessment of their work efficiency. Because of that the possibilities to improve the processes of coal quality control in the interests of fuel-chain targets common to mining companies are not used rationally. This situation is more acute in the Far North, primarily in the northern regions of Yakutia. This is due to the extreme combination and negative effect of a number of geographic, climatic, mining-geological and logistical factors. The fulfillment of all operations here, from prospecting of deposits and ending with the distribution of energy received from coal, is complicated, which leads to a significant rise in price. Approaches to improving the work of open-pits as effective elements of coal supply chains, applied in practice, geotechnological and organizational solutions in the planning and management of mining operations can and should be adjusted on the basis of developed proposals, including additional coal preparation in the opencast mines for deliveries. Their realization allows increasing consumer properties of coal shipped to the market with a reduction in its total qualitative and quantitative losses.

1. Introduction

Various strategies and programs in Russia declare that every individual is guaranteed high-quality supply of energy resources and energy security, including hard-to-reach areas of the Fare East, Siberia and European North [1].

Actual monitoring informs of some adverse factors connected with the current supply of isolated territories of Russia with fuel and power resources (FPR). The costs of FPR purchase and shipment, heat and electricity generation and distribution tend to persistent growth. Fuel is delivered not always in time, there are periodic failures which result in force majeure. Total quantitative and qualitative loss of basic materials along the complex and often irrational supply chains is high. Available relatively inexpensive local coal, including high-grade products, is misused. Coal consumption drops at the concurrent degradation of useful quality of fuel. Mines are subject to closure, or decrease output with increasing cost per 1 t of production; economic performance worsens.

Aimed to improve the situation, by the Decree of the President of the Russian Federation (May 2018), the RF Government is entrusted to ensure stable energy supply to consumers in remote and isolate areas by 20204 based on distributed generation. Solution of this complex problem, despite special attention (market in a greater or lesser degree) renewable sources of energy, assumes wide use of coal alongside with hydrocarbons and for a long time. Coal is a relatively cheap type of fuel; it is produced on site or delivered from other regions in amount sufficient to meet the actual needs given efficient approaches to FPR supply.

For the Far North of Russia, including the comprehensively studied central and northern regions in Yalutia, solution of the said problem is of critical importance. As was shown earlier [1–3], this is connected with the impact of the unique combination of extreme natural, climatic, geological, infrastructural and economic factors which greatly complicate operation of the current and future chains of production, supply and use of coal. As case stands, an emphasis should be laid on the exposure of steady decline in the role of mines situated immediately in the regions of FPR consumption and having low and very low capacity, as a rule, as well as on the identification of background of higher attractiveness and sustainability of operating and new mining companies in the hard-to-reach areas based on development of the appropriate technological and organizational measures.

2. Underground and surface coal mines strategy

Approaches used to obtain relevant estimates can be reduced to some basic provisions included as the key components in the chains to coal supply to consumers in the analysis of performance of mining companies.

Underground and surface coal mines supplying FPR to the hard-to-reach regions operate in conditions of tough fuel competition. Interests of oil, gas condensate, diesel fuel and natural gas suppliers are lobbied by large and strong business. Regulatory bodies and corporate national giants are interested in a wide "innovative" use of more expensive and market-determined sources of energy (mini atomic power stations or plants driven by renewable energy of wind and sun are less reliable in the regions under discussion). The Greens accentuate ecological impact of coal combustion products. At the same time, they conceal littleness of this impact due to small coal consumption and low content of toxic elements in the majority of coal deposits in the region. Advanced technologies of solid fuel preparation and combustion, as well as waste management can ensure efficiency comparable with the other FPR types even on small-capacity plants.

Unlike large coal mining companies, small business is highly dependent on internal risks. For this reason, in order to preserve technological and economic sustainability, small companies have to invest considerable money in preventive control [4]. For example, according to the present regulatory documents, open pit mine planning and operation requirements are identical for all mines irrespective of the capacity. All check-list of design, exploration and various approvals is is subject to compulsory implementation [5]. As a consequence, the period starting from bidding documentation and mining license granting up to commissioning of a mine and its running at full capacity covers from three to five years and longer. In the conditions of the Far North, this period can extend to one–three years more in view of extremely complicated logistics connected with delivery of materials and equipment necessary for construction, modernization and normal performance of a mine. Over such a long period of time, it is highly probable that external difficult-to-predict conditions, under which the decision on mining was made, would change for the worse of the subsoil user.

Coal mining companies should be considered as the basic elements in the supply chain between a coal field and a consumer rather than as self-sustained, independent and exclusive economic market participants. It is expedient to implement vertical or horizontal integration of all chain components in the framework of a common system undergoing dynamic changes and using common approaches to evaluation of operating efficiency. A key figure in this case should the production quantity in the form of kilowatt-hours or mega joules consumed and paid by a final user, or even electric or thermal energy output rather than size of known reserves, or produced and conveyed coal. Local criteria for evaluation of small open pit mine performance should be synchronized with the common criteria for the whole system so that to eliminate and prevent current ore probable contradictions.

For revealing evident and hidden (actual and potential) capacities and risks of coal mining companies when incorporated in the regional market of FPR, it is wise to evaluate alternative, more efficient scenarios of operation under creation of common added value in supply chains. Companies engaged in mineral mining and processing obtain, as a rule, the lowest share in the total economic result of the chain operation [6]. In this situation, along with effectivization, it is a priority task to improve performance of coal mining companies, not only for their convenience but also for the benefit of all participants in the coal supply chain.

A critical backup are the integrated systems of management of coal quality and energy generated from coal. Such systems should target at full-scale improvement of useful qualities of produced and shipped coal, production of grade and calibration fuel, as well as at reduction of total quantitative and qualitative energy loss along the supply chain. Such systems should start with in-depth analysis of mineral potential in the fields under operation or planned for development and finish with distribution of cogenerated thermal and electric energy to a final, including retail, consumer.

With intent to evaluate actual and future role of coal mining companies in FPR supply to the hard-to-reach areas, review positive and negative experience as well as to reveal present difficulties and shortcomings, information on trends in performance of most small open pit mines in the Far East, Siberia and other regions of Russia has been collected from various sources and analyzed (Figure 1). The list of selected mines includes round 50 operating or recently closed coal mining companies (mostly open pits with ROM coal capacity up t0 0.3 Mt/yr, situated in localities with undeveloped transport infrastructure and supplying products first all to local markets.

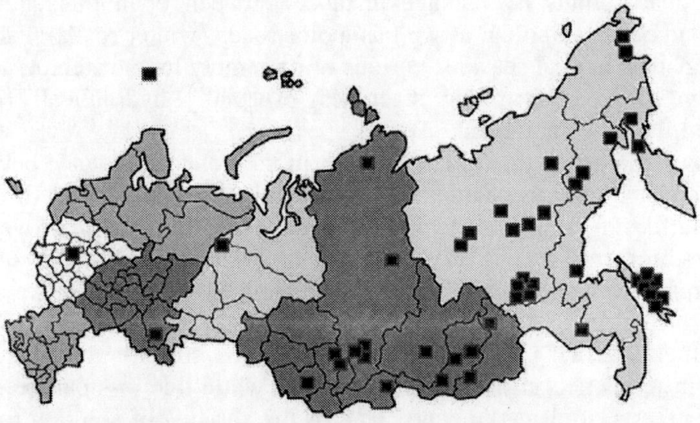

Figure 1. Main small coal mines in Russia.

Table 1. Coal production in central and northern Yakutia, thousand tons.

Open pit mine	Year						
	2011	2012	2013	2014	2015	2016	2017
Bituminous coal							
Dzhebariki-Khaya (underground mine up to 2017)	367	381	559	320	542	398	157
Zyraynsky	243	200	171	160	124	151	148
Kharbalakh	85	108	116	132	129	134	162
Liginite							
Kangalass	126	134	141	131	145	170	160
Kirov	18	18	19	20	23	25	23
Kempendyai	15	16	21	21	25	26	25
Total	854	857	1027	784	988	904	675

Open pit coal mines in Yakituia are reviewed in more detail. Out of 11 6 mines run in the late size years and produced 0.7–1.0 Mt of coal yearly (table). ROM coal was meant for shipment, with numerous rehandling points and long-term delivery periods (to 2–2.5 years), huge unavoidable quantitative and qualitative loss much higher that standards allowed. Percent of logistics and transportation expenditures in the structure of coal cost was and still is very high (to 70–85%). The open pit mines under analysis feature also high committed fixed cost per one produced ton of coal.

Based on the analysis of operation of coal mining companies as components in the fuel supply chain, some basic conclusions have been drawn and are presented below.

The optimistic forecasts made in view of available potential background of sustainable performance and anticipated increase in the number and role of small coal mines in Russia [7, 8] and its separate regions [9, 10] failed on some reasons, including those of subjective nature.

There is high dependence of capacity of open pit mines on demand for coal, which is crucially governed by the government buying. Produced coal output, given indirect cost control, is insufficient to support sustainable operation. Companies persistently run short of assets renewal and technical upgrading to increase labor productivity.

When determining annual regimes of overburden removal and coal extraction, open pit mines should take into account periodicity of demand for coal due to winter and summer seasons. Furthermore, in view of remoteness of many consumers and underdeveloped infrastructure, the impact exerted on annual coal outputs by haulage distance between open pits and users is very high. Schedules of river and ocean shipping, as well as motor roads (winter roads) mismatch, which greatly complicates coal logistics. In coal seasons, periods of extremely low air temperature take place rather often, and operation of modern mining machines, especially hydraulically driven, having wide technological capabilities, is complicated.

Lack of care towards coal quality control results in the nonconformance between coal properties, consumer expectations and process characteristics of combustion equipment. It concerns, for the first turn, the purpose, uniformity and preservation of fuel. Possibilities for improvement in operation of open pit coal mines in terms of delivery are used inefficiently. At stages of mine planning and operation, the features of formation and alteration of quantitative and qualitative characteristics of coal produced are nearly neglected. As a result, efficiency of coal utilization in boiler installations exercising mostly fuel-bed firing, drops [2]. Supply chains suffer appreciable, much higher than standard, total loss in terms if coal quality and quantity, while this loss can be considerably reduced given coordinated interests of all participants. One of the variants of solution to this problem can be introduction of auxiliary process circuits at open pit mines to prepare coal for further use. For instance, production of graded coal and its delivery in flexible containers to very remote regions.

3. Conclusions

1. Review of small coal mining companies as the elements of fuel delivery to hard-to-reach areas has offered the time history of their technological and economic changes. The obtained information is intended to be used to adjust operating conditions so that the mines can be efficient and sustainable for a long time, at expanded limits of production and use of coal as common and relatively inexpensive type of FPR.

2. Low-capacity open pit coal mines running for the benefit of users in hard-to-reach and energetically isolate regions should be considered as the main but not unique links in supply chains which, in their turn, should function with regard to harmonization of interests of all participants.

3. The significant source for improving efficiency of supply chains is shaping and application of special systems of coal quality control. These systems must ensure useful qualities of coal for boiler houses such that to find compromise between mineral potential of coal fields, capabilities of modern efficient mining technologies and admissible economic and ecological consequences.

References

[1] Gavrilov VL, Tkach SM, Batugina NS and Fedorov VI 2014 Geotechnological aspects of coal mining in the Arctic region in Yakutia *J. Fundament Appl. Min. Sci.* Vol 1 pp 79–85 (in Russian)

[2] Tkach SM and Gavrilov VL 2017 Effect of georesource–consumer process flows on coal loss in

energy supply of the Polar regions in *Yakutia IOP Conf. Series: Earth and Environmental Science* 53 (2017) 012017. DOI:10.1088/1755-1315/53/1/012017

[3] Batugina NS, Gavrilov VL and Shepeleva EG 2017 Small-scale coal mine in Polar regions of Yakutia: State and Prospects J. ECO Vol 2 pp 134–145

[4] Kuklina MV 2009 Features of business risks in small coal mining enterprises *Byull. Irkutsk. GTU* Vol 1 pp 108–111

[5] Lapaev VN and Markova IV 2014 Problems of ensuring investment attractiveness of small open pits for the extraction of common minerals *Gornyi Zh. Kazakhstana* Vol 8 pp 38–40

[6] Nikitenko SM, Goosen EV, Pakhomova EO and Kolevatova AV 2017 The added value chains as an instrument of economy development in the region of raw-material orientation *Fundament. Issledovaniya* Vol 10-2 pp 375–380

[7] Imametdinova RH and Liventsev VV 1998 Small business and its place in the structure of Russian coal industry *GIAB* Vol 2 pp 74–77

[8] Melekhin EC and Zinoviev AA 2004 Measures to create competitive conditions for operation of small and medium-size mineral mines *Mineral. Resursy Rossii. Ekonomika Upravlenie* Vol 5-6 pp 68–71

[9] Petrov NA (Ed.) 2010 *Energy Strategy of the Republic of Sakha (Yakutia) for the Period up to 2030* Yakutsk–Irkutsk (in Russian)

[10] Tarasko OA 2010 Current condition and development problems of coal mining on the Kamchatka Peninsula *Gornyi Vestnik Kamchatki* Issue 1(11) pp 67–78

IOP Conf. Series: Earth and Environmental Science **262** (2019) 012019 doi:10.1088/1755-1315/262/1/012019

Highly efficient hard rock-breaking tool for mining machines

BL Gerike*, VI Klishin and PB Gerike

Federal Research Center of Coal and Coal Chemistry, Siberian Branch, Russian Academy of Sciences, Kemerovo, Russia

E-mail:*gbl_42@mail.ru

Abstract. Under consideration is the experience gained in design and manufacture of tools for mining machines meant for operation in uncontrollable conditions. It is shown that equipping the mining machines with disk tools reduces specific consumption of the tools, specific power-input and dusting during breakage of hard rocks.

1. Introduction

Experience of longwall mining equipment operation in off-standard conditions (complex structure coal seams, ore bodies and placers) [1–5] shows that rock destruction consumes much energy and results in high wear of tools [6]. An alternative to series-produced cutting tools can be shearing discs commonly used on various heading and shearing machines [7–9].

The earlier research [10] yields that for shearing discs mounted on shearer drums for breaking hard rocks, the most efficient mode is low-cycle force-impact destruction when stress state is preliminary created and repeatedly used during the passes of the disc tools. Moreover, in this case, it is possible to cut down destruction energy by 15–20 times as compared with the free shearing mode, which allows series-produced cutting machines to be used in extraction of minerals with ultimate compression $\sigma_c \leq 140$ MPa.

2. Bench tests results

As a result of bench tests on breaking of different rock lithotypes by shearing discs, the operating envelope of the tools (cut width t_c and penetration depth h), ensuring minimum energy input, is determined, and arrangement recommendations are made of shearing drums, based on which an aggregate equipment concept is developed.

For the analysis of various arrangement layouts of shearing drums and selection of efficient operation modes of shearers with regard to their energy and force performance, a simulation model of load formation on a shearing tool is developed [10, 11]. The source data of the model are the sequence of load impulses formed on separate tools with regard to variable cross-section of chips, the parameters of the impulses are determined in accordance with the experimentally found laws of distribution; the output of the model is a random discrete sequence of instantaneous values of power taken by the shearer motor.

Examination of various possible design layouts of the mounting group from the viewpoint of reliable secured rotation of the shearing disc, especially if arranged on the goaf side of the drum, has resulted in picking out the mounting group embedded in the body disc rolling cutter, it is also found that the critical factors that govern efficiency of the mounting group are the tribotechnical characteristics (roughness of adjoining surfaces, clearance, friction coefficient, etc.). Based on the analysis of influence exerted by finishing operations on the quality of surface of parts and, eventually, on the joint friction coefficient, technical requirements for the mounting group are formulated.

From the laboratory investigation data, the geometrical parameters of the shearing disc tool such that ensure efficient destruction and sufficient tool strength are determined (Figure 1).

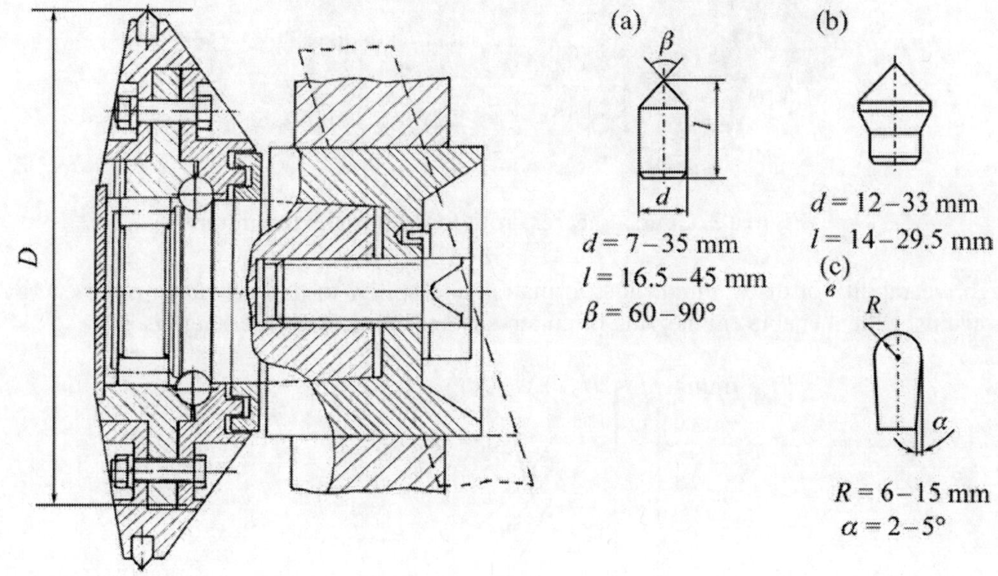

Figure 1. Design of a disc tool with various structure reinforcement insert: (a) conical; (b) mushroom; (c) ellipse.

3. Production tests results

The production tests of prototype shearing discs in cutting coal with dirt bands and partings, metamorphized quartz vein and permafrost placer, as well as in extraction of sulphide ore from thin lodes [13, 14] prove validity of the recommendations used as the framework for shearing disc design and show that:

—destruction energy in coal with fine-graine sandstone partings and hard inclusions makes H_W = 0.7–1.0 kWh/m^3;

—destruction energy in quartz metasomatites (very brittle rocks) is on average H_W = 0.85–1.25 kWh/m^3 and maximum 4.5 kWh/m^3 in extraordinary operation mode;

—specific energy of destruction of permafrost placer alluvium (viscous rocks) makes on average H_W = 2.8–3.5 kWh/m^3 and maximum 4.1 kWh/m^3;

—specific dust formation halves as compared with rock cutting by series-produced tools;

—specific wear of shearing disc tools totals 8 discs per 1000 m^3 of broken rocks;

—TBF of the mounting group is 800–1000 m^3 of broken rocks.

Good results obtained in destruction of viscous and strong permafrost placer rocks made the basis for designing the cutting tool for underground kimberlite extraction without blasting. Kimberlite ore of higher strength ($\sigma_c \geq 50$ MPa) and viscosity ($\chi \geq 0.14$) [15, 16] can be cut by moon miners advantageous for:

—selective slicing of mineral reserves;

—simultaneity of cutting and haulage operations during displacement of cutting tool on the face;

—high mobility and maintainability in limited reserves bodies;

—infinitely variable control of the cut layer thickness.

The cutting drums of boom miners should be equipped by shearing discs (Figure 2) using the scheme implemented on prototype machine TM-D25 [17].

IOP Conf. Series: Earth and Environmental Science **262** (2019) 012019 doi:10.1088/1755-1315/262/1/012019

Figure 2. Cutter–shearer drum for kimberlite ore mining.

High wearability of discs, minor dust formation and ability to transfer much more energy to the face as against cutters enable engineering of cutter–shearer drums with shearing discs.

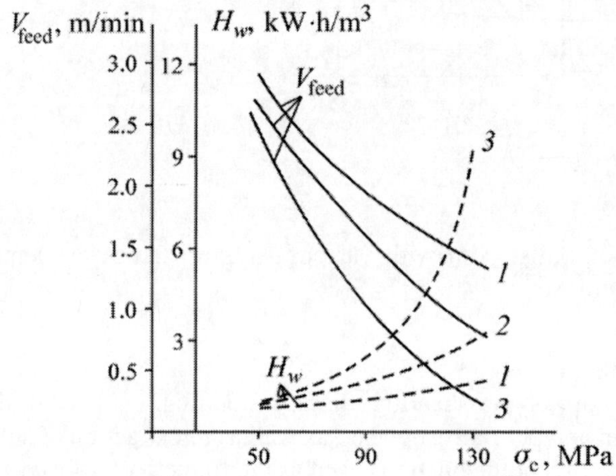

Figure 3. Modeling results of drum performance with shearing discs versus rock strength and brittleness χ: *1*—$\chi = 0.025$; *2*—$\chi = 0.125$; *3*—$\chi = 0.225$ (V_{feed}—feed rate of drum).

The modeling result of the drum performance with the shearing disc tools in destruction of rocks of various strength and brittleness in Figure 3 prove that the range of application of the shearing discs implemented low-cycle force-impact destruction is much wider than the tangential rotary picks have.

References

[1] Korshunov AN, Nesterov VI, Khorshok AA et al 1977 Testing a cutting drum with disc cutters *Ugol'n. Mashinostr.* Issue 4 pp 1–4

[2] Korshunov AN, Revskii FD, Nesterov VI et al 1984 Experience of extraction of sand-and-clay beds with hard inclusions *Gornyi Zhurnal* No 4 pp 6–9

[3] Khoreshok AA, Mamet'ev LE, Tsekhin AM et al 2013 *Manufacture and Operation of Rock-Breaking Tools of Mining Machines* Tomsk: TPU (in Russian)

[4] Kudlai ED 1985 Parameters of mechanical destruction of permafrost rudaceous rocks *Kolyma* No 10 pp 4–6

[5] Tangaev IA 1986 *Energy Intensity of Mineral Mining and Processing* Moscow: Nedra (in Russian)

[6] Prokopenko SA and Shlapakov PA 2015 Increasing safety and efficiency of rock breaking in mines *Vestn. Nauch. Tsentra Bezop. Rabot Ugol'n. Prom.* No 2 pp 27–31

[7] Loguntsov BM 1976 Development of heading machines in 1971–1976 *Mineral and Petroleum Machine Engineering. Science and Technology Summary* Moscow: VINITI Vol 4 pp 7–98 (in Russian)

[8] Johannessen O 1995 *Hard Rock Tunnel Boring. Project Report 1–94* NTH, University of Trondheim, The Norwegian Institute of Technology, Trondheim

[9] Rostami J and Ozdemir L 1993 A new model for performance prediction of hard rock TBMs *RETC Proceedings* Boston MA pp 793–809

[10] Gerike BL, Klishin VI, Gelike PB 2016 A new index of rock-breaking tool efficiency *Journal of Mining Science* Vol 52 No 3 pp 481–486

[11] Su O and Akcin NA 2011 Numerical simulation of rock cutting using the discrete element method *International Journal of Rock Mechanics and Mining Sciences* Vol 48 No 3 pp 434–442

[12] Lizunkin VM, Krylov EI, Gerike BL and Lizunkin MV 2002 RF Patent No 2187640 RU, E21C25/16 Cutting Tool of Mining Machine *Byull. Izobret.* No 27

[13] Lizunkin VM, Volov ES, Kravtsov EM et al 1989 Shearer mining testing in ore mine *Gornyi Zhurnal* No 2 pp 36–38

[14] Lizunkin VM, Gerike BL and Utsyn YuB 1999 *Mechanized Underground Mining of Thin Hard Ore Bodies* Chita: ChitGU (in Russian)

[15] Shreiner LA 1949 *Hardness of Brittle Bodies* Moscow: Gostoptekhizdat (in Russian)

[16] Gogotsi GA 1973 Determination of brittleness of refractories tested for heat resistance 1186–1189 Strength of Materials Springer

[17] Gerike BL, Filatov AP, Gerike PB and Klishin VI 2006 Concept of rock-breaking working element of an underground kimberlite ore mining machine *Journal of Mining Science* Vol 42 No 6 pp 610–616

IOP Conf. Series: Earth and Environmental Science **262** (2019) 012020 doi:10.1088/1755-1315/262/1/012020

Vibration analysis of electromechanical equipment of mining shovels

PB Gerike* and VI Klishin**

Institute of Coal, Federal Research Center for Coal and Coal Chemistry, Siberian Branch, Russian Academy of Sciences, Kemerovo, Russia

E-mail:* am_besten@mail.ru, **klishinvi@icc.kemsc.ru

Abstract. Some results of the scientific study into methodological approaches to development of a group of common diagnostic criteria suitable for the assessment of actual state of electromechanical equipment of electric mining shovels are presented. One of the objectives of this scientific research is to prove feasibility of implementing short-term degradation models in the framework of the preventative maintenance system. The second target is to develop a methodology for creating an algorithm of a common diagnostic criterion suitable for identifying and assessing risk of defects in equipment of mining machines. It is shown that the comprehensive approach to diagnostics based on vibration parameters opens wide opportunities for timely detection of defects in equipment of mining machines, including those at the stage of initiation. Implementation of this study results and the use of the developed criteria will improve management efficiency of maintenance of mining equipment and minimize probability of failure of electric shovel equipment.

1. Introduction

In coal mining, one of the key factors to govern operating efficiency is actual condition of equipment. The situation when faulty equipment remains in operation even if it endangers health of mine personnel can only be changed with introduction of qualitatively new maintenance and repair management. Transition of of mines having outdated planned maintenance system to improved servicing of machinery, without which it is impossible to run safely no mining, haulage or processing equipment, will need representative databases on parameters of vibration, and will require to reveal characteristics of serviceability limit state of shovels, construct degradation prediction models and to develop as set of criteria to detect defects of electromechanical accessories of mining machines.

It is possible to meet the requirements and to create background for introduction of active machinery maintenance based on the actual condition with determination of main regularities of change in electromechanical equipment condition based on the analysis of polyharmonic waves generated during machine operation.

2. Development of common diagnostic criterion for damage risk evaluation

This study is specific for the first attempt made to try to generalize diagnostic data for more than 10 years-long period of observations, as well as to develop on this base a package of diagnostic signs and rules suitable to solution of actual scientific problems such as elaboration of common criteria for actual condition of active mechanisms by vibration parameters which allow finding the limit state ranges for electromechanical equipment of single-bucket shovels. Furthermore, by means of case story

IOP Conf. Series: Earth and Environmental Science **262** (2019) 012020 doi:10.1088/1755-1315/262/1/012020

analyses of vibration parameters demonstrated by electromechanical equipment of EKG and ESH shovels, this paper, unpretending to be offer omnitude, discusses improvement of rating methodology for parameters of mechanical oscillations. The studies rest upon an integrated approach to analyzing parameters of vibration, which enables versatile and useful information on the change in technical state of an active complex mechanical system [1, 2].

This research uses a sampling on 30 single-bucket electrical shovels (models EKG-5A, EKG-8I, EKG-10, EKG-12.5, etc) and 8 draglines ESH 10/70, ESH 6/45, ESH 15/90). The diagnostic measurements were taken in the framework of the industrial safety expert evaluation of engineering devices in operation in hazardous industry branches. The research findings make it possible to draw a conclusion that diagnostics of electrical equipment of EKG- and ESH-type electric shovels should involve an integrated approach to vibration characterization based on the data of Saint-Petersburg-extended frequency-range spectral analysis, excess and envelope analysis. Moreover, sometimes. The listed methods are better to be added with the analysis of acceleration/run-down of an aggregate and with the wavelet transform of initial vibro-acoustic signal. Such combination of diagnostic methodologies is an optimal package and provides maximum valued information with regard to the type and operating mode of equipment (Figures 1–4 illustrate analyses of vibration parameters, where A is the vibration acceleration, m/s^2; V is the vibration velocity, mm/s; f is the frequency, Hz.

The basis of the the integrated approach to diagnosing mining machine accessories by vibration parameters is the spectral analysis as the most flexible tool. The diagnostic properties may then be formalized [3] for further application in development of algorithmic codes for automated diagnostics of defects or common diagnostic criteria suitable for prediction of change in engineering condition of complex mechanical systems [4, 5].

The review of scientific publication shows that models currently available on the software market for modeling change in conditions of mining machine equipment for further prediction based on the analysis of vibration parameters are deficient. Majority of such models are meant to predict actual condition of individual assembles of electromechanical equipment of machines and mechanisms (for instance, a modeling object can be a swivel head reducing gear bearing or a dragline pull winch bearing). As a consequence, the application area of these models is extremely small and trials are not always sufficient [6–8].

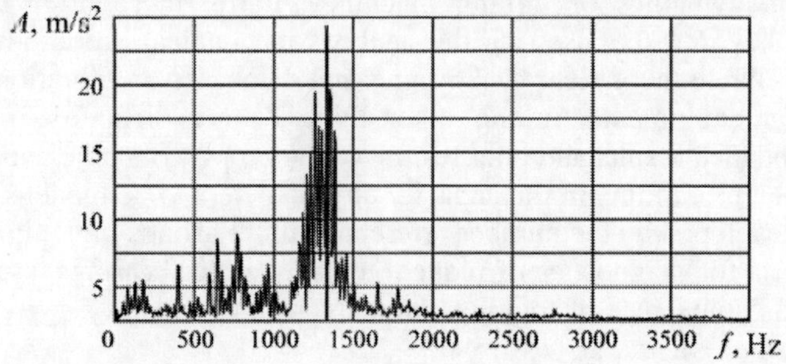

Figure 1. Numerous defects of low-group house generator bearing on ESH 10/70 shovel.

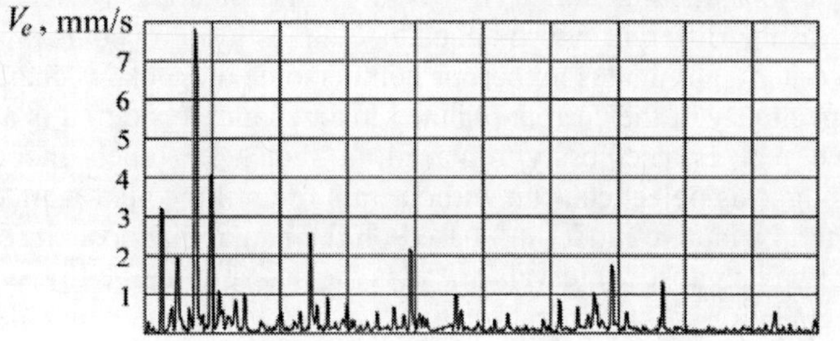

Figure 2. Unfastening of motor from swing gear on ESH 6/45 shovel.

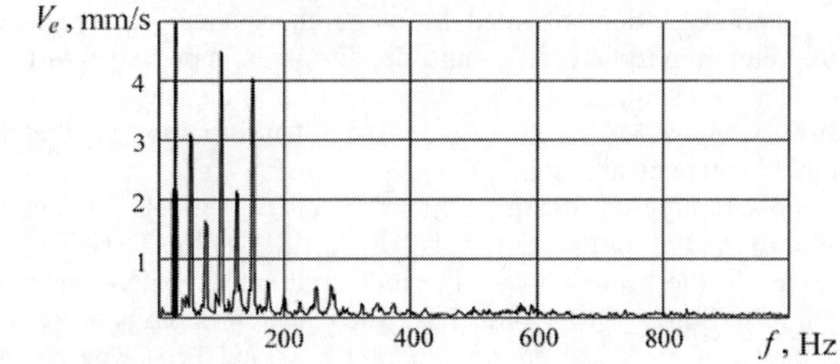

Figure 3. System rigidity failure, tolerable rotor unbalance and misalignment of generator group shaft line on EKG-5A shovel.

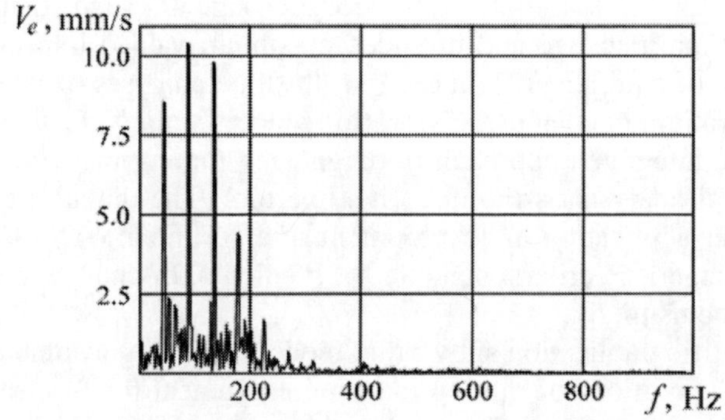

Figure 4. Intolerable misalignment of reducing gear shaft and exposed gear train on EKG-4.6B shovel.

Limitations of the current predicting models are explained by a number of causes, in particular, by lack of knowledge on dynamics of mining machines. Furthermore, it is impossible to develop prognostic models of key defects based on the analysis mechanical vibrations of electromechanical equipment without a reliable theoretical background and databases on vibration parameters of single-type electromechanical equipment. Among other things, more difficulty is brought by complex interpretation of the obtained results and poor testing of the proposed mathematical models. In cases of electric shovels for open pit mining, in the capacity of parameters to be modeled, it is most efficient to use common criteria developed by the number reference failure groups of equipment under diagnostics (individual singe criteria for diagnostics of reduction trains, roller bearings, system rigidity violation, electrical nature defects, unbalance of rotating elements, misalignment of aggregates and various type and design connector sleeves).

Thus, each of the common diagnostic criteria to estimate condition of equipment of electric shovels will be unique, considering that is is based on the outcome of the integrated analysis of vibration parameters using different vibro-acoustic diagnostics procedures, while the process of assessment with such criteria eliminates numerous cumbersome diagnostics characters and rules.

To date, due to specificity of the current planned maintenance system, it is absolutely unimportant where a nondefective part is needlessly replaced; a factor of much more concern is reliable information that the diagnosis object can run without failure until the next maintenance or not. In such conditions, the short-term adaptive modeling, based on the actual diagnosis information and aimed to predict not more than two diagnostic intervals ahead is the best alternative [8].

Modeling of degradation of technical devices is better to base on adaptive algorithms for processing of updated data of diagnostic measurements. Such model parameters are adapted to jumping input conditions, which is required for short-term prediction of change in actual condition of a mechanical system. Majority of scientific studies into a common criterion of equipment condition bas on vibration parameter limit themselves to rolling bearings [9–11], which is reflective of an extremely limited application range of the current algorithms. The ranges of this methodological approach can only be pushed using many common criteria which will enable diagnosis and prediction of degradation for the whole mechanical system rather than for its individual assemblies and elements. Efficiency of such criteria is to be high at all life stages of defects—from initial to moment prior crash stop.

Aside from other things, such factors as sensor position errors, ill-prepared measurement surface, signal recording errors due to bad condition of connections in the sensor–cable–apparatus system, some structural or kinematic features of measurement objects (hard-to-reach measurement points, high alternating impact loads, presence of low-speed structural elements, short operation period, etc.), low temperatures and some other underlying conditions can limit the application area of the common diagnostic criteria. An optimal solution to this problem is addition of the criterion algorithm with filtration algorithms aimed to clean an outlet signal from "unnecessary" components which complicate the initial data analysis.

A first-order condition of any forecast model is information on state estimation ranges beyond which the classification of mechanical system conditions alters; put it otherwise, it is required to possess sound criteria of limit state, suitable to estimate development of defects in elements of diagnosable equipment. In as much as the proposed model uses the common diagnostic criterion per reach group of defects, the required minimum information to cover the actual state of a complex mechanical system can be obtained from the analysis of vibration parameters generated during the system operation. For example, development of basic criteria of limit state can effectively use one or some of the characteristics listed below:

—total level of vibration velocity and vibration acceleration in standard or expanded bandwidth, as well as square root mean of vibration velocity in octave bandwidth, including rotor frequency [12];

—frequency bands, sequences or peak values of harmonics, reflective of the same-nature defect (for example, harmonic sequences of system rigidity violation [$0.4 f_R$–$24 f_{R\max}$] or motor and reducer misalignment [f_R…4…6 f_R]), or a set of significant components of gearing or teeth in expanded bandwidth to 10 kHz by parameters of vibration acceleration and vibration velocity;

—swing of vibratory displacement within a range of 10–1000 Hz;

—similarity of signals, in particular, determined from the spectrum of envelope or wavelet transform of vibro-acoustic signal.

The implemented analysis of diagnostic data obtained on active electromechanical equipment of electric shovels has allowed grouping all defects intrinsic to this class mining machinery into seven packages (misalignment, rigidity violation, rotation parts unbalance, bearings, tooth gears, sleeves and electrical nature defects), which all in all cover more than 120 diagnostic characteristics by vibration parameters. Some of them are formalized for convenient use in development of automatic control algorithm for complex systems by parameters of generated mechanical vibrations [3, 13]. Earlier studies often limited themselves to formalization of signs based on the analysis of vibration parameters of bearings, which are the most comprehensively investigated object of vibro-diagnostics, and no interest was paid to creation of common diagnostic criteria for other elements of technical devices.

The modeling results assure that for the short-term prediction of actual degradation of a complex mechanical system, the adaptive modeling techniques offer high effectiveness of prediction [3, 8]. The adaptive exponential mathematical models using in short-term prediction of 1–2 diagnostic periods ahead (approximately one calendar month) have appeared to be the best in meting the set objectives.

An integral element of the common criterion algorithm is the circulating frequency refinement procedure intended to improve accuracy of calculations and efficiency of diagnostics. Such procedures are widely used in software designs and in creation of highly detailed spectral masks. Within this research, we used the circulating frequency refinement procedure put forward in [8], which proved efficient workability in the conditions of uncertain maximum of circulating frequency.

Furthermore, in order to assess remaining life of a complex mechanical system, which is an open pit mine shovel, it is required to distinguish informative trends in the deterministic component of initial signal for the short-term prediction of equipment degradation. A common criterion best suited to assess condition of complex technical devices should be developed with regard to multi-dimensional space of diagnostic properties using step-wise scalarization algorithms and segmentation of groups of devices with different intensity of defects; in this case the number of the criteria should coincide with the number of groups of defects. An integral part of the algorithm for development of common criteria is the clipping sub-program to clean initial diagnostic data from "unnecessary" information [14]. The number of the clipping sub-programs should be equal to the number of potential defects on the equipment under inspection; all such sub-programs are based on detection and selection of informative components of certain nature (e.g., defects of rolling bearings, tooth gears, connector sleeves, etc.) and, thus, on elimination of all other components usually by means of filtration. Each sub-program includes circulating frequency refinement procedure which, in its turn, is based on finding signal components having maximum amplitudes in low-frequency range. The algorithm [3, 8] applied in the framework of the described research has proved high efficiency in the analysis of vibration acceleration spectra of rolling bearings, where maximum amplitudes if significant harmonics a priori cannot belong to the circulating frequency (e.g., Figure 1), which is the determinant factor for the successful implementation in the framework of development of common diagnostic criteria for estimation of actual state of mechanical systems.

The coal industry historically was and is yet an object of elevated hazard and high risk of emergencies [15]. Thus, evidently, it is required to continue development of the procedure for creating algorithms of common diagnostic criteria to estimate condition of electromechanical equipment of open pit mine shovels. The present research findings allow stating that introduction of maintenance service based on actual condition of equipment in coal mines in Russia is impossible without finding the solution to the problem of common diagnostic criteria of equipment condition and mathematical degradation models capable to predict the change in actual conditions of complex mechanical systems, which is directly dependent on volume and quality of diagnostic information of vibration parameters for wide range of mining machines. Implementation of the methodological approach proposed in this paper to develop common diagnostic criteria and degradation modeling for estimation of complex equipment condition can already be used in minimization of failure probability of mining machines, improvement of safety in open pit mining, as well as in reduction of injury and accident rate due to operation of out-of-order equipment.

3. Conclusions

1. The analytical results of the vibration parameters generated in operation of open pit mine shovel accessories have been generalized and used as the basis for classification and grouping of this equipment defects as well as for formalization of diagnostic criteria applicable for development of algorithm for automatic control of complex systems by frequency of diagnostic properties.
2. The authors have proved expedience of practical application of short-term equipment degradation prediction within the current planned maintenance system of operating mines as this enables forecasting changes in equipment conditions for one–two diagnostic periods ahead.
3. The methodology has been developed for the algorithm of a common diagnostic criterion for detection and evaluation of damage (defect) risk in equipment of mining machines.

References

[1] Gol'din AS 1999 *Vibration of Rotating Machines* Moscow: Mashinostroenie (in Russian)
[2] Klyueva VV (Ed) 2005 *Nondestructive Testing: Handbook* Moscow (in Russian)
[3] Gerike PB 2015 Finding consistent pattern of change in technical state of dynamic equipment of draglines based on the analysis of mechanical vibration parameters *Izv. UGGU* No 4 pp 71–78

[4] Pozhidaeva V 2005 Determining the roughness of contact surfaces of the rolling bearings by the method of shock pulses *World Tribology Congress III* Washington, D.C. USA
[5] Tse P, Peng Y and Yam R 2001 Wavelet analysis and envelope detection for rolling element bearing fault diagnosis. Their effectiveness and flexibilities *Journal of Vibration and Acoustics* Vol 123 pp 303–310 DOI: 10.1115/1.1379745
[6] Krakovskiy YuM 2006 *Mathematical and Software Evaluation of the Technical State of Equipment* Novosibirsk (in Russian)
[7] Barkov AV and Barkova NA 2004 *Vibration Diagnostics of Machines and Equipment. Vibration Analysis: Handbook* Saint-Petersburg (in Russian)
[8] Sushko AE 2007 Dedicated mathematical and software support for automatic diagnostics of complex systems *Candidate of Engineering Sciences Thesis* Moscow: MIFI (in Russian)
[9] Bently DE and Hatch CT 2003 *Fundamentals of Rotating Machinery Diagnostics* ASME Press
[10] Balducchi F, Arghir M and Gaudillere S 2014 Experimental analysis of the unbalance response of rigid rotors supported on aerodynamic foil *bearings Proceedings of ASME Turbo Expo 2014: Turbine Technical Conference and Exposition GT2014*, Düsseldorf, Germany. Available at: http://proceedings.asmedigitalcollection. asme.org.
[11] Skeinik R and Petersen D *Automated Fault Detection via Selective Frequency Band Alarming in PC-Based Predictive Maintenance Systems* CSI, Knaxville, TN 37923, USA
[12] Liu G and Parker R 2008 dynamic modeling and analysis of tooth profile modification for multimesh gear vibration *Journal of Mechanical Design* Vol 130 pp 121402–13. DOI: 10.1115/1.2976803
[13] Gerike PB 2015 Vibroanalysis of dynamic aspiration system equipment at processing plants in Kuzbass *Vestn. Nauch Tsentra Bezop. Rabot Ugoln. Prom.* No 4 pp 73–78
[14] Klishin VI, Zvorygin LV, Lebedev AV and Savchenko AV 2011 Safety Problems and New Technology of Underground Coal Mining Novosibirsk (in Russian)
[15] Lukyanov AV 1999 Classifier of Vibrodiagnostic Symptoms of Defects in Rotary Machines Irkutsk (in Russian)

IOP Conf. Series: Earth and Environmental Science **262** (2019) 012021 doi:10.1088/1755-1315/262/1/012021

Selecting parameters of air-percussion tunneling facility

VP Gileta[1] and YuV Vanag[2]*

[1]Chinakal Institute of Mining, Siberian Branch, Russian Academy of Sciences, Novosibirsk, Russia
[2]Novosibirsk State Technical University, Novosibirsk, Russia

E-mail:*pmx08@mail.ru

Abstract. The paper discusses vibratory displacement of two-mass vibrating system with air-powered impact force generator. In the first stage of studies, the effect of air leakage between chambers on efficiency of the impact force generator in percussion pricking and ramming, with much short travel of the body of the generator than its piston is analyzed. In the second stage, the influence of clearances on the rate of horizontal self-displacement of the vibrating system with impact force generator is examined at the proportional travels of the body and piston. The effect exerted by compressed air leaks between the work chambers in clearances of swivel joints is evaluated. The research results can be used in design of special pneumatic percussion drive for suction headers.

1. Introduction

In Russia and in the world, trenchless pipeline laying commonly uses vibarion–percussion method and air-driven machines. The process of hole-making is based on the effect of vibro-displacement of a vibrating system under asymmetrical resultant force of internal and external forces. This method needs no equipment for generation of static forces, construction of bearing walls, anchorage, etc. Air-driven machines used as generators of driving force for vibro-displacement of vibrating systems are highly reliable, small in size and allow reduced volumes of preparation, earth works, assembly and reclamation at cut down transportation expenditures [1–3].

Holes with a diameter to 300 mm are made by vibro–percussion pricking, i.e. by soil compaction. Many years-long experience shows that such holes remain stable (do not break down) for a long time (sometimes several years). This allows inserting a carrier pipe and, if necessary, casing directly after completion of hole making [2, 4].

Table 1. Specification of series-produced air hammers.

Parameter	IP 4605	IP 4603	SO 134
Length, mm	1500	1550	1720
Diameter, mm	95	130	152
Weight, kg	50	100	150
Blow energy, J	80	220	500
Blow frequency, Hz	5.4	6.2	4.2
Rated compressed air pressure, MPa	0.6	0.6	0.6
Air flow rate, m^3/s	0.073	0.088	0.133

Holes with a diameter more than 300 m are made by ramming. In this case, a steel pipe is driven in soil by an air-powered machine and soil plug is taken out of the pipe by self-propelling suction header [1, 2]. The operating device of the suction header is a receiving or bypass core barrel, the drive unit, depending in the size of the pipe to be cleaned from soil, may be one of the series-produced air hammers: IP 4605, IP 4603 SO 134 (table 1).

The feature of the hole-making vibrating systems with the air-powered impact force generator is that these are the self-oscillating systems [5, 6]. Oscillations are sustained by the oscillating elements of the system. Vibratory displacement of such systems with air-powered impact force generators is studied insufficiently.

Compressed air leakage in clearances of swivel joints has considerable effect on vibratory displacement efficiency of oscillating systems and on parameters of the air-powered impact generators: energy A and frequency f of blowing, and on compressed air flow rate Q. In this regard, air leaks between chambers should be taken into account in calculations and design of a pneumatic–mechanical mechanism. For instance, neglected air leakage between the power and exhaust chambers in a brand-new pneumatic percussion machine—pulsator [7]—resulted in incomplete clearing of the power stoke chamber—under-exhaust—which destabilized operation of the pulsator and ended with scaling down of investigations aimed to design a very powerful and economic percussion mechanism.

2. An air-driven hammer design and work parameters

The present research focused on influence exerted by air leakage between chambers on energy efficiency of impact force generators in trenchless laying of pipes by the methods of vibropercussion pricking and ramming.

As an impact force generator, we study an air-driven hammer with valve-less air distribution and one controllable chamber. Figure 1 shows the the layout and design model of air hammer IP 4603. The impact force generator has a two-mass oscillating system: m_1 is a striking piston, m_2 is the generator body rigidly connected with the pipe or core receiver of the suction header. The piston and the body together with an air-distribution element shape the controllable (back stroke) *1*, main (instroke) *2* and exhaust *3* chambers. Compressed air is fed from the feed line under pressure p_f via control channels and clearances $S_{1\text{-}2}$, $S_{1\text{-}3}$, $S_{2\text{-}3}$ to the chambers and is released from them to the atmosphere p_a. The masses m_1 and m_2 are affected by the internal active driving forces of cyclic nature governed by the type and setting of the impact force generator. The mass m_2 is additionally influenced by the external friction F_{fr}.

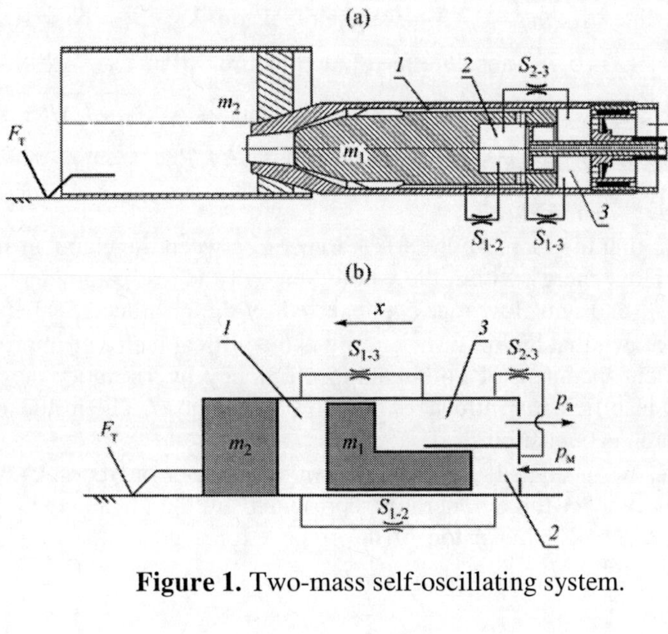

Figure 1. Two-mass self-oscillating system.

The working chambers of the impact force generator have variable volumes due to mobility of the machine components. Air always leaks in clearances in friction pairs. Clearance is an inherent attribute of a swivel joint. The body of the air hammer is a working tool, its length is approximately 15 times its diameter (table). Clearances between the chambers *1*, *2* and *3* (Figure 1) in the air hammer IP 4603 are rather large to eliminate jamming of mobile parts in case of low axial stiffness of the hammer.

The numerical analysis of the effect of air leakages between the air hammer chambers in clearances of swivel joints was implemented at the following assumptions:

—the leakages were evaluated in terms of areas of the joint clearances;

—the clearances were analyzed independently, i.e., the value of air leakage between one pair of the chambers was considered at the constant level of air leakage (areas of clearances) between the other chambers.

The calculations were carried out in two stages. In the first stage, we studied effect of clearances on the impact force generator performance in pricking and ramming. Here, the point is that travel of the hammer body is much shorter than the travel of the piston (less than 3%) [8].

Figure 2 depicts stage I modeling results of the impact force generator performance. The areas of clearances between chambers are given explanatory subscripts, e.g. S_{12} is the clearance area between the controlled *1* and the main *2* chambers.

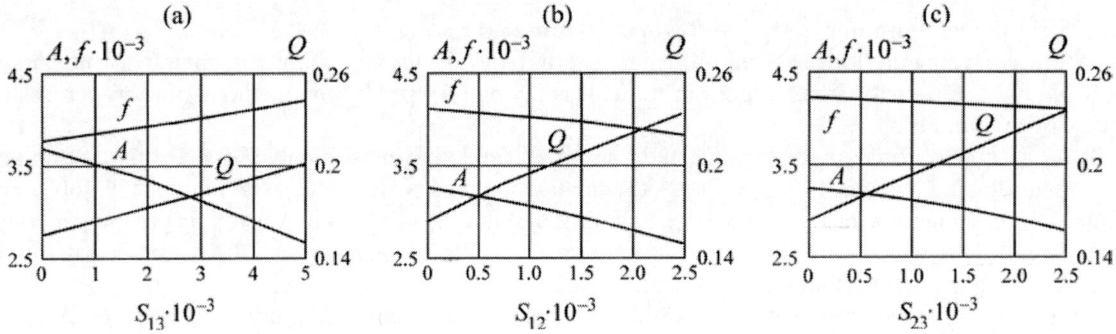

Figure 2. The work parameters of impact force generator versus value of clearances between chambers.

The calculations involved dimensionless values. The main units were: $l_e = d_M = 32\cdot10^{-3}$—diameter of the air feed line, m; $m_e = 37.5$—weight of piston, kg; $S_e = S_1 = 103.9\cdot10^{-4}$—area of controlled chamber, m^2; $p_M = 0.7$—absolute feed line pressure, MPa; $T_M = 290$—temperature in the main chamber, K; $R = 8.31$ J/(mole·K)—universal gas constant; $t_e = \sqrt{(m_e l_e)/(S_e p_M)}$—unit time, s; $A_e = (p_M S_e l_e)/2$—unit blow energy, J; $Q_e = (p_M S_e l_e)/(R T_M t_e)$—unit air flow rate, kg/s; $f = Q/(A_e q_N)$—unit blow frequency, Hz.

It is seen in Figure 2 that the increase in the clearances between the chamber results in higher air consumption and lower blow energy within the whole range of the test parameters. The highest effect on the unit blow energy *A* and unit flow rate *Q* is exerted by the clearance S_{23} between the main and exhaust chambers. It is known that higher blow energy is the critical factor of improvement in the pipe penetration rate, i.e., in enhancement of hole-making efficiency by vibratory percussion [1, 2]. The area of the clearance S_{23} insufficiently influences the blow frequency. *f*. The influence of the clearances S_{12} and S_{13} on *f* of the piston is opposite.

In the second stage, we analyzed the influence of clearances on the self-displacement of the vibrating system with the impact force generator represented by the air hammer IP 4603 as well. In this case, the travels of the body and piston of the impact force generator were proportionate. The external friction force was applied to the body (mass m_2) and represented dry friction. The calculations were in dimensional form.

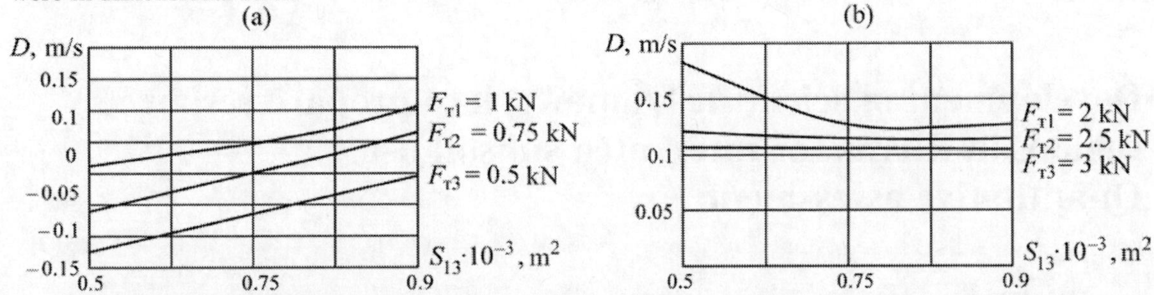

Figure 3. Vibratory displacement rate of impact force generator as function of clearance between working chambers.

The increase in the clearances between the chambers accelerates vibratory displacement of the two-mass system only in the field of insignificant frictions which are less than 30% of recoil of the impact force generator (Figure 3a). This is a signature of this system and is to be taken into account is design and operation of a suction header as in real-time operation, the friction fore F_{fr} of the impact force generator with pipe is low—1 kN, and the generator, instead of advancing in the pipe, will displace backward or even stop. In the field of external frictions exceeding the 30% threshold of the generator recoil, the air leaks between the chambers have insignificant influence of the penetration rate of the vibrating system (Figure 3b).

3. Conclusions

The authors have evaluated influence of compressed air leakage in clearances of swivel joints on performance of valve-less air-powered impact generator with one controlled chamber and on the vibratory displacement rate of the two-mass vibrating system.

The increase in the area of clearances between the working chambers accelerates vibratory displacement of the two-mass vibrating system in the field of insignificant frictions, which are less than 30% of recoil of the impact force generator in the real-time operating conditions. These research findings can be used in design of a special air percussion machine to drive impact force generators.

References

[1] Gurkov KS, Klimashko VV, Kostylev AD et al 1990 *Pneumatic Punchers* Novosibirsk: IGD SO AN SSSR (in Russian)
[2] Gileta VP and Smolyanitsky BN 1998 The use of pneumatic impact machines for making underground passes *Izv. Vuzov. Stroit.* No 11-12 pp 73–76
[3] Kershenbaum NYa and Minaev VI 1984 *Horizontal and Vertical Percussive Hole-Making* Moscow: Nedra (in Russian)
[4] Rybakov AP 2006 *Fundamentals of Trenchless Technologies (Theory and Practice)* Moscow: Stroyizdat (in Russian)
[5] Gileta VP, Smolyanytsky BN, Petreev AM and Tetenov EV 1997 Displacement of a self-excited two-mass system with external dry friction *Journal of Mining Science* Vol 33 No 3 pp 247–252
[6] Gerts EV and Kreinin GV 1975 *Pneumatic Drive Design: Reference Guide* Moscow: Mashinostroenie (in Russian)
[7] Tupitsyn KK 1980 *Air-Powered Impact Machines with Pneumatic Pulsators* Novosibirsk: IGD SO AN SSSR Preprin No 2 (in Russian)
[8] Vorontsov DS, Petreev AM and Smolyanitsky BN 2005 Driving of boreholes in soil by an impact device with annual tool Journal of Mining Science Vol 42 No 2 pp 138–142

IOP Conf. Series: Earth and Environmental Science **262** (2019) 012022 doi:10.1088/1755-1315/262/1/012022

Development of scientific framework to prepare resource regions in Russia for integrated subsoil use: Quantitative assessment

EV Goosen[1,2]*, ES Kagan[1,2] and SM Nikitenko[1]

[1]Federal Research Center for Coal and Coal Chemistry, Siberian Branch of the Russian Academy of Sciences, Kemerovo, Russia
[2]Kemerovo State University, Kemerovo, Russia

E-mail: *egoosen@yandex.ru

Abstract. The degree of preparedness of Russia's resource regions for integrated subsoil use and transition to a balanced development model is revealed. The quantitative assessment has shown that regions dependent on natural reserves are the "locomotives of growth" and, for the most part, have medium and high economic potential. At the same time, the organization of mineral mining must be accompanied by a gradual transition to integrated development of subsoil, which includes efficient use and processing of natural resources. An important aspect of comprehensive exploitation of mineral reserves will be the localization of production in the regions capable to create conditions for the development of internal market and an interconnected economic space in Russia.

1. Introduction

Exhaustion of the traditional growth sources in resource regions dictates finding of a new model of sustainable development. Integrated subsoil use and efficient mineral wealth management are the main sources of balanced advancement. The scientific world has no common perception of resource regions. In the English-language literature, such regions are discussed as the *resource curse* or the *resource abundance* understood as the phenomenon of slower pace of development in countries holding ample natural resources as against countries which lack the environmental assets [1, 2]. Nonetheless, some researchers show that specialization of resource regions is in many ways dependent on the initial economic and institutional conditions as well as on the right-on regional policy accounting for both general regulations and local context [3–5]. The second approach is more sound and relevant, to this paper authors' opinion.

Resource regions have a special path of development, which is a stable replication of an enclave dual economy. This singularity shows itself as formation of a supernormal, isolated, highly productive and export-oriented sector represented by large vertically integrated mining and allied companies in a region. The export-oriented mining and processing sectors of industry concentrate increasingly much more resources, become and remain *engines* of economy in Russia, and, thus, govern, the economic development and place of the country in the global division of labor [6, 7]. Permanent and stable *replication* of enclaves rich in natural and mineral resources ends with extensive inefficient utilization of natural wealth and causes considerable environmental damage [7, 8]. The problem of supersession of the extensive economic development model in Russia is connected with the preparedness and of resource regions to pass on the sustainable development path [9].

2. Features of the region development in Russia

Russia is composed of regions extremely different in social and economic characteristics. For example, in terms of the research and development cost share in the gross products, mismatch of some regions reaches 143–148 times, while in terms of the application of the advanced production technologies, this gap exceeds 1000 times [10]. For this reason, it is important to find regions best prepared for transition to the sustainable development model and, moreover, to specify range of available development paths in every specific case. This requires rejecting a space-neutral approach and switch on to a place-based approach [11–13]. On the other hand, introduction of local context in the analysis of resource regions inevitably raises question on the limits and methods of its inclusion in economic research. It is unallowable that the place-based approach results in complete abortion of search for any general regularities. A way out is the type science which makes it possible to take into account qualitative and quantitative characteristics of resource regions.

Building typology (typological analysis) is a method of studying complex geographical and socio-economic objects with a view to revealing relevant uniform and distinct groups of objects characterized by a set of criteria or properties [14]. In this connection, we have analyzed data on 35 resource regions in Russia, selected within the previous researches [15].

For the economic potential evaluation of a region, index K_1 is calculated as a ratio of GDP per capita in the region to average GDP per capita in Russia. In the case under analysis, K_1 ranged from 0.26 (Republics of Ingushetia and Chechnya) to 11.2 (Nenets Autonomous Region).

Resource-dependence on regions was defined by the share of mining sectors in the regional GDP. The resource-dependence index is the index K_2 calculated as a ratio of mining industries in the regional GDP to a shape of the mining industries in GDP of Russia. In the discussed analysis, K_2 ranged from 0.107 (Ingushetia) to 6.05 (Khanty-Mansi Autonomous Region).

Primarily, evaluation criteria were represented by linguistic variables (LV) with term-sets: economic potential {Te1—low, Te2—moderate, Te3—high) and resource-dependence {Tr1—weak, Tr2—moderate, Tr3—high}. The membership functions (MF) of the terms were the functions of triangular and trapezoid forms. Such MFs are advantageous for the least information required for their determinations as compared with other functions [16].

Table 1. Exact and fuzzy evaluation criteria per regions.

Region	Economic potential index K_1				Resource-dependence index K_2			
	Exact	T1	T2	T3	Exact	T1	T2	T3
Kursk	0.675	0.416	0.584	0	0.766	0.053	0.947	0
Nenets	11.24	0	0	1	6.01	0	0	1
Karelia	0.753	0.155	0.845	0	0.735	0.1	0.9	0
Ingushetia	0.261	1	0	0	0.107	1	0	0
Tatarstan	1.069	0	1	0	1.927	0	0.716	0.284
Tyument	1.407	0	0.741	0.259	1.291	0	1	0
Kemerovo	0.697	0.342	0.658	0	2.284	0	0.477	0.523
Krasnoyars Territory	1.273	0	0.908	0.092	1.562	0	0.959	0.041
Tyva	0.338	1	0	0	0.871	0	1	0
Khakassia	0.721	0.263	0.737	0	1.152	0	1	0
Tomsk	0.992	0	1	0	2.631	0	0.246	0.754

Table 1 gives a fragment of the analysis performed for the selected group of resource regions with initial (exact) and fuzziness (fuzzy values) indicators. In the fuzzy analysis, mismatch of objects can be calculated using the notion of confusing similarity of objects in some property. The similarity of objects A and B in a criterion is calculated from the formula:

$$d_{AB} = \frac{1}{2}\sum_{i=1}^{k} \left| \mu_{iA}(x) - \mu_{iB}(x) \right|, \tag{1}$$

where k is the number of terms in LV.

IOP Conf. Series: Earth and Environmental Science **262** (2019) 012022 doi:10.1088/1755-1315/262/1/012022

Table 2. Fuzzy clustering of resource regions (territories) by integrated subsoil use preparedness.

Economic potential	Resource-dependence		
	Tr1—weak (1; 0; 0)	Tr2—moderate (0; 1; 0)	Tr3—high (0; 0; 1)
Te1—low (1; 0; 0)	***Readiness condition 1*** Kalmykia (0.956); Ingushetia (1); Chechnya (1); Buryatia (0.671)	***Readiness condition 2*** Trasnabikalia (0.495); Buryatia (0.523); Kalmykia (0.044)	***Readiness condition 3***
Te2—moderate (0; 1; 0)	***Readiness condition 5*** Bashkortostan (0.717); Arkhangelsk (0.763); Volgograd (0.523); Kaliningrad (0.053); Buryatia (0.468); Karelia (0.1); Khabarovsk (0.734)	***Readiness condition 8*** Arkhangelsk (0.237); Volgograd (0.478); Transbaikalia (0.505); Irkutsk (0.56); Kaliningrad (0.25); Krasnoyarsk (0.726); Kemerovo (0.477); Tyumen (0.282); Kursk (0.947); Karelia (0.9); Belgorod (0.926); Murmansk (0.686); Irkutsk (0.78); Astrakhan (0.503); Samara (1); Perm (1); Bashkortostan (0.283); Buryatia (0.329); Tatarstan (0.765); Khakassia (1); Amur (1); Tomsk (0.591); Udmurtia (0.495); Khabarovsk (0.747)	***Readiness condition 7*** Astrakhan (0.497); Irkutsk (0.44); Komi (0.341); Tatarstan (0.549); Orenburg (1); Udmurtia (0.505); Kemerovo (0.523); Krasnoyarsk (0.267); Tomsk (0.845)
Te3—high (0; 0; 1)	***Readiness condition 10*** Khabarovsk (0.253)	***Readiness condition 9*** Belgorod (0.074), Krasnoyarsk (0.733), Magadan (0.285), Murmansk (0.314), Tatarstan (0.451), Tomsk (0.155), Tyumen (0.719), Khabarovsk (0.266)	***Readiness condition 6*** Krasnoyarsk (0.275); Magadan (0.715); Nenets (1); Komi (0.659); Sakha (1); Tatarstan (0.235); Sakhalin (1); Tomsk (0.409); Khanty-Mansi (1); Yamalo-Nenets (1); Chukotka (1)

IOP Conf. Series: Earth and Environmental Science **262** (2019) 012022 doi:10.1088/1755-1315/262/1/012022

Then, the mismatch of two objects in criteria n is found as follows:

$$r_{AB} = \frac{1}{n}\sum_{i=1}^{n} d_{AB}\,. \tag{2}$$

To find which cluster an objects belongs to, the distance between the object to a reference point r_j is calculated.

The grade of membership is found as:

$$Cl_j = 1 - r_j\,. \tag{3}$$

Table 2 reports clustering results of resource regions, and it is seen that one of potential clusters is empty. None of the regions appears in the cluster with weak economic potential Te1 (1; 0; 0) and high resource-dependence Tr3 (0; 0; 1). Only one region—Khabarovsk Territory—belongs in the cluster with high economic potential Te3 (0; 0; 1) and weak resource-dependence Tr1 (1; 0; 0) at the grade of membership 0.253. On the whole, this is reflective of the fact that resource regions in Russia are the the *growth engines*—they mostly have medium and high economic potential. Vacancy of the second cluster implies than the enclave duel economic model gradually exhausts itself, and even high mineral production output is incapable to maintain high economic potential of regions.

In four out of nine clusters, there are clear top-level regions. The majority of the regions (22) concentrate in the cluster with moderate economic potential Te2 (0; 1; 0) and resource-dependence Tr2 (0; 1; 0). Moreover, this cluster accommodates regions with different dynamics and prospects of mining industries. For instance, this cluster embraces regions at the stage of exhaust of the old development model based on extensive mineral mining. These regions are Kemerovo (0.477), Krasnoyarsk (0.726) and Tyumen (0.282). In addition, this cluster holds regions developing new reserves: Khakassia (1), Amur (1), as well as regions with same traces of the model based on integrated subsoil development: Kursk (0.947), Belgorod (0.926), Tatarstan (0.765), Samara (1), Perm (1) and some other. It is interesting that these regions have different grades of membership in this cluster and fall within some neighbor clusters. This fact points at the feasibility for these regions to develop along a wide range of growth paths, which calls for additional quantitative assessment of preparedness of resource regions for transition to the sustained development model based on integrated subsoil management.

Table 3. Integrated subsoil development preparedness quantification of resource regions.

Region	Preparedness	Region	Preparedness
Tyumen	8.718505	Magadan	6.854531
Murmansk	8.314166	Volgograd	6.432392
Belgorod	8.07358	Komi	6.34095
Amur	8	Nenets AR	6
Perm	8	Sakha (Yakutia)	6
Khakassia	8	Sakhalin	6
Samara	8	Khanty-Mansi (Yugra)	6
Krasnoyarsk	7.958581	Chukotka	6
Kursk	7.841851	Yamalo-Nenetsk AR	6
Tatarstan	7.715557	Bashkortostan	5.84921
Karelia	7.701004	Kaliningrad	5.75
Irkutsk	7.560286	Arkhangelsk	5.709743
Astrakhan	7.502865	Transbaikalia	5.029613
Udmurtia	7.495099	Buryatia	3.365686
Kemerovo	7.477396	Tyva	2
Khabarovsk	7.284379	Kalmykia	1.043956
Tomsk	7.24589	Ingushetia	1
Orenburg	7	Chechnya	1

At the second stage of research, the integral index of regional preparedness for transition to sustained development based on integrated subsoil management is calculated. Table 3 offers the quantitative assessment of resource regions by their readiness to multipurpose subsoil use.

3. Conclusions

Evaluation of the economic potential of resource regions has disclosed their unequal preparedness for transition to sustainable development model. The main source of economical growth in these regions is mineral mining and processing industries. The integrated subsoil management conditions transition from the stage of mining and rough processing of minerals to deeper conversion and by-production. The main obstacle for projects capable to eliminate the enclave nature of a resource region economy is, first, readiness of the vertically integrated companies to implementation of such projects, second, technical preparedness of resource regions to create by-products, and, third, availability of infrastructure for such production.

Acknowledgements

This study was supported by the Russian Science Foundation, Project No 16-18-10182.

References

[1] Ross ML 2015 What have we learned about the resource curse? *Annual Review of Political Science* Vol 18(1) pp 239–259
[2] Sachs JD and Warner AM 1995 Natural resource abundance and economic growth *NBER Working Papers* No 5398
[3] Van der Ploeg F 2011 Natural resources: curse or blessing? *Journal of Economic Literature* Vol 49(2) pp 366–420
[4] Papyrakis E and Gerlagh R 2004 The resource curse hypothesis and its transmission channels *Journal of Comparative Economics* Vol 32(1) pp 181–193
[5] Alexeev M and Conrad R 2009 The elusive curse of oil *The Review of Economics and Statistics* Vol 91(3) pp 586–598
[6] Kuleshov VV (Ed) 2017 *Resource Regions of Russia in the "New Reality"* Novosinirsk: IEOIP SO RAN (in Russian)
[7] Kondratiev V 2016 Resource Model of Economic Modernization: Opportunities and Limitations Moscow: IMEMO RAN (in Russian)
[8] Levin SN, Kagan ES and Sablin KS 2015 Regions of "resource type" in the modern Russian economy *Zh. Institutsional. Issled.* Vol 7 no 3 pp 92–101
[9] Trubetskoy KN and Galchenko YuP 2015 Methodology for estimating promising development paradigm for mineral mining and processing industry *J. Min. Sci.* Vol 51 No 2 pp 407–415
[10] Ignatov VG 2009 Asymmetry of the socio-economic development of the regions of the Russian Federation and the main directions of its weakening *Ekon. Vestn. RostovGU* Vol 7 No 2 pp 132–137
[11] OECD 2009 *How Regions Grow. Paris: Organization for Economic Growth and Development*
[12] OECD 2009 *Regions Matter: Economic Recovery. Innovation and Sustainable Growth, Paris, Organization for Economic Growth and Development*
[13] Melnikova LV 2014 Spatially neutral and locally targeted regional policy: problems of choice *Region: Ekon. Sotsiol.* No 1 pp 64–85
[14] Grinchel BM and Nazarova EA 2015 Methodology for constructing typologies of Russian regions in the context of competitive capacity *Ekonomich Sotsial. Peremeny: Fakty, Tendent., Prognoz* No 3 (48) pp 111–125
[15] Kagan ES and Goosen EV 2017 The problems of identification of resource-type regions *IOP Conf. Series: Earth and Environmental Science*
[16] Imanov KD and Alieva KhS 2009 Fuzzy models of determining the level of socio-economic development of regions *Ekonomich. Nauki Sovremen. Rossii* No 3 pp 93–102

IOP Conf. Series: Earth and Environmental Science **262** (2019) 012023 doi:10.1088/1755-1315/262/1/012023

Experimental investigation of crack formation in rock mass under impact fracture by a hard indenter

LV Gorodilov and VP Efimov*

Chinakal Institute of Mining, Siberian Branch Russian, Academy of Sciences, Novosibirsk, Russia

E-mail: *efimov-pedan@mail.ru

Abstract. The technique for determining geometric parameters of cracks formed in rock mass under penetration of a hard tool is presented. The results of rock fracture as impact craters and main cracks formed in a marble block after penetration of a wedge-shaped steel tool are discussed. The photographs of the craters and polished sections of drill cores are given.

1. Introduction

Impact fracture of rocks is a method of wide application in mineral mining in destruction of rocks by hammers, in breaking of oversizes, or in rotary–percussion drilling. Although the process remains of concern as judged by numerous experimental and theoretical investigations undertaken, the problem of quantitative description of impact fracture and efficient calculation modeling for predicting its results is far from being solved up to the resent moment. Intrinsically, failure mechanics has initiated a new school on contact failure named Indentation Fracture in the English-language literature. The difficulty of the research consists in extreme concentration of all processes that result in failure. This means that origination of cracks in the conditions of complex stress state deserves close attention. Rather strongly localized volume of material under an indentation tool, being in plastic state due to high compressive stress, is large relative to an incipient crack in the beginning of the process and turns out to be small at the stage of the main crack development. Consequently, the contact failure models embrace both plastic and elastiplastin deformation of a medium under an indenter, as well as brittle crack formation and propagation. Modeling contact interaction using elastic solutions is classic, and such solutions are available in educational books (Hertz contact, Boussinesq problem). Modeling of plastic contact for a wedge was started in [1] and was extended in [2] with regard to strengthening of material under static indentation. Various aspects of rock fracture under indentation of hard tools are discussed in [3–6].

Models of contact-induced fracture verified experimentally in static or dynamic indentation tests. Results of fracture under indentation of a hard tool (indenter) in rocks are mostly estimated by geometrical parameters of a formed crater. The crater is formed either due to plastic deformation of rocks directly in the zone of contact, or due to rock disintegration while edge cracks develop from the contact zone and expose. It is found [4, 7–10] that a system of cracks originates under contact fracture, and the sizes and orientations of the cracks depend on the properties of rocks and on the parameters of indenter (geometrical sizes, weight, pre-impact velocity). Furthermore, axial cracks under a wedge open and grow under loading, and edge cracks propagate under stress relaxation. Under repeated impacts, the developed cracks can exert considerable influence on development of a new system of cracks. Dimension of the plastic deformation zone can be estimated by measuring the indentation depth (crater) whereas determination of penetrated cracks under the wedge impacts is experimentally

IOP Conf. Series: Earth and Environmental Science **262** (2019) 012023 doi:10.1088/1755-1315/262/1/012023

difficult. When studying formation of cracks under indentation of a tool to small samples of rocks, alongside with parameters of a crater, some information can be obtained by visual inspection (exposure of some cracks on the sample surface), ultrasonic sounding or simply saw cutting. In a full-scale experiment, when the listed approaches are impracticable in view of the large dimensions of samples, it seems acceptable to carry out core sampling at the fracture point and, then, to process cores by sawing or polishing. Such procedure is described in [11]. This study aims to improve the procedure in terms of cores taken at the repeated impact point in a marble block.

2. Description of the experiment scheme and results

The method of impact action on a marble block is in detail presented in [12]. A tool was a blunt wedge with a length of 30 mm and an angle of 60°. Rock mass was represented by a marble block 0.95×0.95×1.5 m 4000 kg in weight. Operation of an impact machine was simulated by a pendulum hammer (Figure 1), which allowed varying blow energy and blow structure by changing weights in the piston–impact tool system. In the tests, piston *4* was brought to a wanted velocity to interact with impact tool *5* pressed to marble block *1*. Then, the pre- and post-impact velocities of the piston were determined using a light sensor composed of two laser–photodiode pairs with an amplifier, arranged immediately at the point of the piston and impact tool interaction to record in turn the moments when beams were intersected by the front face of the piston.

The impact process was studied using vibrator inverters AP 33 installed in the impact tool and in the piston; the signals were transferred from the inverters, via amplifiers Bruel&Kjaer 2651, to AD converter L-Card E-440 and, then, to PC for processing in PowerGraph 3.3. The impact tool was also equipped with variable inductance velocity transducer IDS-2.

Figure 1. Experimental design: *1*—marble block; *2*—supporting frame; *3*—pendant; *4*—piston; *5*—impact tool; *6*—acceleration sensor; *7*—optical velocity sensor; *8*—variable inductance velocity transducer IDS-2.

During the tests, geometrical parameters of craters were measured and cores were taken using diamond-coated drill bits MATRIX with diameters of 33 and 55 mm. Coring leaves a ring slot wit a clearance of 2 mm. Cores were separated from the block using the method of slotting using plastic material, developed at the Institute of Mining, SB RAS [13]. A layer of plasticine 3–5 mm thick was placed on the bottom of the core slot, then a knockout with an end anvil was inserted in the slot, and a blow was delivered on the anvil. As a rule, after that, a transverse crack was formed, and the core was separated along the plane of the slot bottom.

Figure 2 gives photographs of the marble block surface before coring, after core removal and the core taken using the drill bit with the diameter of 30 mm. Then the cores were sawed and polished on the axial plane being perpendicular to the wedge tool plane.

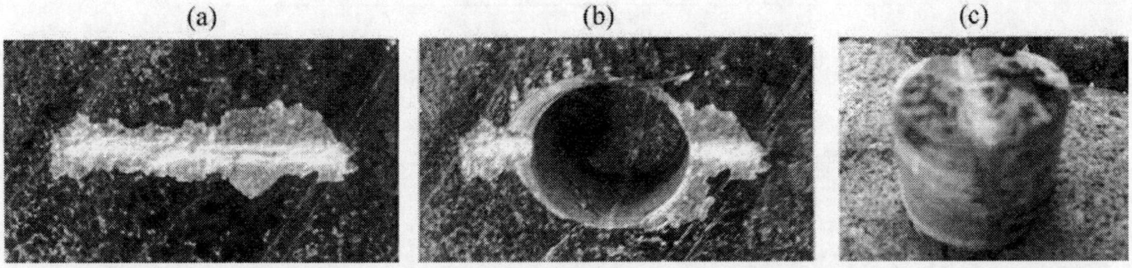

Figure 2. (a) Surface of marble block after wedge indentation; (b) core removal; (c) core.

On the polished surface of the cores, visible are cracks both existing in the block and formed by the wedge tool. Figure 3 offers photographs of the cut sections of the cores, and it is seen that it is rather difficult to determine the length of a crack induced by the impact against the background of the existing fracturing of natural stone.

For example, Figures 3a and 3c show almost an ideal picture—the impact-induced crack merges with none of the natural cracks, and is easy to measure its length. The situation is different in Figures 3b, 3d and 3e—it is very difficult to distinguish a small wedge-induced crack against the main crack exposed on the surface after the impact. In this case, the information recorded during the tests may be helpful. For instance, records of accelerometer installed in the tool display some special features during crack growth.

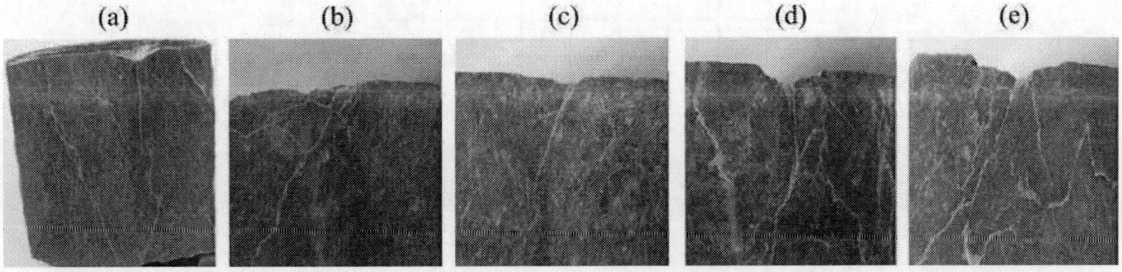

Figure 3. Photographs of polished sections of cores after impacts with an energy, J: (a) 68 (sample blow); (b) 33.6; (c) 68.7; (d) 132.5; (e) 140 (total energy of blows).

Figure 4 presents records of two close-energy blows. By oscillogram *3* of the piston velocity and by oscillogram *1* of the piston acceleration, it is clear that the second blow is more intensive than the first. The fundamental difference of the blow is observable in oscillogram *2* recorded by the acceleration indicator installed on the impact tool. The record of the tool acceleration in penetration is approximately similar in case of both blows. Then, the record represents a so-called "tail" that is a signal generated by a growing crack. Under blow *1*, the "tail" is at the level of noise, which is indicative of the fact that no large crack is generated by this blow. Blow *2* produces a high-amplitude "tail," which implies grow of a large crack under the wedge tool. These records allow a qualitative conclusion to be drawn on initiation or propagation of a wedge indentation-induced crack. This makes it possible to eliminate inference on a long induced crack in case of a low energy blow and to treat such crack as existing, or to assume a short crack as a blow-induced event. Close inspection of the magnified polished sections brings the same conclusion. In the latter case, the color of a fresh crack differs slightly from the old crack color.

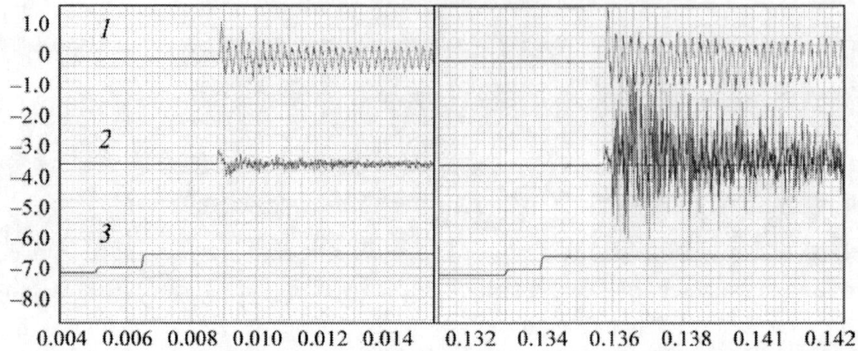

Figure 4. Acceleration oscillograms of piston (*1*), wedge tool (*2*) and light sensor (*3*).

Merging of fresh and existing large cracks (Figure 3a) complicates determination of length of the wedge-induced crack. In this case, the new crack length is assumed as an average length of the cracks measured on the two halves of the core. In Figure 3e, there is one more obstacle for determination of an induced crack length. At a small angle to the core axis (an extension of the blow direction line), two main cracks propagate with opening visible when magnified. In this instance, the crack length was taken as the maximum values of measurements on the two halves of the core.

Geometrical parameters of craters, measured before coring, were used to plot the relationship of the crack length, crater depth and blow energy (Figure 5).

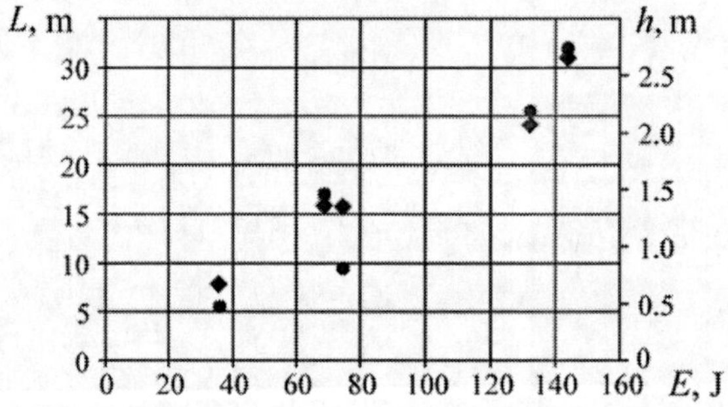

Figure 5. Relationship between crack length, crater depth and energy of blow.

In Figure 5 the characteristics of the external and internal fracture of the marble block under impact. Apparently, there is direct correlation between them, and the scatter of points on the flat is higher for the measured cracks than for the formed craters. Off the general tendency of proportional increase in the fracture characteristics with the blow energy is only a single point at which the total blow energy is ~ 74.4 J. An explanation is that the length of the induced crack in this experiment is assumed to be the thickness of shatters along the impact line as the core has fallen apart during coring. Naturally, this is a rough low estimate. One the objectives of this study is determination of fracture characteristics after multiple impacts and comparison with the fracture characteristics after a unit impact. To this effect, the points on the flat in Figure 5 are grouped by pairs at the two-fold and four-fold energies of unit impact. Nothing can be said about the double blow, for the core has disintegrated. In case of the quadruple blow (total blow energy ~ 144.1 J), fracture efficiency is higher in case of the multiple blows than under unit blow (energy ~ 132.4 J) approximately of the same energy. This proves the conclusion drawn in [8].

3. Conclusions

In the experimental investigation of impact-induced fracture of rocks by a hard indenter, the procedure of core sampling, sawing, polishing and inspection makes it possible to obtain quantitative characteristics of induced cracks—length and path.

The test results show rock fracture under impacts by the piston–wedge tool system is mainly governed by the impact energy transferred to rocks. The integrated recording of the impact action process and fracture parameters inside and outside a rock block offers the required information for proper modeling impact fracture.

Under multiple blows at the same point, fracture efficiency is higher as against unit blow in case that the total energy of multiple blows is equal to the unit blow energy.

Acknowledgements
This study was supported in the framework of Basic Research Program, Project No AAAA-A17-117122090003-2.

References

[1] Hill R, Lee EH and Tupper SJ 1947 The theory of wedge indentation of ductile materials *Proc. Roy. Soc., Lond., Ser. A* Vol 188 No 1013 pp 273–290

[2] Petryk H 1980 Non-unique slip-line field solution for the wedge indentation problem *J. Mech. Applic*. Vol 4 No 3 pp 255–282

[3] Alehossein H, Detournay E and Huangan H 2000 Analytical model for the indentation of rocks by blunt tools *Rock Mech. Rock Engg*. Vol 33 No 4 pp 267–284

[4] Chen LH and Labuz JF 2006 Indentation of rock by wedge-shaped tools *Int. J. of Rock Mech. and Mining Sciences* Vol 43 No 7 pp 1023–1033

[5] Magnenet V, Auvray C, Djordem S and Homand F 2009 On the estimation of elastoplastic properties of rocks by indentation tests International *Journal of Rock Mechanics & Mining Sciences* Vol 46 pp 635–642

[6] Hao Zhang, Haipeng Song, Yilan Kang, Ganyun Huang and Chuanyong Qu 2013 Experimental analysis on deformation evolution and crack propagation of rock under cyclic indentation *Rock Mech. Rock Eng.* Vol 46 pp 1053–1059

[7] Mayrer WC and Rinehart JS 1960 Impact crater formation in rock *L. of App. Phys.* Vol 31 No 7 pp 1247

[8] Paul B and Sikarskie DL 1965 A preliminary theory on static penetration by a ridge wedge into brittle rock material *Transaction of SME-AIME* Dec pp 372–383

[9] Kolesnikov YuV and Morozov EM 1989 *Mechanics of Contact Fracture* Moscow: Nauka (in Russian)

[10] Hao Zhang, Haipeng Song, Yilan Kang and Ganyun Huang 2013 Experimental investigation on deformation and failure of rock under cyclic indentation *Proc. 13th International Conference on Fracture* June 16–21 Beijing, China

[11] Sher EN, Gorodilov LV and Efimov VP 2017 The technique of experimental investigation of crack formation in the fracture of rock mass by solid indenter *Development Trends in Sience and Education: Proc. 29th Int. Conf.* Part 3 Samara: L-Zhurnal NITS pp 22–25 (in Russian)

[12] Gorodilov LV, Efimov VP and Kudryavtsev VG 2016 Experimental research of marble block fracture under action of hammer head–impact tool percussion system *J. Fundament. Appl. Min. Sci.* Vol 3 pp 44–51

[13] Kyu NG 2011 Particular issues associated with fluid fracturing of rocks by plastic materials *Journal of Mining Science* Vol 47 No 4 pp 450–459

IOP Conf. Series: Earth and Environmental Science **262** (2019) 012024 doi:10.1088/1755-1315/262/1/012024

Improvement of drilling efficiency in underground mines in Russia

VN Karpov*, VV Timonin, AI Konurin and EM Chernienkov

Chinakal Institute of Mining, Siberian Branch Russian, Academy of Sciences, Novosibirsk, Russia

E-mail: *karpov@misd.ru

Abstract. Russian and foreign equipment and technologies of drilling are reviewed. The integrated assessment of blasthole drilling energy efficiency is carried out in terms of air-powered hammer P-160-5,5 SHV operation with drilling rig BP100P in a mine. Mechanical, run and volume velocities of drilling are determined using the data of video and acoustic processing of drilling process records. The cost of drilling is evaluated by the integral energy criterion of volumetric destruction of rocks. The analytical estimation of production rate is presented alongside with the methods of improvement. The reasons for the decline in the productivity of drilling operations in domestic mines have been established, and methods for improving it have been proposed. It is shown that energy efficiency and energy source cost should be use as the criteria when selecting drilling machines together with drilling productivity.

1. Introduction

In hard mineral mining, the bulk amount of drilling falls at blastholes which can reach tens and hundreds thousand meters of length annually in mines. In drilling in high-strength rocks, both surface and underground mines widely use rotary–percussive method and down-the-hole air hammers designed in Russia [1].

Methods of drilling in hard rocks with strength higher than 100 MPa feature flexibility in operation, which allows drilling of blast, production, drain, geological, degassing, construction and rescue holes. In drilling-and-blasting, diameters of full well bottom drilling may reach 100–254 mm to a length of 50 m in surface mines and 76–216 mm to a length of 150 m in underground mines. The drilling depth of geological and production wells has grown to 2000 m [2]. The leading role of pneumatic drilling over the other techniques was proved in San Jose mine in rescue of Chilean miners from the depth greater than 700 m [3].

Equipment and technology of rotary percussive drilling are steadily and rapidly developed all over the world. Such companies as Atlas Copco, Sandvic and other design new-generation drilling systems with three automation levels focused at determination of physical–mechanical properties and mineral composition of rocks, as well as air-powered hammers and rock-breaking tools adaptable to physical and mechanical properties of rocks. From 5 to 9 types of drill bits are available for the same size air hammer, etc. Of specific concern is sound power consumption of drilling and training of drilling rig operators to better understand mechanism of rock failure [4–8].

The situation is different in Russia. Series-produced air-powered drilling machines and tools for underground mines fall behind their foreign analogs by 30 years and more. In the meanwhile, the cost of blasthole drilling has risen 2.5 times for the resent 10 years, and productivity of of drilling has dropped 1.5–2 times. Parameters set in the specifications of DTH air hammers differ from the factual performance by 30–50% in terms of capacity and 2–4 times in terms of life, and the cost of drill bit approaches the cost of air hammers due to the use of foreign-manufacture alloys. In drilling in rocks with strength of 140 MPa, domestic hammers powered by low-pressure compressed air (0.5–0.7 MPa), as compared with the foreign high-pressure air hammers (1.6–3 MPa), have 5–10 lower penetration rate, 15–30 shorter life and 10–50 shorter life of drill bits. For these reasons, the drilling rigs of domestic manufacture are gradually pushed by the import models [9, 10]. In connection with this, Russia calls for the integrated R&D projects targeted at improvement of energy efficiency of the pneumatic drilling in underground mining [11].

The problems of drilling productivity effectivization in Russian mines start with the absence of exact data on physical and mechanical properties of rock mass in current drill sites. The many years long experience gained by the members of the Institute of Mining, SB RAS in field works shows that geological exploration data used in open pit and underground mines date back from 20 to 45 years and often fall short of the actual data.

A detail geological characteristic of drilling layouts with mapped data on physical and mechanical properties of rocks in horizontal and vertical cross sections will bring drilling and blasting operations to the next quality level, will enable reasoned adaptation of machines and tools to specific rocks, as well as will assist in organization and evaluation of measures aimed at improvement of drilling efficiency.

Currently, an ample knowledge has been accumulated both on geological structure and physical–mechanical properties of rocks. At the same time, in view of a dim connection between innovations in geomechanics and management of optimization and risks in drilling, it is required to develop a strategy for implementing advanced methods and approaches in planning, design and upgrading of underground mines [12, 13]. Undoubtedly, such a strategy needs decision-making based on the knowledge geology and geomechanics. Model representation of a mineral deposit is a must [14, 15]. Furthermore, the level of safety should be sufficient to minimize all process and financial risks.

To date there are many efficient software packages (mostly, foreign design) on geological modeling and frame representation of mines. On the other hand, the interaction between a geological model and planning and optimization of drilling and blasting is scarcely studied thought there are certain approaches to its determination [16, 17]. The next stage of advancement in software should embrace geological and structural aspects based on analysis of mining geodesy information and drawing. These should be complete intelligence, including geographic location, topographical relief, morphology and lithology of rock mass, tectonics (tectonotype, faults, folds, etc.), jointing, fracturing (main systems of joints, dominant angles and azimuths of dip, jointing density, spacing and lengths).

Yet, availability of data on physical and mechanical properties of rocks is not always a guarantee of optimum drilling mode selected when there no specific recommendations in case of variable geological and geotechnical conditions. Figure 1 illustrates some of the typical problems in operation of foreign machines (Figure 1a) in Sheregesh mine, Gorno-Shorsky division of Evrazruda: oversizes (Figure 1b), excessive wear of expensive rock-breaking tools and deliverables (Figures 1c–1e).

Operation of such equipment results in longer working shifts up to 10 hours, out of which 2–3 hours are taken to troubleshooting in drilling. An often cause of downtime is breaks of energy source feed lines when drilling rigs and load–haul–dumpers travel, which calls for extra measures to ensure occupational safety. Thus, selection of mining systems and equipment should be timely and duly preceded by sound feasibility study.

The energy efficiency of DTH air hammers should be evaluated starting with the hardest-to-drill rocks. A unique local reference of strength and abrasivity is often selected to be ore under mining. For geological composition of enclosing rocks is diverse, they should be divided into groups with quasi-uniform physical and mechanical properties. For example, the Irba iron ore field is composed of 17 types of ores and rocks (Figure 2) [18].

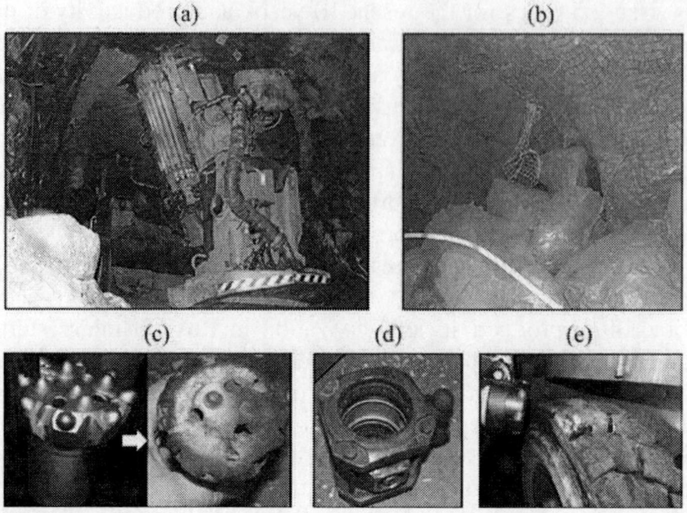

Figure 1. Troubles in operation of foreign equipment in Russia.

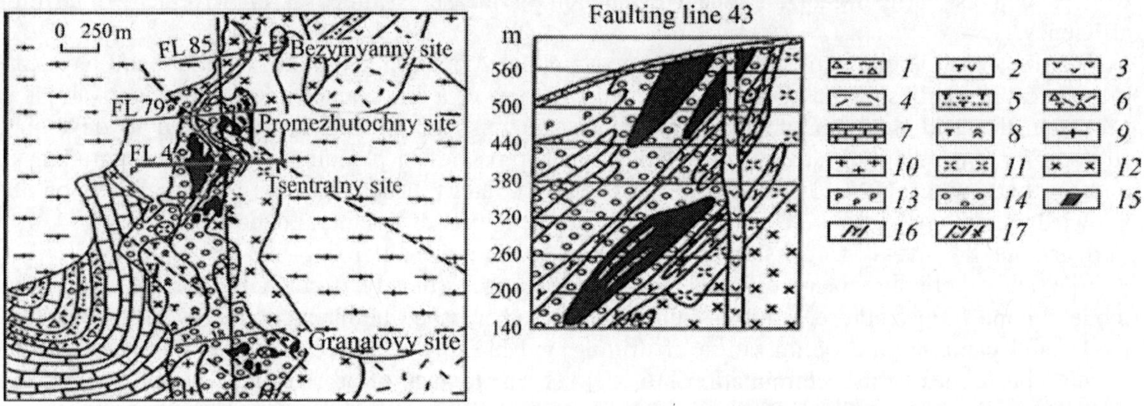

Figure 2. Geological map and profile of the Irba field: *1*—diluvium–alluvium; *2*—andesite porphyritic tuffs; *3*—andesite porphyrite, *4*—felsites, dacite, albitophyre; *5*—sand tuff, tuffoaleurolite; *6*—tuff breccia, tuff conglomerates, acid-composition tuff lava; *7*—marmorized limestone; *8*—orthophyritic and trachitic tuffs; *9*—granosyenite, plagiogranite; *10*—granite; *11*—syenite–diorite; *12*—diorite, gabbro, granodiorite of the Olkhovian intrusive formation; *13*—pyroxene–amphibole and epidote–albite–amphibolic hornfel; *14*—skarn; *15*—magnetite ore; *16*—syenite–porphyrite and orthophyre dykes; *17*—diabase and diorite porphyrite dykes.

When ore and enclosing rocks is grouped according to physical and mechanical properties and correlated with geological maps, it is possible to reasonably selects sites for rate setting and make recommendations on drilling and blasting equipment and technology. In case of additional exploration, the updated information on strength and jointing should duly be introduced on geological maps of drilling areas. Identification of the increased jointing zones along the horizontal and vertical of drilling areas allows anti-accident recommendations on reduction of air hammer power on bottomhole.

2. Evaluation of energy efficiency in blastholes drilling

The technical reasons of drop in the drilling efficiency in mines in Russia are similar in many ways. Below in this paper, the authors perform an integrated evaluation of energy efficiency in drilling blastholes with a diameter of 160 mm and 56 m long by air hammer P-160-5.5 ShV using drilling rig BP100 in an underground mine. The scope of the evaluation embraced the cost of energy source consumption per 1 m of drilling (without regard to water and air consumption by rotation and feed rive system of the rig) and the integral volumetric rock destruction energy criterion *k*, which totally correlates with the cost of energy source in drilling on the bottom. The criterion is calculated from the formulas:

$$k = k_c \eta , \quad (1)$$

where k_c is the criterion factor [19, 20]; η is the air-driven hammer efficiency;

$$k_c = \frac{E_c(t)a}{V_\rho(t)\rho v_\rho^2} , \quad (2)$$

where $E_c(t)$ is the air energy fed to the drilling rig, J; *a* is the estimated coefficient of seismic effect of underground blasts [19]; $V_\rho(t)$ is the volume of broken rock in the control section of drill hole, m^3; ρ is the rock mass density in the drilling site, kg/m^3; v_ρ is the P-wave velocity, m/s.

The air energy fed to the drilling string is given by:

$$E_c = W_c \tau , \quad (3)$$

where W_c is the energy source capacity, Wт; τ is the duration of rotary–percussion drilling in the control section of hole, s.

The value of W_c is found from the formula:

$$W_c = \frac{RT}{\mu} Q_{\mathrm{m}} \ln \frac{p}{p_{\mathrm{atm}}} , \quad (4)$$

where R = 8.31 is the universal gas constant, J/mole·K; T is the absolute air temperature, K; μ is the molecular weight of air, μ = 0.02896 kg/mole; Q_{m} is the mass air flow, kg/s; p is the rated pressure of the air hammer, MPa; p_{atm} is the atmospheric pressure, MPa.

The mass air flow per second is obtained from the relation:

$$Q_{\mathrm{m}} = \frac{Q_{\mathrm{rate}} \rho_{\mathrm{air}}}{60} , \quad (5)$$

where Q_{rate} is the air flow rate, m^3/min; ρ_{air} is the air density at the normal atmospheric pressure, kg/m^3.

The calculation data on magnetite ore are compiled in Table 1; P-wave velocity is assumed as 3100 m/s. the geology of block No. 32 was uncorrelated with the drilling site, and drill cuttings were therefore take for the analysis for additional exploration.

Table 1. Geological characteristic of rocks in block No. 32.

Ore and rock types	Strength, MPa	Mass, kg/m^3	Jointing category	Joint set spacing, m
Grey magnetite ore	150	3950	III/II*	0.8 / 0.3
Pecky banded chlorite–magnetite–granate–epidotic skarn	130	2950	III/I	0.65 / 0.1<
Banded chlorite and sheet–plate shale	110	2720	III/I	0.75 / 0.1<
Pinky–grey syenite	100	2730	III/I	0.7 / 0.1<
Light-green feldsparic porphyre dyke	100	2830	II	0.25
Dioritic porphyre dyke	90	2830	II	0.25

*After the slash there is the jointing category for rocks nearby Vostochny tectonic fault.

Inasmuch as the common integral criterion of drilling is the cost of 1 m drilled, we calculated the cost of air spent by the air-driven hammer. The air cost was assumed 1 Rub/m^3. At the air hammer consumption of 14.9 m^3/min, the cost of the machine operation made 894 Rub/h, or 0.248 Rub/s. The cost of water consumption and air taken by rotation and feed drives was disregarded. Efficiency of the discussed-type air hammers within the pressure limits of 0.5–06 MPa totaled $\eta \cong 0.11-0.12$. In view of the fact that efficiency of the hammer model P-160-5.5 SHV was not determined in bench-test and drilling was carried out within the mine mains pressure limits of 0.48–0.58 MPa, the efficiency was assumed to be $\eta = 0.12$. At W_c = 45.7 kW (T = 295 K, $\rho_{\text{B}} = 1.2466$ kg/m^3, p = 0.58 MPa), capacity of the air hammer model P-160-5.5 SHV is W = 5.5 kW. According to the specifications, this air-driven hammer has the capacity W = 5.5 kW at the pressure of 0.5–0.7 MPa [21]; thus, the boundary conditions at η = 0.12 are set correctly. When p = 0.48 MPa, it is assumed that air flow rate is 14 m^3/min, accordingly, W_c = 38.3 kW and W = 4.6 kW. The minimum energy input of rock drilling is reached at $k = 1–4\cdot 10^{-9}$.

The first meter drilling results were excluded from the calculation, for auxiliary operation connected with collaring were performed. Drilling involved reused drill bit. The mine air feed pressure was 0.58–0.6 MPa always, except for one shift when drilling pressure reduced to 0.48 MPa, which ended with an emergency situation.

Table 2 presents calculated data for a hole 55 m long with the determination internal of 5 m; for the shift with emergency situation in the hole section from 35 to 40 m with the determination interval of 0.75 m; and for the rod section of 0.3 m during drilling of which the emergency took place with the determination interval f 0.3 m. The air consumption cost was calculated for the specified check intervals L_c (index C_{check}) and per 1 m in each interval (C_{air}). The penetration rates V_p were determined based on video and acoustic processing of the drilling process records [20].

Table 2. Calculated energy efficiency of drilling with air hammer P-160-5.5 SHV and drilling rig BP100 for the hole with diameter 160 mm and 56 m long.

Check interval length L_c = 5 m											
L, m	5	10	15	20	25	30	35	40	45	50	55
p, MPa	0.58	0.58	0.58	0.58	0.58	0.58	0.58/0.48	0.58	0.58	0.58	0.58
V_p, m/min	0.057	0.05	0.047	0.042	0.038	0.035	0.023	0.03	0.029	0.025	0.023
k_c, 10^{-7}	1.28	1.47	1.55	1.74	1.93	2.09	2.58	2.37	2.52	2.93	3.09
k, 10^{-8}	1.54	1.76	1.87	2.09	2.31	2.51	3.1	2.84	3.03	3.52	3.71
C_{check}, Rub	1296	1484	1573	1761	1949	2117.5	3116.5	2395	2550	2968.5	3125
C_{air}, Rub/m	259.2	296.8	314.6	352.2	389.8	423.5	623.3	479	510	593.7	625
Check interval length L_c = 0.75 m (35–40 m)											
L, m	0.58	0.58	0.58	0.48	0.48	0.48	0.48	—	—	—	—
p, MPa	0.75	1.5	2.25	3	3.75	4.5	5.25	—	—	—	—
V_p, m/min	0.04	0.038	0.037	0.016	0.018	0.02	0.021	—	—	—	—
k_c, 10^{-7}	1.54	1.62	1.66	3.84	3.41	3.07	2.92	—	—	—	—
k, 10^{-8}	1.84	1.94	1.99	4.61	4.09	3.69	3.51	—	—	—	—
C_{check}, Rub	278.3	293	300	696	618.4	557	530	—	—	—	—
C_{air}, Rub/m	371	390.5	401.1	927.6	824.6	742.1	706.7	—	—	—	—
Check interval length L_c = 0.03 m											
L, m	0.03	0.06	0.09	0.12	0.15	0.18	0.21	0.24	0.27	0.3	—
p, MPa	0.02	0.021	0.019	0.019	0.019	0.014	0.017	0.0125	0.013	0.01	—
V_p, m/min	2.97	2.9	3.11	3.28	3.17	4.2	3.58	4.91	4.47	5.8	—
k_c, 10^{-7}	3.56	3.48	3.73	3.93	3.81	5.04	4.3	5.9	5.36	6.96	—
k, 10^{-8}	21.6	21.08	22.568	23.808	23.064	30.504	26.04	35.712	32.488	42.16	—
C_{check}, Rub	712	695.6	744.7	785.6	761.1	1006.6	859.3	1178.4	1072.1	1391.2	—

Figure 3 illustrates energy efficiency of drilling with air hammer P-160-5.5 SHV beyond the range of $k \in (1 \div 4) \cdot 10^{-9}$. This is typical of low-pressure and high-rate air hammers with low energy performance [21, 22].

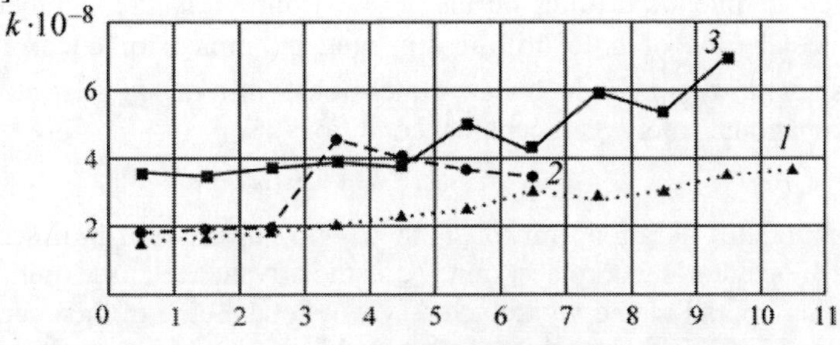

Figure 3. Drilling energy efficiency of air hammer P-160-5.5 SHV by *k*: *1*—hole 55 m long; *2*—hole section of 35–40 m; *3*—emergency section 0.3 m long.

The volumetric rock destruction criterion *k* and the criterion energy factor k_c are identical in terms of rock–machine feedback and correlated with the energy source consumption cost $C_в$. For this reason, it is simple to evaluate cost of compressed air of drilling on the bottom by deviation Δ of *k* or k_c. Figure 4 depicts energy efficiency evaluation by *k*.

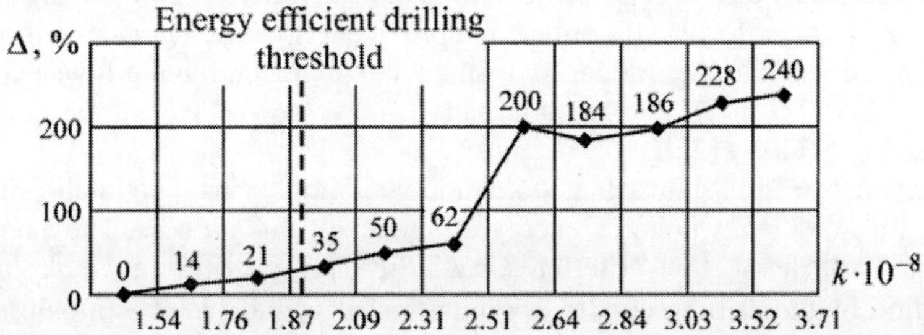

Figure 4. Evaluation of drilling energy efficiency with air hammer P-160-5.5 SHV.

Considering the data of earlier studies focused on evaluation of energy efficiency in air-powered hammer drilling by the direct cost *C*, as well as based on the international experience gained in operation of drilling machines and rock-breaking tools, the energy efficiency range of pneumatic drilling by $k(k_c)$ is Δ = 15–30% [23–25]. The dash line in Figure 4 marks the point where it is required to eliminate the causes of decline in energy efficiency of drilling with air hammer P-160-5.5 SHV. With regard to the specific character of long blasthole drilling efficiency in mines, the range of 30–50% is tolerated in holes having reached 2/3 of their length as multi-trip drilling results in a drop of drilling productivity and in the drilling cost escalation.

The research finds that the major case of lower productivity of long blasthole drilling with the air hammer model P-160-5,5 SHV with a diameter of of 160 mm and drilling rig BP-100 is the loss of percussive capacity of the hammer due to:

—increased water consumption in composition of air-and-water mixture (by a factor of 1.5);
—lack of sealing rings in drill rods;
—use of small-size drill rods ($L = 0.75$ m);
—unavailability of sharpening for tungsten carbide insets of drill bits;
—long operation f drill bits with worn inserts (one drill bit per 3–4 holes).

During the experiments, standard capacity was 7 meters of drilling. Using the data on actual time of drilling the check sections in the hole, the other standard machine capacity is calculated as:

$$H_{\mathrm{st}} = \left(\frac{T_{\mathrm{act}} \cdot i}{t_{\mathrm{act1}} + t_{\mathrm{act2}} + t_{\mathrm{act2}} \cdots + t_{\mathrm{act}n}} \right) k_{\mathrm{cor}} , \tag{6}$$

where T_{act} is the actual time of drilling during a shift, min; i is the number of check sections; $t_{\mathrm{act1}}, \dots t_{\mathrm{act}n}$ is the actual time of hole drilling (in open pit mines) or check sections of holes (in underground mines) with the known linear dimensions, min/m; k_{cor} is a correction factor of standardization (occupational safety, geotechnical conditions);

$$T_{\mathrm{act}} = T_{\mathrm{shift}} - (T_{\mathrm{rt}} + T_{\mathrm{serv}} + T_{\mathrm{pn}}) , \tag{7}$$

where T_{shif} is the overall shift duration, min; T_{rt} is the rigging up and tail-in time, min; T_{serv} is the time of servicing of the drill operator's workplace, min; T_{pn} is the private need time, min.

The time of rigging up, tail in and workplace servicing totals 9.5% of the overall shift duration T_{shift} (420 min), checkup and safing of workplace—10 min; lubrication, fore-actuation and minor repair of drill rig—21 min; rigging up—4 min; cleanup of drill operator workplace—5 min; time for private needs and rest—10 min, $k_{\mathrm{cor}} = 1$.

From (6) the standard machine capacity $H_{\text{в}}$ = 9.73 m, which means that prompt replacement of drill bits after drilling of two holes (100–112 m) with air hammer P-160-5.5 SHV, the minimum increment in the productivity of drilling (other failure causes unremoved) will make 39%. After all revealed trouble causes have been eliminated, the anticipated increase in the drilling productivity will be 12–14 meters at the monthly saving of money in amount of up to 2 million Rubles (at values of 2016). In the period of introduction of drill rigs BP100 and DTH air hammers P-160-5.5 SHV, such productivity per shift was assumed as average [26]. In order to improve energy efficiency of drilling in magnetite ore, it is required to replace this hammer model by a smaller size analog having lower air flow rate and the same capacity as P-160-5.5 SHV. An alternative may be DTH air hammer model P-150S (Q_{air} = 10 m^3/min, W = 6.3 kW) [27].

The implemented research shows that potentiality of improving pneumatic drilling productivity should be analyzed from the viewpoint of energy efficiency of this process. The express-estimation can use the volumetric rock destruction criterion k and the energy criterion k_e. As against the energy efficiency evaluation by the drilling cost C, computation of k_e andи k takes much less time and needs no economic figures though correlates with them (in terms of energy source consumption). Furthermore, the criteria k_e and k enclose physical and mechanical properties of rocks and, unlike C-index, allow exact and reliable determination of drilling energy efficiency over any time interval. The index C lacks that stability, for the cost of energy sources and the other indexes are fluid. It is very difficult to analyze decline or growth of pneumatic drilling energy efficiency in different years without conversion factors for all indexes involved in the calculation of drilling cost.

3. Conclusions

The review of the drilling technologies and equipment shows that introduction of heavy drill rigs of foreign manufacture in Russian mines is often economically inexpedient as the cost of energy sources (electricity, water, diesel fuel) rises essentially in this case.

The integrated assessment of energy efficiency in blasthole drilling in mines in terms of DTH air hammer model P-160-5.5 SHV and drill rig model BP 100 has discovered that the major cause of low productivity is the loss of the impact capacity of the hammer to to increased water flow rate in the air-and-water mixture, lack of sealing rings in drill rods, unavailability of sharpening for tungsten carbide inserts of drill bits, use of short drill rods and long operation of drill bits with worn inserts. From the evidence f the analyzed drilling costs and machine standard capacity calculations, it follows that elimination of the revealed causes will result in the anticipated improvement of drilling productivity not less than by 40%.

The undertaken studies show that the energy efficiency and the energy source cost should be the criteria for selecting drilling equipment alongside with the criterion of drilling productivity. The express-assessment of energy efficiency in drilling with DTH air hammer is advisable to perform using the volumetric rock destruction criterion and the energy criterion.

Development of detailed flow charts, including geological characterization of drilling sites with indicated physical and mechanical properties of rocks along the vertical and horizontal will enable the next level performance of drilling and blasting. Moreover, it will become possible to use the same drilling machines and tools in the drilling site with analogous characteristics and distribute the equipment between scheduled drilling sites with regard to the wanted number of drilling machines and rock-breaking tools.

Acknowledgements
This study was supported by the Russian Science Foundation, Project No 17-17-01282.

References
[1] Emelyanov PM, Easin NN, Zinoviev AA, Semenov LI and Suksov GI 1965 *DTH Hammer Drilling Machnies for Mines* Novosibirsk: SO AN SSSR (in Russian)
[2] Poderni RYu, Zamyshlyaev VF and Prasolov SK 2012 Current structures of drill rigs for open pit mines *Gornye Nauki i Tekhnologii* No 12 pp 100–113
[3] Nell Lukosavich 2010 Chilean miner rescue: The drilling team that defied all odds *World Oil* Vol 231 No 12 pp 1–6
[4] Marcelo Prado 2013 Autonomy in the Andes *Mining & Construction* No 2 pp 34–35
[5] Josef Pappenreiter 2016 Bohrgigant am Erzberg *Baublatt. Österreich* No 2 pp 50–51
[6] Brian Fox 2013 Insight of automation *Mining & Construction* No 2 pp 32–33
[7] Brian Fox 2011 *Blasthole Drilling in Open Pit Mining* Garland Texas USA: Atlas Copco Drilling Solutions LLC
[8] Li J-G 2018 Intelligent mining technology for an underground metal mine based on unmanned equipment *Engineering* Vol 4 Issue 3 pp 381–391
[9] Oparin VN, Rusin EP, Tapsiev AP, Freidin AM and Badtiev BP 2017 *International Experinece of Underground Mining Automation* Novosibirsk: SO RAN (in Russian)
[10] Smolyanitsky BN, Oparin VN, Denisova EV, Kondratenko AS, Tishchenko IV, Smolentsev AS, Khmelinin AP and Konurin AI 2016 *Modern Long Hole Drilling Technologies and the Path Control Facilities in Soil* Novosibirsk: SO RAN (in Russian)
[11] Kondratenko AS, Timonin VV, Karpov VN and Popelyukh AI 2018 Ways to improve rotary-percussive drilling efficiency *Gornyi Zhurnal* No 5 pp 63–68
[12] Kelly B 2013 Stress analysis for boreholes on department of defense lands in the western United States: A study in stress heterogeneity *Proceedings 38th Workshop on Geothermal Reservoir Engineering* California: Stanford University pp 139–150
[13] Reiter K and Heidbach O 2014 3-D geomechanical-numerical model of the contemporary crustal stress state in the Alberta Basin (Canada) *Solid Earth* No 5(2) pp 1123–1149
[14] Bashkov VI, Shtirtz VA, Eremenko AA and Konurin AI 2015 Assessment of geomechanical conditions of rockburst-hazardous rock mass under large-scale blasting in Gornaya Shoria *J. Fundament. Appl. Min. Sci.* Vol 2 pp 205–213 (in Russian)
[15] Freidin AM, Neverov SA, Neverov AA and Konurin AI 2016 Geomechanical assessment of geotechnology at a project stage of underground ore mining *Gornyi Zhurnal* No 2 pp 39–45
[16] Potvin Y and Giles G 2008 The development of a new high-energy absorption mesh *Australasian Institute of Mining and Metallurgy Publication Series* pp 89–94
[17] Balg C, Roduner A and Geobrugg AG 2013 Ground support applications *Int. Ground Support Conf.* Lungern, Switzerland: AGH University
[18] Eremenko AA, Eremenko VA and Gaidin AP 2009 *Geology and Geomechanical of Iron Ore Deposits in the Altay–Sayan Folded Zone* Novosibirsk: Nauka (in Russian)
[19] Oparin VV, Timonin VV and Karpov VN 2016 Quantitative estimate of rotary–percussion drilling efficiency in Rocks *Journal of Mining Science* Vol 52 No 6 pp 1100–1111

[20] Oparin VV, Timonin VV, Karpov VN and Smolyanitsky BN 2017 Energy-Based Volumetric Rock Destruction Criterion in the Rotary–Percussion Drilling Technology Imporvement *Journal of Mining Science* Vol. 53 No 6 pp 1043–164

[21] Lyukhanov VV 2010 *Down-the-Hole Air Hammer Drill P-160-5.5SH* Yekaterinburg: Mashinstroit. Holding (in Russian)

[22] Karpov VN 2016 Assessment testing procedure for downhole pneumatic hammers under production conditions *J. Fundament. Appl. Min. Sci.* Vol 3 No 2 pp 74–80 (in Russian)

[23] Timonin VV, Alekseev SE, Karpov VN and Chernienko EM 2018 Influence of DTH hammer impact energy on drilling-with-casing systen performance *Journal of Mining Science* Vol 54 No 1 pp 53–60

[24] Bo Presson 2012 How sharp rock drilling tools put money in the bank Mining & Construction No 3 pp 26–27

[25] *Sharpening Tools* Available at: https: http://www.mining-solutions.ru/ product/225

[26] Lyukhanov VV and Alferov SB 2010 Use of drilling tools and equipment produced by Mashinostroitelnyi Holding—A guarantee of drilling efficiency *Gornaya Promyshlennost* No 4 pp 38–42

[27] Smolyanytsky BN, Repin AA, Lipin AA, Alekseev SE et al 2013 *Improving Efficiency and Durability of Pulse-Generating Long Hold Drilling Machines for Rocks* Novosibirsk: SO RAN Issue 43 pp 32–38

IOP Conf. Series: Earth and Environmental Science **262** (2019) 012025 doi:10.1088/1755-1315/262/1/012025

Role of large mineral mining companies in resource regions: Finding tools of balanced interaction

ON Kavkaeva[1], EO Pakhomova[1], EV Goosen[1,2]*, SM Nikitenko[1] and GA Chakhoyan[2]

[1]Federal Research Center for Coal and Coal Chemistry, Siberian Branch, Russian Academy of Sciences, Kemerovo, Russia
[2]Kemerovo State University, Kemerovo, Russia

E-mail: *egoosen@yandex.ru

Abstract. This article is devoted to assessment of weight of large mining companies in resource regions of Russia in terms of Kuzbass. Performance of three Kuzbass companies is analyzed with demonstration of positive and negative economic effects. Obstacles for balanced development of the resource region are revealed. The key problem is interaction between regional authorities and large mining companies, as well as the enclave nature of regional economy. For the balanced development of resource regions, it is suggested to define and fix rules of play by means of a regulatory contract.

1. Introduction

Despite sanctions imposed by Western countries on Russia, the country's largest mining companies are scratching for new opportunities for business growth and expansion. Such opportunities may be offered in the regions the companies base. Resource regions are represented by vertically integrated companies (VIC) from mining and intermediate sectors and are characterized by steady replication of dual economy of enclave nature, i.e. economy with high-production export-oriented sector isolated from the rest of economy. The process of finding tools for development aims to abate impact of VIC on the resource region economy and to ensure VIC–resource region interaction to the benefit of mutual balance advancement.

The impact exerted by large exterritorial companies on their presence area is the subject of study for many Russian and foreign researches, including V.A. Kryukov, V.V. Shmat, V.I. Nefedkin, A.G. Shelomentsev, W.M. Corden, J.P. Neary, J. Saks, A. Warner et al. Mining companies in the fuel-and-energy sector are the largest tax payers and employers; they form vertical coordination and displace horizontal management based on interaction of equal subjects on the market [1]. Moreover, resource regions differ in "make-up" of the leading sector. For instance, the Kemerovo Region relies on the coal mining industry unlike many other regions where the chief role belongs to oil-and-gas companies. Fluctuations in coal mining are mirrored by the regional economy. Here, we should emphasize distinctions of the coal industry as against the oil and gas sector: first, up 97% of coal is produced by large mines; second, almost all coal mines are private companies; third, production is concentrated and labor efficiency grows year by year; fourth, coal export expands while internal consumption remains on the same level; fifth, coking coal is totally subjected to preparation while cumulative processing of power-generating coal is low (30% in 2017).

2. Analysis of activities of coal mining companies in the Kemerovo region

The influence of coal mining companies on regional economy was assessed based on the analysis of their delivery chains. To this end, Russia's three larges coal-mining companies operating in the Kemerovo region were selected: SUEK, SDS Ugol and Evraz.

In 2016 SUEK produced 105.5 Mt of coal, ore 27% of overall coal production in Russia (1st place). Inside the Kemerovo Region, the company produced 37.7 Mt of coal (1st place among the other areas of presence). Aside from Kuzbass, the company operates large surface and underground mines in other regions of Siberia and Far East of Russia [2].

SDS Ugol Holding produced 28.6 Mt of coal in 2016, which made 7% of total coal production in Russia (3rd place). The company operates entirely in Kuzbass.

Evraz produced 22.1 Mt of coal in 2016, which was 6% of overall coal production (5th place in Russia). The company's output in the Kemerovo region was 22.5 Mt [3].

Below, we present the supply chains of the companies, their types and effects on the regional economy.

SUEK is a vertically integrated company; it holds control over the whole cycle—from coal extraction, preparation and shipment to marketing all over the world via network of the manufacturer's representatives (Figure 1).

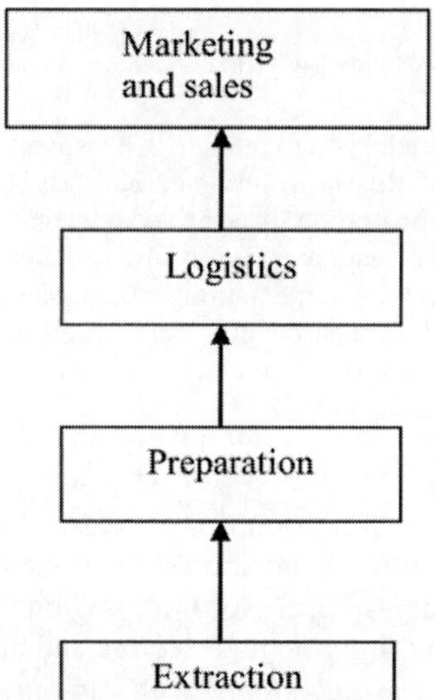

Figure 1. Supply chain of SUEK.

SUEK possesses port and railway infrastructure to deliver products to consumers in Russia as well as in the Atlantic and Asia–Pacific countries. The global supply chain of SUEK has adverse effect on the areas of the company's presence as it shapes enclaves.

SDS Holding is a political conglomerate formed by way of acquisition of effective different-branch business using managerial assistance rather than through the business development. In the resultant supply chain, the company has a control center in all divisions (Figure 2). SDS Ugol carries out extraction, preparation and haulage of coal; some part of coal goes to SDS Energo. Thus, this supply chain represents coal assets within a conglomerate. This variant is the most productive for the region.

Evraz is a vertically integrated mining and metallurgical company. It possesses its own reserves of iron ore and coking coal to meet the internal requirements nearly in full. The supply chain of Evraz is coal assets within a metallurgical company (Figure 3).

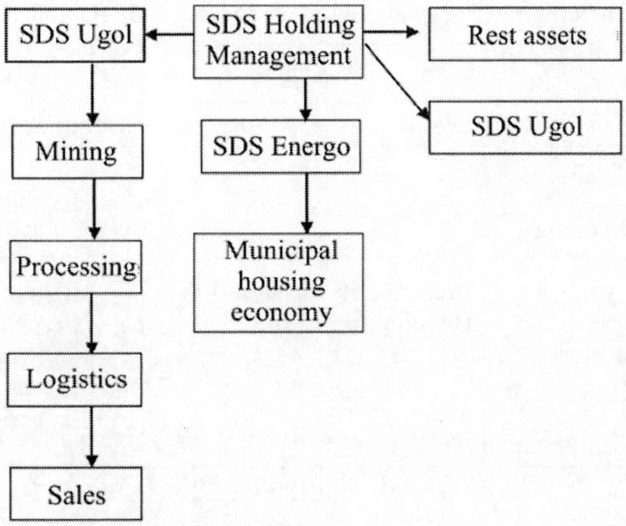

Figure 2. Supply chain of SDS Ugol.

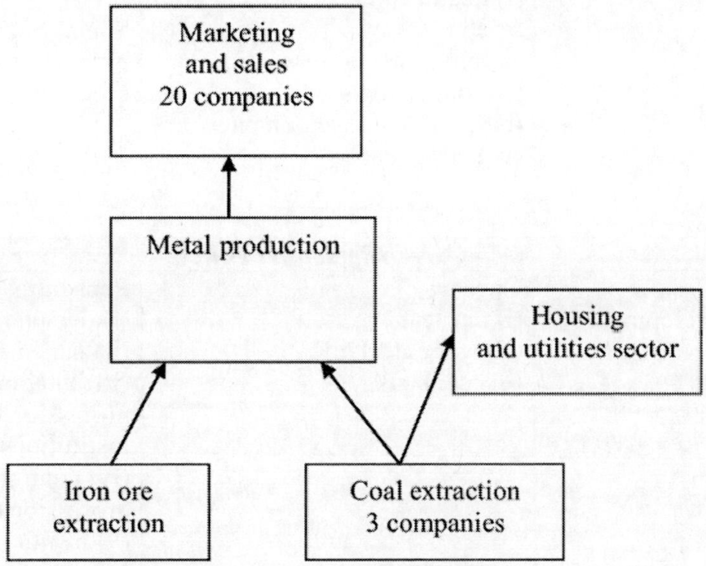

Figure 3. Supply chain of EVRAZ.

The latter supply chain is the most admissible variant for the regional economy as the company creates markets for its coal divisions.

Apart from negative impact, large- exterritorial companies have positive potential. The potential is generally represented by four blocks: (1) nature preservation/ecology; (2) social responsibility; (3) career; (4) cooperation agreement with territorial, regional and local authorities (table 1).

In terms of Evraz company, Health, Safety and Ecology block includes: (i) strategy; (ii) health and safety; (iii) environment; (iv) international safety standards.

Charitable Projects block is (i) children; (ii) youth; (iii) education; (iv) environment; (v) sports.

Our Fellows block is comprised of (i) personnel development strategy; its principles are: competitive wage for the best productivity, 100% personnel engagement in development programs, in-house specialists to take new vacancies; (ii) Evraz youth; (iii) professional advancement.

Table 1. Interaction between large coal mining companies and regional authorities in Kuzbass [2–4].

Evraz	SUEK	SDS Holding (SDS Ugol)
Ecology/Nature Preservation/Occupational Safety		
Health, safety and ecology: (i) strategy (ii) health and safety (iii) environment (iv) international safety standards	**Included in Sustainable Development block**	**Ecology:** (i) ecological management/philosophy (ii) environmental impact (iii) biodiversity preservation (iv) control blasting (v) basic aspects of activities (iv) contacts
Social Responsibility/Social Policy		
Charitable projects: (i) children (ii) youth (iii) education (iv) environment (v) sports	Sustainable development (i) health and occupational safety (ii) environment protection (iii) our fellows (iv) regional development (v) social reports	**Social policy:** No sections and data in this block
Career/Personnel Policy		
Our fellows: (i) personnel development strategy (ii) Evraz's youth (iii) professional development	**Career:** (i) our team (ii) work at SUEK (iii) contacts	**Personnel policy:** (i) personnel monitoring (ii) personnel reserve (iii) interaction with all-level professional education institutions (iv) continuing education and professional improvement (v) health monitoring (vi) Counseling subprogram
Cooperation Agreements with the Kemerovo Region (included in the blocks above)		
None	+ 2018	+ 2018

Sustainable development at SUEK involves: health and safety, environmental protection, personnel. Regional development (SUEK implements social programs in 48 towns in various regions of Russia) and social reports (reports on basic activities and performance of the company towards production, economic, social and ecological objectives are published since 2016).

At SDS Ugol, Ecology block includes ecological management/philosophy; environmental impact evaluation; biodiversity preservation and environment monitoring under blasting. SDS Ugol is an official partner of the draft program of UN, Global Ecological Foundation and Russia's Ministry of Nature: Biodiversity preservation objectives in policy and development programs of the energy sector of Russia. The company's personnel policy involves personnel monitoring, formation of personnel reserve, communication with professional education institutions at all levels (for instance, the Company has negotiated a long-term contract with the Kuzbass State Technical University for with the target training of specialists, continuing education and professional improvement), health monitoring and Counseling subprogram. The Company's Social Policy is presented without explanations.

SUEK and SDS Ugol concluded cooperation agreements with the Kemerovo Region in 2018. Furthermore, SUEK has cooperation agreements negotiated with the Khabarovsk and Krasnoyarsk Territories in 2017.

Thus, all three VICs possess potential to handle the problem connected with the negative impact on the area of presence. On the other hand, fluctuating policy relative to the resource regions disables a balanced decision. Underdevelopment of the institutional environment with an attempt to offer a universal transition approach without regard to signatures of each resource region and unpreparedness of development institutions to function efficiently provokes the opportunist behavior of both business and regional authorities.

The authors propose a way out by introducing a regulatory contract in the area of power engineering. The initiator was the Ministry of Finance of Russia by elaborating in partnership with the Russian Union of Industrials and Businessmen the Law on Protection and Encouragement of Investment in the Russian Federation and on Introduction of Alteration in Some Legislative Acts of Russia [5].

It is suggested to alter the public and private partnership format. The governmental agencies will present their projects for investors in the road show format in compliance with the generally accepted business standards (NPV, IRR) and other financial and economic indices. Such approach will provide a "filter" to pick up the most financially attractive projects.

The principal singularity of the investment protection and encouragement agreement (IPEA) lies in the stability of the project implementation conditions. According to the draft, since IPEA enters into force and up to its expiration, the company implementing the project is not subjected to any amendments in standards and regulations (such as international contracts, federal laws, legislations of subjects of the country, etc.) enforced after the effective date of the given agreement, which regulate production or any other economic activities under the project implementation and make the project conditions worse [6]. The regulatory contract format reduces investor's risks by means of the guaranteed governmental protection from changes in the rules of play in the course of project implementation.

The agreement can be concluded in case of investment of not less than 3 billion rubles and will preserve all fiscal and regulatory terms for the investor for the period of 6 years. Upon investment of more than 30 billion rubles, the guarantee of no institutional changes extends for the period of 12 years. It is worthy of mentioning that participants of the agreement are the Russian Federation, a private investor and a subject of the Russian Federation inside which the project is to be implemented.

3. Conclusions

In this manner, the balanced interaction between large mining companies and regional authorities is constrained by economic privacy of a region, vector of development of the region and its preparedness to alter the model of development. The background may become the regulatory contract capable to fixate the rules of play in the region for a long term in order to avoid opportunism. In this case, there appear conditions for new value-added production projects (based on NPV) [7] and reduction of resource dependence [8].

Acknowledgements

The study was supported by the Russian Science Foundation, Project No 16-18-10182.

References

[1] Nafedkin VI Economic power of large corporations and regional development Available at: http://institutiones.com/general/2902-ekonomicheskaya-vlast-krupnyx-korporacii-i-razvitie-

regionov.html

[2] *SUEK 2017 Annual Report* Available at: http://www.suek.ru/upload/iblock/bfb/SUEK_AnnualReport2017_RUS.pdf

[3] SDS Ugol Official Website: http://sds-ugol.ru

[4] Evraz Official Website: http://www.evraz.com/ru/

[5] Law On Protection and Encouragement of Investment in the Russian Federation and on Introduction of Alteration in Some Legislative Acts of Russia. Available at: http://www.evraz.com/ru/

[6] Shokhin AN Coming to an agreement on wholy new business and authority interafction format. Available at: http://xn--o1aabe.xn--p1ai/viewpoint/view/1008

[7] Nikitenko SM and Goosen EV 2017 Chains of value added as an instrument for the development of the Kuzbass coal industry *ECO Journal* No 9(519) pp 104–124

[8] Pakhomova EO, Goosen EV and Nikitenko SM 2017 New institutes for development and territorial arrangement in resource type regions *Proceedings of IV All-Russian Symposium on Regional Economy* pp 50–54

IOP Conf. Series: Earth and Environmental Science **262** (2019) 012026 doi:10.1088/1755-1315/262/1/012026

Modeling coal deposits in hard-to-reach regions of Yakutia

EA Khoyutanov* **and VL Gavrilov****

Chersky Institute of Mining of the North, Siberian Branch, Russian academy of Sciences, Yakutsk, Russia

E-mail: *khoiutanov@igds.ysn.ru, **gvlugorsk@mail.ru

Abstract. The article presents brief characteristics of the Kharbalakhs and Nadezhda deposits currently being mined. Exploration data on their specific structure, the variability of the geometric parameters of the seams, the statistical heterogeneity of the ash content in coal are reported. Frame and block models of the main coal seams are constructed based on the compiled database with use of mining and geological information systems. The paper considers a number of methods for interpolation of coal grade at different density of the exploration network. The spatial distribution of ash content in coal seams is mapped. The use of the obtained results can increase the validity of the decisions made at the planning, mining and coal preparation stages.

1. Introduction

Although coal is being currently replaced by alternative kinds of energy, it yet remains and will be the main source of heat generation in the Republic of Sakha (Yakutia) [1]. This sparsely inhabited region features underdeveloped infrastructure and severe climatic conditions, unfavorable for mineral mining.

The operating and promising coal mine in Central and Northern Yakutia have, as a rule, considerable reserves and resources of various kind and grad solid fuel. When erroneous or inefficient decisions are unpardonable, it is absolutely indispensable to use the available information about these georesources to maximum effect. Such decision-making at the stages of mine planning, design and operation would improve performance of both mines and deposit–consumer chains.

Among other things, this objective can be reached by analyzing mine planning and operation processes using advanced software programs and mining geology information systems (MGIS). The geo-information analysis aims to study properties, arrangement, structure and interaction of objetcs and phenomena in space. To date, it is difficult to image an efficient mine without using such tools as 3D modeling and applications to the effect of production planning and control [2–4], information technologies of mining process supervision meant for optimizing expenses, time and productivity [5], geo-information systems to handle problems connected with spatial analysis and exploration of different objects[6], or ensuring interactive access to data mining tools on Internet servers [7], geostatistical estimation of spatial patterns of qualitative and quantitative data, determination of optimal size and weight of samples [8]; creation, operation and supervision of data bases on sampling and qualimetry of minerals and spatial information [9, 10], etc.

2. Kharbalakh and Nadezhda coal fields

With regard to the desired improvement of managerial decision-making at mining and allied companies, we are going to discuss in more detail general approaches to comprehensive analysis of georesources based on upgrading of information analysis systems in terms of two coal fields

(Kharbalakh in Central Yakutia and Nadezhda in Zyryansky Coal Basin). These approaches can be later on used to create estimation procedures for appraisal of quality and quantity of mineral reserves in other deposits, including information and analysis support of coal quality management along the mining and delivery chains.

Kharbalakh coal field occurs in the Low Aldan area of the Lena Coal Basin [11, 12]. Commercial-value coal seam Karierny features wide area and is comprised of two coal bands. The seam is from 6.5 to 9.8 m thick; the upper band thickness is 3.1–4.5 m, and the lower band thickness varies from 2.7 to 4.7 m. The seam is consistent, with distinct partition by a dirt band from 0.15 to 1.3 m thick. Coal has medium ash content, is low sulfur and belongs to grade D (long-flame). The remaining reserves in-place intended for extraction by the open pit mining method make 3.5 Mt at the annual output at the level of 150–170 thousand tons. According to the guidelines of the Federal Agency for Mineral Resources of Russia, the deposit is put in difficulty group II. The operating open pit mine supplies coal to populated areas of the central region of Yakutia. For the latest 10 years, open pit mine production has totaled 1.2 Mt of coal at the loss of 5.2%.

Stripping involves blasting and dozing. Owing to thin overburden layer and a dirt band, coal is extracted by four benches: overburden, two coal bands and the parting. ROM coal quality is off any control. Coal is only crushed on a crushing-and-screening plant, which allows dirt and high ash rocks enter the outlet product. Coal is hauled to boiler houses and public sector by trucks at the distinct seasonality of consumption. In view of the demand for more resources, the adjacent area of the open pit mine is subjected to additional exploration.

Nadezhda coal site of Zyryansky open pit min is a part of Zyryansky Coal Basin at the Kolyma River. Commercially valuable are five coal seams of grade ZH (fat) with total in-place reserves of more than 9 Mt. Coal is low-ash, highly calorific, low sulfur and caking. The open pit mine field is divided into three sites suitable for independent development. The minimum thickness of five seams of complex and simple structure is 2 m (upper band 32v—1 m). dip angles vary from 5 to 10° in the north and from 25 to 30° in the south. The average thickness of seams 13, 18 and 22 is from 1.0 to 6.6 m. The overlying seam 32 is consistent, has complex structure and ranges from 4.0 to 12.8 m in thickness. A coal band to 2.7 m thick splits off from this seam in the south. The top-most seam 32v (1.0–3.4 m) has simple structure and is relatively consistent.

Coal is accessed by ramps cut in the center and on sides, with encircling haul roads and an exit to the river dam and a main road. Side strips of the upper bench are removed by dozers. Further stripping is proceeded by the truck-and-shovel scheme with loaders. In addition, drilling-and-blasting and benching by shovels and dozers are involved until a productive seam is exposed. Loose rocks are loaded by shovels to trucks, or are pushed by dozers to mined-out void or to dumps. A stripping bench height varies from 6 to 10 m.

Stage-wise exploration, planning, design and mining assumes implementation of some series–parallel, interconnected and mutually complimentary procedures ensuring persistent improvement of our knowledge on the georsource at hand. The relevant experience accumulated at the Institute of Mining of the North, SB RAS, allows rising to a wholly new level of efficient and resource-saving subsoil management.

First, necessary data are collected from various sources with maximum coverage of information on previous stages of exploration and operation (geology, climate; standards and design data; coordinates of exploring opening, geophysical studies, coal sampling with characteristics of quality and structure of seams; various technical and economic indexes; fuel haulage routes; resource consuming points, etc.). Then, after preliminary consideration in the framework of a uniform information space creation, the source information is digitized and formalized with creation of databases (with state registration if required) on individual deposits.

The next stage is processing of the data using mathematical statistics methods and software support, for instance, Microsoft Excel, with a view to shaping general representation of basic characteristics of the deposits (sites) and their variability (table).

At the general compatibility of the statistical characteristics, the lower band of Karierny seam in Kharbalakh field has the lowest ash content at the high coefficient of variation. In Nadezhda field, seam 32n has the highest ash content and low CV.

Nextly, to get deeper insight into the variation of geometrical parameters of coal seam and coal quality, 2D mapping is implemented (using Golden Software Surfer). The accomplished interpolation and visualization of general ash content across the area of seam 32n in Nadezhda field shows relatively uniform zones of different size and values (Figure 1).

Basic statistical characteristics of total ash content of Kharbalakh and Nadezhda coal.

Characteristics	Kharbalakh, Karierny seam		Nadezhda				
	Coal band						
	Upper	Lower	32v	32n	22	18	13
Average	15.7	14.5	15.1	19.6	15.2	17.5	17.7
Standard deviation	4.6	5.7	4.8	4.2	4.8	5.8	4.2
Coefficient of variation	29.4	39.5	31.5	21.5	31.9	32.9	23.4
Excess	−0.4	−1.3	0.1	−0.5	0.6	−1.2	−0.7
Asymmetry	0.5	0.2	0.7	0.1	1.1	−0.2	−0.4
Holes	23	37	64	72	87	84	71

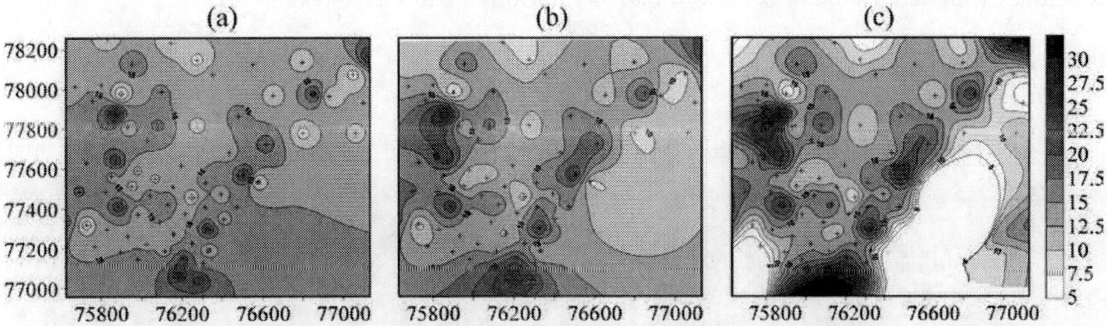

Figure 1. Map of total ash content contour lines for seam 32n of Nadezhda field: (a) inverse distance weighting; (b) krigging; (c) Shepard method.

Finalized presentation of coal deposits in the context of sound resource-saving subsoil management is implemented using MGIS as the basic analysis, design and modeling tool.

Review of the key foreign and Russian program products using various information sources (publications, web sites of software designers, user companies, conferences, workshops, webinars, personal communication, application of license and demo-versions) shows that ideologies of architecture and functioning of these systems are mainly alike. The choice of a specific MGIS is often subjective. Available information on efficient application of a software product is of local nature, and the product cannot be automatically used at another deposit or a company without adjustment to its specificity (scale, mineral production output, complex structure, sales market, financial assets, etc.).

The present paper authors posses MGIS systems MINEFRAME [13]) and Micromine [14]. Both systems allow integrated decision-making with 3D modeling, multilateral appraisal, planning, design and optimization of mining operations at a various degree of detailing.

Using the MGIS data, 3D frame and block models of the discussed coal fields are constructed (Figure 2), and ash contents of coal seams are assessed.

IOP Conf. Series: Earth and Environmental Science **262** (2019) 012026 doi:10.1088/1755-1315/262/1/012026

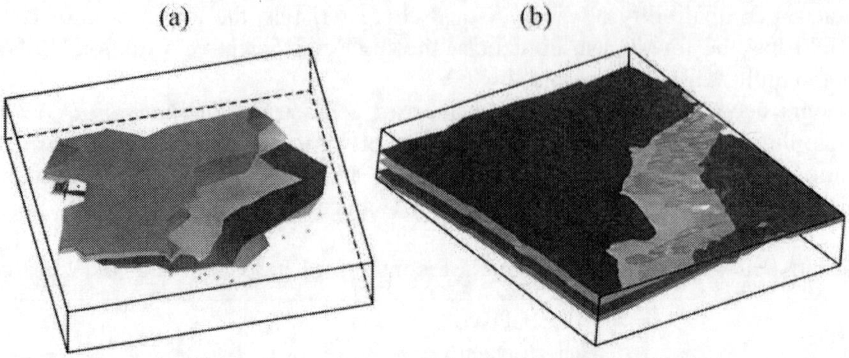

Figure 2. Frame models of Kharbalak and Nadezhda fields: (a) Mineframe; (b) Micromine.

In the course of work, for comparison, many variants of the models were constructed using different methods of geostatistcs, mainly, inverse distances and krigging, with various sets of parameters, to study nature and patterns of spatial distribution and quality of coal reserves. Coal quality mapping involved such procedures as coarsening of grid to estimate influence of exploration grid density on delineation accuracy of coal reserves by useful quality; testing of different methods of bridging data gaps on coal sampling in seams; 2D and 3D modeling of coal seams using geostatistics techniques with delineation of zones of relatively uniform quality coal (Figure 3).

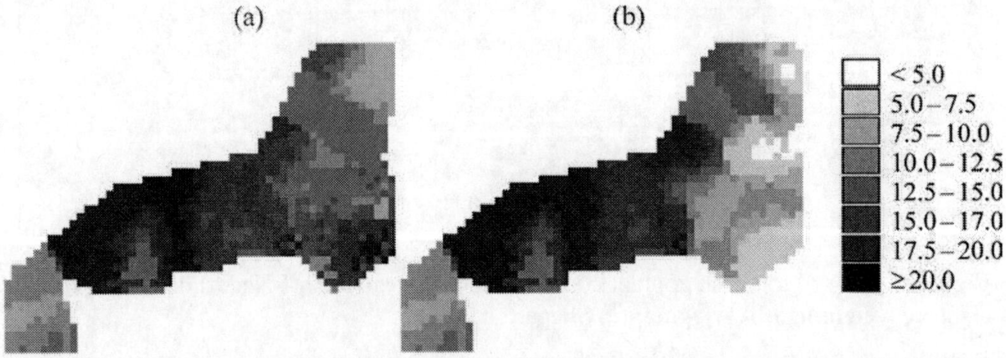

Figure 3. Ash content by block models of the lower band in Kareierny seam of Kharbalakh field in various density exploration grids: (a) actual; (b) coarse.

3. Conclusions

1. The result of the stage-wise analysis of coal seam parameters and properties in accordance with the proposed approach is the additional knowledge on the georesource. This knowledge is a more reliable information background for determination of order and sequence of mining in extraction panels with regard to quality of coal; for studying geotechnologies with a view to optimizing their parameters; for improvement or creation of quality management systems with regard to the interests of final consumers—boiler houses in hard-to-reach regions.
2. Depending on coal quality requirements imposed by consumers, complexity and unique nature of coal deposits, as well as objectives, goals and capabilities of subsoil users, it is possible to use any course of geoinformation–based analysis of deposit structure, including the proposed and discussed approach.

References

[1] Gavrilov VL, Ivanov VG, Kugaevskiy AA and Rakhleev PD 2017 The state and prospects of production and consumption of coal in the Central regions of Yakutia *GIAB* Vol 11 Special Issue 24 pp 422–430

[2] Zhao H, Bai R and Liu G 2011 3D modeling of open pit based on AutoCAD and application *Procedia Earth and Planetary Science* No 3 pp 258–265

[3] Yanbing W, Lixin WU, Wenzhong S and Xiaomeng L On 3D GIS spatial modeling *ISPRS Workshop on Updating Geo-spatial Databases with Imagery & The 5th ISPRS Workshop on DMGIS* pp 237–238

[4] Zu XF, Hou WS, Zhang BY, Hua WH and Luo J 2012 Overview of three-dimensional geological modeling technology *IERI Procedia* No 2 pp 921–927

[5] Miladinović Milena, Čebašek Vladimir and Gojković Nebojša 2011 Computer programs for design and modelling in mining *Underground Mining Engineering* No 19 pp 109–124

[6] Böhner J, Bock M, Selige T and Köthe R 2014 *Geographical Information Systems—Applications to Soils BT— Reference Module in Earth Systems and Environmental Sciences* Elsevier https://doi.org/https:// doi.org/10.1016/B978-0-12-409548-9.09107-7

[7] Yuhendra 2016 Development of web based GIS mapping for geothermal resources information system *ARPN Journal of Engineering and Applied Sciences* Vol 11 No 11 pp 7349–7353

[8] Dimitris Drymonitis 2015 Some mineral exploration problems that can be solved by mathematics and geostatistics *Procedia Earth and Planetary Science* No 15 pp 747–753

[9] Whateley MK 2002 Measuring, understanding and visualising coal characteristics—Innovations in coal geology for the 21st century *International Journal of Coal Geology* Vol 50 No 1–4 pp 303–315. https://doi.org/10.1016/S0166-5162(02)00122-2

[10] Ogunbadewa EY 2012 Developing natural resources database with Nigeriasat-1 satellite data and geographical information system *Egyptian Journal of Remote Sensing and Space Science* Vol 15 No 2 pp 207–214. https://doi.org/10.1016/j.ejrs.2012.04.002

[11] Moyakunov EV and Galenchikov SA 2017 *Real-time Coal Reserves Appraisal in Kharbalak Field of Jan 1, 2017: Report* Yakutsk: Tele Vol 1 (in Russian)

[12] Khoyutanov EA, Gavrilov VL and Fedorov VI 2018 Modeling additional exploration and development of Kharbalakh coal field *Geology and Mineral Reserves of the North-East Russia: VIII All-Russian Conference* Yakutsk Vol 2 pp 407–410 (in Russian)

[13] Lukichev SV 2017 Institute's experinece in creation of software for mining technology *GIAB* S23 pp 19–31

[14] *Reasonability of GIS Introdution in Coal Mines* 2017 http://www.micromine.ru/micromine-integration-in-coal-enterprise

Innovative technologies for thick coal seam mining on the basis of powered roof support with controlled coal discharge

VI Klishin

Institute of Coal, Federal Research Center of Coal and Coal Chemistry, Siberian Branch, Russian Academy of Sciences, Kemerovo, Russia

E-mail: klishinvi@icc.kemsc.ru

Abstract. A new trend for designing powered supports for thick coal seam mining with controlled coal discharge is substantiated. Numerical modeling for gravity flow of disintegrated rocks and the flow interaction with the support section is performed. A test bench is developed, and the laboratory studies into coal discharge process are carried out.

1. Introduction

New technologies for mining thick flat and steep coal seams use use force of rock pressure for physical destruction of coal, which needs machines with extra capabilities. Many countries investigate and justify mining technologies for thick coal seams with discharge of softened top coal [1–6]. The new proposed design of a powered roof support equipped with controlled coal discharge to armored face conveyor possesses advantages and eliminates drawbacks of the known analogs (Figure 1) [7, 8].

(a) (b)

Figure 1. (a) Longwall top coal caving technology in mining thick flat coal seams; (b) general view of powered roof support with controlled discharge of caved coal to armored face conveyor.

In this regard, it is important study the process of caved coal discharge. One of the most popular methods used to model deformation of geomaterials is the method of discrete element (DEM) for the first time put forward in [9]. This method allows determining stress state evolution in broken rock mass both at pre-limiting stage of loading and in transition to shearing localization and failure. At the

IOP Conf. Series: Earth and Environmental Science **262** (2019) 012027 doi:10.1088/1755-1315/262/1/012027

present time, DEMe enjoys wide application in solving geomechanical problems [10–12]. In DEM a real medium is replaced by a set of discrete particles with the assigned laws of interaction between them. This method faces no difficulties in solution of problems with high (finite) deformations and rotations. Thus, DEM is a principal alternative to classical methods based on conventional concepts of continuum mechanics. In the framework of DEM, gravity flow of coal discharge to AFC via feeder was modeled numerically [13].

2. Numerical and laboratory modeling

The experimental design is depicted in Figure 3. The figures marks the model elements of the powered roof support unit: capping *1* with a length of 2.5 m; immobile vertical gate *2* with a height of 1.75 m to limit velocity and volume of granular material flow; feeder *3* with a length of 3.575 m inclined at 12° to horizontal plane; discharge opening *4* with a width of 2 m, shut at the stage of initial flow configuration formation; side shields *5* spaced at 1 m.

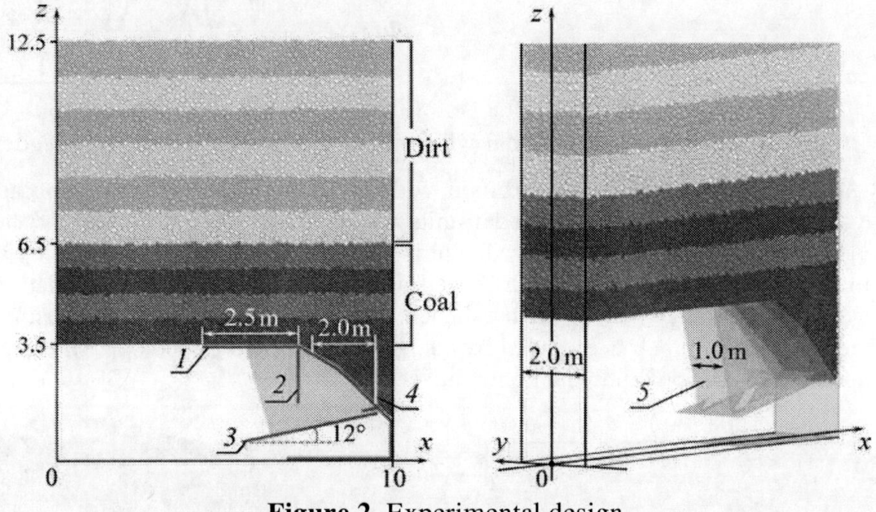

Figure 2. Experimental design.

The influence of the feeder design on the flow behavior was studied in two tests. In the first test, the feeder surface was smooth, and in the second test, it was corrugated with step plates 1.15 long arranged in parallel to horizon plane. The inclination of the feeder was 12° in both tests. When broken rock mass reached equilibrium, the discharge opening was opened and the material started flowing unit the next equilibrium state was reached owing to the vertical gate after the feeder was deactivated. The pressures of the flow on the gate and on the feeder, P_d and P_f, respectively, were recorded and plotted (Figure 3)

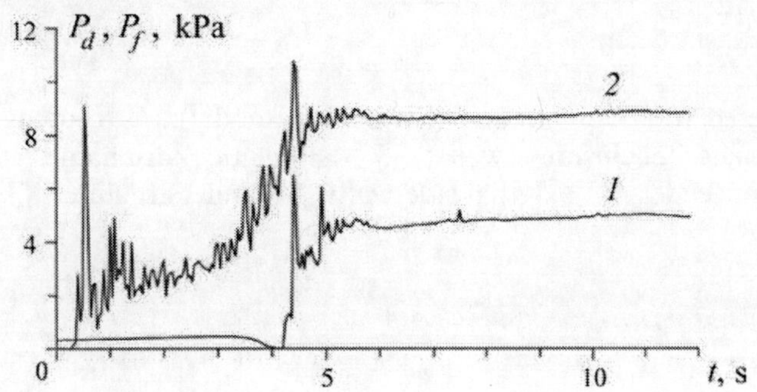

Figure 3. Pressure of granular material flow on the gate (*1*) and deactivated feeder (*2*).

The actuated feeder, due to contact friction between the particles and the particles and the feeder, makes the material move. The broken material flow reaches steady state in some time. The mass flow rate of coal, M_c, in this mode within 45 s is described in Figure 4. With the smooth feeder, M_c is 47 kg/s at the total discharged coal mass of 2193 kg (Figure 4a). With the corrugated feeder, these indexes are 32 kg/s and 1500 kg, respectively (Figure 4b).

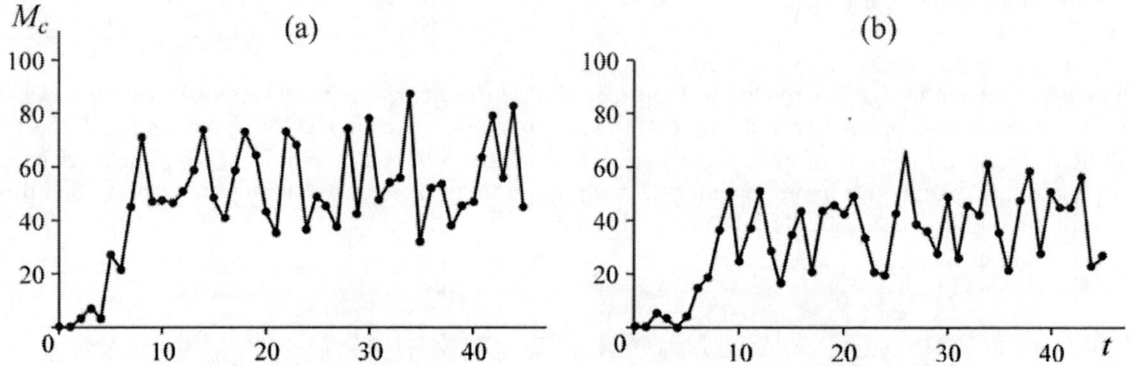

Figure 4. Mass flow rate of coal with (a) smooth and (b) corrugated feeder.

Twenty models of powered roof support units with advancing facilities were manufactured for the tests (Figure 5). The total length of the model units was equal to the inner width of the transparent testing box (1600 mm) (Figure 6). Discharged coal is transported from the face zone by a belt 65 mm wide and 4 mm thick, mounted in front of the base frame of the laboratory test installation. The belt is driven by a gear motor arranged under the test bench. The belt capacity is adjusted using a frequency converter. The tension of the belt is executed by a tensioner mounted on the opposite side of the testig box and composed of a tension drum and tightening screws.

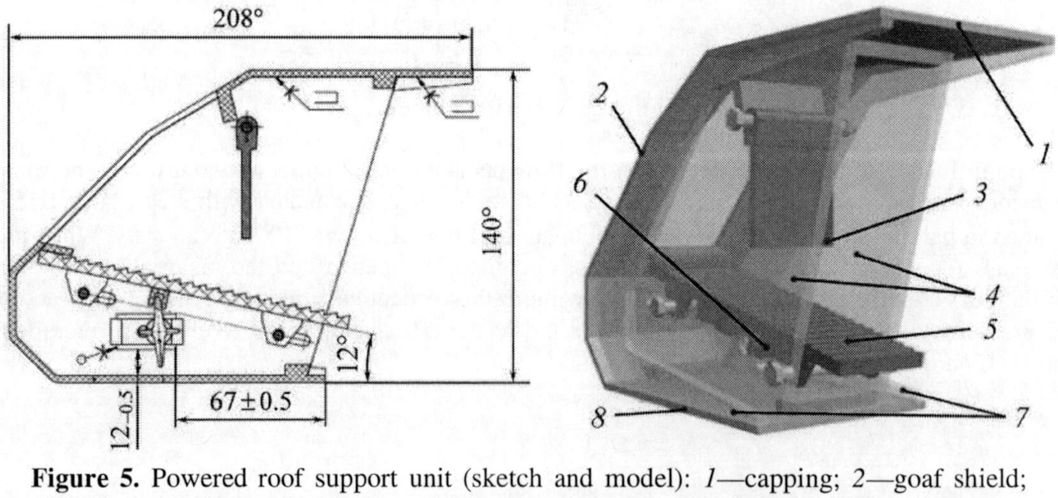

Figure 5. Powered roof support unit (sketch and model): *1*—capping; *2*—goaf shield; *3*—gate; *4*—inner enclosure walls; *5*—automatic discharge facility (feeder); *6*—servomotor of feeder; *7*—external side walls; *8*—understructure.

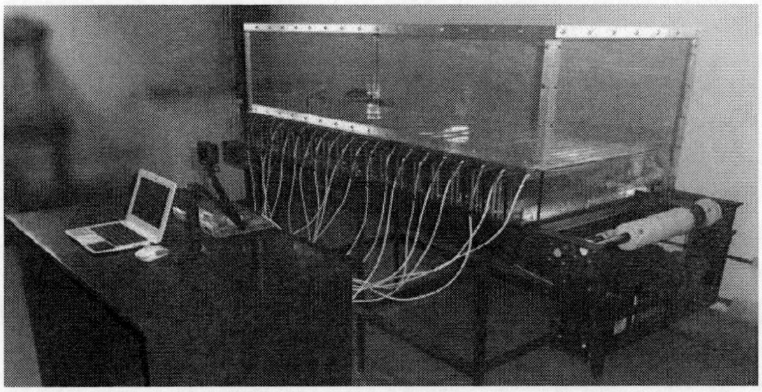

Figure 6. Laboratory test bench.

The control system of the discharge flow includes instrumentation and software making it possible to vary feeder parameters (frequency, amplitude) and operating modes of roof support units (individual, grouped). The test bench is also equipped with photo and video recording devices.

This test bench allows experimentation on coal discharge from the powered roof support units operating in various regimes: individual and grouped, wavy and full-length.

Coal is simulated by gravel painted black, faction 5–15 mm; dirt is modeled by grey-pinky marble chips 13–15 in size. For better visualization of flow process, coal layer is parted by a layer of white gravel particles 5–10 mm in size (Figure 7).

Figure 7. Laboratory testing of discharge flow in different operating modes.

3. Conclusions

1. The presented DEM-based model of gravity flow of broken rock mass in the longwall top coal caving technology includes all process stages: creation of initial equilibrium state of rock mass; discharge of mineral to feeder of powered roof support unit; operation of the feeder and material discharge in accord with the process chart of the support.
2. The cyclic operation of the feeder is the cause of the cyclic change both in the mineral flow and in the load imposed by the flow to the elements of the powered roof support unit during discharge. The configuration of the feeder surface has considerable influence on the process parameters under analysis.
3. The developed model of the powered roof support unit is equipped with the discharge control system composed of instrumentation and software making it possible to vary the feeder parameters (frequency, amplitude) and the unit operation mode)individual, grouped). The designed test bench is fitted with photo and video recording devices. The tests of the discharge flow process on the test bench shows conformance between theoretical knowledge and experimental results.

Acknowledgements
The study has been supported by the Ministry of Education and Science of the Russian Federation in the framework of the Federal Targeted Program on R&D in Priority Areas of Advancement in the Science and Technology of Russia for 2014–2020, Topic: Development of Efficient Robotic Longwall Top Coal Caving Technology and Equipment (Unique Identifier RFMEFI60417X0173).

References
[1] Melnik VV and Sushchev VA 2009 Technology of longwall top coal caving in mining thick flat coal seams *GIAB* No 5 pp 198–201
[2] Alireza Jabinpoura, Alireza Yarahmadi Bafghib and Javad Gholamnejad 2016 Application of Vibration in Longwall Top Coal Caving Method *International Academic Journal of Science and Engineering* Vol 3 No 2 pp 102–109
[3] Kumar R, Singh AK, Mishra AK and Singh R 2015 Underground mining of thick coal seams *International Journal of Mining Science and Technology* Vol 25 Issue 6 pp 885–896
[4] Hebblewhite BK 2005 Status and prospects of underground thick coal seam mining methods *The 19th International Mining Congress and Fair of Turkey, IMCET2005* Izmir, Turkey
[5] Guo J, Ma L, Wang Ye and Wang F 2017 Hanging wall pressure relief mechanism of horizontal section top-coal caving face and its application—A case study of the urumqi coalfield, China *Energies* 10 (9) pp 1371
[6] Unver B and Yasitli NE 2006 Modelling of strata movement with a special reference to caving mechanism in thick seam coal mining *International Journal of Coal Geology* Vol 66 Issue 4 pp 227–252
[7] Klishin VI, Shundulidi IA, Ermankov AYu and Soloviev AS 2013 *Mining Technology for Thick Flat Coal Seams with Dischare* Novosibirsk: Nauka (in Russian)
[8] Klishin VI, Fokin YuS, Kokoulin DI and Kubanychbek B 2007 *Thick Coal Mining with Powered Support with Adjustable Coal Discharge* Novosibirsk: Nauka (in Russian)
[9] Cundall PA and Strack ODL 1979 A distinct element model for granular assemblies *Géotechnique* Vol 29 No 1 pp 47–65
[10] Klishin SV, Lavrikov SV, Mikenina OA and Revuzhenko AF 2018 Discrete element method modification for the transition to a linearly elastic body model *Journal of Physics: Conference Series* 973(1) 012008
[11] Klishin SV, Lavrikov SV and Revuzhenko AF 2017 Numerical simulation of abutment pressure redistribution during face advance *AIP Conference Proceedings* 1909 020086
[12] Klihin SV, Klishin VI Opruk GYu 2013 Modeling coal discharge in mechanized steep and thick coal mining *Journal of Mining Science* Vol 49 Issue 6 pp 932–940
[13] Klihin SV, Klishin VI and Opruk GYu 2018 Discrete element modeling of gravity flow of broken rocks in the technology of longwall top coal caving *Naukoemk. Tekhnol. Razrab. Ispolz. Mineral. Resurs.* Vol 4 pp 80–85

IOP Conf. Series: Earth and Environmental Science **262** (2019) 012028 doi:10.1088/1755-1315/262/1/012028

Development of device for interval-by-interval hydraulic fracture

VI Klishin[1]*, GYu Opruk[1], DI Kokoulin[2] and AL Tatsienko[3]*****

[1]Institute of Coal, Federal Research Center for Coal and Coal Chemistry, Siberian Branch, Russian Academy of Sciences, Kemerovo, Russia
[2]Chinakal Institute of Mining, Siberian Branch, Russian Academy of Sciences, Novosibirsk, Russia
[3]Kirov Mine, SUEK-Kuzbass, Leninsk-Kuznetsky, Russia

E-mail:*klishinvi@icc.kemsc.ru; **konstruktor430@yandex.ru; ***tatcienkoal@suek.ru

Abstract. Under consideration is the experience gained in design and manufacture of tools for mining machines meant for operation in uncontrollable conditions. It is shown that equipping the mining machines with disk tools reduces specific consumption of the tools, specific power-input and dusting during breakage of hard rocks.

1. Introduction

Low-permeability zones decelerate tunneling speed and entail high face outputs due to small volume of gas drainage work and under-efficiency of drainage mechanisms in many mines in Russia. In highly productive longwalls in gas-bearing coal seams, the in-seam drainage technology has found widest application [1–3]. By estimates, the current gas drainage techniques applied in Russia extract from 20 to 30% of total methane released [4–7]. Surface-based gas drainage is connected with the known difficulties detailed in [8]. Further improvement in safety, environmental security and efficiency of underground mining in coal mines with high gas content is inextricably connected with development of special measures and techniques aimed to stimulate gas recovery from high-stress coal [9–12].

Gas recoverability in high-stress coal seams can be enhanced by improving coal permeability using multi-stage hydraulic fracturing (Figure 1); a variant of this approach is described in [13, 14]. Multistage hydraulic fracturing is used to shorten and intensify pre-mine gas drainage process. In a coal seam, a set of hydrofractures of assigned configuration is created for partial unloading of rock mass and for making filtration channel for water (dehydration) or gas drainage.

Hydraulic fracturing improves permeability of coal faces by means of initiation or opening and elongation of natural cracks. The fractures created during directional multistage hydraulic fracturing can reach a length of a few tens of meters and, when merged, considerably improve permeability of gas drain hole [15].

The hydraulic fracturing tool (Figure 2) is advanced in the hole in the column of high-pressure pipes (Figure 3) using a drilling rig.

2. Results of practical application of the hydraulic fracturing method

In Yubileinaya Mine in Kuzbass, pilot multistage hydraulic fracturing was carried out in conveyor drive 16–19 in an experimental hole with a length of 145 m and diameter of 93 mm (Figure 4).

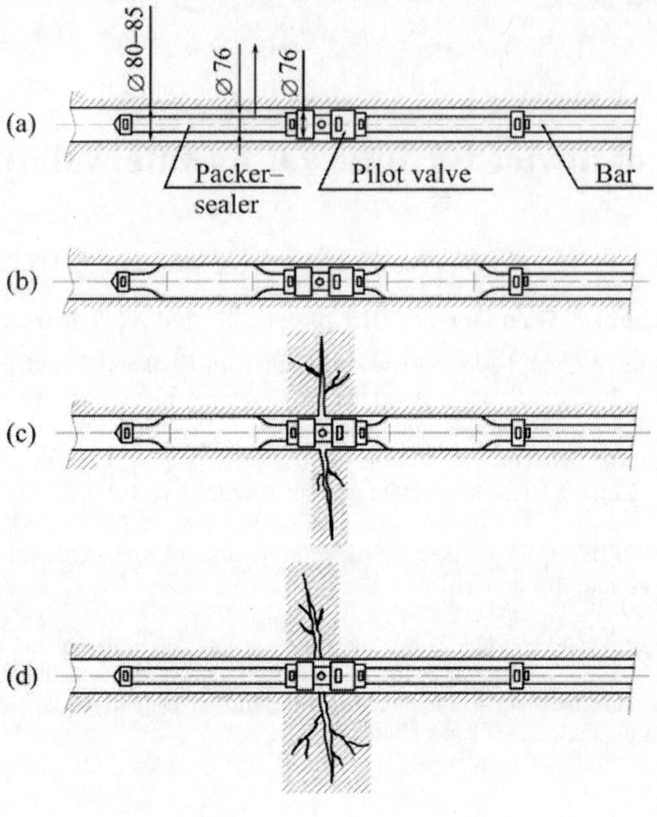

Figure 1. Process chart of multistage hydraulic fracturing of coal: (a) inlet of packer in hole; (b) sealing of hole; (c) hydraulic fracturing of coal; (d) unsealing of hole and advance of packer.

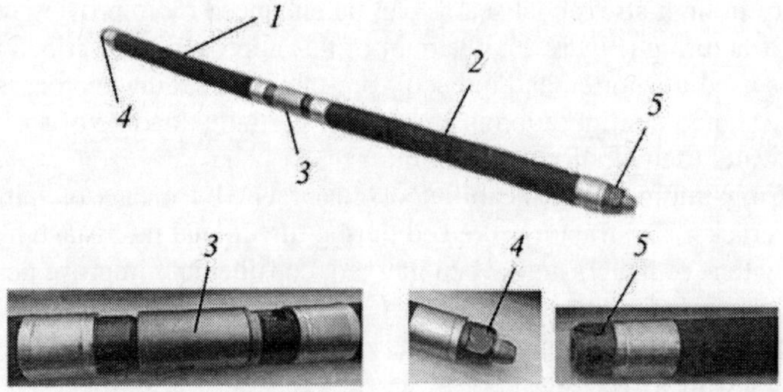

Figure 2. Hydraulic fracturing tool: *1*, *2*—elastic inflatable hoses; *3*—inter-packer valve, *4*—shat-off nose-piece, *5*—connector sleeve.

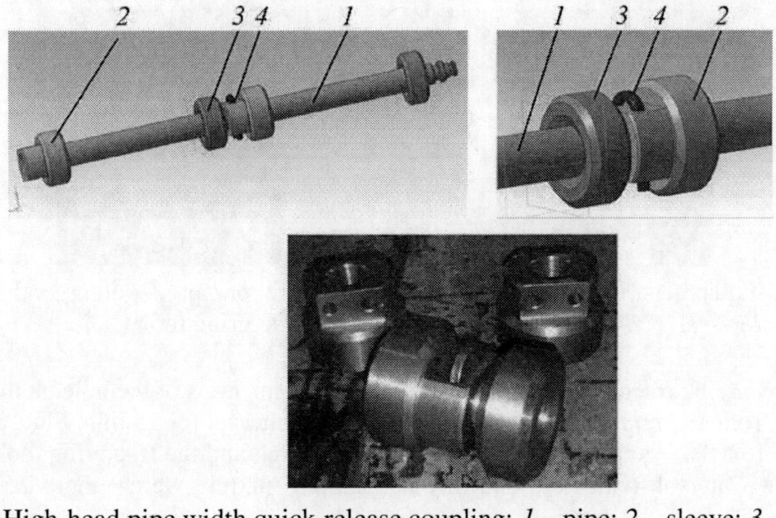

Figure 3. High-head pipe width quick-release coupling: *1*—pipe; *2*—sleeve; *3*—nipple; *4*—clip.

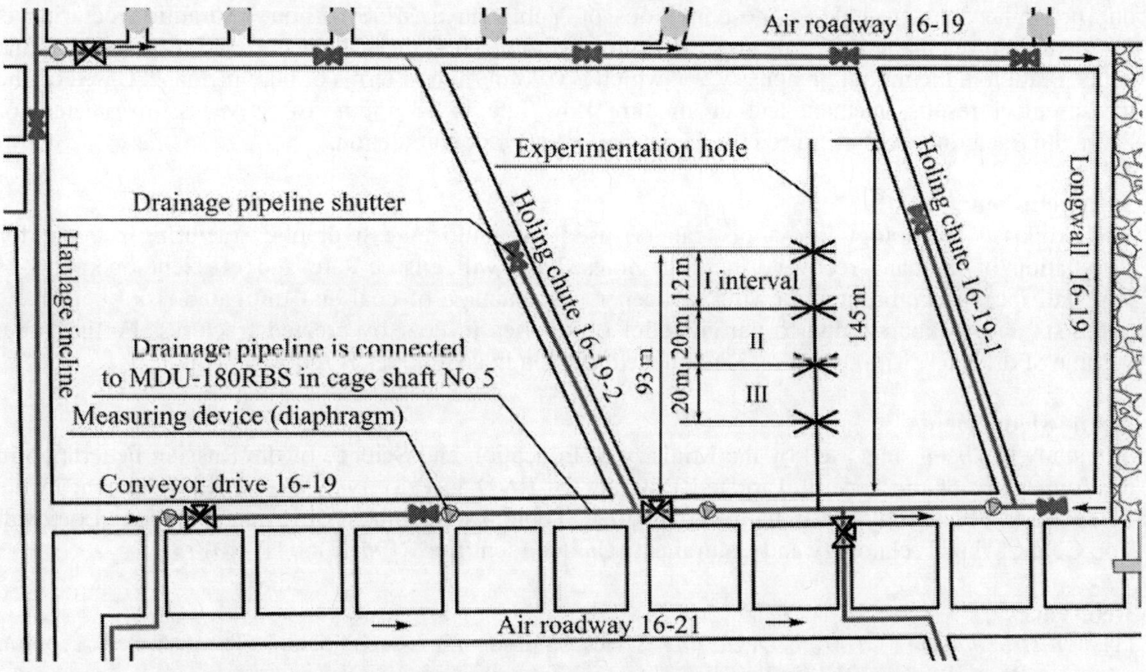

Figure 4. Experimental hole for multistage hydraulic fracturing: Layout.

The coal seam 16 is from 1.48 to 1.78 m thick, with average thickness of 1.60 m. Natural gas content is 11.8–25.0 m^3/t. The maximum depth of mining in longwall 16-19 is 730 m. The seam is rockburst-hazardous down from the depth of 150 m and outburst-hazardous down from the depth of 325 m. Before the experiment, the high-head equipment and instrumentation were assembled and installed (Figure 5).

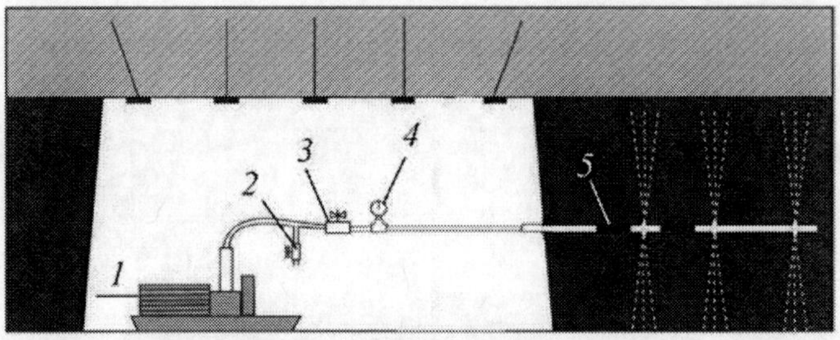

Figure 5. Equipment configuration: *1*—high pressure pump; *2*—dump valve; *3*—high-pressure screw-valve; *4*—manometer; *5*—hydraulic fracturing tool.

Multistage hydraulic fracturing was implemented in four intervals of the hole, at the distances of 93, 81, 61 and 41 m from the mouth. Fluid was injected in each interval for 15 min. Fluid injection pressure was no less than 16 MPa. As the high-head pipes were removed and the fracturing tool was displaced to the next fracturing interval, both fluid and methane-and-air mixture were jump-wise ejected from the hole.

After completion of experimentation, when water overflow stopped, the hydraulic fracturing hole was connected vacuum degassing network. From the monitoring, the average methane flow rate per one pre-mine drainage hole in the conditions of Yubileinaya Mine without hydraulic fracturing is 0.073 m^3/min, methane flow rate from the pilot hydraulic fracturing hole is 0.315 m^3/min. Methane concentration in the mixture reaches 72%, which is 0.23 m^3/min in terms of pure methane. Based on the measurement results, methane-and-air mixture flow rate is 13 l/min. Effectiveness parameters of hydraulic fracturing were measured for 55 days after the work completion.

3. Conclusions

The proposed technology and tools can be used for multistage hydraulic fracturing of coal for stimulation of methane recovery in drain holes. This will ensure safe and efficient operation of longwall faces at high outputs owing to deeper gas drainage of coal and mitigated risk of methane outbursts and rockbursts due to partial relief of stresses in coal by created fractures. Furthermore, volume of drainage drilling and toxic methane emission in coal mines is reduced.

Acknowledgements

The study has been supported by the Ministry of Education and Science of the Russian Federation in the framework of the Federal Targeted Program on R&D in Priority Areas of Advancement in the Science and Technology of Russia for 2014–2020, Topic: Development of Efficient Robotic Longwall Top Coal Caving Technology and Equipment (Unique Identifier RFMEFI60417X0173).

References

[1] *Instruction on Cal Mine Degassing* Series 05 Issue 22 Nauch.-Tekhn. Tsentr Issled. Problm. Prom. Bezop. 2012 (in Russian)

[2] Trubetskoy KN, Ruban AD and Zaburdyaev VS 2011 Justification methodology of gas removal methods and there parameters in underground coal mines *Journal of Mining Science* Vol 41 No 1 https://doi.org/10.1134/S1062739147010011

[3] Ruban AD, Zaburdyaev VS, Zaburdyaev GS and Matvienko NG 2006 *Methane in Russian Coal and Ore Mines: Forecast, Extraction and Use* Moscow: IPKON (in Russian)

[4] Nozhkin NV 1979 *Preliminary Coal Degassing* Moscow: Nedra (in Russian)

[5] Slastunov SV 1996 *Preliminary Degassing and Methane Extraction from Coal Beds* Moscow: MGU (in Russian)

[6] Malyshev YuN, Khudin YuL, Vasilchuk MP et al 1997 *Methane-Bearing Seam Development in the Kuznetsk Coal Field* Moscow: Akad. Gorn. Nauk (in Russian)

[7] Puchkov LA, Slastunov SV and Present GM 2002 Prospects of commercial coalbed methane recovery *GIAB* No 6 pp 6-10

[8] Sikora P, Smyslov D and Pletner O 2008 Peculiarities of coal seam preliminary degassing by means of borehole drilling from the surface *Glukauf* No 1 pp 39–45

[9] Shumilov VA, Akselrod SM and Shumilov AV 2014 *Downhole Geophysical Survey In Coalbed Methane Prospecting and Production* Perm: PGNIU (in Russian)

[10] Klishin VI, Kokoulin DI, Kubanychbek B and Durnin MK 2010 Coal seam softening as a method of methane release stimulation *Ugol* No 4 pp 40–42

[11] Klishin VI, Opruk GYu and Tatsienko AL 2017 Technology and means of a coal seam interval hydraulic fracturing for the seam degassing intensification *IOP Conf. Series: Earth and Environmental Science* 53 012019

[12] Rodin RI and Plaksin MS 2016 Features of coalbd gas permeability stimulation *Vestn. Nauch. Tsentra Bezop. Rabot Ugoln. Prom.* No 1 pp 42–48

[13] Klishin VI and Kokoulin DI 2013 RF Patent No 2472941 Method of Hydraulic Fracturing of Coal *Byull. Izobret.* No 2

[14] Klishin VI, Kokoulin DI and Klishin SV 2012 Downhole Device for Rock Hydraulic Fracturing RF Patent No 123064 *Byull. Izobert.* No 35

[15] Chernyi SG, Lapin VN, Esipov DV and Kurnakov DS *Methods for Crack Initiation and Propagation Modeling* Novosibirsk: SO RAN (in Russian)

Investigation of coal discharge modes on brassboards of powered roof support unit

VI Klishin[1]*, EL Varfolomeev[1], IL Borisov[1] and SV Klishin[2]**

[1]Institute of Coal, Federal Research Center of Coal and Coal Chemistry, Siberian Branch, Russian Academy of Sciences, Kemerovo, Russia
[2]Chinakal Institute of Mining, Siberian Branch, Russian Academy of Sciences, Novosibirsk, Russia

E-mail: *klishinvi@icc.kemsc.ru, **sv.klishin@gmail.com

Abstract. Physical and analytical modeling of caved top coal discharge by adjustable feeders on armored face conveyor during underground mining of thick flat seams has been performed. The process of top coal caving is studied using laboratory installation with the feeders operating in wave and areal modes. Based on 3D discrete element method (DEM), gravity flow of coal and its interaction with the powered roof support unit is modeled numerically.

1. Introduction

In the world practice, there are to prevailing technologies for thick flat coal seams: slicing and full-height cutting with top coal caving. Longwall slicing features high loss of coal and endogenous fire hazard. Therefore, the second-listed technology is more preferable.

Advantages of LTCC technology are described below:

—coal is extracted full-out, which contributes to high concentration of mining and reduction in coal loss and endogenous fire risk;

—development heading, as well as the cost of construction and support is reduced 1.5–2 times;

—the force of rock pressure and the effect of top coal self-caving are utilized, which lowers energy expenditure of coal mining;

—expenses of longwalling equipment and face haulage machines are cut down;

—expenditures connected with assembly and disassembly of transportation facilities, electric equipment, pipelines as well as precautions, etc. are lowered;

—coal production cost is reduced as output per face jumps at a comparatively minor increase in the number of longwall personnel.

There two variants of LTCC: with coal discharge to the front or rear-mounted armored face conveyor. In the first variant, the discharge opening is located nearby the face, which allows shorter roof support units but top coal is in this case underprepared for self-caving due to small distance from the roof to the canopy. For this reason, even in soft coal, it is required to undertake additional loosening. Furthermore, insufficient coal discharge is accompanied by considerable dusting, which elevates operation riskiness. In the second variant, the conditions of coal discharge to rear-mounted AFC are favorable for deformation and fracture of top coal. On the other hand, it is required to increase dimension of the roof support units and to mount the rear AFC available. Thus, the structure of roof support is complicated and it is necessary to include a rehandling facility at the longwall-and-conveyor drift junction, which creates additional maintenance difficulties [1–4].

These technological variants have a common drawback—the caved coal flow size is limited and, thus, the loss and ash content of coal increase, especially in coarse-size flow. These shortcomings can be eliminated with coal discharge along the whole length of longwall, from all support units simultaneously. In this case, the coal and rock interface caves in concurrently, which allows an area–controllable discharge [5].

Rock flow is adjusted by means of forced–controlled discharging by feeders included in the support units. For implementation of these technologies, the optimized approached to treatment of top coal to ensure physical possibility of coal discharge have earlier been justified: vibro-seismic treatment and multi-stage hydraulic fracturing.

The proposed structure of powered roof support with adjustable coal discharge to AFC is designed with regard to geomechanical behavior of coal and roof rocks, embraces advantages and eliminates disadvantages of the known designs (Figure 1).

Inclusion of feeders for discharge and loading of caved top coal in the structure of powered roof support is a new trend in creation of high-productive technologies in the coal mining industry. In this regard, it is of the current concern to undertake both fundament and applied research into laws and consistent patterns governing rational parameters of powered roof supports with forced-controllable coal discharge.

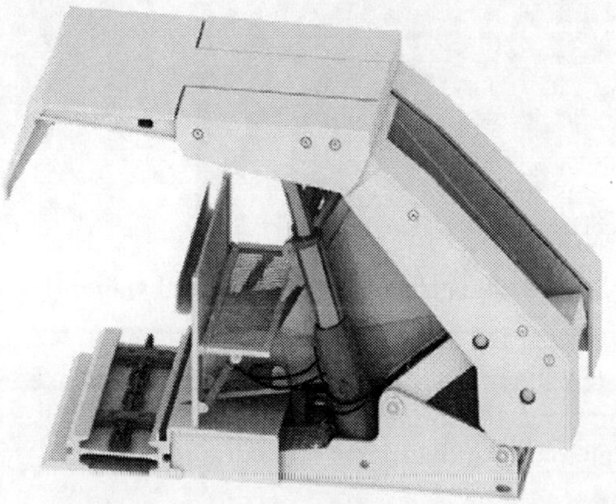

Figure 1. General view of powered roof support unit with adjustable flow of caved top coal to AFC.

2. Laboratory experiment

The main approach to study laws of flow is observation and recording of flowing mass simulating coal with interbeds. For the investigation of controlled caved top coal discharge to AFC during thick coal mining, a laboratory installation [6, 7] as well as the lab test program and procedure [8] were developed.

The lab-test installation allows simulating the process of thick coal cutting with top coal caving and discharge to face area. The body of the installation is a box structure with transparent sides and bottoms made of organic glass 10 mm thick. The transparent walls of the box enable visual observation of flow behavior.

Twenty brassboard units of roof support with relocating devices are placed in the box until their shielding touches the back wall of the box. The brassboard units are made at a scale of 1 : 25 relative to the full-size unit of a powered roof support. Total width of the brassboard unit equals the inner thickness of the transparent box (1600 mm).

Coal flow is simulated by black painted gravel fraction 5–15 mm; dirt is modeled by grey-pinky marble chips 13–25 mm in size. For visualization, the coal-simulating material is parted by a layer of white gravel fraction 5–10 mm (Figure 2a).

IOP Conf. Series: Earth and Environmental Science **262** (2019) 012029 doi:10.1088/1755-1315/262/1/012029

The discharge control system is instrumentation and software capable to measure characteristics of feeder (frequency, amplitude) and operating modes (individual, grouped). The installation is equipped with photo and video recording tools.

The investigations were carried for geological conditions as follows: coal seam thickness—7.5 m; active roof thickness—4 m; cutting layer thickness—3.5 m; top coal thickness—4 m.

Coal was discharged by sequentially actuated feeders on the 1st to the 8th units with their velocity decreasing in the same sequence. When the coal–rock interface approached the discharge openings of the 1st and 2nd units (Figure 2a), these feeders were deactivated and the feeders of the 9th and 10th units were switched on. Then, the units were advanced. The advance step was 32 mm, which corresponded to the advance of the real unit of powered roof support for 0.8 m. After three advances, the discharge process was restarted. All in all, three advances and discharges were performed. Total length of advance made 5.6 m (in real scale terms). The final stage of the laboratory experiment is demonstrated in Figure 2b.

(a) (b)

Figure 2. Wavy discharge flow: (a) initial stage; (b) final stage.

The feeders were actuated sequentially, with the velocity decrease toward the 10th unit. That was the wavy mode of discharge flow. From the 11th to 20th unit, we observed the full-length flow, with all feeders operating simultaneously and at the same velocity (Figure 3).

(a) (b)

Figure 3. Full-length discharge: (a) initial stage; (b) final stage.

Alongside the wavy and full-length flows, the laboratory experimentation procedure provided individual (by each unit one-by-one) and grouped (by a group of 3–5 units simultaneously) modes of discharge.

3. Discrete element method-based modeling

Together with lab tests of controllable coal discharge, numerical modeling of gravity flow of precrushed rocks was carried out in order to find consistent patterns in interaction between the powered roof support units and the discharge material flow. The 3D modeling used the discrete element method. In a space *Oxyz* a domain was set as a parallelepiped bounded by planes directed along the coordinate axes. The length and width of a box filled with granular material are 10 and 2 m, respectively. The fill height is 12.5 m (Figure 4a). The gravity vector is directed along the axis *Oz* vertically downward. The coal seam thickness is 7.5 m, the powered roof support units, 3D modeled in Figure 4b, is 3.5 m high.

(a) (b)

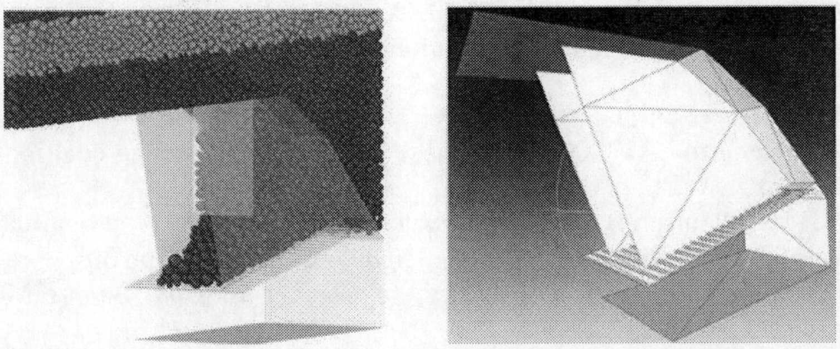

Figure 4. (a) Crushed rock discharge; (b) 3D model of powered roof support unit with gravity discharge.

The model elements of the powered roof support have the following parameters: canopy is 2.5 long; the immobile vertical flow gate meant to limit velocity and volume of granular material flow has a height of 1.75 m; the feeder is 3.575 m long and inclined at 12° relative to horizontal plane; the discharge opening with a width of 1 m is closed at initial stage of flow shaping; the distance between the side shields is 1 m. The feeder surface has corrugation in the form of step plates 0.15 m long arranged in parallel to horizontal plane.

The selected physical parameters of coal and dirt particles (elasticity modulus, Poisson's ratio, density) are typical of coal deposits in Kuzbass. The calculation of forces at the particle contacts took into account dry friction, rolling resistance and viscosity.

The domain was filled with discrete elements using a dynamic algorithm imitating movement of particles under the action of a pre-set external force (gravity). In order to reach maximum density of the packing, friction between the particles and between the particles and the box walls was withdrawn from that stage of modeling. Finally, the equilibrium packing of granular material was represented by a population of discrete elements with total mass of coal particles—58 t and dirt—195 t. Then, the angles of the contact sliding friction, $\varphi_{ij} = 30°$, rolling resistance, $\psi_{ij} = 30°$, between the particles were fixed. When the crushed rock mass reached equilibrium, the discharge opening was opened, and the material started moving until the next state of equilibrium. The latter was ensured by the immobile flow gate and switched-off feeder. Later on, the feeder was set in reciprocating motion (full travel 0.3 m, vibration period 2 s). Owing to friction at the contacts between the particles and feeder, the material was dragged in motion. The crushed material flow became steady-state in some time. The total flow rate of coal in that mode of discharge for 45 s of testing made on average 32 kg/s; the total mass of discharged coal was 1500 kg.

4. Conclusions

As a result of the laboratory simulation and numerical modeling, it has been found that the design of the powered roof support with discharge openings equipped with adjustable capacity feeders enables metered flow of caved top coal to AFC. As a consequence, this allows coal discharging through opening in a number of the support units simultaneously, which provides full-length or wavy discharge, which prevents coal dilution and minimizes losses.

The presented DEM-based model of gravity flow of fragmented rocks in coal mining by the longwall top coal caving technology takes into account basic stages of the process: creation of initial equilibrium state of rick mass; discharge of coal to the feeder of a unit of powered roof support; operation of the feeder and coal discharge in accord with the process chart of the powered roof support.

Acknowledgements
The study has been supported by the Ministry of Education and Science of the Russian Federation in the framework of the Federal Targeted Program on R&D in Priority Areas of Advancement in the Science and Technology of Russia for 2014–2020, Topic: Development of Efficient Robotic Longwall Top Coal Caving Technology and Equipment (Unique Identifier RFMEFI60417X0173).

References

[1] Belov VP and Remezov AV 2008 Experinece of thick getly dipping coal mining *Ugol* No 1 pp 11–14
[2] Alehossein H and Poulsen BA 2010 Stress analysis of longwall top coal caving *International Journal of Rock Mechanics and Mining Sciences* Vol 47 No 1 pp 30–41
[3] Salamatin AG 1997 *Underground Mining of Thick Flat Coal Seams* Moscow: Nedra (in Russian)
[4] Shadulidi IA 2001 *Geomechanical Substantiation of Powered Roof Support Parameters forMining Thick Flat Coal Seams with Top and Interbed Coal Caving* Kemerovo: Kuzbassvuzizdat (in Russian)
[5] Kubanychbek B and Opruk GYu 2010 Full-length coal discharge control *Naukoemk. Tekhnaol. Razrab. Ispolz. Mineral. Resurs.* pp 10–15
[6] Klishin VI, Varfolomeev EL, Borisov IL, Kokulin DI and Khudyntsev EA 2018 Development of a mechanized supports model and laboratory installation for simulating controlled sublevel caving *Naukoemk. Tekhnaol. Razrab. Ispolz. Mineral. Resurs.* No 4 pp 199–203
[7] Kizilov SA, Nikitenko MS and Khudintsev EA 2018 Scheme of management of technical means to ensure the controlled sublevel caving on the face conveyer *Naukoemk. Tekhnaol. Razrab. Ispolz. Mineral. Resurs.* No 4 pp 313–316
[8] Klishin VI, Varfolomeev EL, Borisov IL and Malakhov YuV 2018 Substantiation of the program and methodology of laboratory studies of controlled sublevel caving *Naukoemk. Tekhnaol. Razrab. Ispolz. Mineral. Resurs.* No 4 pp 196–199

Influence of energy characteristics of vibropercussion source on efficiency of soil plug removal from casing pipes

AS Kondratenko, AM Petreev, AYu Primychkin and AS Smolentsev

Chinakal Institute of Mining, Siberian Branch, Russian Academy of Sciences, Novosibirsk, Russia

E-mail: *kondratenko@misd.ru

Abstract. This article describes the method of virbopercussion ramming of an open-ended steel pipe in soil. The study focuses on interaction between the pipe and internal soil plug by the law of dry friction. The test bench for the effect exerted by impact impulses on displacement of soil plug is presented. The reduction in the friction force under the simultaneous static influence on the soil plug and vibropercussion load on the pipe is evaluated. The influence of energy characteristics of vibropercussion source on the efficiency of soil plug removal from the pipe is determined.

1. Introduction

Trenchless laying of underground utility lines is actually an efficient alternative to shoveling and sometimes is the only way of achieving the goal. The method of vibropercussion driving of open-ended pipes finds increasingly wide application these days [1]. Efficiency of this method is connected with high rate and accuracy of penetration in soil cut by the front edge of the driven open-ended pipe [2]. Soil enters the pipe drive under dynamic impacts, gets compacted and eventually forms a plug. After that new batches of soil cannot enter the pipe and are pressed into walls of the hole, which increases radial deformation in surrounding soil body while the pipe penetrates at far smaller rate. Thus, removal of soil plug during pipe driving is an important factor governing efficiency of the technology [3].

2. Soil plug removal: laboratory test results

In this connection, the method of vibropercussion driving with soil plug removal by batches has demonstrated practical success [4]. Figure 1 shows schematic of the technology with pipe ramming by pneumatic hammer.

In this technology, to the tailpiece of the pipe *1*, via a separate line component *2*, a fluid *3* is fed. As soon as a soil plug *4* is formed, the fluid is fed under pressure into the pipe through a hole *4*. The fluid cuts off a batch from the plug and displaces it toward the back end of the pipe *1* where the soil batch is ejected through a discharge opening to the pit *7*. The pipe ramming process continues until the soil plug is formed again. The ramming and plug removal cycles are repeated until the pipe comes in to the inlet pit [5]. It has been found under production conditions that owing to soil plug removal by batches, it is possible to accelerate ramming, reduce energy consumption of the process, and improve degree of mechanization and accuracy of pipe installation.

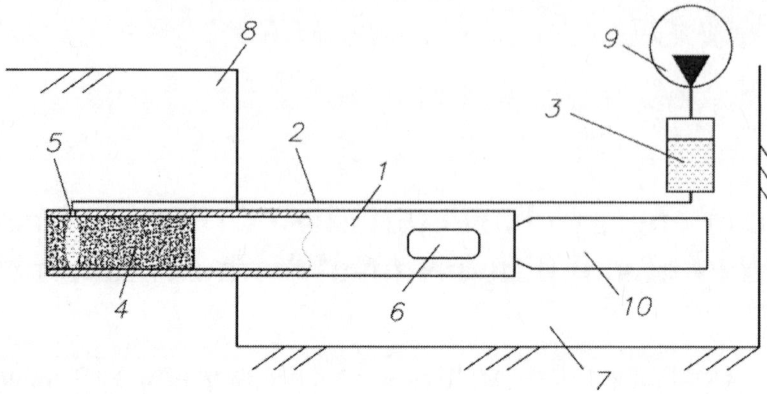

Figure 1. Pipe ramming with soil plug removal by batches: *1*—driven pipe; *2*—separate line component; *3*—ejecting fluid container; *4*—soil plug; *5*—feed hole for ejection fluid; *6*—discharge opening; *7*—working pit; *8*—soil body; *9*—compressor; *10*—pneumatic rammer.

In vibropercussion ramming, additional factors to influence separation and displacement of soil plug are vibration of the pipe and adjacent soil as well as elastic deformation of the pipe walls under impacts [6, 7]. It is known that internal and external friction coefficients can be reduced under the action of alternating loads [8, 9]. However, these processes are studied insufficiently, and their favorable effect on removal of soil plug by batches needs to be analyzed additionally using physical analogs.

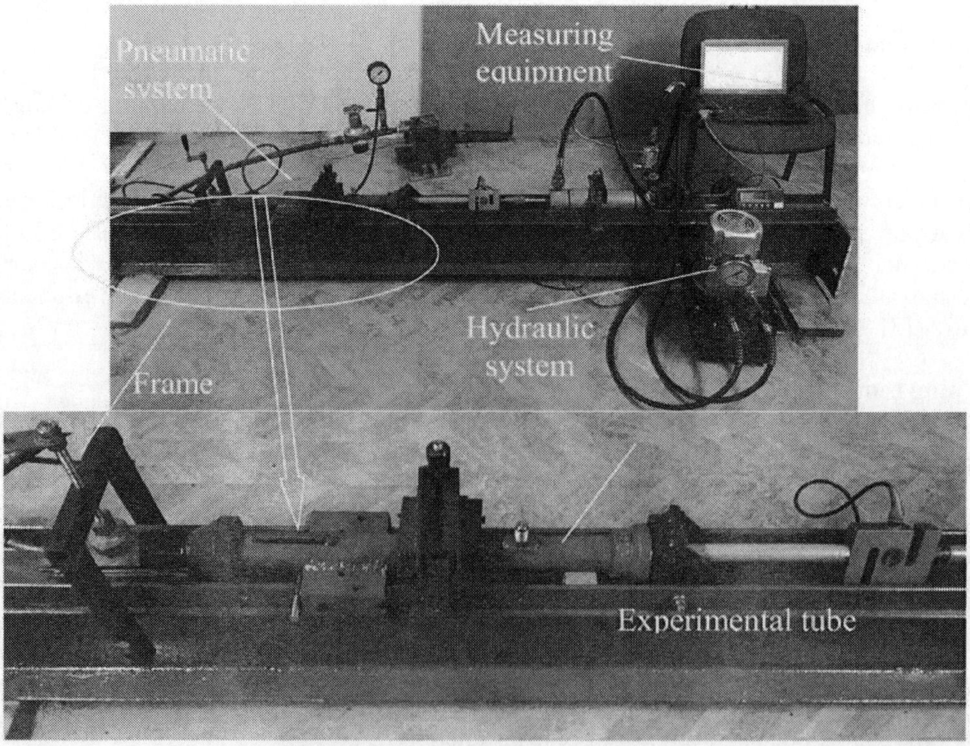

Figure 2. Test bench: general view.

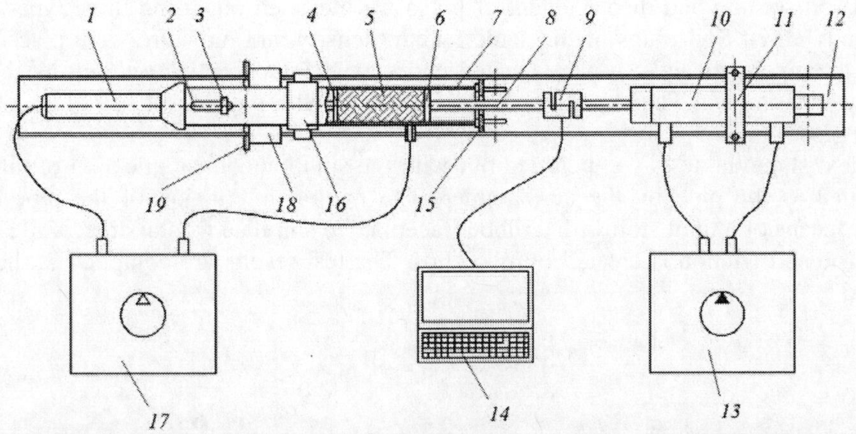

Figure 3. Test bench schematic: *1*—air hammer; *2*—visual control channel; *3*—sliding element; *4*—measurement piston; *5*—soil batch; *6*—loading piston; *7*—pipe; *8*—rod; *9*—force sensor; *10*—cylinder; *11*—frame; *12*—hinged support; *13*—pumping station; *14*—computer; *15*—limit stop with a faceplate; *16*—clip; *17*—compressor.

Figures 2 and 3 present the lab test bench designed to determine how the impact load influence efficiency of soil batch displacement. The thick metal frame of the test bench is made of I beams. The pipe with a diameter of 48 mm is pressed to the frame by clips with rubber inside to crate friction force variable depending on the pressing force. Inside the pipe, a cylinder rod is placed. The cylinder is hinged to the frame. The force sensor mounted on the rod records the drag of a soil batch relative to the pipe and sends the data to a computer. At the opposite end of the pipe, an air hammer is arranged (piston weight is 0.5 kg, impact frequency is 8–28 Hz, blow energy is 1.8–5 J). For convenience of observation over the soil batch motion, an axial channel is made in the pipe, with an output sliding element rigidly connected with the piston. The equipment set includes air compressor and pumping station to power the hammer and cylinder, respectively.

For the repeated experimentation, soil plug was man-made. In order that the test conforms with a real process, density and moisture of a soil plug formed in the same length pipe in ramming in natural soil were measured and then used to form artificial soil plugs (Table 1). In each test series, a soil plug was formed by compaction of filled soil under axial force of 4 kN. For all types of soil, the length of the plug was 150–155 mm.

Table 1. Testing results.

Experiment	Clay		Loam		Sandy clay	
Content of clay particles, %	62		39		21	
Batch length, mm	155		150		150	
Density. kg/m^3	1960		1840		1860	
Moisture, %	14		13		13	
Type of loading	Statics	Vibro	Statics	Vibro	Statics	Vibro
1	3.0	1.8	2.76	1.53	2.5	0.38
2	2.8	2.0	2.35	1.4	2.8	0.44
3	2.8	2.0	3	1.26	2.6	0.32
4	2.4	1.7	2.6	1.29	2.6	0.36
5	2.7	1.9	2.5	1.47	2.7	0.34
Average	**2.7**	**1.9**	**2.5**	**1.38**	**2.6**	**0.37**

The tests on starting and displacement of plugs were carried out using three types of soil (clay, loam and sandy clay). Soil plugs of the same length, density and moisture were placed at the same distances to the pipe ends and ten were ejected under axial force applied by the rod; the moving-off resistance was determined. The experiments were repeated with ejection of soil plugs by compressed air fed in the pipe.

At the next stage the tests were carried out with the simultaneous application of vibropercussion loading applied to the pipe by the air hammer. The natural interaction of the pipe and soil was simulated by means of a limit stop with a rubber faceplate to simulate frontal drag, while the resistance due to pipe and soil friction is created by two clips. The test results are compiled in the table and in Figures 4 and 5.

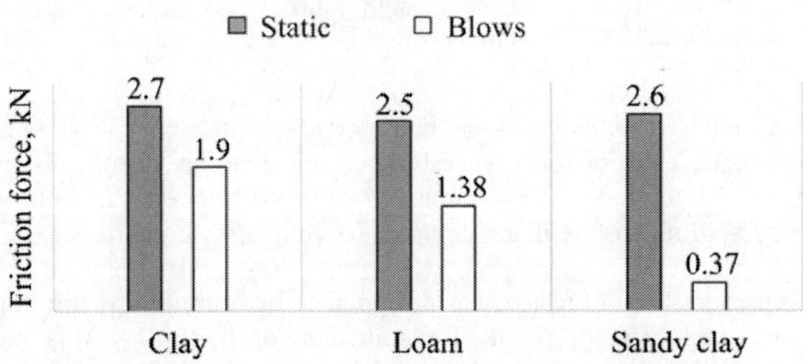

Figure 4. Frictional resistance of soil batch.

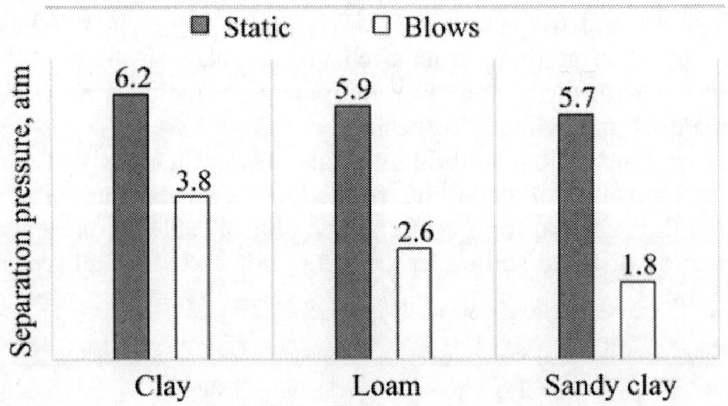

Figure 5. Compressed air separation of soil batch.

Furthermore, the influence of energy characteristics of the vibropercussion rammer on efficiency of soil plug removal was assessed. The vibropercussion source was chosen to be Typhoon air hammer designed at the Institute of Mining, SB RAS.

The tests of the blow energy effect were carried out using loam type of soil. The blow energy was adjusted by variation of energy source pressure in the feed line, and using different cross-section air bleeds installed in the feed line of the back run chamber (Figure 6).

The tests show that the friction between the soil plug and pipe walls reduces with increasing blow energy. The curve of the friction force as function of the unit blow energy of air hammer is presented in Figure 7.

IOP Conf. Series: Earth and Environmental Science **262** (2019) 012030 doi:10.1088/1755-1315/262/1/012030

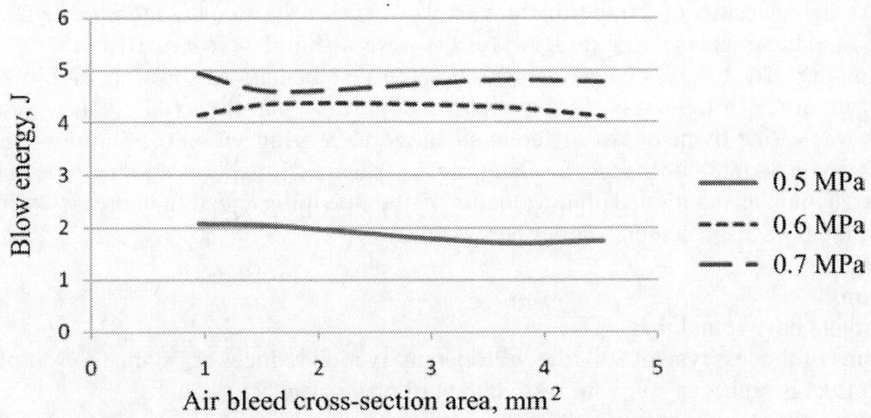

Figure 6. Air hammer blow energy versus flow area of air bleed.

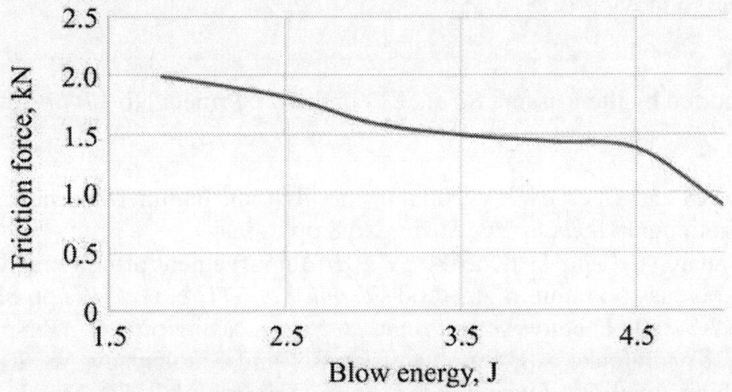

Figure 7. Friction force and blow energy curve.

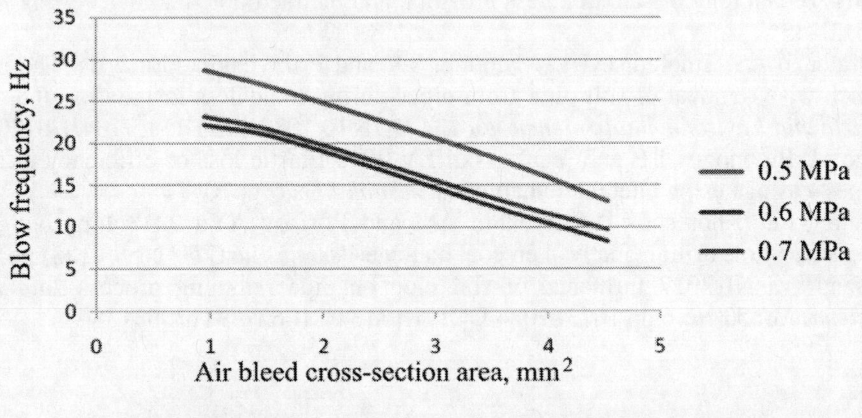

Figure 8. Blow frequency versus flow area of air bleed.

In the experiments with loam, the influence of blow frequency of the air hammer on the friction force was evaluated under constant energy of blows at 4.5 J, the blow frequency was varied from 8 to 28 Hz using air bleeds with different flow areas installed in the feed line of the back run chamber of the hammer.

After averaging results of 5 experimental series of each test, no evident dependence between the friction force and the impacts was found. All results were within the error of 10%.

The damping effect was studied using the same test bench. Absorption of blow energy was regulated by the force of tightening of the clips *16* (Figure 3), and the axial displacement of the pipe under blows was varied by means of different stiffness rubber ring set on the damping unit *15* (Figure 3). The tests revealed no influence of damping on the friction between the pipe and soil plug. However, it should be mentioned that velocity of the soil plug ejection increases with the higher amplitude of the pipe displacements under percussions.

3. Conclusions

Thus, the authors have found that:

1. In terms of the test type of soil, the vibropercussive lod reduces resistance to soil plug moving – off and this effect is higher in soil with less content of clay particles.

2. As the unit blow energy of the vibropercussion rammer is increased from 1.8 to 5 J, the moving-off resistance reduces 2 times.

3. The variation in the blow frequency and damping parameters has no considerable influence of the soil plug resistant to moving-off.

Acknowledgements

The study was supported by the Russian Science Foundation, Project No 17-77-20049.

References

[1] Kondratenko AS and Chervov VV 2006 Typhoon-70 air hammer and new method of soil plug removal from pipe *Mekhanizats. Stroit.* No 8 pp 8–12

[2] Gileta VP, Vanag YuV and Tishchenko IV 2016 Imporvement of hole making efficiency by the vibration–percussion rammin gmethod *Vestnik KuzGTU* No 6 (117) pp 82–89

[3] Kondratenko AS 2008 Features of soil removal from a casing pipe *GIAB* No 7 pp 327–331

[4] Danilov BB, Kondratenko AS, Smolyanitsky BN and Smolentsev AS 2017 Improvement of pipe pushing method *Journal of Mining Science* Vol 53 No 3 pp 478–483 DOI: 10.1134/S1062739117032391

[5] Kondratenko AS, Timonin VV, Abirov AA, Gosmanov MK, Esenov BU and Zharkenov EB 2014 Technology of safe trencless hrozintal and inclined hole making *Vestnik KuzGTU* No 1 pp 40–45

[6] Kondratenko AS, Smolentsev AS, Timonin VV and Primychkin AYu 2018 The perspective of batch-wise removal of soil plug from pipes during trenchless installation *IOP Conf. Series: Earth and Environmental Science* Vol 134 DOI :10.1088/1755-1315/134/1/012029

[7] Gromov I, Belonogov LB and Yankovskii LV 2010 Elastic loss of efficiency in intsallation of pipes–casings using pneumatic hammers *Vestnik PGTU* pp70–74

[8] Mametiev LE, Khoreshok AA, Nekhin AM and Borisov AYu 2018 Improving effiiency of degassing hole driling and broken coal haulage *Vestnik KuzGTU* No 1 (123) pp106–112

[9] Aleksandrova NI 2017 Influence of soil plug on pipe ramming process *Journal of Mining Science* Vol 53 No 6 pp 1073–1084 DOI: 10.1134/S1062739117063138

IOP Conf. Series: Earth and Environmental Science **262** (2019) 012031 doi:10.1088/1755-1315/262/1/012031

Engineering solutions for casing drilling in pre-mine drainage

AS Kondratenko and IO Shakhtorin*

Chinakal Institute of Mining, Siberian Branch, Russian Academy of Sciences, Novosibirsk, Russia

E-mail: *dodysya@yandex.ru

Abstract. The authors discuss application of the casing while drilling technology in pre-mine drainage. The currentness of this technology and its advantages are proved. The scope of the discussion embraces operating principle of the equipment used in casing drilling. A new method for batch removal of drill fines from large-diameter boreholes is described. The economic efficiency of the technology is evaluated in terms of drainage holes drilled using rig SANDVIK DE-880 in Kirov mine at SUEK-Kuzass.

1. Introduction

Drilling and casing in difficult ground conditions is a challenge both to mining industry and civil construction. Instable rocks constitute risk of sloughing in wells and, as a consequence, impractibility of casing [1, 2]. Furthermore, drilling equipment can be lost in wells and never pulled out [3, 4].

The casing drilling systems are intended for drilling wells and holes in difficult ground conditions (boulder beds and cobble deposits, water inflows, etc). Systems of pneumatic percussion drilling are expedient to be used in casing drilling in ground conditions favorable for DTH hammers [5, 6].

To this effect, a DTH hammer is equipped with a special rock-breaking tool of original design which allows drilling diameter larger than casing diameter. During drilling, a casing pipe goes down the hole by gravity (Figure 1a). Vibration of the rotary–percussive mechanism facilitates the casing pipe driving by reducing the force of friction at the rock–casing interface. After drilling is completed, dimensions of the rock-breaking tool can be diminished to the internal diameter of the casing, and the tool is freely pulled out of the well (Figure 1b). If necessary, drilling can be continued using a standard cylindrical tool (Figure 1c) [7, 8].

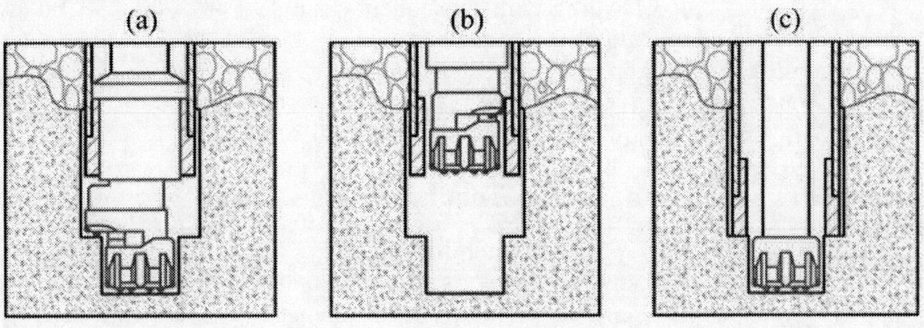

Figure 1. Operation of the hammer drill bit: (a) casing drilling; (b) removal from the well; (c) drilling without casing.

IOP Conf. Series: Earth and Environmental Science **262** (2019) 012031 doi:10.1088/1755-1315/262/1/012031

2. Casing drilling for methane drainage

According to safety regulations for gas outburst-hazardous coal mining, the mines use the method of pre-mine drainage with holes drilled from ground surface to a coal seam. SUEK-Kuzbass mines practice the method of rotary drilling SANDVIK DE-880 multi-purpose drill rig. First a well is drilled, then different diameter casing is installed with cementing of annular space.

According to specifications of drainage hole drilling from ground surface to longwall no. 24-60 in Boldyrevsky coal seam in Kirov Mine, design of the casing string has the parameters as shown in Figure 2a. Hole drilling is followed by different diameter casing and cementing of annular space.

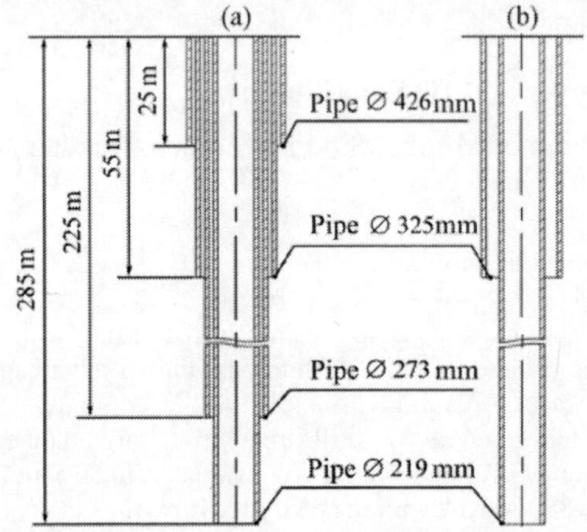

Figure 2. Well design: (a) according to drilling specifications for Kirov Mine; (b) casing drilling technology.

The drilling process has four stages in conformity with the number of casing pipes. The telescopic step-wise design of a well is governed by the risk of wedging of a pipe during its installation after the drilling tool is pulled out. As a rule, the length of each step is determined based on the exploration data, and transition to smaller diameter drilling is undertaken in the zone of change in properties of rocks.

Application of the casing drilling technology in this specific case will allow considerable reduction in the time of drilling and casing operations, cost saving due to elimination of two connecting pipes (Figure 2b) as well as abridgement of upward/downward operations and cementing volume.

Table 1. Durations of drilling and volumes of drilled solids.

Hole depth, m	Casing pipe diameter, mm	Hole diameter, mm	Volume of drilled solids, m^3	Drilling capacity, m/shift*	Drilling duration, shift*
Drilling with subsequent casing					
25	426	490	4.71	5–15	1.67–5
55	325	393.7	3.65	8–20	1.5–3.75
225	273	304.5	12.37	10–25	6.8–17
285	219	244.5	2.82	20–40	1.5–3
			Total: 23.55		**Total: 11.47–28.75**
Casing drilling					
55	325	348	5.23	9–22	2.5–6.11
285	219	240	10.40	21–41	5.61–10.95
			Total: 15.63		**Total: 8.11–17.06**

*Work shift is 12 h.

In order to evaluate economic efficiency, let us compare durations of drilling and volumes of drilled solids in drainage hole design using the casing while drilling technology and in drilling with subsequent casing (Table 1).

It is seem from the Table 1 that the casing drilling technology shortens the period of drilling by 30–40% and reduces the volume of drilled solids by 34%. The fact that the two-step hole drilling needs drilling equipment of two rather than four standard sizes is speaks in favor of the efficiency of the proposed approach.

Casing drilling in subsurface soils and weak rocks can go without drilling machines. The casing pipe is driven in soil by pneumatic hammers in this case [9]. Vibropercussion drilling technique can drive pipes with a diameter from 102 to 3600 mm to a depth to 120 m. Material of pipes is limited to steel as a rule, due to considerable forces transferred to the pipe while driven. This method is applicable in wide range of rocks—from sand, silt and plastic clay to weathered rocks, chalky clay and schist [10].

When a pipe is subjected to dynamic driving, rocks enter inside the pipe, compact gradually and form a plug. As a consequence, new portions of rocks cannot enter the pipe and are pressed in the hole walls, which intensifies radial deformation in surrounding rock mass, and the pipe is penetrated in rocks at a much lesser rate down to complete cessation of the process [11, 12]. Thus, timely removal of soil plugs from pipes is an important factor governing efficiency of the whole technology [13].

In the conditions of vertical pipe driving, it seems logic and wise to remove soil plug from the pipe by batches without auxiliary equipment. The implementation of such technology in pipe driving using pneumatic hammer is depicted schematically in Figure 3 [14].

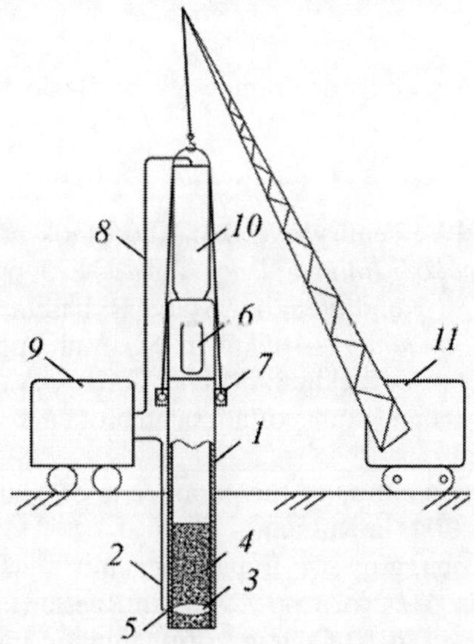

Figure 3. Pipe driving with removal of soil plug by batches: *1*—driven pipe; *2*—air line; *3*—air gap; *4*—soil; *5*—compressed air supply hole; *6*—adapter with discharge hole; *7*—tightening device; *8*—air supply hose; *9*—compressor; *10*—air hammer; *11*—lift crane.

In this manner, compressed air is supplied to bottom hole end of the pipe *1* via a separate line *2*. As soon as a soil plug *4* is formed, compressed air is fed under pressure inside the pipe through the hole *5*. The air cuts a batch from the plug, generates am air gap *3* and displaces the soil batch upward the pipe *1* to be ejected to the pit through the discharge hole of the adapter *6*. This sequence of operations is repeated in the course of the soil plug formation until the project depth of the pipe penetration is reached. Soil batches can be removed from the free end of the pipe but this needs the air hammer to be disassembled.

The whole process of batch-wise removal of soil plug can be divided into three stages: formation of a soil batch, separation of the soil batch from the plug and displacement of the batch along the pipe up to removal. It is important that the soil batch is compact in the zone of cutting and fills the whole cross-section of the pipe. This is required to create and maintain compressed air pressure sufficient to cut and displace the soil batch to be then removed. At the same time, it is clear that is the plug is over-compacted, it can be impossible to cut a batch, or in case of loose soil, an air pocket can be formed, which will result in the pressure drop in the yielding medium. A signal of the plug formation and removal is a drop in the penetration rate. The other factors influencing cutting and removal of soil plugs is vibration of the pipe in adjacent rock mass and elastic deformations of the pipe walls under impacts, which can lower the drag force by 30–90% [12, 15].

Replacement of the large-diameter drilling with subsequent casing in weak rocks by impact driving eliminates the risk of loss of drilling equipment or wedging of the casing pipe due to damage or erosion of hole walls. Furthermore, impact driving in the starting interval of a vertical hole excludes expenditures connected with the larger diameter drilling tools and operating costs of drilling rigs at this stage.

3. Conclusions

The application of the casing drilling technology in drainage hole drilling from ground surface will reduction of drilling and saving of drilling cost due to increased productivity of the process and shortened period of upward/downward operations. The methods proposed in this paper will make it possible to avoid accidents connected with the hole wall sloughing and, as consequence, wedging of casing pipes in holes.

Acknowledgements

This study has been performed within the framework of Basic Research Program No AAAA-A17-117122090003-2.

References

[1] Lekontsev YuM, Sazhin PV, Temiryaeva OA, Khoreshok AA and Ushakov SYu 2013 Two-side sealer operation *Journal of Mining Science* Vol 49 No 5, pp 757–762

[2] Patutin AV, Timonin VV, Kondratenko AS and Rybalkin LA 2014 Integrated deep-borehole study of coal beds *J. Fundament. Appl. Min. Sci.* Vol 1 pp 23–26

[3] Mametev LE, Khoreshok AA, Tsekhin AM and Borisov AYu 2018 Increasing the efficiency of drilling degassing wells and transportation of broken coal *Vestnik KuzGTU* Vol 1 pp 106–112

[4] Shakhtorin IO 2017 Determination of the parameters of a submersible hammer *Interexpo GEO-Siberia* Vol 2 pp 199–204 (in Russian)

[5] Shakhtorin IO 2017 Debugging of impact machines using modern software software *J. Fundament. Appl. Min. Sci.* Vol 1 pp 72–77 (in Russian)

[6] Repin AA and Alekseev SE 2010 Creation of pneumatic impact tools for work on the elevated pressure of the energy carrier *Fundamental Problems of Geo-Environment Formation under Industrial Impact: Int. Conf. Proc.* Vol 3 pp 324–329 (in Russian)

[7] Petreev AM and Primychkin AYu 2016 Ring-type elastic valve operatin in air hammer drive Journal of Mining Science Vol 52 No 1 pp 135–145

[8] Primychkin AYu 2017 Ring-type elastic valve in the model of a pneumatic percussion machine *J. Fundament. Appl. Min. Sci.* Vol 4 No 1 pp 39–44 (in Russian)

[9] Chervov VV and Kondratenko AS 2006 Investigation of operating modes of balanced sealer *Mekhanizats. Stroit.* No 8 pp 8–12

[10] Schrank JS, Havekost MD and Njoloma SM 2009 Pipe ramming under three active railroad lines in difficult soil conditions International *No-Dig Show 2009* The North American Society (NASTT) and the International Society for Trenchless Technology (ISTT) Toronto Paper F-2-05 pp 1–10

[11] Gileta VP, Vanag YuV and Tishchenko IV 2016 Increasing the efficiency of drilling wells with the method of vibro-impact punching *Vestnik KuzGTU* Vol 6 pp 82–89

[12] Anderson WK, Thomas JL and van Leer B 1986 Comparison of finite volume flux vector splittings for the Euler equations *AIAA J* Vol 24 No 9 pp 1453–1460

[13] Möller M, Charypar D and Gross M 2003 Particle-based fluid simulation for interactive applications *Proceedings of the 2003 ACM SIGGRAPH, Eurographics Symposium on Computer Animation* Aire-la-Ville pp 154–159

[14] Monaghan JJ 1992 Smoothed particle hydrodynamics *Annual Review of Astronomy and Astrophysics* Clayton pp 543–574

[15] Weiss JM, Maruszewski JP and Smith WA 1999 Imlicit solution of preconditioned Navier–Stokes equations using algebraic multigrid *AIAA J* Vol 37 No 1 pp 29–36

IOP Conf. Series: Earth and Environmental Science **262** (2019) 012032 doi:10.1088/1755-1315/262/1/012032

Mechanism of physisorption of collectors in activation of non-sulfide mineral flotation by metal ions

SA Kondratiev and DM Tsitsilina

Chinakal Institute of Mining, Siberian Branch, Russian Academy of Sciences, Novosibirsk, Russia

E-mail: kondr@misd.ru

Abstract. The authors address the problem of activation of non-sulfide minerals by metal ions. A brief analysis of modern ideas about the mechanism of activation of rutile, ilmenite, and fluorite by copper sulfate or lead nitrate is given. It is shown that modern thermodynamic ideas about the mechanism of activation incompletely agree with the accumulated experimental and practical information on flotation with the use of activators. A new mechanism of mineral flotation activation based on the physically adsorbed species of collectors is proposed. The proposed mechanism considers the kinetics of the formation of flotation contact and the influence exerted by the collector–activator associates stabilized by the collector anions.

1. Introduction

The research into the influence exerted by salts of polyvalent metals on flotation of oxides, silicates and salt-forming minerals using radioactive isotopes shows that cations of metal in many cases activate flotation at low consumption of salts. According to [1], salt consumption should "equal the amount capable to attach to mineral surface owing to the increased consumption and attachment strength of an anion collector. The higher consumption of salt suppresses flotation as the collector is bound in pulp and oleates of polyvalent metals, possessing no collecting abilities, are generated. Even in small amount, ferric chloride suppresses cassiterite" (Figure 1).

2. Non-sulfide minerals activation by metal ions: a review

Collectability of hardly dissolvable oleates of different metals in flotation of nonsulphide minerals is highlighted in [2]. The outcome of flotation by finely dispersed fresh precipitates of oleates of polyvalent metals is equivalent to flotation by initial carbonic acid.

In [3] the data on flotation of fluorite by tagged precipitates of ^{14}C-bearing calcium tridecylate (see Table 1). The precipitates had different dispersiveness. The collector consumptions was equivalent to 300 g/t of sodium tridecylate, pine oil consumption was 40 g/t, and S : L = 1 : 14.

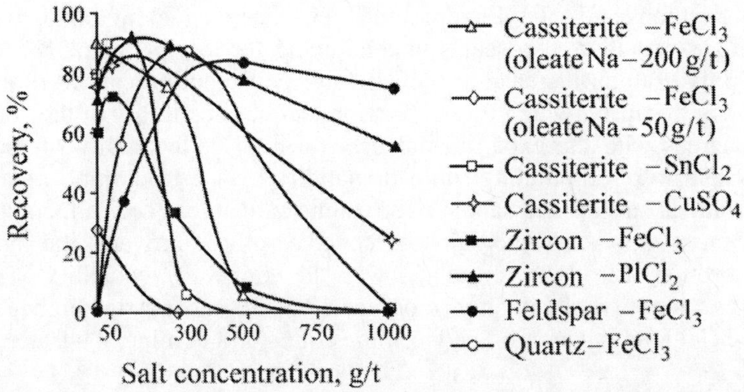

Figure 1. Effect of concentration of salts of polyvalent metals on mineral recovery by sodium oleate [1].

Table 1. Calcium tridecylate precipitate flotation of fluorite [3].

Collector	Recovery, %	Adsorption of tridecylate		
		$\times 10^{-10}$ mole/cm^2	% of initial	pH
Sodium tridecylate solution	93.8	4.860	74.6	6.01
Finely dispersed fresh precipitate of calcium tridecylate	92.0	5.550	85.5	6.30
Finely dispersed precipitate of calcium tridecylate after aging for a few hours before flotation	88.2	5.175	78.6	6.10
Coarsely dispersed precipitate after mixing concentrated solutions of sodium tridecylate and calcium chloride	78.1	4.285	66.0	6.30
Coarsely dispersed precipitate of calcium tridecylate in mixing in smaller volume	79.2	4.730	72.8	6.61
Dried precipitate of calcium tridecylate	11.0	0.93	14.3	6.4

It follows from the table that precipitates are not waste of reagent but possess almost the same collectability as the reagent solution. The authors of [3] suppose that precipitates of fat acid salts exert water-repellent action due to adhesion to the mineral surface or as a result of adsorption of fat acid anions, which shifts equilibrium towards further dissolving of precipitate. This explanation of high collecting ability is assumed as unsatisfactory. First, precipitates attach to mineral surface insufficiently strong for the flotation aggregate to survive in the turbulent flows of slurry and, second, precipitates are represented as separate micro-particles than the cover on the surface of mineral to be recovered. The micro-particles create extra chemical heterogeneity on the mineral.

Rutile flotation with salicyl hydroxamic acids is studied in [4]. The activator was lead nitrate. The authors suppose that hydroxyl lead species $Pb(OH)^+$ interact with the Brønsted centers of Ti–OH and form surface compounds of Ti–O–Pb^+ on rutile. The surface compound has facilitating effect of adsorption of $C_7H_7NO_3$ and floatability of rutile. Attachment of the agent is governed by chemosorption and ion–electrostatic interaction. It is emphasized that the maximum change in ζ-potential of the mineral surface at the consumption of $Pb(NO_3)_2$ of $2.5 \cdot 10^{-4}$ mole/l and acid of $5 \cdot 10^{-4}$ mole/l is observed in the range of pH of pH 6.3. In the same pH range, the maximum recovery of rutile is noticed. Rutile floatability with salicyl hydroxamic acid at the concentration of $5 \cdot 10^{-4}$ mole/l drops when the lead nitrate concentration is raised above $2.5 \cdot 10^{-4}$ mole/l. The authors explain the drop in floatability by competition of hydroxyl ions and salicyl hydroxamic acid ions for making bonds with the lead on the mineral surface. This explanation also seams unconvincing for low pH of 6.3. Another cause of the drop in flotation ability with an increase in the activator consumption can be formation of lead and salicyl hydroxamic acid in slurry, which means fixation and inaccessibility of the collecting agent. The latter statement disagrees with the experimental data on high collectability of precipitates (see the table) [3]. The mechanism of stimulating rutile floatability by the Coulomb interaction between the positively charged mineral surface and salicyl hydroxamic acid anions seems incompletely realistic. In the range of low concentrations of activator, the conclusions are proved—the increased shift of ζ-potential due to adsorption of lead ions correlates with the improved floatability of the mineral. In the range of the higher consumption of activator (more than $2.5 \cdot 10^{-4}$ mole/l), the higher shift of ζ-potential and the forces of the Coulomb interaction between the collector and mineral surface brings no positive effect.

The authors connect the mineral floatability strongly with the hydrophobic behavior of its surface and pay no attention to the kinetics of formation of flotation contact between air bubble and mineral particle. For this reason, the lead nitrate is assumed as a conditioner of $Ti–O–Pb^+$ compounds which give positive charge to the mineral surface.

An increase in the floatability of ilmenite with lead ions is illustrated in [5]. The inactivated ilmenite micro-flotation with oleate was efficient in the pH range of 6–8. The maximum adsorption was observed at pH 4 and it was higher than adsorption in the pH range of 6–8, i.e. in the range of the maximum recovery of ilmenite. Adsorption of the collector gradually decreased with growing pH. In the optimum range of pH, the collector derivative $(RCOO)_2H^-$, by the authors' opinion, has the highest effect of the floatability of the mineral.

After addition of lead ions, both adsorption of the collector and floatability of ilmenite grew. The rate of adsorption increased from $8.6 \cdot 10^{-7}$ to $11.3 \cdot 10^{-7}$ mole/m^2, and the mineral recovery rose from 44.08 to 74.18%. The optimum range of pH remained the same. The authors think that lead ions increase the energy of the chemical bonding between oleate-ions and ilmenite. As a result of treatment of ilmenite with oleate, by the data of X-ray photoelectron spectroscopy, the bonding energy of Fe2p at the surface of ilmenite changed by 0.3 eV; after treatment by lead ions and oleate—by 0.4 eV. The chemical shift of other metals in the ilmenite lattice was unaltered. The authors suppose that the increase in ilmenite floatability after addition of lead is governed by the change in the energy of the chemical bonding between oleate and ferrum ions. Cations of titanium and calcium are the places of physical sorption of the collector.

In the optimum pH range of 6–8, i.e. in the range of the maximum floatability, lead-activated ilmenite changes the sign of ζ-potential of the surface from positive to negative (Figure 2). The mineral floatability remains maximum, which means that flotation properties of the mineral are thanks to both physisorption and chemisorption.

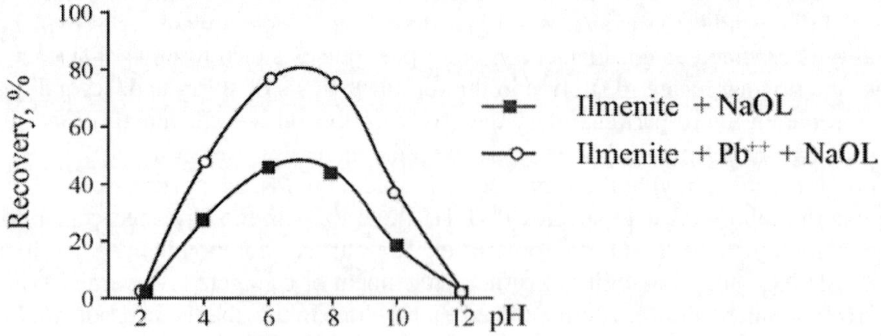

Figure 2. Recovery of ilmenite versus pH. Concentration: sodium oleate—$1 \cdot 10^{-5}$ M, lead nitrate—$1 \cdot 10^{-4}$ M [5].

IOP Conf. Series: Earth and Environmental Science **262** (2019) 012032 doi:10.1088/1755-1315/262/1/012032

No explanation is given to which properties of the associates $(RCOO)_2H^-$ are connected with the high floatability of the mineral. Nor the authors disclose the cause of low floatability in the acidic range of pH.

Effect of one- and bi-valent salts (NaCl, $Ca(Cl)_2$) on floatability of artificial malachite with sodium oleate has been investigated in [6]. The medium pH is chosen as 9.5. It is found that an increase in the sodium chloride concentration from $1\cdot10^{-6}$ to $1\cdot10^{-3}$ M results in a gradual rise in the floatation ability of malachite treated in solution with the reagent concentration of $2\cdot10^{-5}$ M. The curve of the malachite floatability with sodium oleate versus the concentration of Ca^{2+} ions is an extremum function. At $Ca(Cl)_2$ concentration higher than $3\cdot10^{-5}$ m, the mineral floatability drops. The authors explain the gradual increase in the floatability with sodium ions by the compression of the double electric layer of bubbles and the drop in the floatability with calcium ions at the concentration higher than $3\cdot10^{-5}$ M by their adsorption at the hydrophobizated surface of malachite and by the increase in its hydratation capacity. Formation and consolidation of calcium oleate precipitates is beyond the scope of the discussion.

Influence of ionic force of solution of adsorption of sodium oleate at the surface of apatite is studied in [7]. The topography and phase contrast of apatite crystal before and after treatment by the reagent solution are estimated. It is found that after treatment, the surface roughness grows from 25.3 to 288 nm. The authors think that calcium dioleate in the form of associates precipitates at the mineral surface and makes it better floatatable. However, the authors leave unspecified the treated surface properties that contribute to flotation. Formation of coarse associates 288 nm in size implies that these properties have no connection with the increased hydrophobic behavior of the mineral surface.

The review of the experimental studies into flotation of rutile, ilmenite, fluorite and other minerals shows that the results are explained by the mechanism of physisorption. Precipitate is represented by dispersed particles the structural units of which are the metal and collector compounds and the size is stabilized by the collector anions. These particles attach to the hydrophobic mineral surface in accord with the law of balancing of polarities of the contact media and, for this reason, cannot reduce the surface wettability. Upon the rupture of interlayer, such particles can function as a removal of kinematic constraint of the particle–bubble flotation contact. The increase in the concentration of metal and approach to stoichiometric ratio of reaction components results in enlargement of particles of the colloid system. In this case, colloids contain coarse dispersed particles with their surface activity relative to gas–water interface less than the surface activity of colloids with fine dispersed particles. The spreading rate of colloids with the coarse dispersed particles is also lower than the spreading rate of colloids with the fine dispersed particles. Precipitated on the mineral surface from the solution, such colloids fail to exert considerable influence on removal of water from the interlayer. The drop in the floatability with the increase in the flow rate of the activator [4, 5] is connected with the described phenomenon. Flotation of fluorite with the tagged precipitate of calcium tridecylate illustrated the influence of the precipitate coarseness of the mineral recovery [3].

3. Conclusions

Activation of flotation of non-sulphide ore has been discussed in the framework of thermodynamics of flotation event. Formation of precipitates of collectors with multi-valent metals and associates is understood as additional hydrophobization of mineral surface. The drop in the floatability under concentrations higher than a threshold value is explained by the hydrophilization of the recovered minerals with hydroxides irrespective of pH of the medium. It is shown that the increased floatability of minerals with the heavy metal ions can be explained in the framework of the kinetics of flotation event. The collector–metal ion precipitation removes the kinetic constrain of the bubble–particle flotation contact.

Acknowledgements

This study was carried out in the framework of R&D project, Registration No. AAAA-A17-117092750073-6.

References

[1] Polkin SI 1987 *Beneficiation of Rare and Precious Metal Ores* Moscow: Nedra (in Russian)
[2] Eigeles MA 1964 *Elements of Flotation of Nonsulphide Minerals* Moscow: Nedra (in Russian)
[3] Bogdanov OS, Podnek AK, Kainman VYa amd Yanis NA 1959 Theory and technology of flotation *Mekhanobr Institute Trasnactions* Isue 124 (in Russian)
[4] Li H, Mu S, Weng X, Zhao Y and Song S 2016 Rutile flotation with Pb^{2+} ions as activator: Adsorption of Pb^{2+} at rutile, water interface *Colloid and Surfaces A: Physicochemical and Engineering Aspects* No 506 pp 431–437
[5] Chen P, Zhai J, Sun W, Hu Y and Yin Z 2017 The activation mechanism of lead ions in the flotation of ilmenite using sodium oleate as a collector *Mineral Engineering* Vol 111 pp 100–107
[6] Choi J, Choi SQ, Park K, Han Y and Kim H 2017 Flotation behaviour of malachite in mono-and di-valent salt solutions using sodium oleate as a collector International *Journal of Mineral Processing* Vol 167 pp 35–41
[7] Paiva PRP, Monte MBM, Simao RA and Gaspar JC 2011 In situ AFM study of potassium oleate adsorption and calcium precipitate formation on an apatite surface *Mineral Engineering* Vol 24 pp 387–95

IOP Conf. Series: Earth and Environmental Science **262** (2019) 012033 doi:10.1088/1755-1315/262/1/012033

Sublevel caving under protection of ore-and-barren rock cushion during transition from open pit to underground mining

AI Konurin, SA Shchukin, SA Neverov* and AA Neverov

Chinakal Institute of Mining, Siberian Branch Russian, Academy of Sciences, Novosibirsk, Russia

E-mail: *nsa_nsk@mail.ru

Abstract. These studies aimed to evaluate an efficient method of ore mining under open pit bottom during transition from open pit to underground mining. In the transition period and further mining with the sublevel caving method, it is proposed to use a protection in the form of an artificial large-block ore-and-barren rock cushion (capping) created by partial caving of benches and bottom of open pit. The analytical calculations and modeling performed for optimizing parameters of geotechnology and ore discharge provide indexes of ore recovery completeness and quality.

1. Introduction

In modern conditions of intensification of mineral production at increasingly deeper levels, it is a common practice to transit from open pit to underground mining. In this connection, it is highly important to develop and introduce safe methods for recovery of ore reserves under open pit bottom with further efficient transition to deeper levels.

Operating mines all over the world have accumulated ample experience of mining under open pit bottom using systems with caving and backfilling. Mines that have efficiently implemented the hybrid mining method include Zyryanovsky, Gaisky, Sheregesh mines, etc, in Russia, krivoi Rog basin mines in Ukraine, Questa in USA, Finsch and Bellezane in South Africa, Vammala in Finland, Hemlo and Kidd Creek in Canada, and many other [1–4].

The hybrid mining features high probability of pitwall rock fall in underground mine. The disruptive effect of the rockfall can be prevented using cut-and-fill systems, or by creating a safety pillar between the open pit and underground mines [5–7]. These measures make mining expensive and dependent on backfill material supply. In addition, considerable ore resources are left in the protective pillar.

In this regard, it seems efficient to mine ore reserves under open pit bottom by the system with sublevel caving. The stoping face is continuously regenerated, and ore is discharged under caved rocks, which enables sufficient stability of stopes and controllability of mining operations [8–12]. Caved rocks offer a protection layer to counteract probable rock fall from open pit to mined-out stoping void, as well as contribute to better ventilation in underground mine. Furthermore, sublevel caving method features simple preparation of rock mass for stoping, high productivity, low cost, adaptability of single-type machines to heading and extraction, and is efficient in medium-grade ore bodies [13]. Disadvantages of sublevel caving include high loss and dilution variable in wide ranges of 7.8–15.5 and 9.0–27.0%, respectively.

In such conditions of hybrid mining, the backbone of engineering solutions aimed to ensure efficient sublevel caving at the stage of transition from open pit to underground mining is improvement of ore recovery completeness and quality factors, i.e., reduction of ore loss and dilution. The major avenues towards this objective are:

—optimizing SCL method parameters to fit with specific deposit (sublevel height, cavable layer thickness);

—modifying SLC design (shaping stoping zone as ellipse, trapezoid or rhomb).

This study aimed to find rational parameters and improve ore recovery indexes during sublevel caving under open pit bottom using a large-block protection capping made of broken ore and barren rocks of the open pit bottom and walls.

2. Mining under open pit bottom

In sublevel caving under open pit bottom, the ore body is split into sublevels with a height h = 15.5–28.5 m (Fig. 1). Broken ore is drawn from sublevels via ore chutes to haulage level and, then, to skip hoisting shaft. As a rule, stoping is divided into two stages: the first stage is extraction or ore reserves adjoining open pit (sublevels 1 and 2); the second stage comprises the bulk of the deposit (deeper sublevels).

In sublevels 1 and 2, coarse drilling pattern is designed so that to size open pit bottom and walls, and blasting is performed so that to produce coarse fragmentation to create a large-block ore-and-barren rock cushion [14]. Broken or is either never or partly drawn (not more than 50%), and sublevel caving is continued under protection of the cushion.

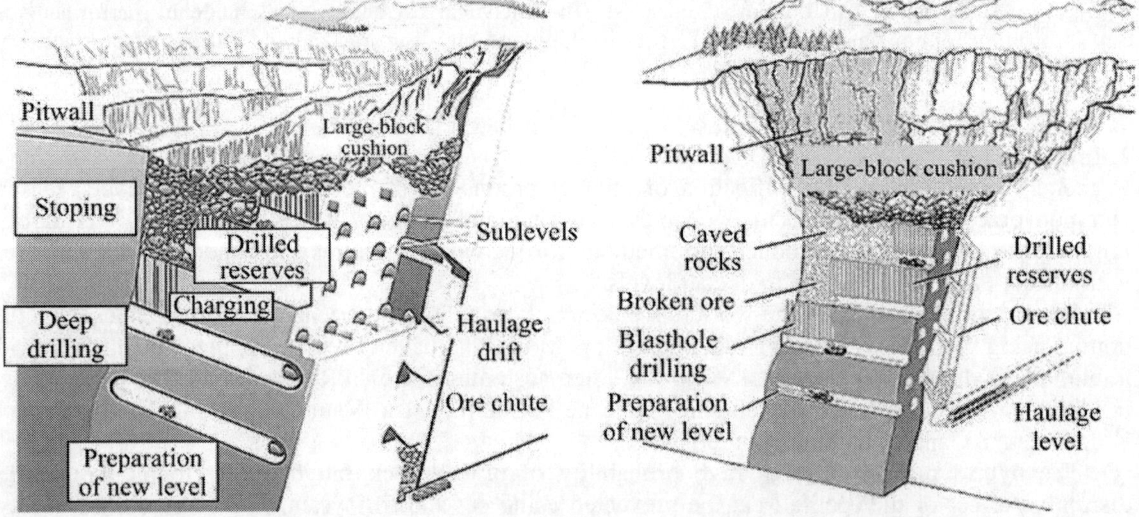

Figure 1. Examples of ore recovery under open pit bottom by sublevel caving.

On sublevels, extraction panels are formed by means of cutting drill-and-haulage drifts every 12–18 m, as well as air connections and ore chutes Figures 1 and 2). Ore is broken in layers of definite thickness and drawn from drill-and-haul drifts. Ore drawing amount in each broken layer is to be chosen based on conditions of the protection cushion integrity and maximum sustainability of initially set parameters.

During the first stage sublevel caving, the ore-and-barren rock cushion can gradually accumulate at the footwall of the ore body and get thinner at hanging wall. In order to keep initial thickness of the cushion at the hanging wall, a few last layers f broken ore under the hanging wall should be never or limited discharged so that some ore and rocks is left to renew the cushion. Accordingly, in case of ore extraction immediately under open pit bottom, ore drawing modes should be set based on the preservation of the initial parameters of the ore-and-barren rock cushion (capping), while in deeper level mining—based on the minimum penetration of barren rocks from the cushion and rational reasoning of ore loss and dilution.

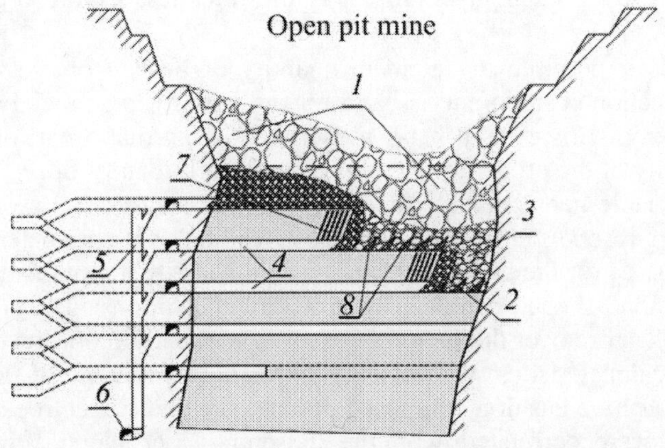

Figure 2. Sublevel caving under open pit bottom: *1*—large-block ore-and-barren rock cushion; *2*—broken ore; *3*—caved rocks; *4*—drilling-and-haulage drift; *5*—ore and rock chute; *6*—haulage level; *7*—blastholes; *8*—remaining ore.

The proposed technology eliminates caving of enclosing rocks onto the large-block capping. At the same time, during sublevel stoping, it is possible that pitwalls and exposed rocks of the handling wall and footwall, or undermined top ore partly cave and accumulated on the large-block cushion. Barren rocks can penetrate from this layer into the cushion and result in an increase in ore dilution later on. The nature and condition of the barren rock layer on the cushion should be persistently monitoring, and this is a mandatory requirement of the proposed technology to be efficiently applied. According to mining practice in Zyryanovsky and Gaisky mines, granulometric composition of the large-block ore-and-barren rock cushion should be composed of fraction ranging between 0.4 and 1.2 linear size of a stope. As sublevel caving is advanced, depending on safety of the protection pillar, condition of sidewalls in mined-out voids and mining safety requirements, partial caving of enclosing rocks in the hanging wall is admissible.

Lumped calculation of the protective cushion pillar at different depths in the worst caving conditions [13] show that ore recovery at a depth of 30–120 m below open pit bottom, the minimum thickness of the protective cushion is 18.7–48.4 m.

3. Rational parameters of mining system

Transition from open pit to underground mining is possible in stable enclosing rock mass. Sublevel caving with ore drawing from drilling-and-haulage drifts can only be safe under a protective cushion made of broken ore and barren rocks. Thickness of this protective cushion should never change in the course of mining, which is ensured by adhering ration mode of ore discharge.

In case that the protective cushion is composed of broken ore, the advised mode of ore drawing is preserving equal volumes of ore discharge and ore reserves in a layer:

$$V_{\text{layer}} = K_{\text{loose}} V_{\text{draw}}, \tag{1}$$

where V_{layer} is the volume of ore reserves in a layer, m^3; K_{loose} is the loosening coefficient of broken ore, m; V_{draw} is the volume of ore drawing, m^3.

Equality (1) is the framework to determine height of an ore draw point (shaped as an ellipsoid):

$$H = \sqrt{\frac{3S_{\text{layer}}\, t_{\text{layer}}\, K_{\text{loose}}}{\pi\rho}}, \qquad (2)$$

where S_l is the area of the broken layer, m^2; t_l is the thickness of the broken layer, m; ρ is the flowability index, m.

Relation (2) makes it possible to calculate quantity of broken ore left in a stope to create a constant-thickness protection cushion and is valid for ore with average metal content over the deposit (found out experimentally). However, it happens during stoping that a part of ore is cut under caved rocks due to discontinuity of the protection layer (Figure 2, right-hand side).

No matter how variable geotechnical conditions are, the major criteria of efficient mining is the maximum recover of ore reserves at the minimum cost. The rational parameters of the sublevel caving system are calculated based on the analytical studies with some assumptions made. It is assumed that ore is extracted in diamond-shaped panels. The protective cushion (capping) is created by means of partial shrinkage of broken ore over the period of extraction of certain portion of ore reserves.

The ore recovery characteristics were calculated for the conditions of ore drawing under caved rocks and under the protective cushion composed of large ore and barren rock blocks. The draw point parameters (Figure 3) were correlated with the loosening coefficient, flowability index, ratio of sublevel height to spacing of drilling-and-haulage drifts and with the layout of sublevel stopes.

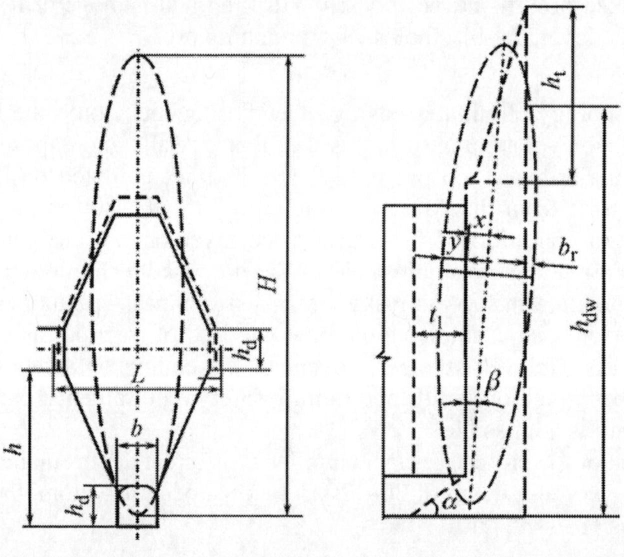

Figure 3. Analytical model for determining ore recovery characteristics and rational parameters of sublevel caving system: h—sublevel height, m; h_d—height of draw point, m; b—width of draw point, m; L—spacing of drilling-and-haulage drifts (width of extraction panel); H—height of the draw point ellipsoid, m; t_l—height of broken layer, m; h_{dw}—draw point working height, m; h_t—minimum height of tongues, m; b_r—with of enlacing rock layer in the draw ellipsoid, m; x—deviation of the ellipsoid axis from ore body, m; y—width of segment cut from draw ellipsoid by ore body, m; α—slope of discharge, deg; β—incline of longer ellipsoid axis to vertical plane, deg.

Efficient parameters of the hard ore mining system were selected with respect to relative net profit per 1 t of in-place reserves. Variable operating cost of mining was chosen subject to the change in the diamond-shape panel parameters and the associated specific volumes of development and heading operations. Drilling-and-blasting ad other operating costs of ore mining and processing were assumed to be constant. The calculation algorithm of ore recovery characteristics and the value of profit per mining variants were, respectively, based on the ore drawing patterns under caved rocks from draw points shaped as ellipsoids of revolution (Figure 3) and on the parametric modeling of structural elements within the mining system (sublevel height spacing of drilling-and-haulage drifts and thickness of broken ore layer).

The calculation results made it possible to determine efficient parameters for sublevel caving in diamond-shaped panels in different conditions of ore drawing at the minimized ore loss and dilution:

$$t_{\text{layer}} \approx \sqrt{\frac{2}{3}\frac{h}{\sqrt[3]{1-R}}\rho}, \quad (3)$$

where R is the limit dilution in the last ore draw portion, unit fractions,

$$L = 2\left(\frac{h}{\text{tg}f}+b\right)+\frac{b}{4}, \quad (4)$$

$$f = \text{arctg}\left[\frac{2h}{3\left(\sqrt{h\rho}-\frac{b}{2}\right)}\right]. \quad (5)$$

A reasoned ratio of sublevel height to broken ore layer thickness is governed by the latter parameter, ranges as 3–6 and is on average 4–5.

Sublevel caving under protective ore-and-barren rock cushion, owing to extraction of "cleaner" ore as against ore drawing under caved rocks, allows expanding the net of stopes by 15–20%. By the criterion of relative profit in this case, the optimized variants are the panel width to sublevel height ratios ranging from 0.85 to 1.5 (Figure 4).

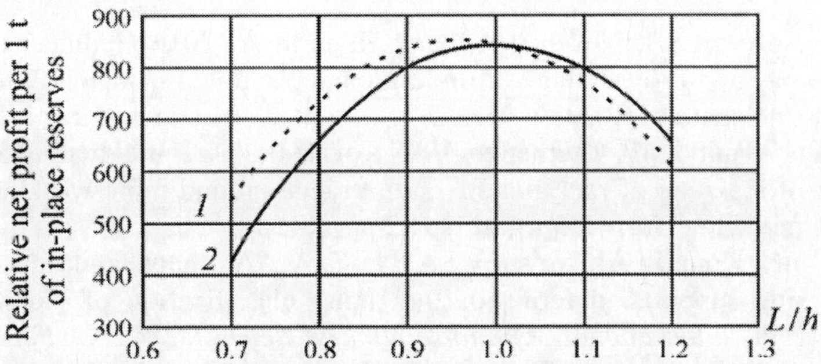

Figure 4. Net profit per 1 t of in-place reserves versus ratio of stoping caving system parameters: *1*—ore drawing under protection ore-and-barren rock cushion; 2—ore drawing under caved rocks.

The calculations have determined efficient mode of ore drawing under protection of ore-and-barren rock cushion. The required thickness of the cushion is maintained by drawing its part equaling the remaining part (11–22%) after discharge of basic reserves from the extraction panel (78–89%). This measure makes it possible to maintain the thickness of the large-block cushion for the whole period of of safe mining at minimized cost.

4. Conclusions

The implemented analytical studies have determined optimized ratio between the mining system parameters for the stage of transition from open pit to underground mining and efficient modes of ore drawing to preserve continuous thickness of large-block ore-and-barren rock cushion and the required mining safety. The key recommendations on application of the proposed technology say that:

—ore extraction panels should be shaped as diamonds at most escribed into ellipsoids of revolution;

—coarseness of fraction of the protective ore-and-barren rock cushion should be set not less than 0.4–1.2 of maximum linear size of a drilling-and-haulage drift;

—volume of ore drawing should be 40–50% of broken rock volume from panels in the first and partly second sublevels immediately under open pit bottom and 100% from the lower lying sublevels.

In the course of underground mining, the proposed parameters of sublevel caving system will be continuously updated with regard to the change in the depth of mining and in the flowability index, which nonlinearly diminishes with the growing depth of mining due to compaction and compression of rocks.

Acknowledgements
This study was supported by the President of Russian Federation, Project No MK-6827.2018.5.

References
[1] Farafonov VP 1984 Experience of hybrid mining in Gaisky deposit *Gornyi Zhurnal* No 4 pp 13–15
[2] Jalonen M, Forster D et al 2013 *Global Gold & Precious Metals* Merrill Lynch 09 April USA
[3] Barnov NG, Eremenko VA, Kondratenko AS and Timonin VV 2015 Substantiation of parameters of geotechnology for mining primary corundum deposits in difficult working conditions in upland areas *Gornyi Zhurnal* No 11 pp 42–47
[4] Neverov AA, Konurin AI, Shaposhnik YuN, Neverov SA and Shaposhnik SN 2016 Geomechanical substantiation of sublevel-chamber system of developing with consolidating stowing *16th International Multidisciplinary Scientific Geoconference—SGEM 2016: Science and Technologies in Geology, Exploration and Mining* Vol II Albena, Bulgaria pp 443–450
[5] Freidin AM, Neverov SA, Neverov AA and Konurin AI 2016 Geomechanical assessment of geotechnology at a project stage of underground ore mining *Gornyi Zhurnal* No 2 pp 39–45 DOI: 10.17580/gzh.2016.02.08
[6] Lushnikov VN, Sandy MP, Eremenko VA, Kovalenko AA and Ivanov IA 2013 Method of definition of the zone of rock massif failure range around mine workings and chambers by numerical modeling *Gornyi Zhurnal* No 12 pp 11–16
[7] Shaposhnik YuN, Konurin AI, Neverov SA, Neverov AA, Shaposhnik SN 2016 Justification of mine working supports in terms of the rating classification of Norwegian Geotechnical Institute *16th International Multidisciplinary Scientific Geoconference—SGEM 2016: Science and Technologies in Geology, Exploration and Mining* Vol II Albena, Bulgaria pp 519 – 526
[8] Balg C, Roduner A and Geobrugg AG 2013 Ground support applications Int. *Ground Support Conf.* AGH University, Lungern, Switzerland
[9] Reiter K and Heidbach O 2014 3-D geomechanical-numerical model of the contemporary crustal stress state in the Alberta Basin (Canada) *Solid Earth* No 5(2) pp 1123–1149
[10] Gunzburger Y and Magnenet V 2014 Stress inversion and basement-cover stress transmission across weak layers in the Paris basin, France *Tectonophysics* Vol 617 pp 44–57
[11] Hofmann H, Weides S, Babadagli T, Zimmermann G, Moeck I, Majorowicz J and Unsworth M 2014 Potential for enhanced geothermal systems in Alberta, Canada *Energy* No 69 pp 578–591
[12] Potvin Y and Giles G 2008 The development of a new high-energy absorption mesh *Australasian Institute of Mining and Metallurgy Publication Series* pp 89–94
[13] Imenitov VR, Abramov VF and Popov VV 1983 *Localization of Voids in Underground Ore Mining* Moscow: Nedra (in Russian)
[14] Timonin VV and Karpov VN 2016 Assessment of rock failure process under percussion–rotary drilling from the viewpoint of nonlinear geomechanics *J. Fundament. Appl. Min. Sci.* Vol 2 No 3 pp 172–176 (in Russian)

IOP Conf. Series: Earth and Environmental Science **262** (2019) 012034 doi:10.1088/1755-1315/262/1/012034

Compressive vacuum resistance to percussion hammer in downhole vibration source

AO Kordubailo*and AYu Primychkin

Chinakal Institute of Mining, Siberian Branch, Russian Academy of Sciences, Novosibirsk, Russia

E-mail: Kordubaylo_ao@mail.ru

Abstract. The paper focuses on downhole vibration source used for wave generation in an oil stratum in order to increase oil recovery and stimulate hydrocarbon flow in a well. Simulation results of pre-shock compressive vacuum resistance to a hammer in an electromagnetic shock unit in a downhole vibration source are presented. The effect of axial hole in a hammer on induced compression is analyzed. Experimental data on changes in coil electromagnetic traction are presented depending on axial hole diameter in the hammer. Based on the effect of hammer hole diameter on compressive vacuum resistance and coil electromagnetic traction the present research findings can be of great use in designing an electromagnetic shock unit.

1. Introduction

The present-day problem of prime importance in the oil and gas industry concerns escalation of hydrocarbon production rates. The oil recovery factor for conventional oil and gas production technologies in Russia reaches 0.4, the essential portion of oil reserves remains in the interior of the Earth. The novel processes for oil recovery improvement (ORI) are in demand in order to involve the residual oil reserves into production. Since the 1970s the permanent research and pilot-full-scale tests on development of vibration-wave processes designed to improve oil recovery of wells have been advanced in Russia, the idea of the process rests on the application of the directed vibration-wave effect on the productive strata. The generated elastic oscillations cause variations in structure of macro- and micro-fractures in an oil reservoir and ways of repressuring medium percolation, thus involving the earlier intact oil reserves into active oil flow contour. Moreover, it is established appreciable acoustic wave influence on oil viscosity reduction, as well as suppressing capillary and adhesive forces hampering the displacement of hydrocarbons from porous structure of a stratum. At present the extensive many-years theoretical and experimental experience justifies the efficiency and promising perspectives of ORI wave process [1–5].

2. Numerical and experimental investigation results

At the Institute of Mining SB RAS the research and development work is conducted on investigation into a downhole electromagnetic impulse vibration source to generate the wave effect on oil reservoir for last three years. The research installation consists of electromagnetic reciprocating percussion motor and a hydraulic power unit, intended for transformation of axial hammer impacts into radial power impulses and their transmission to downhole production string.

In [6] the researchers prove the feasibility to design an electromagnetic drive based on long-stroke magnetic systems with ratio of magnetizing coil length to core radius of more than 15. Increase in this ratio leads to increase in a stroke and hammer velocity; it seems the single way to higher impact energy in diametrically confined well space. However increase in motion velocity induces the compressive vacuum resistance because of air presence in hermetic volume. The laboratory tests of the pilot electromagnetic percussion unit at modes with/without sealed cases revealed that in operation with a one-piece hammer in a sealed case the compression generated under the bottom end of the hammer in a working stroke tends to reduce a hydraulic pressure impulse generated by a hammer impact in a power element by 15–20 % as compared to impulses generated in non-pressurized case. The similar test with a hammer with axial hole of 23 mm in diameter did not indicate the compression effect on pressure impulse in the power component. .

From above the necessity becomes obvious to establish interrelation between the axial channel diameter in the hammer and the magnitude of compression in a cavity under it, so the targets of the present reseach are:

• to study the effect of axial hole diameter in a hammer on compression induced under hammer in a working stroke;

• to determine the minimal diameter of hole in the hammer at which the stabilized pressure difference is reached between cavities under and over the hammer, when the hammer is moving at pre-impact velocity;

• to estimate the effect of axial hole in a hammer on electromagnetic coil thrust;

• to work out recommendations on designing an electromagnetic percussion unit with account for test results of the present research.

The virtual model describing the air pressure growth under the lower end at the final stage of the working stroke of the hammer was developed in software system SimulationX to study the effect of hole diameter in the hammer on compression. The computation scheme of the model is presented in Figure 1: hammer *1* of 14 kg in mass and 48 mm in external diameter, moving with a zero gap in guide *2*; compression cavity *3*, which volume is equal to a sum of volumes of air pockets of a real machine; pressure meter *4*; pipe *5*, connecting compression cavity *3* with cavity *6* over hammer. Pipe *5* has internal diameter equal to diameter of the hole in hammer and length equal to hammer length 1030 mm.

Assumptions made in the model are:

—under consideration is only pre-impact interval of hammer motion h_h, equal to 200 mm. Growth of compression in the initial hammer motion is not considered;

—by the moment when the hammer reaches the interval under consideration its pre-impact velocity v_h, is 5 m/s. The hammer continues motion by inertia.

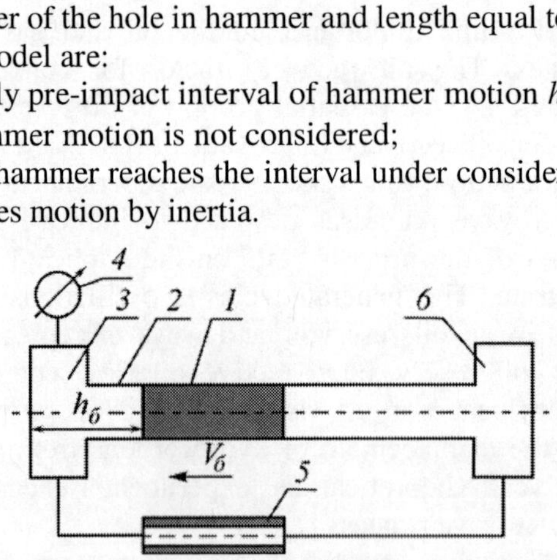

Figure 1. Calculation scheme to evaluate air resistance to hammer motion: *1*—hammer; *2*—guide; *3*—compression capacity; *4*—pressure meter; *5*—pipe; *6*—cavity over hammer.

In the numerical tests the internal diameter of pipe *5* was preset, the pressure growth in cavity *3* in the interval during which the hammer *1* advanced at distance h_h. Then the internal diameter of pipe *5* was increased stepwise with 4 mm stroke and the test was repeated. The initial diameter of the hole was preset at 4 mm and the cross-section is preset equivalent to a real gap in hammer–guide conjunction.

The oscillograms are obtained for different diameters of holes reduced to the same coordinate plane for illustration purposes (Figure 2a). Figure 2b demonstrates the relationship of the maximum pressure in the compression cavity versus hole diameter in a hammer.

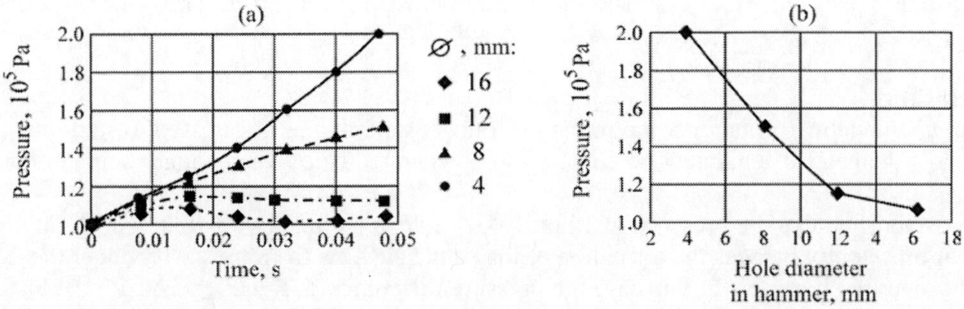

Figure 2. (a) Oscillograms for pressure in compression cavity and (b) dependence of the maximum pressure vs. hole diameter in the hammer.

By the analytical data on oscillograms the minimum 12 mm hole diameter in the hammer is sufficient to form a stable air flow from the compression cavity to a cavity over the hammer at constant difference in pressures. Thereto, difference in pressure is equal to $0.13-0.15\cdot10^5$ Pa, and the average resistance force is no more than 25.3 N.

Reduction in hole diameter (Figure 2b) leads to sharp growth of compression. In the case with one-piece hammer the average resistance at the calculation interval is 99 N. Furthermore, in the calculation interval it fails to gain the stable air flow under the constant pressure difference (oscillograms do not reach horizontal asymptote). This feature indicates that pressure will continue growing far to the beginning of the calculation interval during the entire working stroke of the hammer. Therefore, the model representation of these modes is not correct. Actual pressure values can be really higher than calculated ones.

Increase in hole diameter causes a cyclic air flow through it; this statement is proved by wave shape of pressure oscillogram. Thereupon, the time interval when difference in pressures between cavities is close to zero and accompanied with low or even zero air flow rate, thus indicating irrationality to make a hammer hole of more than 12 mm in diameter.

The hole diameter also affects the thrust of electromagnetic coil in addition to resistance acting on the hammer. This relationship relates to reduction in active cross-section of the anchor because of the hole [7].

The laboratory results of pilot electromagnetic hammer prototype tests give thrust characteristics of electromagnits operating with three hammer types: one-piece, with hole diameters of 13 and 22 mm. The thrust characteristics of the hammer having 48 mm external diameter at working current values of 20 A and 30 A are reported in Figure 3.

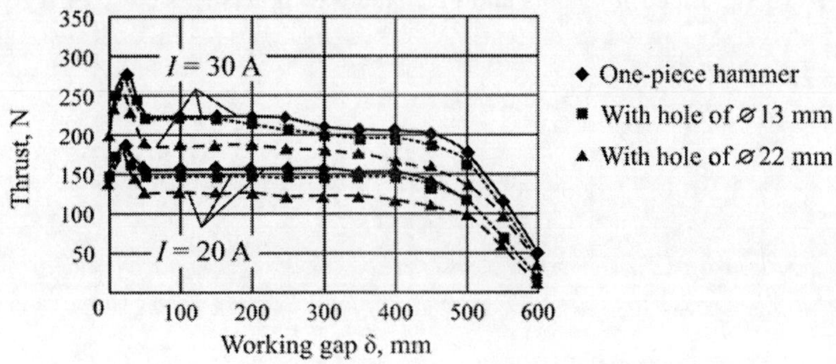

Figure 3. Experimental thrust characteristics of electromagnets in operation with different-type hammers.

By experimental data, the electromagnet thrust lowers with increase in hole diameter in the hammer. For hammer with 13 mm hole diameter the drop of the thrust amounts to no more than 3–6% in the complete working gap as compared to that for a one-piece hammer. This proves insufficient effect on the impact energy as compared to the compression resistance, when one-piece hammer is moving.

3. Conclusions

SimulationX modeling of hammer motion at pre-impact time moment was used to study the influence of axial hole diameter in a hammer on compression, generated below the hammer being in the working stroke.

It is established that the presence of 12 mm axial hole in hammer with 48 mm external diameter at 5 m/s motion velocity induces the formation of the stable air flow from the cavity under the hammer to the cavity over the hammer. In this case the pressure difference does not exceed $15 \cdot 10^5$ Pa, and head resistance is 25.3 N.

The electromagnet thrust in operation with hammer of 48 mm external diameter and 12 mm axial hole diameter is by 3–6 % lower than that for a one-piece hammer. Such reduction in thrust is admissible with account for multiple reduction in compression in the cavity under the hammer.

When designing the percussion unit, it is recommended:

—to reduce volume of compression cavity;

—to provide overflow air chute between cavities over and under the hammer as a hole in the hammer of 12 mm in diameter or as specially made cuts on surface of coils; the cross-section area of a cut should be equivalent to cross-section area of the hole.

References

[1] Oparin VN, Simonov BF, Yushkin VF et al. 2010 *Geomechanical and Technical Provisions for Enhanced Oil Recovery with Vibro-Wave Technologies* Novosibirsk: Nauka (in Russian)

[2] Oparin VN and Simonov BF 2010 Nonlinear deformation-wave processes in the vibrational oil geotechnologies *Journal of Mining Science* Vol 46 No 2 pp 95–112

[3] Simonov BF, Cherednikov EN, Serdyukov SV et al 1998 The technology of volumetric wave influence on oil and gas deposits to increase hydrocarbon output of oil reservoirs *Neftayn. Khoz.* No 4 pp 42–44

[4] Dyblenko VP, Marchukov EYu, Tufanov IA et al 2012 *Wave Technologies and Application in Oilfields with Hard Recoverable Resources* Moscow: RAN (in Russian)

[5] Savchenko AV, Simonov BF and Cherednikov EN 2014 Influence of hydropulse effects on the bottomhole zone of a productive formation *J. Fundament. Appl. Min. Sci.* Vol 1 No 1 pp 285–288 (in Russian)

[6] Simonov BF, Neiman VYu and Shabanov AS 2017 Pulsed linear electromagnetic drive for a downhole vibration source *Journal of Mining Science* Vol No 1 pp 118–126

[7] Slivinskaya AG 1972 Electromagnets and Permanent Magnets Moscow: Energiya (in Russian)

Modern trends of development in mining industry in the Kyrgyz Republic

KCh Kozhogulov* and DA Bespalov**

Institute of Geomechanics and Development of Subsoil, National Academy of Sciences, Bishkek, Kyrgyz Republic

E-mail: *ifmgp@yandex.ru, **divorses@mail.ru

Abstract. Modern trends are considered for developing mining industry in the Kyrgyz Republic under conditions of transition to the JORC system in evaluating mineral deposits.

1. Introduction

Mining industry is one of the most important sectors in development of the Kyrgyz Republic and makes a weighty contribution of 11% to GDP of the country. By estimates quoted in the National Strategy for Mining Industry, gold makes around 87% of total cost of mineral reserves, coal–10%, and the rest is mostly copper and silver [1]. According to official figures, on the national balance currently there are gold deposits (proved reserves 430 thou t) tin (208 thou t), tungsten (144 thou t), rare earth metals (51 thou t), aluminium (349 Mt), coal (over 1 Bt), antimony (157.5 thou t), mercury (40 thou t), molybdenum (2358 thou t) and some other minerals [2]. One of the largest deposits, Kumtor has produced 339.6 t of gold over a period of 1997 to 2016, and holds proven and inferred reserves of 241 t [3].

To date many companies invest in the mining sector of Kyrgyzstan. The top investor is Canada with the capital expenditures of 48% (Kumtor). Capital is also invested by China (14%), Great Britain (7%), Germany and Kazakhstan (5% each), Russia and Switzerland (4% each), and USA (3%) [5].

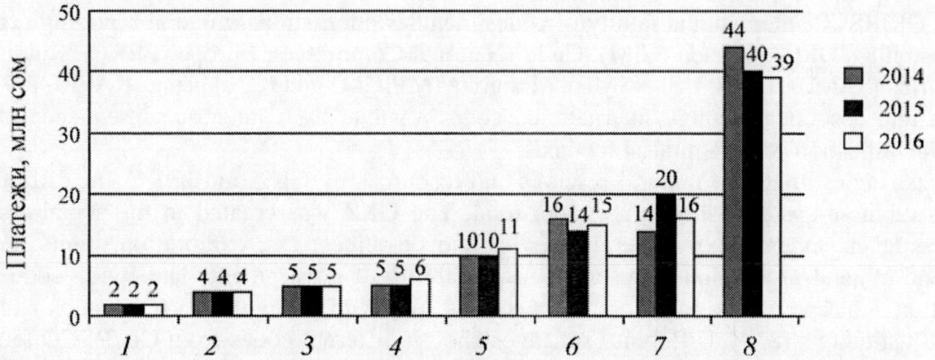

Figure 1. Nontax and tax payments per industry branches (excluding state and social insurance): *1*—oil and gas; *2*—financial services; *3*—professional occupation; *4*—construction; *5*—communication; *6*—trade; *7*—nonferrous metals; *8*—other industries. Payments are given in million Kyrgyzstan som.

In 2016 gold mining was carried out at the deposits Makmal, Terekkan, Solton-Sary, Bozymchak and Taldybulak Levoberezhny. Concentrate produced at Ishtamberdi, Dzhamgyr and Karakazyk deposits is shipped for processing outside Kyrgyzstan [5].

The highest budget receipts in the form of nontax and tax payments were from the nonferrous metal industry although its share in GDP was merely 5.05% in 2016 (Figure 1) [6].

2. Development Strategy for Mining Industry of the Kyrgyz Republic

The National Sustainable Development Strategy for Kyrgyzstan is the critical document to shape the growth strategy of the country, as well as the lines and priorities of economic, political ad social advancement. The Strategy includes mining industry in the list of five strategic sectors of economy, and development of the mining industry is totally entrusted to the private sector due to the public finance deficiency. The Strategy highlights that favorable investment climate requires clear, simple and transparent legislation to be elaborated by the government.

In 2014 the consortium of local experts prepared a draft mid- and long-term Development Strategy for Mining Industry of the Kyrgyz Republic for 2015–2035 with comprehensively planned implementation. This document is overlooked by reformers of the mining sector at the present time. Regarding legislative and subordinate acts, mineral resources management is regulated by a series of regulatory legal documents, including the crucial Law on Subsoil (2012) and its provisions, Law on Environment (1999), Law on Ecological Expertise (1999), Tax Code (2008) and Land Code (1999). The government accounts management is subjected to regulation by common regulatory legal acts which fall within government finance administration, in particular, Budget Code of the Kyrgyz Republic (2016).

Numerous federal and local regulatory bodies participate in development and implementation of subsoil management policy. Accordingly, the functions and responsibilities of each body should be set distinctly, with duly arranged coordination between them for efficient communication and consistency of acts. In some case, regulatory bodies can be given new functions. For instance, local authorities are often underactive in development of normative acts, especially if the point is on income distribution, or reclamation and preservation of mining industry assets.

The mining industry, in order to increase its weight in economy of Kyrgyzstan, urgently requires transition to innovative international codes of evaluation of mineral reserves. The improvement of innovation-based competitive ability of the mining and processing industry of the country calls for transition from evaluation of mineral resources and reserves by CRIRSCO family of codes to JORC system. This problem was raised on the governmental level but unfortunately was not put into motion.

The CRIRSCO international family of codes includes international mineral reporting organizations from Australia (JORC), Canada (CIM), Chile (National Committee), Europe (PERC), Russia (NAEN), South Africa (SAMCODES), USA (SME), Mongolia (MPIGM) and Kazakhstan (KAZRC) [7]. It should be taken into account that these international codes regulate the contents and sequence of reporting without identification of the appraisal methods.

Comparing estimates of mineral reserves and resources by GKZ and JORC, the difference in the purposes of these systems should be understood. The GKZ was created in the social economy for governmental management of mineral reserves and resources; i.e., exploration, mine planning and design and mineral production were under responsibility of government. The JORC code aims offer information and reporting to investors. This code is subjected to regulation from stock exchange and financial institutions [8, 9]. Different exchanges choose different codes from CRIRSCO family. Put it otherwise, the aim of the GKZ reporting is registration of mineral reserves and resources on the governmental balance, while the JORC system provides exchanges and investors with public reports. The difference of the aims brings different formats and approaches.

The GKZ and JORC codes are based on such criteria as: goal, materiality, competence, transparency, independence, input data evaluation, exploration approaches, appraisal, report authors, approval.

Regarding *Materiality*, both systems provide complete information for validation of mineral resource and reserve estimation.

Competence makes distinctions between areas of responsibility of participants. The GKZ estimation is the team-work result without personal responsibility, the JORC Code imposes responsibility on the competent reporting person.

From the viewpoint of *Transparency*, JORC Code goes ahead as the competent person, plainly and with no omission (and no disorientation) offers complete and significant information in whole. The GKZ system only provides information required for registering reserves and resources on governmental balance, while ill-transparent and complex expertise is often subjective, which results in investment errors. As for the criterion of *Independence*, the GKZ system experts are not independent as they serve interests of the government; in conflict of interests, the independent consultant of the JORC Code naturally wins. Regarding *Input Data Evaluation*, both system approaches are in many ways the same, with the collegial evaluation in GKZ and QA/QC analysis by a competent person in JORC.

Exploration and Geological Appraisal in JORC are based on geological features and industrial standards, are nor strictly formalized, while rest on regulations in GKZ although final decision is made by executors.

JORC reporting is implemented by competent persons, and GKZ system engages subsoil users or consultants. The *Report Approval* procedures differ greatly. The JORC report is approved by a competent person, while the GKZ procedure is more formal and extended. The GKZ experts should be professionals in the area of evaluation / audit, with no less than 5 years-long experience of operation at the specific-type deposits. JORC's competent persons bear full responsibility for the reporting, whereas GKZ experts take on the joint responsibility, which results in worse quality of reporting, in buck-passing and in shifting responsibility to government.

By pre-estimates, Kyrgyzstan should right now think on introduction of the CRIRSCO family codes, in particular, JORC Code toward innovation and efficiency in mineral mining. It is the time to begin training independent experts—competent persons and to purse consistent investor policy. As a consequence, potential risks will be reduced for international investors and their interest in mineral mining in Kyrgyzstan will be raised.

3. Conclusions

In summary, the emphasis should be laid on:

—complete removal of doubling of public authorities in the mining sector by means of reformation and optimization;

—early transition to the new innovative system of estimating mineral reserves and resource, namely JORC Code, which will enhance investment inflow, reduce risks of international investors, improve estimation transparency and abate corruption.

In the future, the influence of the mining sector on the economic advance of Kyrgyzstan will grow, and the JORC Code will help an innovative breakthrough in this sector.

References

[1] *The Impact of Kumtor Gold Mine on the Economic and Social Development of the Kyrgyz Republic* Working Paper No 32, 2015 Available at: http://www.big.kg/full-publications/mining-industry-kyrgyzstan-problems-and-prospects/ru/
[2] http://www.big.kg/full-publications/mining-industry-kyrgyzstan-problems-and-prospects/ru/
[3] *Kumtor Gold Company* https://www.kumtor.kg/en/
[4] http://www.big.kg/full-publications/mining-industry-kyrgyzstan-problems-and-prospects/ru/
[5] *KAZ Minerals* https://www.kazminerals.com/ru/
[6] *Report on the Results of the Work of the State Agency for Geology and Mineral Resources under the Government of the Kyrgyz Republic (Successor to the State Committee of Industry, Energy and Subsoil Use of the Kyrgyz Republic) for 2016*

[7] *Report on the Results of the Work of the State Agency for Geology and Mineral Resources under the Government of the Kyrgyz Republic (Successor to the State Committee of Industry, Energy and Subsoil Use of the Kyrgyz Republic) for 2016*
[8] Seredkin M and Urbisinov S some issues of comparison of mineral resources *Ore Reserves According to the Codes of the CRIRSCO Family and Reserves/Resources by the GKZ System*
[9] *Australian Guidelines for the Assessment and Classification of Coal Resources 2014* IMC Montan translation http://www.imcmontan.ru/files/kodex.pdf

IOP Conf. Series: Earth and Environmental Science **262** (2019) 012036 doi:10.1088/1755-1315/262/1/012036

Gas explosion hazard in underground coal mining in Kuzbass

EA Kozlovsky[1], GN Sharov[1], AE Kontorovich[2], GI Gritsko[2], FA Kuznetsov[3], MV Kurlenya[4]*, VA Kovalev[4], VI Rostovtsev[4], IM Belozerov[5], VA Tchernook[5], VA Minin[6] and NYu Vashlaeva[7]

[1]Institute of Geological-Economical Problems, Russian Academy of Natural Sciences, Moscow, Russia
[2]Institute of Gas-Oil Geology and Geophysics, Siberian Branch, Russian Academy of Sciences, Novosibirsk, Russia
[3]Institute of Inorganic Chemistry, Siberian Branch, Russian Academy of Sciences, Novosibirsk, Russia
[4]Chinakal Institute of Mining, Siberian Branch, Russian Academy of Sciences, Novosibirsk, Russia
[5]Novosibirsk Branch of the State Specialized Design Institute, Novosibirsk, Russia
[6]Institute of Geology and Mineralogy, Siberian Branch, Russian Academy of Sciences, Novosibirsk, Russia
[7]Kemerovo Region Administration, Kemerovo, Russia

E-mail: *kurlenya@misd.ru

Abstract. The data on gas compositions are obtained in a Kuzbass mine by the Novokuznetsk Detached Force of Militarized Mine-Rescue Unit. The analysis of 1200 mine air samples, as for instance, in Alarda mine, shows that the extreme values of fire-hazardous gas contents are (% by volume): max 0.8 for hydrogen, max 28.3 for methane, and min 0.8 for oxygen. The conclusions are drawn about possible concept of neutron–proton structure of the core being the source of hydrogen and hydrocarbon-bearing gases releasing from the subsoil of the Earth as well as about potential applicability of chemical technology for removing hydrogen from the gas mixture to increase its explosion safety.

1. Introduction

Media systematically inform on tragedies connected with explosions of combustible gases and mass deaths of people in coal mines both in Russia (Kuznetsk, Pechora, Sother Yakutia coal basins, etc.) and abroad (Kazakhstan, China, New Zealand, Ukraine, etc.). For example, Central mine (Darasun, Chita Regiona) in September 2006, Ulyanov mine in Kuzbass in March 2007, or Raspadskaya mine in Kuzbass as well in May 2010. Unfortunately, this list is to be continued.

Today any more or less educated man would routinely explain explosion of gas mixture in coal mine as a blast of methane-and-air mixture containing airborn coal dust. Is this true? Let us analyze the situation more carefully. Air with sufficient content of oxygen enters underground excavation from ventilation system, the origin of dust is clear too. The source of methane formation is yet obscure.

IOP Conf. Series: Earth and Environmental Science **262** (2019) 012036 doi:10.1088/1755-1315/262/1/012036

2. The content of gases in the mine atmosphere according to the measurement results

We address geo-sciences and physical chemistry. Methane CH_4, as natural combustible gases of propane C_3H_8 and butane C_4H_{10}, currently produced in colossal amounts, are the chemical compound of two elements—carbon C and hydrogen H (sometimes with small additions of other elements—oxygen, nitrogen and sulfur). These are typical hydrocarbons [1, 2]. In a natural variety of hydrocarbons, methane is chemically distinct by being a saturated limit substance molecule of which contains maximum number of hydrogen atoms—4—per one atom of carbon as hydrogen occupier all 4 possible valence bonds of carbon. Naturally, such substance of limit composition, in accord with the chemical laws, can be originated under conditions when one of the compounds is in essential excess—hydrogen in our case [1, 2]. Put it otherwise, methane in any natural gas is a reliable sign of the presence of hydrogen. This thesis was illustrated by the Academician RAS Adushkin who with his scholars found that the composition of combustible gases which heavily emitted in the open pit mine at Udachnaya kimberlite pipe in the south-west of Yakutia was "to 50% hydrogen and the rest was methane" [3]. By the data of ZapSibGeologiya geologica service, gas emissions from jointed dolerite and sandstone in Southern Kuzbass at a depth of 2200 m contain up to 8.5% hydrogen. There are many such examples.

Where carbon is from in coal mines is a rhetorical question as the mechanism of centuries-long circulation of carbon and its transformation, through carbon dioxide (owing to biosynthesis), to coal, among other things—a mineralized product of natural origin. The puzzle is presented by hydrogen which, aside from the known natural circulation mainly through water, has permanently been escaping the Earth in huge amounts (round 1 Bt annually) [4–10] in the form of light gases (hydrogen and methane) for 4.5–5 billion years. (Classical geo-sciences have sadly little opportunity to explain the origin of this hydrogen and only offer a theoretical hypothesis on capture of gas from a once breathing protoplanet cloud). At the present time, escape of hydrocarbons and light gases (hydrogen, helium) from the subsoil is energetically investigated and discussed in the framework of scientific conferences held by the Russian Academy of Sciences (Institute of Petroleum Problems, Schmidt Institute of the Earth's Physics) [7, 8, 10]. It is worthy of mentioning that, owing to the research findings of the Skobeltsyn Research Institute of Nuclear Physics at the Lomonosov Moscow Scate University, Geophysical Center of the RAS and other scientific institutions [11–14], a neoteric hypothesis has been put forward to explain the described phenomena and their infinite horizon [15, 16, 1, 6, 17].

According to this hypothesis, hydrogen, which forms inside the Earth during natural self-decomposition of free neutrons outflowing from the Earth's interior, and which is partly in the transient state of extremely chemically active atomic form [1, 2], continuously hydrogenates, including carbon-bearing components of coal, up to the limiting methane state, and turns into a more stable molecular form. When methane and hydrogen enter mine air, a three-component mixture of methane, hydrogen and air is appears.

It is known that, by explosibility, properties of oxygen or air mixed with hydrogen or methane strongly differ. Figuratively, stoichiometric mixtyre of oxtgen (air) and hydrogen, called a detonating mixture, might be primed by a patch of sunlight. In the meanwhile, a methane-bearing mixture (similarly to household propane–butane in a kitchen) to be fired needs a temperature of a burning match. The temperature difference in initiation of interaction between hydrogen- and hydrocarbon-bearing mixtures is 150–200°C to the disfavor of the latter [1, 6, 2]. For this reason, it seems justified to suppose that combustible gases play different parts in the three-component air+methane+hydrogen mixture. Hydrogen is a detonator (it is primed first), while methane (similarly to coal dust), even at excessive concentration, acts as a fuel, though it definitely can detonate under more complex conditions (concentration, temperature, etc.), and so does a pure methane-and-air mix. This fact supposedly offers a way of improving coal mine safety by maximum possible reduction of hydrogen concentration in the gas mixture. Naturally, this theory needs experimental approval with all testing, project and other procedures. Considering properties of two-component methane-and-air mixture, it would be naïve to expect the the risk of explosion would completely vanish with hydrogen removed from from the gas mixture. Still, a noticeable increase in explosion safety of coal mining is quite realistic [9].

Regarding coal dust, it would be very unwise to treat it as a fuel during experiments. Initially organic, coal (and coal dust, accordingly) cannot but have micro-porous volumetric structure featuring sensible absorption capacity in dry air. As a consequence, that any gas component reaches higher concentrations in bulk or on the outward surface of a coal particle (dust). In this case, the coal particle saturated with hydrogen and/or methane can be a provoker and/or catalyst of an undesirable voluminous process. It is also expedient to emphasize that adsorption properties of coal differ per mines, which would require preliminary absorption tests to be carried out in specific coal mines.

The methane–hydrogen concept of gas explosion hazard in underground coal mining was to be checked, in particular, it was required to find hydrogen in coal mine air. To this end, the comprehensive analysis was undertaken to dig the data obtained in January–April 2011 in operating Alarda mine of Yuzhkuzbassugol. The gas composition of mine air was determined at the Novokuznetsk Detached Force of Militarized Mine-Rescue Unit and kindly provided by second in command of the brigade, General A. P. Eruslanov through the assistance of S. V. Popov, head of Research Laboratory.

In the bulk of the mine air samples, 41.0 % of the samples were take n routinely; 25.7 %—as a mater of urgency; 33.3 %—in an emergency. A specific flow sheet of coal mining was characterized by 1195 samples. The basic data in the sample analysis were the contents of oxygen, methane and hydrogen. The picture to describe gas explosion risk in Alarda mine is presented in Table 1.

Table 1. Gas composition of air in Alarda mine.

Place in the mine	Sampling point	Sampling number			Extreme contents, % by volume			Fire
		Total	Including hydrogen					
		samples	samples	%	min O_2	max H_2	max CH_4	events
Near the actual operation site	Underground	14	10	71.4	13.6	0.4	8.6	—
Gus suction from longwall no. 6-1-14	Ground surface	100	70	70.0	18.6	0.032	2.9	—
Conveyor roadway 3-39; railway incline 3-3	Entrance	105	66	62.9	15.2	0.22	9.8	—
Behind brattices	Underground	545	297	54.5	0.8	0.79	28.3	105
Hole from the mined-out void of longwalls	Ground surface	61	21	34.4	1.7	0.17	19.4	19
Hole no. 8065	Ground surface	107	72	67.3	5.0	0.77	17.9	8
Other	Underground	6	5	83.3	4.6	0.24	15.5	—
Total		938	541	57.7	0.8	0.79	28.3	132

It is seen in Table 1 that:

—hydrogen is present in more than 50 % of the samples;

—oxygen content reduces from natural (≈ 21 % by volume) to 0.8–5 % by volume in some places; the maximum decrease is observed in the mined-out and poorly ventilated void;

—hydrogen concentration grows from 0.032 % by volume (at the gas suction from operating longwall) to 0.79 % by volume (in mined-out void);

—methane content grows from 2.9 % by volume in the longwall to 28.3 % by volume in the mined-out area;

—each 6th sample was taken during gas combustion with the majority of fires (round 80 %) took place in the mined-out unventilated voids behind brattices.

It follows from the aforesaid that the critical situations with approaching overall combustion of gases intervene in mined-out area behind brattices which are unventilated or ventilated via holes.

The available information enables assessment of gases in the mined-out voids. In Alarda mine, there were 63 mined-out voids, in three voids (behind brattices nos. 115, 1187 and 1189), long-lasting fires took place. These voids, according to the database at hand, were under systematic monitoring by the Novokuznetsk independent mine rescue brigade, and 287 samples (53 % of the overall sampling) were taken in them. Out of 63 voids, hydrogen was present in 297 cases (54.5 5 of total number) in 27 voids (42.9 %). The gas composition of air behind brattices in these three critical voids is characterized in Table 2.

Table 2. Gas composition of air behind brattices in critical mined-out voids in Alarda mine.

Brattice no.	Sampling			Extreme contents, % by volume			Fires	
	Total	Including hydrogen						
	samples	samples	%	min O_2	max H_2	max CH_4	events	%
1115	87	40	46.0	1.0	0.02	28.3	20	23.0
1187	113	106	93.8	0.8	0.69	24.4	24	21.2
1189	87	85	97.7	1.4	0.79	24.9	24	27.6
Total	287	231	80.5	0.8	0.79	28.3	68	23.4

It follows from Table 2 data that these three critical voids encompass all extreme values of the analyzed parameters, namely, hydrogen presence in the samples (to more than 90 % of cases), extreme concentrations of oxygen (minimum) as well as hydrogen and methane (maximum) in the samples.

Thus, the information on Alarda mine discloses that:

—hydrogen is present at various concentrations in the majority of the samples taken at different points in the mine;

—the critical mined-out voids behind brattices being damaged or failed can be the source of accidents and emergencies in the mine.

The presented concept of unavoidable origination of fire and explosion hazards in coal mines and the illustrative data obtained in an operating mine proved another outlook of the nature and mechanism of hazardous gas mixtures. The change in vector of activities in the field of coal mine safety from methane to hydrogen, or, more specifically, hydrogen-and-methane, is an absolute innovation and will surely bring new beneficial effects. This certainty ensues from the fact that early in the 21st century, which can be regarded as the age of oncoming hydrogen energy generation, scientists and practitioners in many countries created and successfully operated systems and facilities capable to ensure complete safety of operations with much amounts of hydrogen (power generation, chemistry, metallurgy, space technologies, etc.). It is quite justified to assume that adaptation of such technologies and equipment to methane–hydrogen and other similar gas mixes in underground mineral mining will be one of the top-priority avenues of R&D for many research institutions in mining and the related sciences.

3. Conclusions

The resonance problem of occupational safety improvement in mining runs beyond the territorial borders of west Siberia and Russia as a whole. Accordingly, this problem can and is to be solved only if backed up by a solid foundation provided by science, physical infrastructure and finance with all-support of the government. By now, this problem can be handled by such a mighty scientific institution as the Russian Academy of Sciences represented by its Siberian Branch and, first of all, by the Kemerovo Research Center with the created Coal Science City technologically and financially bolstered by the whole Kuzbass. Successful management and advance in this area requires dedicated administration and research headquarters, an experimental test ground as one or a number of operating mines, as well as an extensive program of time-urgent R&D projects. A special emphasis should be laid on prompt authorization of data obtained by the domestic researches and designer within the framework of these projects.

Innovative approaches and knowledge require that society acknowledges then and undertakes vigorous measures toward their vivid advancement. To be fair, this concept of hydrogen and hydrocarbon formation in coal mines is not equally supported by this paper co-authors. However, this is not an obstacle for the co-authors team in integrating efforts in the analysis of the proposed mechanisms of fire and gas explosion risk in coal mines with intent to find ways of reduction (and, maybe, elimination) of these fatal hazards, which is an absolute priority.

References

[1] *The Great Soviet Encyclopedia* 1970–1978 Moscow: Sovetskaya Entsiklopediya Vols 1–30 (in Russian)

[2] Condensed Chemical Encyclopedia 1961–1967 Moscow: Sovetskaya entsiklopediya Vols 1–5 (in Russian)

[3] Adushkin VV, Kudryavtsev VP and Khazins VM 2006 Hydrogen degassing of the Earth and ozone anomalies Dokl. Akad. Nauk Vol 406 No 2 pp 241–243

[4] Uchitel AD and Gushchin VV 1981 *Vibratory Discharge of Rocks* Moscow: Nedra (in Russian)

[5] Voitov GI 1986 Chemistry and the scale of modern flow of natural gases in different geostructural zones of the Earth *Zhurnal Ross. Khim. Ob-va im. D.I.Mendeleeva* Vol 31 no 5 pp 53–60

[6] Kozlovskiy EA, Belozerov IM, Minin VA and Sharov GN 2011 Gas explosion hazard in underground coal mining *New Ideas in the Earth Sciences: X International Conference Proceedings* Moscow: Extra-Print Vol 2 (in Russian)

[7] Larin VN and Larin NV 2008 Hydrogen degassing on the Russian Platform *Degassing of the Earth: Geodynamics, Geofluids, Oil, Gas and Their Paragenesis—All-Russian Conference Proceedings* Moscow: Geos pp 267–269 (in Russian)

[8] *Degassing of the Earth Geotectonics, Geodynamics; Oil, Gas; Hydrocarbons and Life–All-Russian Conference Proceedings* Moscow: Geos 2010 (in Russian)

[9] Portnov AM 2010 Volcanoes—hydrogen reservoirs *Prom. Vedomisti* no 10–12 (158–161)

[10] Syvorotkin VL 2002 *Deep Degassing of the Earth and Global Catastrophes* Moscow: Geoinformtsentr (in Russian)

[11] Volodichev NN, Kuzhevskiy BM, Nechaev OYu et al 2002 Earth's crust—An active source of neutrons *Vestnik MGU. Fizika. Astonomiya* No 5 pp 69–73

[12] Gorshkov GV, Zyabkin VA et al 1996 *Natural Neutron Background of the Atmosphere and Earth's Crust* Moscow: Atomizdat (in Russian)

[13] Kuzhevsky BM 2011 Gravity of celestial bodies and neutron flows *Nauka v Russii* No 5 (125) pp 12–19

[14] Shestopalov IP and Kharin EP 2006 Time-to-time variability of the relationship between the Earth seismicity and different duration cycles of the Sun activity Gefiz. Zh. Inst. Geofiz. NAN Ukrainy Vol 28 no 4 pp 59–70

[15] Belozerov IM 2008 Nature through the eyes of physicist *International Scientific Journal for Alternative Energy and Ecology (ISJAEE)* No 12(68) pp 8–58

[16] Belozerov IM, Mezentev LN, Minin VA and Mitkin VN 2011 The Earth—an active source of neutrons and hydrogen *State-of-the-Art in the Earth Sciences: Proceedings of the International Conference devoted to V. E. Khain memory* Moscow: MGU pp 211–215 (in Russian)

[17] Emelyanov VS (Ed) 1958 *Condensed Encyclopedia of Atomic Energy* Moscow: Bolshaya sovetskaya entsiklopediya (in Russian)

Strength of inhomogeneous rib pillars

AA Krasnovsky

Chinakal Institute of Mining, Siberian Branch Russian, Academy of Sciences, Novosibirsk, Russia

E-mail: visanta@ngs.ru

Abstract. The author analyzes stress state of a piecewise-homogeneous pillar with a curved interface. Based on the system of singular integral equations, relations are derived for stresses and displacements at the boundary of the pillar and at the interfaces of the pillar components. Examples of numerical implementation are given.

1. Introduction

In underground mines, various pillars serve as support, barriers or walls, i.e. elements intended to ensure safe and regular operation. Rib pillars are a kind of spacer assemblies which interact with enclosing rock mass in independent, conjoint or hybrid modes as practice shows [1].

Modern rock mechanics sets a stress on the analysis of nonuniform bodies. Knowledge on stress state is required to calculate strength, stiffness and stability of mine structures, all components of stresses and displacements at boundaries and interfaces are needed to anticipate possible failure.

A major difference of layered and uniform structures is more complex stress state of the former. It is impossible to predict which components of stresses and displacements can or cannot be neglected, which is finally governed by geometry and position of layers, and their mechanical constants.

The most of calculations in mechanics and engineering are carried out within the framework of elasticity. Mathematical modeling, permissive of a theoretical analysis of the solution, provides the fullest understanding of the behavior of a structure under loading.

2. The problem formulation and solution

This study addresses stress state at the boundary of a piecewise-homogeneous pillar with a curved interface for the primal boundary-value problems without actual finding of complex potentials $\varphi(z)$ and $\psi(z)$ in the domain under analysis [2, 3]. The speculation is based on the system of singular integral equations connecting boundary values of stresses and displacements [4].

Consider a piecewise uniform pillar, the scheme of which is shown in Figure 1. On the border of the presented region, various variants of boundary conditions are assumed within the framework of the three main problems of the theory of elasticity.

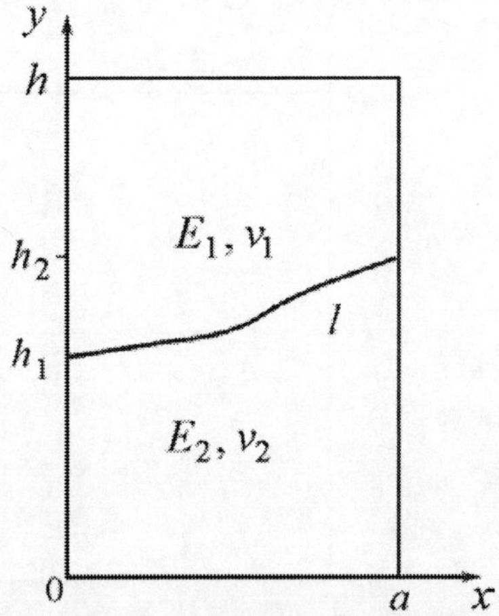

Figure 1. Schematic layout of piecewise-homogeneous pillar.

It is assumed that the interface of the pillar components experiences cohesion:

$$\sigma_n^+ = \sigma_n^-, \quad \tau_n^+ = \tau_n^-, \quad u^+ = u^-, \quad \upsilon^+ = \upsilon^-, \tag{1}$$

i.e. continuity of the normal and shearing stresses and displacements, where the upper sign implies belonging of the parameters to a respective pillar component.

The system of singular integral equations connecting boundary values of all stresses ad displacements in a general form for an arbitrary and simply connected domain [4] contains the function $f(t)$:

$$f(t) = i\int_0^t (X_n + iY_n)ds, \tag{2}$$

where X_n, Y_n—forces at the domain boundary in the lines of the axes x and y, respectively; t point of the boundary; i imaginary unit.

By way of illustration, we analyze a piecewise-homogeneous pillar (Figure 2) with the boundary $\Gamma = \Gamma_1 + \Gamma_2$ ($\Gamma_1 = \Gamma_{11} + \Gamma_{12} + \Gamma_{13} + \Gamma_{14}$, $\Gamma_2 = \Gamma_{21} + \Gamma_{22} + \Gamma_{23} + \Gamma_{24}$) and parabolic interface.

The boundary conditions are set as follows:

$$\sigma_n = \sigma_0 = -1, \ u = 0 \quad \text{at the faces } 0 \le x \le a \text{ and } y = h,$$

$$\sigma_n = 0, \ \tau_n = 0 \quad \text{at the faces } 0 \le y \le h, \ x = 0 \text{ and } x = a, \tag{3}$$

$$u = 0, \ \upsilon = 0 \quad \text{at the faces } 0 \le x \le a \text{ and } y = 0,$$

where σ_n, τ_n are the normal and shearing stresses; u, υ are the horizontal and vertical displacements.

Based on [4–6] with regard to the condition (1), the relations are written to connect the boundary values of displacements and the function $f(t)$ for the pillar components and at their contact line l.

IOP Conf. Series: Earth and Environmental Science **262** (2019) 012037 doi:10.1088/1755-1315/262/1/012037

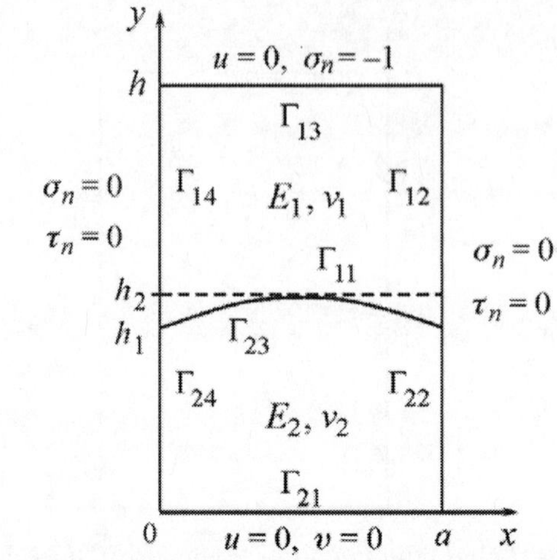

Figure 2. Analytical model of piecewise-homogenous pillar.

For the numerical implementation of the resultant system of equations, we go to dimensionless values such that the length dimension is related to the pillar width and the stress dimension is related to the value of stresses at the computational domain boundary.

The calculations presented in this paper were carried out at $h = 6$, $a = 1$, $\nu = 0.25$ for:

$$1)\ E_1 = 10^4,\quad E_2 = 3E_1;\quad 2)\ E_2 = 10^4,\quad E_1 = 3E_2, \tag{4}$$

as well as at different positions and shapes of the interface.

Figure 3 demonstrates deformation of the pillar boundary in variants (4) at $h_1 = 2$. The results are zoomed across the width for demonstrativeness.

The test case was a one-dimensional pillar with the interface represented by a parabola and a straight line. The calculated results agreed along the whole boundary of the pillar in both scenarios.

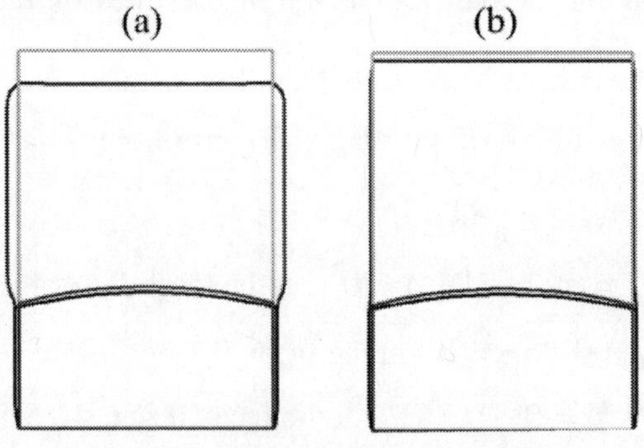

Figure 3. Deformation of pillar boundary: (a) $E_1 = 10^4$, $E_2 = 3E_1$; (b) $E_2 = 10^4$, $E_1 = 3E_2$.

Figure 4 depicts calculations for the horizontal and vertical displacements u, υ at the boundary of the upper pillar part. The boundary $\Gamma_1 = \Gamma_{11} + \Gamma_{12} + \Gamma_{13} + \Gamma_{14}$ is unfolded into a straight line. Curves *1–3* stand for the scenarios of $E_2 = 3E_1$, $E_1 = E_2$ and, $E_1 = 3E_2$, respectively.

The results are reflective of a complex structure of the stress state and the dependence of the solution on many parameters. The analysis of the functions (2) provides insight into the stress state both at the interface and the other branches of the boundary. To this effect, it is sufficient to differentiate the obtained results for the real and imaginary parts of *f*. The comparison of the results with the case when $h_2 = h_1$ (i.e. l is a horizontal straight line) shows that even a minor increase in h_2 causes difference in the solution, which is the greatest in the neighborhood of the points $(0, h_1)$ and (a, h_1) (Figure 2).

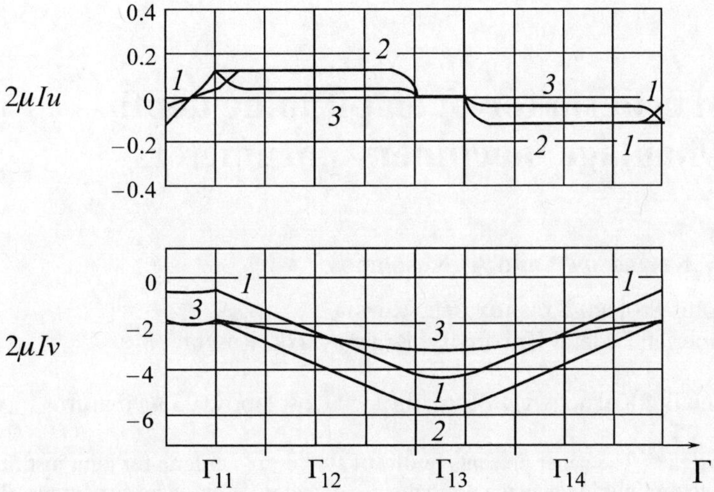

Figure 4. Displacements at the boundary of upper part of pillar.

3. Conclusions

The obtained relations and their implementation offer relevant information on all components of the normal and shearing stresses and displacements at the whole boundary of the pillar and at the interface of its components without factual determination of the stress state on the domain under study. The integral form of the solution allows both analyzing the solution itself and varying the involved parameters in order that solution possesses the wanted properties. The numerical implementation algorithm has been constructed. The obtained results have been analyzed.

References

[1] Borisov AA 1980 *Mechanics of Rocks and Rock Masses* Moscow: Nedra (in Russian)
[2] Muskhelishvili NI 1968 *Some Basic Problems of the Mathematical Theory of Elasticity* Moscow: Nauka (in Russian)
[3] Muskhelishvili NI 1968 *Singular Integral Equations* Moscow: Nauka (in Russian)
[4] Mirenkov VE 1978 Relation between stresses and shifts at the periphery of a working *Journal of Mining Science* Vol 14 No 3 pp 251–254
[5] Mirenkov VE 2007 Contact problems in rock mechanics *Journal of Mining Science* Vol 43 No 4 pp 370–381
[6] Mirenkov VE and Krasnovsky AA 2014 Identification of mechanical characteristics and interfaces of composite samples *J Fundamental and Applied Mining Science* Vol 1 No 1 pp 225–230 (in Russian)

Justification criteria for open pit mine depth and mining/haulage machinery parameters

DV Kuznetsov[1]* and AI Kosolapov[2]**

[1]Polus Project, Krasnoyarsk, Russia
[2]Siberian Federal University, Krasnoyarsk, Russia

E-mail: *KuznetsovDV@mail.ru, **Kosolapov1953@mail.ru

Abstract. The paper presents results of studies into criteria for joint justification of open pit mine depth and mining/haulage equipment parameters. New indexes of mineral reserves appraisal are obtained, and a classification of mineral reserves by value is proposed. In terms of a specific deposit, the relations are given for prompt determination of shovel bucket and dump truck capacities, and for justification of rational pit depth under conditions of unstable market of mineral reserves.

1. Introduction

Justification of ultimate pit depth limit is of special concern in ore mining. It defines basics of open pit mining and its key technical and economic indicators. At the same time, solution of this problem depends on many factors and requires complete and precise calculations with many complex questions explained. Under worsening natural and engineering conditions of mining, or in economic instability, it is necessary to estimate open pit mine dynamic, characteristics of modern mining and haulage machinery and to predict possible risk and new capabilities.

In Russia, the ultimate pit depth limit is conventionally set based on open pit mining without loss. The limit and boundary stripping ratios are determined and they should obey the condition that:

$$H \in K_b \leq K_{\lim}, \tag{1}$$

where H is the pit depth, m; K_b is the perimeter stripping ratio, m^3/t (t/t); $K_{\lim}$ is the limit stripping ratio, m^3/t (t/t).

The boundary stripping ratio is found as the ratio of rock volume increment to ore volume increment with increase in the ultimate pit depth limit by one bench, and the limit stripping ratio is calculated based on unit production cost of mineral:

$$K_{\lim} = \frac{C_{al} - C_{mp}}{C_{st}}, \text{ m}^3\text{/t}, \tag{2}$$

where C_{al} is the allowable cost per 1 t ore, Rub/ton; C_{mp} is the cost of mining, concentration and metallurgical processing of 1 t ore and other costs, excluding stripping, Rub/t; C_{st} is the cost of stripping, Rub/m^3.

The allowable cost per 1 t ore is found from the formula:

$$C_{al} = \frac{PC(1-d)R}{1+E_{п}}, \text{ Rub/t}, \tag{3}$$

where P is the market price of the mineral, Rub/t; C is the useful component content of ore, g/t; d is the ore dilution coefficient, unit fractions; R is the useful component recovery in processing, unit fractions; E_{pr} is the profit rate, unit fractions.

The economic criterion of profit in open pit mining can by given by:

$$E\,\mathrm{Pr} = P \cdot C \cdot R(1-n) - (C_{opm} + C_{st} \cdot K_b + C_{proc} + C_{\exp})\frac{1-n}{1-d}, \text{ Rub/t,} \quad (4)$$

where n is the ore loss coefficient, unit fractions; C_{opm} is the cost of open pit mining, Rub/t; C_{st} is the cost of stripping, Rub/m^3; C_{proc} is the cost of processing of 1 t ore, Rub/t; C_{exp} is the general and other expenses, Rub/t.

In case the profit calculated from (4) is higher or equal to zero, the ore reserves within the open pit mine limits are assumed as economic reserves minable and processable with the available advanced technologies and equipment. When the profit value is under zero, the ore reserves are assumed as uneconomic reserves as their mining is economically inexpedient or technically and technologically impossible although they can be assumed economic reserves in the future [1]. Justification of economic reserves in accord with the effective classification [2] takes into account categorization of ore reserves in terms of the extent of their exploration: categories A, B, C_1 and C_2.

At the further stages of mine planning, design and operation, neither volume not quality estimates are changed. During mining of a deposit, some portion of total reserves, depending on the degree of there preparation for mining, is grouped into prepared, accessed and ready for mining reserves.

The prepared reserves are the reserves that can be involved in primary process flow charts before extraction-and-loading; accessed reserves are the reserves with transport access required for extraction and haulage; ready for mining are the accessed reserves ready for extraction, loading and haulage directly from rock mass, or after pre-treatment by ripping or blasting [3].

The present paper authors think that this approach needs system amendments. This is mostly connected with the justification of the threshold limit of economic efficiency of reserves, with regard to the cost behavior as open pit mine depth is increased, inflation and possible price variation, including the given mineral price.

2. Justification criteria for open pit mine parameters

The review of open pit mining and mineral markets shows that for the latest 15–20 years, some open pits have reached the depth more than 500 m while their annual rock mass output has more than doubled in the conditions of time-varying demand for minerals and price level [4]. Moreover, geographically, open pit mining drifts to the regions with severe climate, which complicates mining technologies. Open pit mines are limited in terms of size as their depth is increased, which results in nonuniform development of the open pit mine area [5]. One the other hand, progression of mining and haulage machine engineering with increased capacity more than 1.5 times offers some technological advantages [6, 7].

The increase in capacity of open pit mining machines was gradual and conditioned mainly by reequipment of open pit mines, while the prices set by the machine manufacturer as well as market prices of of minerals jumped depending on currency rates. For example, gold price as per data of the Central Bank of Russia was 689 Rub/g (859 \$/oz), in 2008, 1201 Rub/g (1230 \$/oz) in 2010, 1436 Rub/g (1396 \$/oz) in 2013 and 2362 Rub/g (1260 \$/oz) in 2018. The average purchase coat of import equipment, e.g. manufactured by Caterpillar or Komatsu, at the elevation of dollar rate relative to rubble in 2014–2015 from 33 to 73 Rub in Russia, jumped by 1.8–2 times.

All these changes required multiple refreshed justifications of depth and other related parameters of open pit mines at different stages of planning and design, and governed high significance of assessment of technological advance rate in mineral mining relative to the rate of deletion of mineral reserves.

In this regard, the present paper authors have attempted to propose new justification criteria which are believed to be capable to enhance reliability and promptness of decision-making after being adapted to the specific deposit conditions. The background of the approach is feasibility of rating of any mineral reserves by value based on preset classification and possibility of finding ration open pit mine depth based on parameters of employed mining and haulage equipment, as well as its purchase and operation cost dynamics.

The proposed expression for the mineral value V is:

$$V = \frac{P \cdot C \cdot R \cdot (1-n)}{(C_{opm} + C_{st} K_b + C_{proc} + C_{\exp}) \dfrac{1-n}{1-d}} . \quad (5)$$

The joint analysis of the formulas (4) and (5) shows that: when $E\text{Pr} \geq 0$, $V \geq 1$ and when $E\text{Pr} < 0$, $V < 1$. On this basis, the value rating of mineral reserves is proposed: $V < 1$—no value (uneconomic); $1 < V < 1.25$—low-value (poor); $1.25 < V < 1.5$—medium value (low-grade); $1.5 < V < 1.75$—of value (ordinary); $V > 1.75$—high-value (high grade, rich).

In view that the common criteria of absolute and relative efficiency of investment projects are currently net present value *NPV* and internal rate of return *IRR*, they are correlated with the above characteristics as follows:

$$NPV = f(E\text{Pr}), \quad IRR = f(V).$$

Calculation of these relations in optimization of open pit mine limits will offer more exact dynamics of open pit mining, allow planning of extraction volumes and improve ability of a company to response to different unstable situations on the mineral market.

The values of *NPV* were calculated from the known formula:

$$NPV = \sum_{t=1}^{T} (R_t - Ex_t) \frac{1}{(1+E_{\text{av}})^t}, \quad (6)$$

where T is the estimation period, years; R_t is the income at a t-th calculation step, Rub; Ex_t is the expenses at this step, Rub; E_{co} is the industry-average rate of discount, %.

The values of *IRR* were set as the positive rate E_{av} such that *NPV* is zero. Assuming that the pit depth H would satisfy that condition under most unfavorable conditions, the optimal value domains of the studied parameters were predicted.

The expenses Ex_t in (6), embracing both operating costs and investments, for the main variants of mining and haulage machines available on the market in Russia were calculated directly. The correct evaluation of these expenses is one of the critical and labor-consuming stages in justification of economic efficiency of open pit mining. The calculations need much information which is always available. The expenses should be forecast for the long term comparable with operation life of equipment.

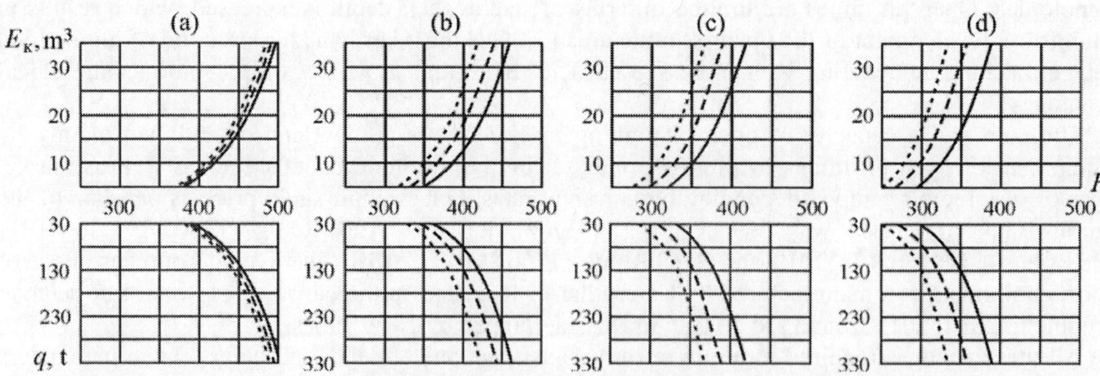

Figure 1. Effect of open pit depth H on average bucket capacity E_{buc} of electric shovels manufactured by IZ-KARTEKS, UZTM, and on capacity q of dump trucks manufactured by Caterpillar, Komatsu, Terex, Hitachi, Liebherr in mining of ore reserves of different value V: (a) 1; (b) 1.25; (c) 1.5; (d) 1.75.

The calculations were performed for open pit mines operating in rock mass with different physical and mechanical properties and in different climatic conditions. The rate of discount E_{av}, reflective of the inflation level, was assumed in the range from 10 to 20%. The values of the other indexes in (4)–(6) and their possible combinations were set such that to ensure the mineral value V in the range from 1 to 1.75. Some results are depicted in the Figure 1.

The presented relations are valid for a round open pit mine, column-like ore body and rock mass having ultimate compression strength to 120 MPa, density of 2.85 t/m^3, average structural block size of 1 m, and agree with the research findings in [8, 9]. Alongside shovels and dump trucks, the basic equipment package includes drill rigs and dozers. The number of buckets unloaded to a truck body is from 3 to 7.

3. Conclusions

It follows from the analysis of the obtained results, that the open pit mine depth is tightly connected with both value of ore and parameters of mining and haulage machines. Technical excellence of modern mining machines and their elevated capacity makes it possible to reach high productivity and expand open pit mine limits at higher economic efficiency.

The propose criteria enable justification of mining technology and equipment at the extended classification of mineral reserves.

References

[1] Trubetskoy KN, Kaplunov DR et al 2006 *Mining, Dictionary of Technical Terms* Moscow: IPKON RAN (in Russian)

[2] *Classification of Deposits and Predicted Hard Mineral Resevres* Moscow: Min. Prirod. Resurs. RF Order No 40 as of 7 March 1997 (in Russian)

[3] Rzhevsiy VV 2010 *Open Pit Mining. Technology and Full-Scale Mechanization* Moscow: Lbrikom (in Russian)

[4] *Global Mining Industry* Moscow: Gornoe delo 2005 (in Russian)

[5] Kuznetsov DV and Kosolapov AI 2018 Dynamics of mining equipment concentration in deep open pit mines in the north *Izv. Vuzov. Gornyi Zh.* No 1 pp 22–29

[6] Kuznetsov DV, Kosolapov AI and Malofeev DE 2016 *Complexes of Equipment for Open-Cut Mines in Severe Climatic Conditions* Deutschland: LAP Lambert Academic Publishing

[7] Burt C and Caccetta L 2013 *Equipment Selection for Surface Mining: A Review* Australia: Department of Mathematics and Statistics http://www.optimization-online.org/DB_FILE/2013/04/3831.pdf

[8] Camm TM 1991 *Simplified Cost Model for Prefasibility Mineral Evaluation* Bureau of Mines Information Circular

[9] Matsko NA 2002 Methods for assessment and management of dynamic accessability of mineral reservs *Doctroal Theses* Moscow: IPKON RAN (In Russian)

IOP Conf. Series: Earth and Environmental Science **262** (2019) 012039 doi:10.1088/1755-1315/262/1/012039

FEM simulation of deformation of self-stress rock mass with disjunction nearby a tunnel

SV Lavrikov*, OA Mikenina and AF Revuzhenko

Chinakal Institute of Mining, Siberian Branch, Russian Academy of Sciences, Novosibirsk, Russia

E-mail:*lvk64@mail.ru

Abstract. In the work the constitutive equations which allow to describe the structurally inhomogeneous rock mass taking into account the property of rocks to accumulate and release elastic energy, are applied. The finite element algorithm and programs for solving plane boundary-value problems are developed. The authors solve a problem on deformation of rock mass with a disjunction surrounding a tunnel. It is shown that driving of horizontal tunnel nearby the disjunction can provoke the accumulated elastic energy release, which affects stress-strain state of rock mass around the tunnel

1. Introduction

One of the priority trends in the modern geomechanics is concerned with the studies into internal structure of soil and rock. In the mechanics of soil and granular material, the granular structure governs internal friction, cohesion and dilatancy. During deformation, such properties may give rise to an increase in the pressure applied to enclosing structures, as well as to localization of shears and generation of isolated slip surfaces [1–3]. In the mechanics of rocks, the occurrence of discontinuities and the block structure determine essential nonlinearity and anisotropy of rock mass behavior [4–7].

One of the important properties of rock mass is the capacity to accumulate energy of external forces in the form of internal self-balancing stresses. This property is connected with the hierarchical block structure of rocks, having numerous equilibrium states [8]. Some areas in rock mass can accumulate energy and later, under certain conditions, to release it, i.e. such areas function as energy sources. The energy release process can either be stable relaxation or unstable disaster [9–11].

There exist different approaches to taking into account the hierarchical structure and internal self-balancing stresses of rock mass. Among them there are, for instance, the mathematical apparatus and models with internal variables [12–15], the approach based on the methods of the non-Archimedean mathematical analysis [16], the discrete element method [17], etc. In this study we used the approach from [12] and the mathematical model from [15] to analyze numerical deformation of rock mass surrounding a horizontal tunnel

2. Mathematical model

We choose a two-scale model of self-stress rock mass [12, 15] (plane deformation). The microscale elastic particles (grains) occur at the points of a square lattice. The pore space is filled with a binding material having elastic characteristics other than the properties of the grains (Figure 1). In-between the particles plastic shears develop by the nonlinear law, including stages of strengthening, softening and residual strength. The diagram of microslip between grains, connecting tangential microstress t_{12} and microshears ε_{ij}^{R} (in the general case, it is assumed that $\varepsilon_{12}^{R} \neq \varepsilon_{21}^{R}$), is shown in Fig. 2 (piece-wise linear approximation).

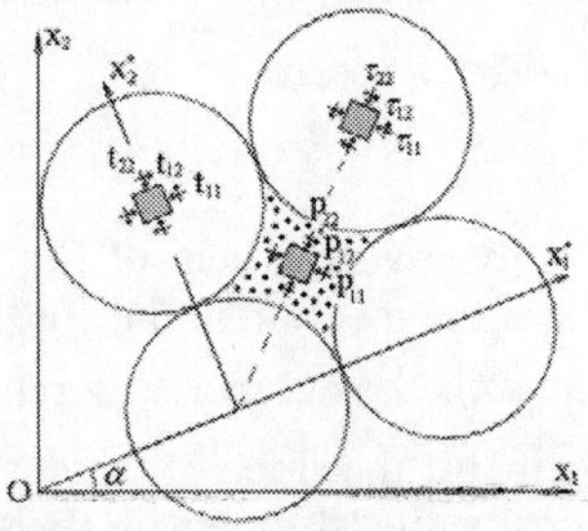

Figure 1. Internal microstructure of packing.

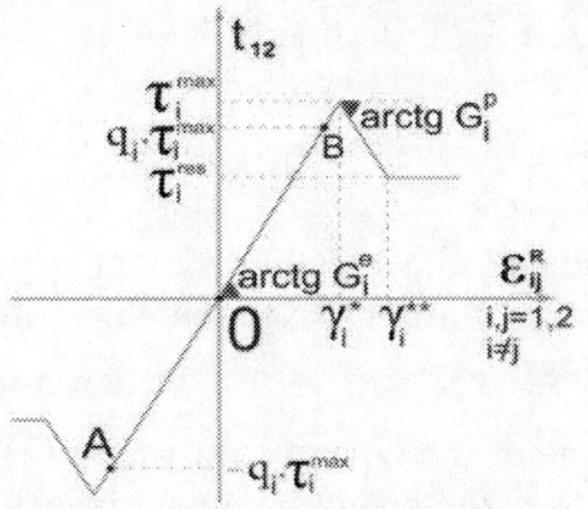

Figure 2. Diagram of inter-grain slip.

The conditions for the slip between grains are approximated by the piecewise-linear diagram (Fig. 2), and for the increments in the microstresses and microshears are given by

$$\Delta\varepsilon_{12}^{R} = \Delta t_{12} / G_1^s, \quad \Delta\varepsilon_{21}^{R} = \Delta t_{12} / G_2^s. \qquad (1)$$

The parameters G_1^s, G_2^s in Eq. (1) arethepresetmoduliofslipforeachfamilyofcontacts. The moduli G_i^s are defined by the preset constants $\gamma_i^*, \gamma_i^{**}, \tau_i^{\max}, \tau_i^{res}$, where $i = 1,2$ is the number of the family of contacts $G_i^s = \begin{cases} G_i^e, & 0 \le \gamma_i < \gamma_i^* \\ -G_i^p, & \gamma_i^* \le \gamma_i < \gamma_i^{**} \\ 0, & \gamma_i^{**} \le \gamma_i \end{cases}$.

The constitutive relations at the macroscale of the model connect the increments of the macrostresses $\Delta\sigma_{ij}$ with the increments of the macrostrains $\Delta\varepsilon_{ij}$:

$$\begin{pmatrix} \Delta\varepsilon_{11} \\ \Delta\varepsilon_{22} \\ \Delta\varepsilon_{12} \end{pmatrix} = W \cdot \left((T+R)^{-1} + 2(T+P)^{-1} \right) \cdot W^{-1} \cdot \begin{pmatrix} \Delta\sigma_{11} \\ \Delta\sigma_{22} \\ \Delta\sigma_{12} \end{pmatrix}. \qquad (2)$$

The coefficients in Eq. (2) depend on the microproperties of grains (matrix T), pore-filling material (matrix P), moduli of contact slip between grains (matrix R) and on the angle α (see Fig. 1), governing the orientation of regular grain packing in the Cartesian coordinates (matrix $W = W(\alpha)$).

The forces appearing in grains cause deformation of the grains and inter-grain shearing, which results in the deformation of the pore structure. Compression of the pore space is balanced by the tangential forces at the inter-grain contacts, and, in this way, even at zero external stresses, the internal self-balancing stresses can be considerably high. The condition of compatibility of micro- and macroparameters make it possible to connect the macrostresses σ_{ij}^* (in the coordinates of regular grain packing, see Fig. 1) and microstresses of grains t_{ij} and pore-filling material p_{ij} in the form of:

$$\sigma_{ij}^* = t_{ij} + 2m \cdot p_{ij}, \; i, j = 1,2 \qquad (3)$$

where the dimensionless value $0 < m < 1$ acts as a parameter of areal porosity [12].

The conditions (3) mean that one and the same macrostresses σ_{ij}^* can agree with the multitude different stress states of grains t_{ij} and pore-filing material p_{ij}. In this regard, zero macrostresses do not imply the absence of the internal microstresses. The microstresses in grains and in pore-filling material may have different signs and may completely balance each other.

The constitutive relations (2) in combination with the Cauchy relations, connecting strains and displacements, is closed by the equations of equilibrium (in terms of increments):

$$\frac{\partial \Delta\sigma_{11}}{\partial x_1} + \frac{\partial \Delta\sigma_{12}}{\partial x_2} + \Delta X_1 = 0, \quad \frac{\partial \Delta\sigma_{12}}{\partial x_1} + \frac{\partial \Delta\sigma_{22}}{\partial x_2} + \Delta X_2 = 0. \tag{4}$$

The formulated quasi-static problem is numerically solved with a help of finite element method by steps of loading with regard to the changing properties of the medium in the course of deformation: $\sigma_{ij}^{k+1} = \sigma_{ij}^k + \Delta\sigma_{ij}^k$, $\varepsilon_{ij}^{k+1} = \varepsilon_{ij}^k + \Delta\varepsilon_{ij}^k$, where the superscript k is the number of iteration.

The analysis of the type of the system (2), (4) shows that if the slope of the descending branch of the microslips at the inter-grain contact is less than a critical value (depends on the shear modulus and Poisson's ratio of grains, μ^t, ν^t, and pore-filling material, μ^p, ν^p), the system of equations is elliptical. In this case, deformation is a stable process of microstrength loss without jumps. The examples of the numerical calculations in the given range of parameters are discussed below.

3. Numerical modeling results

The computational domain is set in the neighborhood of a horizontal tunnel with an arch cross-section (Fig. 3). The initial macrostress state is chosen as the linear distribution: $\sigma_{22}^0 = -\gamma(H - x_2)$, $\sigma_{11}^0 = \xi\sigma_{22}^0$, $\sigma_{12}^0 = 0$, where γ is the specific weight of rocks, H is the depth of the tunnel occurrence, ξ is the lateral earth pressure coefficient. Alongside with the macrostresses, it is required to determine initial distribution of microstresses such that satisfy the relations (3). We select distribution of microstresses as function of the parameters q_i: $t_{11}^0 = 0{,}5\sigma_{11}^0$, $t_{22}^0 = 0{,}5\sigma_{22}^0$, $t_{12}^0 = \pm q_i \tau_i^{\max}$, $p_{ij}^0 = (\sigma_{ij}^0 - t_{ij}^0)/2m$, $i, j = 1{,}2$ and consider two different problems. In the first problem, it is set that in the whole computational domain the shear microstresses of grains and, accordingly, pore-filling material, are zero: $q_i = 0$, $i = 1{,}2$. In the diagram of shearing (see Fig. 2), this state corresponds to the point O. The second problem includes a disjunction nearby the tunnel (grey-colored zone in Fig. 3), and it is assumed that the shear microstresses of grains in the disjunction make a certain portion of the critical value, e.g. $q_i = 0{,}9$, while $q_i = 0$, $i = 1{,}2$ in the surrounding rock mass. This state in the shearing diagram conforms with the points A or B depending on the sign of the shear microstress (see Fig. 2). In this manner, one and the same initial macrostresses are set in the two problems, and the difference is the distribution of the microstresses of the structural elements.

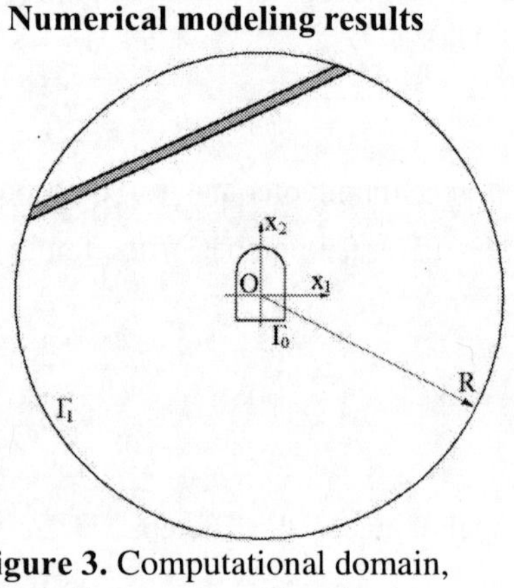

Figure 3. Computational domain, disjunction is grey-colored.

The boundary values are set as a sequential reduction in stresses at the tunnel boundary: $\Delta\sigma_n|_{\Gamma_0} = -\Delta d^k \cdot \sigma_n^0$, $\Delta\tau_n|_{\Gamma_0} = -\Delta d^k \cdot \tau_n^0$, where $\Delta\sigma_n$, $\Delta\tau_n$ are the increments in the normal and shear macrostress, respectively; σ_n^0, τ_n^0 – initial normal and shear macrostresses; Γ_0 is the boundary of the tunnel; $0 \le \Delta d^k \le 1$ is the increment in the dimensionless loading parameter at a k-th step. The condition $\sum_k \Delta d^k = 1$ means complete relieving of the tunnel boundary. The external boundary Γ_1 is assumed immobile: $\Delta u_i|_{\Gamma_1} = 0$, $i = 1{,}2$.

The input dimensionless parameters (all values of the dimension of stresses are related to the value of the maximum shear microstress $\tau_1^{\max}$ at the peak of the diagram in Fig. 2) are: $\mu^t = 3{,}75 \cdot 10^3$, $\nu^t = 0{,}2$, $\mu^p = 0{,}625 \cdot 10^3$, $\nu^p = 0{,}3$, $\gamma_1^* = \gamma_2^* = 0{,}5 \cdot 10^{-3}$, $\gamma_1^{**} = \gamma_2^{**} = 10^{-3}$, $\tau_1^{\max} = \tau_2^{\max} = 1$, $\tau_1^{res} = \tau_2^{res} = 0{,}5$, $\gamma H = 33{,}75$, $\Delta X_1 = \Delta X_2 = 0$, $\xi = 0{,}42$, $m = 0{,}5$, $\alpha = \pi/4$, $H/R = 100$. The calculations are continued till complete relieving of the tunnel boundary

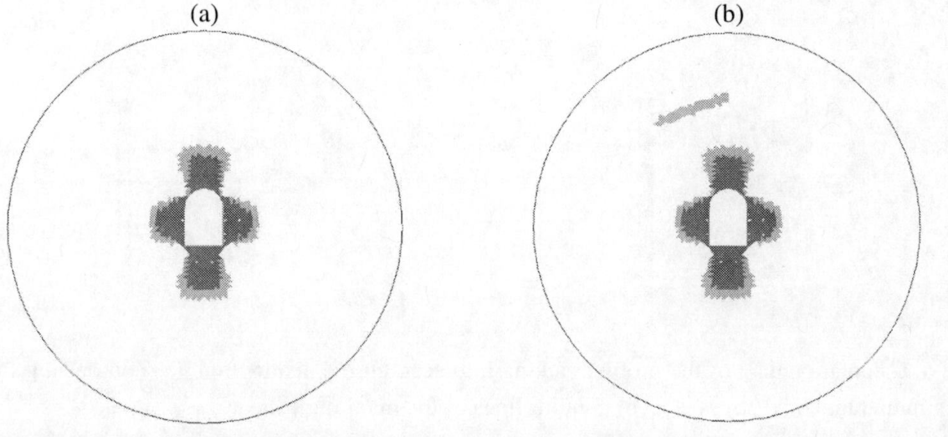

Figure 4. Calculated zone of plastic deformation: (a)without disjunction, (b) taking into account the disjunction.

Figures 4 and 5 depict the results of the first problem solution without the fracture. Figure 4a illustrates evolution of plastic deformation zones (light-grey are the zones of local softening—the descending branch; dark-grey are the residual strength zones—horizontal branch in Fig. 2). Figures 5a and 5b show contour lines of the maximum tangential stress $\tau_0 = 0{,}5\sqrt{(\sigma_{11} - \sigma_{22})^2 + 4\sigma_{12}^2}$ and maximum shear $\gamma_0 = \sqrt{(\varepsilon_{11} - \varepsilon_{22})^2 + 4\varepsilon_{12}^2}$, respectively. Apparently, plastic shears initiate from the natural stress raisors at the tunnel boundary and propagate in rock mass predominantly in the vertical direction (since the preset gravity stresses exceed the tectonic stresses).

Figures 4b and 6 present the results of the second problem solution with regard to the disjunction. It is seen (Figure 4b) that during deformation the material inside the disjunction zone relatively rapidly passes to plastic state and exerts influence on the stress state of surrounding rocks (Figure 6a and 6b).

On the whole, under the same initial macrostresses, the release of the accumulated energy in a relatively narrow section of the disjunction in the course of the tunnel driving induces macroshear of the fracture edges and the related change in the macrostress state of surrounding rock mass. The influence of the disjunction will even more increase in case of unstable deformation, when the release of the accumulated elastic energy will have the uncontrollable dynamic behavior.

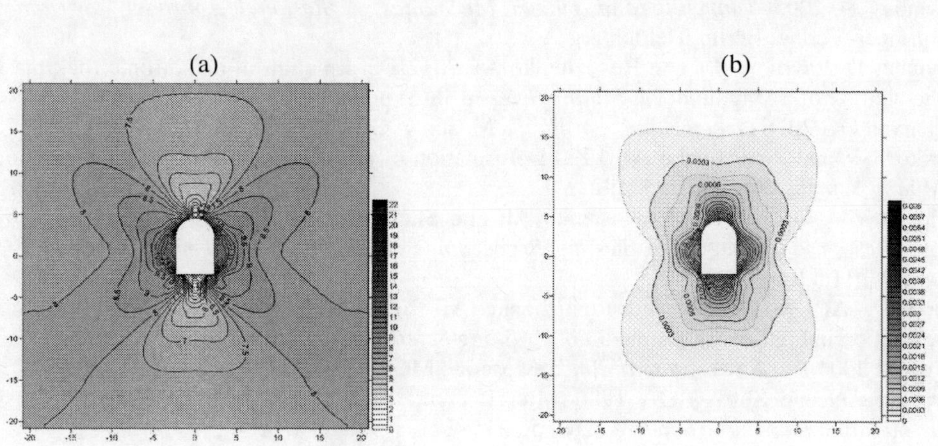

Figure 5. Calculation data of the problem without the disjunction: (a) contour lines of the maximum tangential stress τ_0, (b) contour lines of the maximum shear γ_0.

IOP Conf. Series: Earth and Environmental Science **262** (2019) 012039 doi:10.1088/1755-1315/262/1/012039

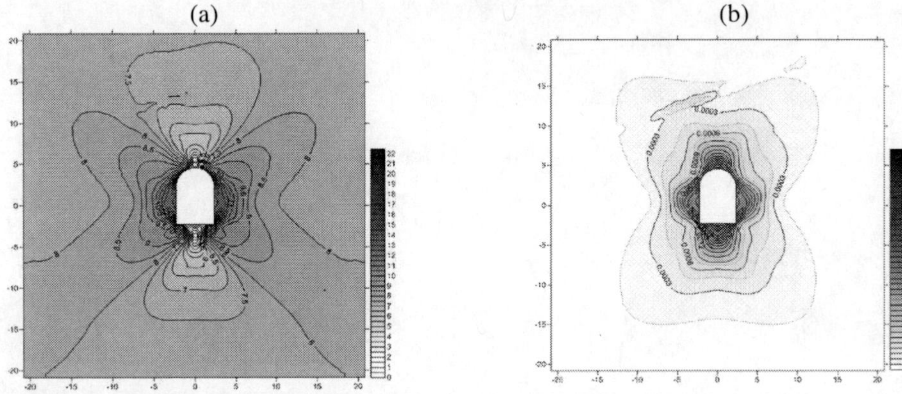

Figure 6. Calculation data of the problem taking into account the disjunction: (a) contour lines of the maximum tangential stress τ_0, (b) contour lines of the maximum shear γ_0.

On the whole, under the same initial macrostresses, the release of the accumulated energy in a relatively narrow section of the disjunction in the course of the tunnel driving induces macroshear of the fracture edges and the related change in the macrostress state of surrounding rock mass. The influence of the disjunction will even more increase in case of unstable deformation, when the release of the accumulated elastic energy will have the uncontrollable dynamic behavior.

4. Conclusions

The implemented approach allows the problems on deformation of structurally inhomogeneous geo-materials to be solved with regard to internal self-balancing stresses.

Driving of a tunnel nearby a disjunction can result in relief of internal self-balancing stresses, which affects stress state of surrounding rock mass.

Acknowledgements
The study was supported by the Russian Science Foundation, Project No 16-17-10121

References

[1] Gudehus G 2001 *Physical Soil Mechanics* Springer-Verlag, Berlin, Heidelberg
[2] Wu W, Bauer E, Niemunis A and Herle I 1993 Visco-hypoplastic models for cohesive soils In Kolymbas (Ed) *Modern Approaches to Plasticity* pp 365–383 Elsevier Amsterdam
[3] Borja R. Song X, Rechenmacher AL, Abedi S and Wu W 2013 Shear band in sand with spatially varying density *J. Mechanics and Physics of Solids* 61 pp 219–234
[4] Kolymbas D 2005 *Tunneling and Tunnel Mechanics. A Rational Approach to Tunneling* Springer-Verlag, Berlin, Heidelberg
[5] Kolymbas D, Lavrikov SV and Revuzhenko AF 2012 Deformation of anisotropic rock mass in the vicinity of a long tunnel *J. Mining Science* 48(6) pp 962–974
[6] Kocharyan GG 2016 *Geomechanics of Faults* Moscow: GEOS (in Russian)
[7] Lavrikov SV and Revuzhenko AF 1990 Deformation of a blocky medium around a working *J. Mining Science* 26(6) pp 485–492
[8] Kurlenya MV, Oparin VN, Revuzhenko AF and Shemyakin EI 1987 Some features of rock mass response to near-range blasting *Proceedings of the USSR Academy of Sciences* 293(1) pp 67–70 (in Russian)
[9] Bobryakov AP, Kosyh VP and Revuzhenko AF 2015 Trigger initiation of elastic energy relaxation in high-stress geomedium *J. Mining Science* 51(1) pp 10–16
[10] Moroz AI 2004 *Self-Stress State of Rocks* Moscow: MGGU (in Russian)
[11] Stavrogin AN and Shirkes OA 1986 Aftereffect in rocks caused by preexisting irreversible deformations *J. Mining Science* 22(4) pp 235–244
[12] Revuzhenko AF 2000 *Mechanics of Elastoplastic Media and Nonstandard Analysis* Novosibirsk: NGU (in Russian)

[13] Lavrikov SV and Revuzhenko AF 1994 Model of deformation of pillars with consideration of the effects of energy storage and weakening of the material *J. Mining Science* 30(6) pp 533–542
[14] Lavrikov SV 2003 A method of increase in bearing strength of rock mass around a mine working *J. Mining Science* 39 (5) pp 444–452
[15] Lavrikov SV 2010 On stress-strain state calculation of softened block mass nearby a mine working *Physical Mesomechanics* 13(4) pp 53–63
[16] Lavrikov SV, Mikenina OA and Revuzhenko AF 2008 Rock mass deformation modeling using the non-Archimedean analysis *J. Mining Science* 44(1) pp 1–14
[17] Klishin SV and Revuzhenko AF 2017 3D discrete element approach to the problem on abutment pressure in a gently dipping coal seam *IOP Conf. Series: Earth and Environmental Science* 84 012003(2017)

IOP Conf. Series: Earth and Environmental Science **262** (2019) 012040 doi:10.1088/1755-1315/262/1/012040

Laboratory tests of the packer sealing elements

YuM Lekontsev OA Temiryaeva and PV Sazhin*

Chinakal Institute of Mining, Siberian Branch, Russian Academy of Sciences, Novosibirsk, Russia

E-mail: *Pavel301080@mail.ru

Abstract. The test bench to study sealing elements made of elastomers is designed and manufactured. The relationships of their geometric parameters vs. axial load and probable reasons for jamming of sealing devices under packing in wells are established. The solutions to improve performance of sealants (packers) are worked out and proposed.

1. Introduction

Application of the directed and interval hydraulic fracturing tends to grow at Kuzbass coal mines. We consider some factors affecting the performance of hydrofracturing in a rock mass, disregarding the detailed description of the process, which, no doubt, contributes to higher safety of underground mining operations.

Hydrofracturing in wells drilled in a rock mass can be realized, given that a section of this well is properly sealed [1–4]. Special facilities are available to guarantee this condition. They are inflatable sealers of Taurs-type and mechanical sealers capable to vary its external diameter under axial compression [5]. The imperfection of an inflatable sealer is a relative small increment of the external diameter (8.0 mm), this drawback constricts the application scope and does not enable to realize the internal hydrofracturing. An additional point is that its low axial stiffness does not allow its mounting in an upward well and at depth of more than 20 m. Thus, under conditions of extended wells and interval hydrofracturing it is reasonable to employ rigid-body sealers and sealing elements made of elastomers. They are less sensitive to mechanical effects and more adaptable to variable conditions of interaction with well surface. However interaction of elastomers with host rocks in wells has not got thorough investigation yet. All the above complicates selection of an elastomer material considering mining and geological environment and a high risk for sealer jamming after hydrofracturing.

2. Laboratory tests of the elastomer specimen

The test bench was designed and manufactured to test elastomer sealers on axial compression and stability (Figure 1). The test bench consists of frame *10*, collet clamp *2*, sliding carriage *3*, balance wheel *4* with screw *5*. Sealer case end *1* is rigidly locked by collet clamp *2*, the other end is dressed with running-fit sleeve *7*. Dynamometer *6* is mounted between sleeve *7* and screw *5*. Elastic elements *8* of a sealer are mounted on case *1* and loaded by axial compression through sleeve *7* by screw *5* and balance wheel *4* in its rotation. Ball *9* centralizes force transmitted from screw *5* to sleeve *7*. Diameter of elastic element *8* is changed by means of clock-type indicator *11* at three points: in center, and on ends of the element.

Table 1 reports test data on polyurethane specimen with Shore hardness of 40 units. Figure 2 demonstrates basic dimensions of the test specimen made of elastomer and its loading scheme. The

IOP Conf. Series: Earth and Environmental Science **262** (2019) 012040 doi:10.1088/1755-1315/262/1/012040

left end of the specimen supports on a plain washer and the right end supports on a conical washer. Loading was cancelled after the test specimen lost stability.

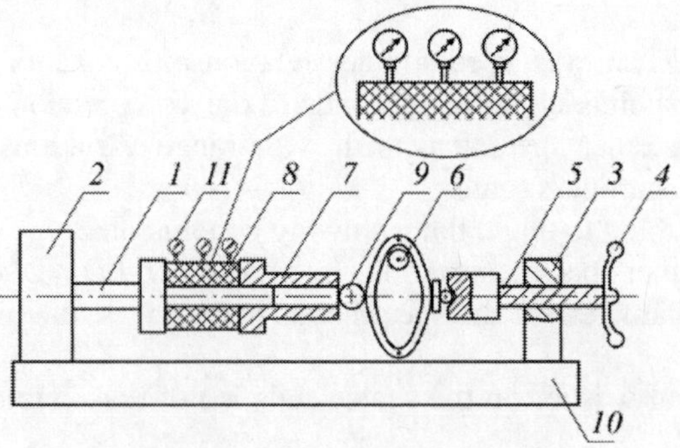

Figure 1. Test bench scheme.

Table 1. Tests of the elastomer specimen.

Plunger stroke, mm	Specimen length, mm	External diameter d_{ext}, mm			Increment Δd, mm			Dynamometer readings, kg
		left end	center	right end	left end	center	right end	
0	60	31.5	31.7	31.5	—	—	—	180
4	56	32.5	33.4	33.1	1	1.7	1.6	550
8	52	34.1	34.8	35.8	2.6	3.1	4.3	950
12	48	34.6	34.5	43.3	3.1	3.1	12.8	1000
16	44	Stability loss						1020

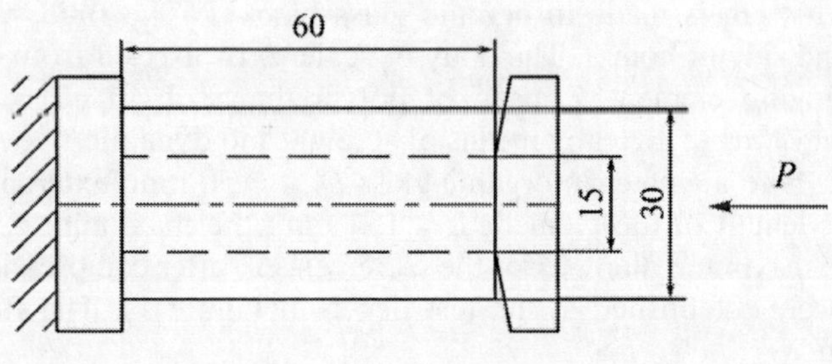

Figure 2. Scheme of elastomer specimen of 40 units in Shore hardness.

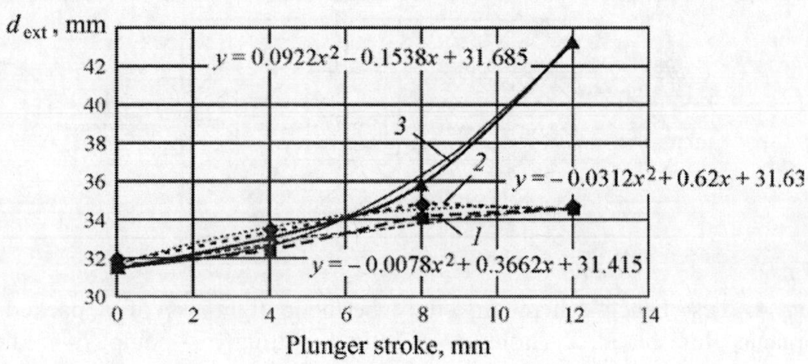

Figure 3. Relationship of the elastomer specimen diameter variations d_{ext} at three points versus plunger stroke: *1*—left end; *2*—central section; *3*—right end

It is obvious from Table 1, the largest increment of diameter in the middle section amounts to Δd = 3.1 mm. Thereto, the right-side end of the specimen supporting on conical washer increased in diameter by Δd = 12.8 mm.

The experimental evidence was used to plot relationships, one of which illustrates variation in specimen diameter at three points along length under axial compression (Figure 3). Similar tests were conducted on 15 other elastomer specimens with wide-range variations in geometric parameters and stiffness properties in Shore hardness scale.

Abundant database enabled to make the following conclusions:

—stability of an elastomer specimen depends on ratio $K = L / D \leq 2$, where L is specimen length; D, d is external and internal diameters of the specimen. The less it is, the higher stability of the specimen under axial compression is;

—application of conical washer on the sealer ends is not reasonable, as this enhances probability to "jam" a sealer in a well;

—to apply axial compression exceeding the calculated stability of the specimen to this specimen is not admissible. After stability loss the specimen loses symmetrical shape and fails to bear uniformly against well walls.

Next, the research concerned evaluation of the effect of axial elastomer specimen compression on hermetic tightening of a well section under the supplied fluid pressure.

In [6] the calculation of hermetic tightening between packer body and a well wall is made under simplified assumption running as: a ring-shaped space is filled with an incompressible mass of a sealing element, like a fluid. However, it is found in laboratory tests that the filling of a ring-shaped gap with a sealing material is not sufficient for the hermetic level needed to realize hydrofracturing. After the axial compression and contact with a well wall the sealing element requires additional pressing to gain its density value, corresponding to the maximum pressure of the fluid supplied to the area of hydrofracturing. Otherwise, the working fluid can affect the sealing element and compress it along its external diameter and initiate leaking from the sealed section. This statement is verified by tests executed at the test bench in Figure 4.

Hole *2*, imitating a well, is made in organic glass block *1*. The fluid was supplied by pump *5* through fitting *4* to blind side of hole *2*. The body of sealer *6* is inserted from the opposite side; sealer *6* contains elastic-springing element *7* made of polyurethane of 40–60 u. Shore hardness. Axial compression of element *7* was realized by means of screw *9* and dynamic sleeve *8*.

The input data are: hole diameter in organic glass D_{hole} = 50 mm; external diameter of the elastic element d_{ext} = 43 mm; length of the element L = 154 mm; the maximum compression stroke up to stability loss is 40 mm; maximum diameter of the increment Δd after compression—from 6.8 to 7.2 mm (two last parameters were established at the test bench in Figure 1). The fluid (3% emulsion) was supplied by hand pump *5*.

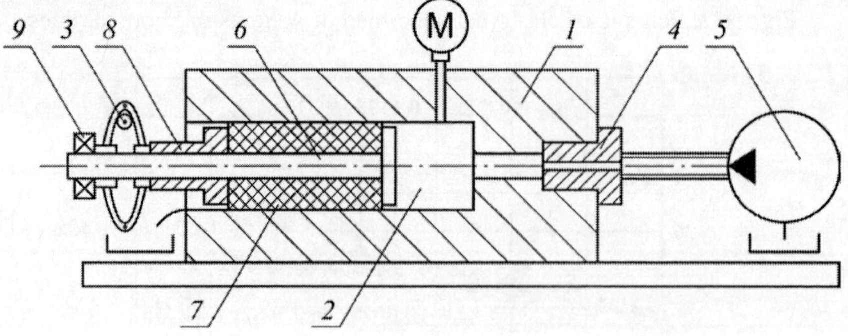

Figure 4. Test bench scheme to study hermetic tightness of a packed well section: *1*—organic glass block; *2*—hole; *3*—a ball; *4*—fitting; *5*—pump; *6*—sealer; *7*—elastic-springing element; *8*—dynamic sleeve; *9*—screw.

Test sequence:

—screwing screw *9* into screw section of sealer *6* through sleeve *8* provides compression of sealing element *7* up to contact of its external diameter with walls of hole *2*.

—pump *5* supplies the fluid through fitting *4* into hermetized area.

—successive screw motion stroke is 5 mm (compression of the test specimen is held up to stability loss).

—ring-shaped inserts made based on epoxy resin were used to vary diameter of the hole D_{hole} in organic glass to size of 48 and 46 mm.

The test bench results cited in Table 2 indicate that the sealing element after axial compression up to stability loss (< 40 mm) is pressed to walls of a conditional well, but pressurization failure is initiated at pressure of 7.5 MPa. Axial compression exceeding critical stability (> 40 mm) does not increase hermetization level. The highest hermetization level is observed in test no. 3 where axial compression of the sealing element by 35 mm led to pressure higher than 18 MPa. Further increase in pressure was constrained by strength of the body made of organic glass.

Table 2. Experimental results.

D_{hole}, mm	Parameter	Shore hardness 60A							
50	P_{max}, MPa	0.3	1.5	4.8	5.2	6.2	7.5	7.5	7.5
	Screw stroke, mm	15	20	25	30	35	40	45	45
48	P_{max}, MPa	1.5	5.1	6.8	8.0	9.0	12.0	—	—
	Screw stroke, mm	15	20	25	30	35	40	—	—
46	P_{max}, MPa	5.0	8.2	12.0	14.0	18.0	—	—	—
	Screw stroke, mm	15	20	25	30	35	—	—	—

The above sequence was followed in tests on sealing specimens Nos. 2–5. Parameters of the test specimens are, mm: No. 2—$D_{ext}\times d_{int}\times l$ = 30×15×60; No. 3—$D_{ext}\times d_{int}\times l$ = 30×15×30; No. 4—$D_{ext}\times d_{int}\times l$ = 40×15×160; No. 5—$D_{ext}\times d_{int}\times l$ = 70×40×170.

3. Conclusions

A series of test on different length and external-diameter sealing elements at the test bench revealed that the maximum hermetical tightening in a well is feasible under the following conditions: axial compression (linear value) should not exceed critical stability value; the initial ring-shaped gap between internal diameter of a well and external diameter of a sealing element should be by 5–8% less than the maximum increment of diameter increase under the free axial compression. Given that axial compression of the sealing element exceeds stability limit, in most cases the sealing element used to be jammed in the well, viz., loss of the sealer without improved hermetic tightness.

The efficient sealing elements should be of high elasticity stability along with high-level stiffness. Soft high-elasticity specimens of less than 40A hardness tend to leak to all irregularities of a well and to produce high level of hermetic tightness at the well section at their application point, but because of their insufficient elasticity they remain in the compressed state after axial loading is cancelled. This imperfection can lead to loss of such sealing elements.

References

[1] Klishin VI, Zvarygin LV, Lebedev AV, Savchenko AV 2011 Safety *Problems and New Technologies of Underground Coal Mining* Novosibirsk 524 pp

[2] Chernov OI and Kyu NG 1996 Experimental study of oriented rupture of solids by high

viscosity fluid *Journal of Mining Science* Vol 32 No. 5, pp. 362–367
[3] Kurlenya MV, Zvarygin LV, Serdyukov SV 1999 Control of longitudinal hydraulic fracturing of wells *Journal of Mining Science* Vol 35 No 5 pp 445–454
[4] Lekontsev YuM, Sazhin PV 2014 Technology of directional hydraulic fracturing of rocks to control hard-to-collapse roofs in treatment faces and degassing of coal seams *Journal of Mining Science* Vol 50 No 5 pp 914–917
[5] Lekontsev YuM and Temiryaeva OA RF Patent No 143091 Device for rock fracturing *Bull. Izobret.* 2014 No 19
[6] Ropay VA, Zaichenko VA 2012 Estimation of errors in determining the modulus of elasticity of rubber in compression associated with the failure to take into account changes in the size of the sample during the test *Sb. Nauch Rabot Nats. Gorn. Univer.* No. 39 pp 51–60

IOP Conf. Series: Earth and Environmental Science **262** (2019) 012041 doi:10.1088/1755-1315/262/1/012041

Results of research and development of vibration machines for mining

SYa Levenson*, LI Gendlina and VM Usoltsev

Chinakal Institute of Mining, Siberian Branch Russian, Academy of Sciences, Novosibirsk, Russia

E-mail: *lev@misd.ru

Abstract. The paper presents results of the bulk material vibration displacement studies, the results being the basis for the creation of vibrotechnics for the mining industry. The analysis is given of the capabilities and technical characteristics of the machines being created in comparison with that of the existing technique similar in operation principle and purpose, advantages of the machines are shown.

1. Introduction

In mining and construction, it is required to handle large volumes of bulk multi-component materials having different physical and mechanical properties. As a rule, these mixtures include clayey components which considerably increase cohesiveness when wet. Process lines provide temporal storage of such materials in accumulation bins. The bins are unloaded by gravity or using feeders, for instance, vibro-feeders which reliably operate in mines, make it possible to improve capacity of discharge opening and prevent from bridging. However, handling of wet coherent rocks can face problems connected with adhering or freezing over of materials on load-carrying surface of feeders.

Many institutions engage in engineering of vibration machines for loose bulk materials [1–5] both in Russian and abroad, such as Vibrotekhmash Research and Production (Russia), Electrovibromachine Tskhvinvali Woks (South Osetia), AViTEQ Vibrationstechnik GmbH (Germany), Schenckprocess (Germany), leading European company and Shanghai Zenith Mining and Construction Machinery Co. (China). It should be mentioned that these companies offer standard design vibro-machines, with rigid active element fixated on heavy frame by elastic supports to ensure directional vibrations. Disadvantages of these machines are high metal and power consumptions, high dynamic loading on supports and small area of vibrational impact. These machines can only handle non-adhesive materials, which is specified in technical documents.

2. Vibro-feeder characteristics and schematic layout

For the same purposes, the Institute of Mining, SB RAS, has designed vibro-feeders Sibiryachka (VDPU-4TM) and Volna [6]. Vibro-feeder VDPU-4TM (Figure 1) is composed of a welded load-carrying element (platform) *1*, the unloading end of which is connected to two inertia vibration exciters V-4,5 with circular driving force, coupled by an elastic sleeve. The platform is affixed to the discharge excavation floor by supports *3* and *4*. The platform section between the supports oscillates laterally, and broken rocks flow under vibration and by gravity on the sloped load-carrying surface to a truck or a conveyor. This feeder has simple design, but slope of the active element to horizon to ensure the flow should not be less than 15–18°, and this requirement complicates the process control.

Shortages of this machine are large weight (more than 4 t), high power consumption and difficult assembly–disassembly.

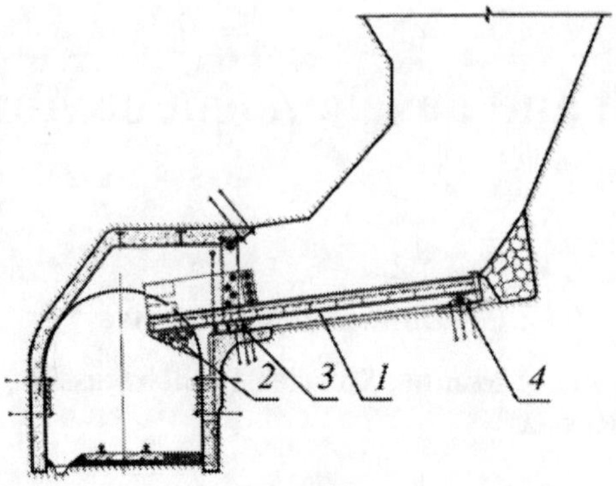

Figure 1. Schematic layout of Sibiryachka vibro-feeder in underground excavation: *1*—active element; *2*—vibration exciter; *3* and *4*—support.

Despite these disadvantages, Sibiryachka reliably operates in difficult conditions in Russian mines and is series-produced. This feeder is run in Evrazruda mines in Gornaya Shoria in ore drawing and ore chutes before loading rocks to haulage vehicles.

Vibration machines with elastic active element (vibro-belt) drastically differ from the other vibro-machines. This class feeders offer simple structure, small metal consumption and reliable operation under gobs under under subzero temperatures. Material is moved by an active element which is a thin metal oscillating sheet [7]. These machines are free from amortization, need no thick foundation and are driven by electric inertia or pneumatic vibro-exciters designed at the Institute of Mining, SB RAS.

Figure 2 depicts schematically Volna-4P feeder in ore drawing from extraction panel, and the Table 1 gives specifications of vibro-feeders Sibiryachka and Volna-4P.

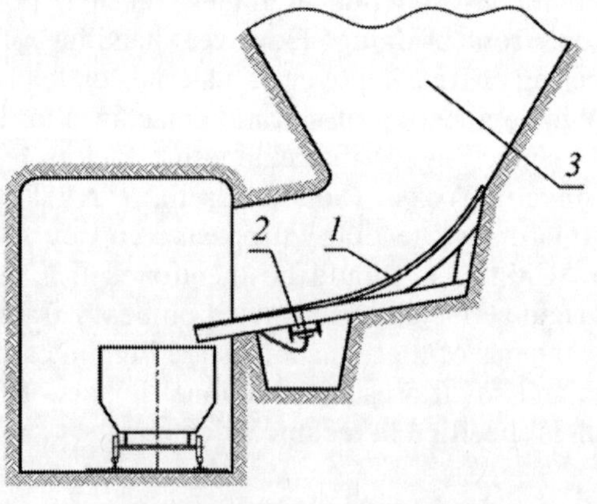

Figure 2. Schematic layout of Volna-4P vibro-feeder in underground excavation: *1*—active element; *2*—vibration exciter; *3*—accumulation bin.

Examination of loose rock discharge by vibro-belts show that curved active element of of the feeder creates more favorable conditions for rock flow. Application of such feeders in ore chutes, designed in conformity with the theory of loose material flow in narrowing and nonsymmetrical channels developed at the Institute of Mining, SB RAS, allows elimination of bridging of material at intersections of vertical and horizontal excavations.

Volna vibro-feeders ensure reliable discharge of loose materials from different reservoirs (ore chutes, bins) and loading to haulage machines in mining and construction.

Table 1. Specifications of vibro-feeders.

Description	Sibiryachka (VDPU-4TM)	Volna-4P
Dimension, mm:		
length	6015	3950
width	1200	1590
height	620	1040
Vibration exciter	electrical inertia	Napor-3M pneumatic
Number of vibration exciters	2	1
Driving force, kN	total 90	35.6
Vibration frequency, Hz	25	40
Slope, deg	15–20	15
Weight, kg	4000	1800
Capacity, t/h	Not less than 300	Up to 500

This class feeders ensure continuous and uniform flow of low-cohesive material. Difficulty arises in haulage of wet material with high content of clay, which entails bridging. It is possible to improve situation by enlarging the zone of vibration impact. This is especially important for machines with elastic active element as in this case, due to loss owing to external friction, vibrations damping greatly in both directions from vibro-source. As a consequence, amplitude of vibrations at the loading end of the active element can tens times less that at the vibration exciter, which decelerates flow rate of material and lengthens the period of material stay on the vibrating surface. As a result, loose material compacts at the loading end of the feeder. Since vibrations transferred to material at the loading end are much lower than at the vibro-exciter, a small portion of rocks directly above the unloading end of the feeder is only discharged, which results in chimney formation and in under-use of volume of accumulation bins.

In order to eliminate this disadvantage of vibration feeders with elastic active elements, their design was modified. In the new design, the active element of low bending stiffness is suspended on elastic supports installed normally to the end sections of the load-carrying surface. Each support has one its end rigidly connected with the active element and the other—with the frame, and takes bending moment. Thanks to such design, longitudinal oscillations of the active element are possible, which allows essentially higher vibrations of the loading section of the vibrating facility.

Bending stiffness of supports is selected in conformity with the admissible sag of the active element under maximum static load. On the assumption of maximum sag of the active element at the vibro-source arranged symmetrically relative to the supports, the relationship between the support bending stiffness W, design factors of the active element and its allowable sag f_{alwb} under maximum static load has been found and experimentally proved:

$$W = \frac{1}{3} N l_0^3 \left(R \sin\frac{l}{2R} - \frac{l^2}{8 f_{\text{alwb}}} \sin\frac{4 f_{\text{alwb}}}{l} \right)^{-1},$$

where l_0, l are the lengths of the support and active element; R is the curvature radius of the active element; N is the force applied on the support at the maximum static load on the active element.

The allowable sag value is set at design stage as per condition that the slope of the unloading section of the active element under maximum static load should not exceed 7–10°.

3. Experimental tests results

Figure 3 shows the experimental curves of the maximum dimensionless vibration velocity (V, V_s—vibration velocities at any point of the active element and at the vibro-source, respectively) of the active element without supports and with supports of different bending stiffness (curves *1* and *2*, *3*, respectively). The origin is matched with the loading end of the active element of the vibro-feeder, and

IOP Conf. Series: Earth and Environmental Science **262** (2019) 012041 doi:10.1088/1755-1315/262/1/012041

the point of the maximum vibration velocity is the driving force application point. The measurements are taken at the beginning of discharge, when the load on the active element is maximal.

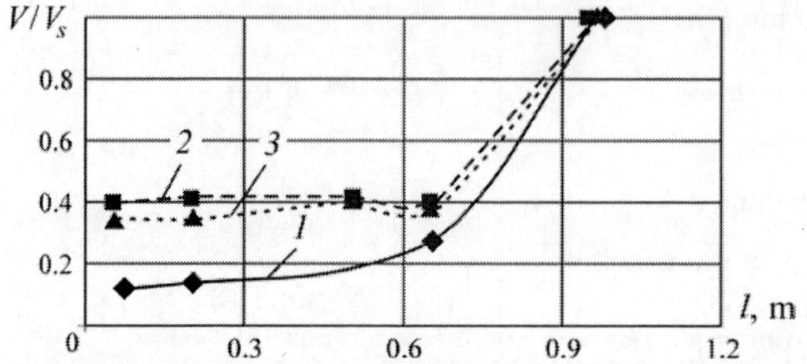

Figure 3. Distribution of dimensionless amplitude of vibration velocity along loaded active element without supports (*1*); and with supports with bending stiffness of 2.31 kN·m^2 (*2*) and 5.77 kN·m^2 (*3*).

These results show that the new design offers essentially higher (3.5–4.0 time) oscillation amplitude of the loading end of the active element, and, as a consequence, coherent material flow accelerates at this end, which creates conditions for sufficiently steady-state and uniform flow along the whole transporting surface of the vibration facility.

The research results on vibration discharge of coherent materials were used in design of a new vibro-feeder (Figure 4). The active element *1* made of steel sheet 5 mm thick is suspended on the frame *2* using elastic elements fixated at the ends of the active element, which enables longitudinal vibrations of the feeder and higher oscillations of the loading end. The active element and the frame, which is a box, form a reservoir with the bottom represented by the active element. The vibration source is inertia electric vibration exciter.

Figure 4. Development model of virbo-feeder for loose and coherent materials: *1*—active element; *2*—frame.

Specifications of the feeder:

Dimension, mm:	
length	1500
width	664
height	1050
Discharge opening area, m^2	0.27
Weight, kg	160
Vibration frequency, kHz	35
Driving force, N	2400
Vibro-exciter capacity, W	950

Under the specified driving force, discharge capacity of loam with clay content of 10% and moisture content of 13.5% makes 550–560 t/h.

For application in open pit mining, the feeder should be not shorter than 6 m, but the length is limited as vibrations damping. This shortage is eliminated by distributing driving force along the active element by means of installation of two vibration exciters at certain spacing as well as by ensuring their synchronous rotation.

The experimental research and analysis of dynamics of vibro-system [8] including elastic active element and inertia vibration exciters using physical models show that such system develops steady-state stable synchronous regime of vibration sources based on the phenomenon of synchronization [9]. The results of this research have become the framework for engineering feeder for vibrational dumping.

For dumping using dump trucks with capacity up to 300 t, the Institute of Mining, SB RAS, has designed a self-propelling vibrational stacker with the load-carrying element composed of three vibrating transportation facilities [10].

In the dumping technology with vibrational facilities, the stacker is arranged so that the unloading end of the load-carrying element is beyond the boundary of the dump level. The stacker length ensures unloading of dump truck beyond a possible sliding triangle. The distance from the dump level edge to the unloading point is set according to the operational safety standards. Broken rocks are unloaded from a dump truck onto receiving table of the load-carrying element and moved by vibration down the slope of the dump level.

Economic efficiency of the proposed technology was evaluated in comparison with the dozer stacker in the conditions of Lomonosov Mining and Processing Plant of Severalmaz Co. It is found that in operation with dump trucks with capacity of 90 t, the vibro-stackers with annual productivity of 3.8 Mm make it possible to reduce the number of dozers by 67%, to diminish the number of dumps and levels as well as to enhance operating safety. The dumping capacity can be increase 2.5 times, while operating and capital costs are cut down by 43 and 54%, respectively. The use of one self-propelling vibrating stacker allows the annual economic efficiency round 8 million rubles.

Currently, the pilot vibrating stacker has been manufactured for operation with dump trucks MAZ-5551. Efficiency of the stacker has been proved in the condition of a testing ground.

4. Conclusions

Design features of the vibration machines engineered at the Institute of Mining, SB RAS, are suitable for rock discharge from extraction panels and ore chutes in difficult conditions of mines.

Based on results of vibro-discharge tests, the designed structural layout of the feeder reduces damping of oscillations at the loading end of the active element of the feeder. The vibro-feeder designed by this layout implements uniform discharge of loose and coherent materials.

Application of stackers equipped with vibrating facilities with elastic active element improves safety and economic efficiency of mining.

References

[1] Blekhman II 2016 Effects of vibratory displacement in mineral processing: Useful applications and emergency situations *Resource Saving and Environmental protection in Mineral Mining and Processing—Plaksin's Lectures 2016: Int. Conf. Proc.*, Saint-Petersburg pp 65–67 (in Russian)

[2] Batrakov DV 2013 Ore drawing by vibrating feeders in iron ore mines of the Krivoy Rog Basin *Vestn. Krivorozh. Nats. Univer.* 34(1) pp 104–108

[3] *Electric Vibrating Feeders* Available at: http://www.mtspb.com//production_current.php?id=4&id_group=22

[4] *Vibratory Rock Screens* Available at: http://www.generalkinematics.com/mining/proddesc.cfm/productid/94

[5] *Vibrating Feeder* Available at: http://www.break-day.com/ru/vibrating_feeder.htm

[6] *Vibration Handling and Loading Unit VDPU-4TM*. Available at: http://npgm.ru/images/com_acatalog/vdpy-4tm.pdf
[7] Levenson SYa, Gendlina LI, Eremenko YuI, Kulikova EG and Morozov AV 2008 Vibratory equipment for underground and open cast mining *Equipment for Mineral Mining and Petroleum Produciton: IV Int. Conf Proc.* Yekaterinburg, pp 34–38
[8] Molotilov SG, Vasil'ev EI, Kortelev OB, Norri VK, Levenson SYa, Gendlina LI and Tishkov AYa 2000 *Intensification of Handling and Transportation Operations in Open Pit Mines* Novosibirsk: SO RAN (in Russian)
[9] Blekhman II 1971 *Self-Synchronization of Dynamic Systems* Moscow: Nauka (in Russian)
[10] Levenson SYa and Gendlina LI 2014 Safe dumping equipment *Journal of Mining Science* Vol Vol 50 No 5 pp 938–942

Justification of new construction diagrams of vibrating feeders for loose material discharging from hoppers

SYa Levenson, LI Gendlina and EG Kulikova*

Chinakal Institute of Mining, Siberian Branch, Russian Academy of Sciences, Novosibirsk, Russia

E-mail: *shevchyk@ngs.ru

Abstract. The paper present construction diagrams developed at the Chinakal Institute of Mining to expand capacities of vibrating feeders with low bending stiffness for discharging various difficult-to-transport materials, including cohesive, sticky and hot. The authors justify support elements for fixing the vibrating body on the frame and its equipment with auxiliary elements (chains, elastic seals and self-closing gate).

1. Introduction

Vibrating feeders designed at the Vibrotechnique Laboratory of the Chinakal Institute of Mining are intended for discharge of loose materials from hopper. The active part of these machines put on a support frame at an angle 10–15° relative to horizontal line and represents a thin metal sheet effecting bending vibrations under the action of exciting force. For creating the best conditions of loose material transfer, the vibrating sheet can be curved [1]. The vibration source can be an elastic or pneumatic exciter. Possible spills are eliminated using flexible rubber seals.

Unlike the machines with the stiff active part [2–4], the new vibro-feeders feature simple design, easy assemblage and maintenance as well as various modes of vibro-transport [5].

However, when a feeder operates under a hopper, vibrations greatly attenuate along the distance from the application point of exciting forces to the loading point of the feeder. The attenuation has no influence on transport efficiency in case of materials with low cohesion whereas during conveying of coherent and adhesive materials, high resistance forces arise in the material and at the feeder–material interaction zone. As a result, material adheres to the vibrating surface, and bridging is possible. Furthermore, rubber seals used to eliminate spills prevent from operation of the feeders with hot loose materials [6].

2. Vibrating feeders efficiency improvement

Currently, some engineering solutions are proposed, which enable using vibrating feeders with low bending stiffness for transporting loose, coherent, adhesive and hot materials.

Vibrations at the loading point of the feeder can be intensified by means of removal of friction between the carrying surface and the support. To this effect, it is proposed to the active part of the feeder is suspended on support elements with one end of which is connected with the active part and the other end is fixed on the support frame.

Such supports may be stiff elements (for instance, rods [7]) set in unstable equilibrium and elastically fixed. In this case, the exciting force of the vibration source generates bending vibrations of the active part; vibrations reach contact zones with the support elements and deflect them from the initial position. The support elements each time reset equilibrium owing to elasticity. Such support elements ensure shuttle motion of the active part relative to the support frame, with variable forms and intensities of vibrations both at the loading and unloading points of the feeder. The values of vibrations at the loading points become commensurable with the value of bending vibrations at the unloading point in the zone of the vibration exciter attachment, which allows creating extra shearing force, overcoming resistance of material flow on the feeder surface and maintaining uniform flow of material along the transporting surface.

Figure 1 demonstrates two construction diagrams of a feeder with such support elements. In order to recover initial position, the element s can be hinged with spring-assistance on the frame (Figure 1a), or set in elastic holders fixed on the frame (Figure 1b).

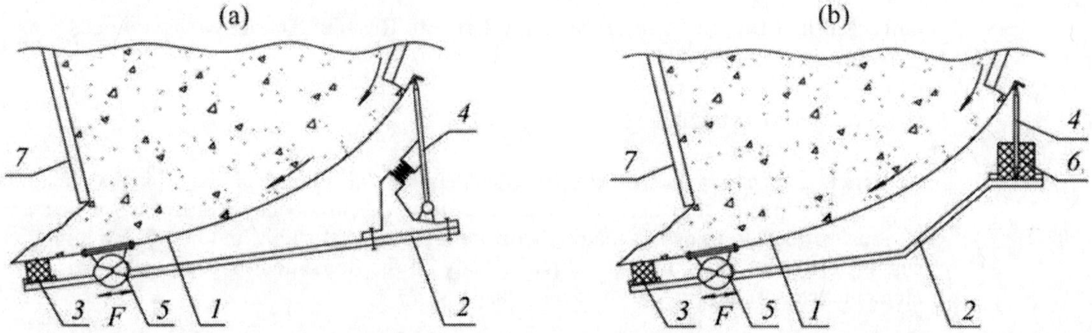

Figure 1. Construction diagram of vibrating feeder with low-bending stiffness active part connected with the frame by support elements: (a) and (b) alternatives of loading point of the feeder; *1*—active part; *2*—frame support; *3*—elastic support element at unloading points\; *4*—stiff support element in unstable equilibrium at the loading point; *5*—vibration exciter; *6*—elastic holder; *7* —hopper.

The support can be elastic (e.g., flat springs) and bend. When suspended, the active part under the action of load can sag. The resultant tension force is transferred to the support elements and bends them. The support elements keep the active part strained owing to elasticity [8–10]. As there is no friction between the loaded active part and the frame, vibrations are intensified at the loading point, and coherent material is moved at the same rate along the length of the transporting surface.

The influence of stiffness of the support elements on the efficiency of vibro-discharge of coherent material from hopper was studied experimentally.

3. Numerical and experimental investigation results

The physical experimentation used a test bench composed of a vibro-feeder and a receiving container. The binder was sand clay composed of particles 0.005–0.1 mm in size, with clay content of 10%. Parameters of vibrations were determined a measurement and computation equipment. The time of discharge of measured volume of the coherent material was recorded.

The numerical modeling of the dynamics of feeder characterized with low bending stiffness was carried out using the finite element modeling in ANSYS environment.

The research has found that the amplitude of transverse vibrations at the loading point is independent of the stiffness of the support elements. The amplitude of longitudinal vibrations grows with the reduction in the stiffness. According to the calculation data, the maximum increase in the vibration amplitude is observed when the total stiffness of the support elements at the loading and unloading points, EI_{Σ}, is less than 40 kN·m^2. On the other hand, the physical tests show that given such stiffness, the active part of the feeder sags greatly, the slope of its unloading point is decreased and, as a consequence, the discharge capacity is reduced by 1.3–1.6 times [11]. Thus, parameters of support elements are limited by the allowable sag of the active part. The observations over the coherent material discharge reveal that the use of elastic support elements in the structure of vibro-feeder enables uniform decrease of the material level in the volume of the hopper.

Aimed to prevent adhesion of material to the surface of the active part, it is proposed to equip the latter with the flexible movable elements, e.g. chains or ropes freely placed on the active part of the feeder and fixed at the loading point (Figure 2) [12].

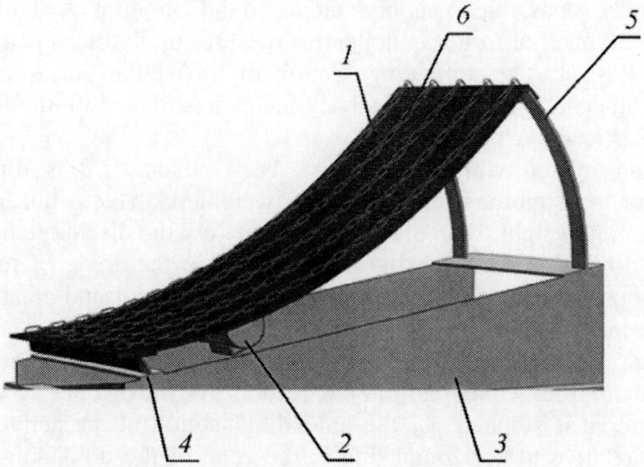

Figure 2. Vibrating feeder: *1*—active part; *2*—vibration exciter; *3*—support frame; *4*, *5*—elastic support elements at the unloading and loading points; *6*—flexible movable elements.

After the vibration exciter is actuated, coherent material is moved on the active part, tenses elements *6* in the line of unloading and presses them to the transporting surface. Under bending vibrations of the feeder, the contact between it and elements *6* is cyclically broken and recovered. At the same time, material on the vibrating surface is mixed, and the adhered material layer is decomposed, which improves efficiency of the vibro-feeder operation with adhesive materials.

With hit materials, it is possible to use heat-proof elastic (e.g., metal) plates (Figure 3) [13]. Each plate has a small length. The plates are arranged in two layers. In each layer, the plates are rigidly fixed in conjunction with each other and bent in the direction of active part *1*. The bottom layer plates *5* freely lie on the feeder surface and the top layer plates *4* cover the conjunctions of the bottom player plates.

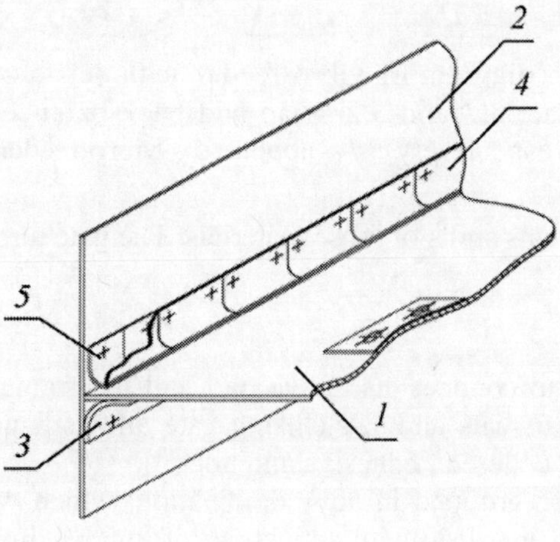

Figure 3. Construction diagram of vibro-feeder with heat-proof seals: *1*—active part; *2*—walls of frame; *3*—elastic support elements; *4, 5*—elastic sealing plates to prevent spills.

The seals have small lengths and are not connected rigidly; thus, even a mild force is sufficient for their elastic deformation. Bending vibrations generated in the active part under the actuation of the vibro-exciter make the plates bend too, but their contact with the transporting surface of the feeder is unbroken. Arrangement of the plates in two layers securely prevents spills, making the plates heat-proof allows using the feeder in transportation of hot materials. As the active part of the feeder is suspended on the frame, it is possible to supply cold air to the vibration exciter for cooling.

Spontaneous flow of material from the hopper in case of no vibrations can be prevented by a self-closing gate (Figure 4) [14]. The unloading section of the active part should be arch-wise bent downward. The gate composed of a flap and a back-balance is hinged on the frame at the level of the unloading end of the elastic active part.

When off-load (unactuated vibration exciter), back-balance *3.2* is directed vertically down (position *a*) so that the force moment created by its weight relative to the hinge exceeds the force moment created by the flap weight. Flap *3.1* closes completely the discharge hole of the hopper *4* and prevents unprompted flow of material on the active part *1* at the angle of repose. When the vibro-exciter *5* is actuated, loose material is moved along the active part *1* and creates pressure on the self-closing gate flap with increasing the force moment relative to the hinge. The gate turns toward the discharge side such that the flap goes from the position *a* to the position *b* and becomes an extension of the unloading point of the active part of the feeder. As the exciter is unactivated, the forced discharge of loose material is stopped. As the unloading section of the active part is arch-wise bent downward, its incline relative to horizontal line in the zone of the unloading end is higher than the angle of repose of loose material. Therefore, loose material moves by gravity and leaves totally this section of the active part. The force moment of the back-balance relative to the hinge becomes higher than the force moment of the flap and re-puts the self-closing gate into initial position. The discharge hole is shut.

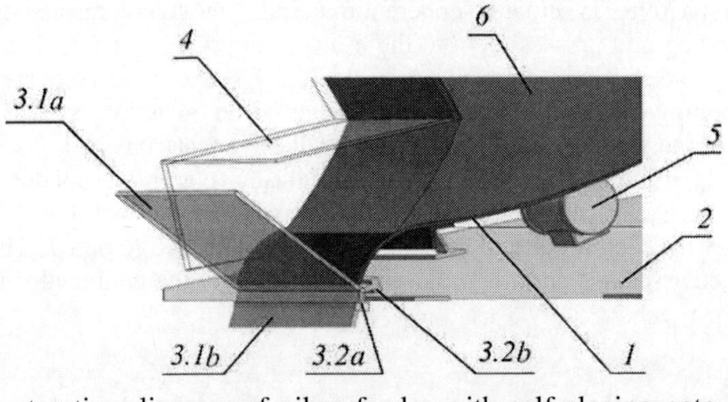

Figure 4. Construction diagram of vibro-feeder with self-closing gate (section view): *1*—active part; *2*—frame; *3.1* and *3.2*—flap and back-balance, respectively (*a* and *b*—gate is open and shut, respectively); *4*—hopper; *5*—vibro-feeder; *6*—loose material.

The self-closing gate prevents spills of loose materials. The gate turns by itself without extra energy intake.

4. Conclusions

The use of support elements to connect the active part and the frame, as well as auxiliary movable flexible elements, elastic metal seals and self-closing gate allows handling of loose and difficult-to-transport materials, including, coherent, adhesive and hot.

The proposed diagrams can be grouped in any combination, which will make it possible to expand capacities of vibrating feeders equipped with active part of low bending stiffness in discharge of geo-materials from hoppers.

References

[1] Zimonin LV 1983 Increasing efficiency of ore discharge using vibrofeeder with the elastic active member *Scientific Foundations for Mechanization in Open and Underground Mining: Collection of Papers* Novosibirsk pp 17–21 (in Russian)

[2] Batrakov DV 2013 Ore discharge by vibrating feeders in the Krivoy Rog iron ore mines *Vestnik Krivorozh. Nats. Univer.* No 34 pp 104–108

[3] Lyashenko VI, Dyatchin VZ and Franchuk VP 2014 Design and introduction of vibrating feeders–screens in the mining industry *Chernaya Metallurgiya* No 5 pp 72–80

[4] Uchitel AD and Gushchin VV 1981 *Vibratory Discharge of Rocks* Moscow: Nedra (in Russian)

[5] Levenson SYa, Gendlina LI and Kulikova GR 2017 Conditions of efficient vibrodischarge of materials in modern technologies of mineral mining and processing *J. Fundament. Appl. Min. Sci.* Vol 4 pp 28–33

[6] Kuzmin AV and Maron FL 1983 *Hanbook on Design of Mechanisms for Hoisting and Transport Machines* Mins (in Russian)

[7] Gendlina LI, Eremenko YuI, Levenson SYa and Kulikova EG 2008 RF Useful Model 77258 Vibrating facility for discharge of loose, coherent and caking material *Byull. Izobret.* No 29

[8] Weaver W, Jr, Timoshenko SP and Young DH 1990 *Vibration Probems in Enginering* John Willey and Sons

[9] Belyaev NM 1976 *Strength of Materials* Moscow: Nauka (in Russian)

[10] Svetlitsky VA 1978 *Mechanics of Flexible Rods and Threads* Moscow: Mashinostroenie (in Russian)

[11] Gendlina LI, Kulikova EG and Usoltsev VM 2017 Effect exerted by characteristics of support elements on p[arameters if vibrations of vibrofeeder *Inter-Expo Geo-Sibir Conference Proceedings* Novosibirsk: SGGA Vol 2 No 2 pp 159–163 (in Russian)

[12] Gendlina LI, Eremenko YuI, Levenson SYa, Kulikova EG and Aleski MYu 2008 RF Useful Model 77259 Vibrating facility for discharge of loose, coherent and caking material *Byull. Izobret.* No 29

[13] Gendlina LI, Levenson LYa, Eremenko YuI, Shevchuk EG and Golubykh SG 2003 RF Useful Model 3514 Vibratory transporting facility *Byuill. Izobert.* No 34

[14] Levenso SYa, Gendlina LI, Kulikova EG and Usoltsev VM 2015 RF Useful Model 156680 Vibrating facility for discharge of loose, coherent and caking material *Byull. Izobret.* No 31

Integrated performance analysis of ventilation schemes for double-line subway tunnel

IV Lugin and EL Alferova*

Chinakal Institute of Mining, Siberian Branch, Russian Academy of Sciences, Novosibirsk, Russia

E-mail: *alferova@mosk.ru

Abstract. The analysis of ventilation schemes for double-line subway tunnel uses criteria of operational economy and efficiency in routine and emergency modes. The longitudinal, transverse and longitudinal–transverse schemes ventilations are considered. It is shown that the most efficient and economic scheme is longitudinal ventilation in the routine mode and longitudinal–transverse ventilation in the emergency mode.

1. Introduction

Ventilation in subway tunnels is to remove harmful off-gases, as well as to maintain the preset air conditions and definite chemical composition. The major characteristics of a ventilation system inlet and outlet air flow rate which is to meet actual standards in the routine mode of a tunnel [1, 2]. In the emergency mode, a tunnel ventilation is to ensure safe evacuation of passengers and personnel from the zone of firing (smoke generation, gas pollution). To this effect, fresh air is fed in the escape to remove smoke and prevent smoke pollution of the other subway areas.

The international experience shows that double-line subway tunnels increasingly more frequently use the system of separate ventilation of tunnels and platforms [3]. The air distribution analysis in subways implies that tunnel ventilation is possible without ventilation chambers to be installed in the tunnel [4]. This considerably reduces capital cost due to decreased volume of underground construction, improves maintainability and accelerates troubleshooting.

2. Analysis of ventilation schemes for a double-line tunnel

Let us discuss variants of ventilation in a double-line tunnel:

—without an air duct—longitudinal ventilation (Figure 1a);

—with an air duct equipped with a baffle to divide fresh and disposal air flows—transverse ventilation (Figure 1b);

—unbaffled air duct (Figures 1c–1e)—hybrid longitudinal–transverse ventilation.

The criteria for the comparison of the listed ventilations include:

—aerodynamic parameters of fans (governed by air flow rate and loss);

—efficiency of operation in the emergency mode;

—operational economy.

The earlier research has found that the design flow rate of disposal air in the routine mode is mostly governed by excess heat in the tunnel. For heat loss in soil has minor influence on the heat balance, the air flow rates are approximately equal in the tunnel design in Figure 1. In the section composed of a tunnel and a platform in the Moscow Metro, the required air flow rate is 117.7 m^3/s [8].

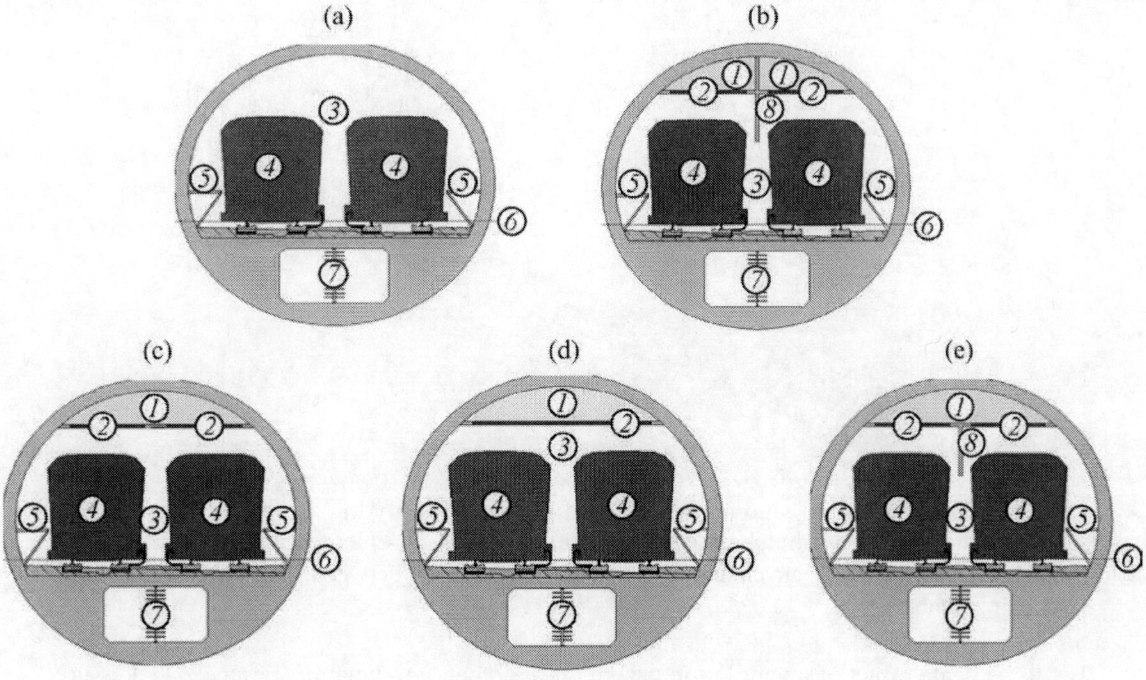

Figure 1. Double-line tunnel designs: *1*—air duct; *2*—hatch; *3*—tunnel; *4*—train; *5*—bench; *6*—level of head of rail; *7*—cable channel; *8*—line shield.

In the emergency mode, e.g., fire in the tunnel, ventilation is strongly required to ensure the design air flow rate in the escape so that to prevent spreading of smoke toward fresh air flow. This allows air flow not to be reversed, which ensures the safe escape. In a double-line tunnel, the highest velocity of the required air flow is 2.07 m/s [1]. Thus, for the tunnel design in Figure 1a, the required air flow rate is 87.6 m^3/s, Figures 1b and 1e—78.7 m^3/s, Figure 1c and 1d—80.9 m^3/s; this means that the air flow rate in the longitudinal ventilation

The design of the tunnel without an air duct in Figure 1a has a number of advantages: it is cheaper as there no expenses connected with the manufacture, installation, debugging, maintenance, etc. of the air duct; the air drag in the tunnel is lower, which brings in lesser blast pressure and electric energy spent for airing. The disadvantage of this design is the implementation of exclusively longitudinal ventilation. In case of emergency, passengers can only escape in the single direction off the fire source, which can result in fatality if cars get inflamed in the middle of a train.

For the passengers to evacuate in both directions from a fire source, it is required that fresh air is fed from both side toward evacuating people while gases of fire and smoke are removed from the scene of fire. Smoke is to be removed via a ventilation channel so that the evacuation routes are free from smoke. In this case, it is necessary to arrange the ventilation channel to ensure the longitudinal–transverse airing of the tunnel. Such tunnel designs are illustrated in Figures 1b–1e. The tunnels of such designs can be used for various purposes [5–7]. The advantages of the tunnel design in Figure 1e (with a line shield) are described and proved in [8].

Figure 2 depicts the longitudinal–transverse ventilation in a tunnel during fire and train stop in it: hatches *1* are opened above the burning car *2*; in the platform on the two sides of the emergency tunnel, air chambers *4* connected with the ventilation channel via valves *5* are switched into the mode of draught to remove smoke. Fresh air is fed from the two sides of the tunnel by air chambers *3* in the mode of air inflow toward evacuating people. This ventilation scheme conforms with the tunnel designs in Figures 1c–1e.

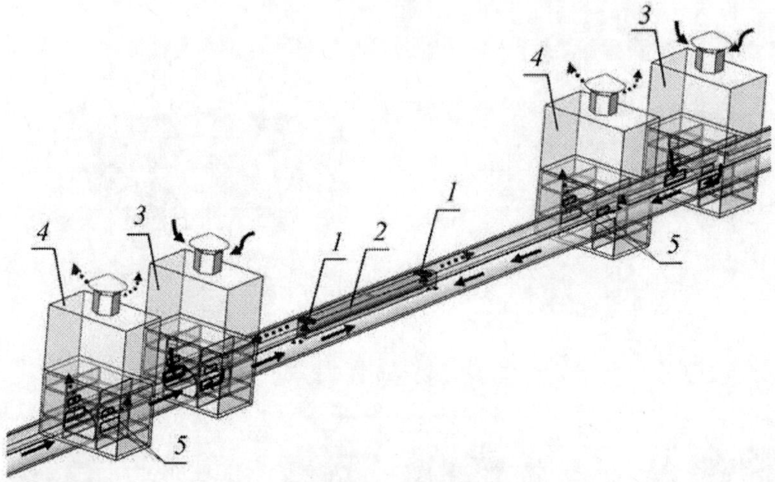

Figure 2. Schematic of smoke removal and fresh air supply in case of car fire and train stop in the tunnel: *1*—hatches to ventilation channel; *2*—emergency car; *3*—feeding air chamber; *4*—removing air chamber; *5*—connecting valves between the ventilation channel and air chambers.

Let us estimate effect of a ventilation design on the other aerodynamic parameter—pressure loss. The pressure loss ΔP in air flow along a long straight pipe are determined from the Darcy–Wesibach equation [9]:

$$\Delta P = \frac{\lambda \rho v^2}{2D_\mathrm{h}} l,$$

where λ is the fricition resistyance coefficient; $\rho = 1.2$ kg/m^3 is the air density in normal conditions; v is the air flow velocity, m/s; l—is the channel length, m; $D_\mathrm{h} = 4F / \Pi$ is the hydraulic parameter of the tunnl, m; F is the cross section area, m^2; Π is the perimeter, m.

The friction resistance coefficient λ depends on the pipe roughness $\mathrm{Re}(k_\mathrm{eqv} / D_\mathrm{h})$, where $k_э$ is the absolute equivalent roughness of the channel surface, mm; it depends on the channel lining material. In a double-line tunnel, $\mathrm{Re}(k_\mathrm{eqv} / D_\mathrm{h})$ is always higher than 500 mm/m, and the friction resistance coefficient is found from the Shifrinson equation for totally rough pipes:

$$\lambda = 0.11\left(\frac{k_\mathrm{eqv}}{D_\mathrm{h}}\right)^{0.25}.$$

Tables 1 and 2 describe pressure loss ΔP in the routine and emergency modes of operation, respectively, in a double-line tunnel 1 km long in case of different tunnel designs and ventilation schemes, as well as for half-tunnel (500 m long) air chambers of different design arranged in the tunnels. The pressure loss in the air chambers was disregarded, for they are similar in all schemes discussed and, according to [10], makes up to 70 % of the fan pressure.

Table 3 compiles values of air capacities in different routine and emergency ventilations (longitudinal, transverse and longitudinal–transverse).

Evidently, the transverse ventilation (Figure 1b) is inefficient due to high loss in friction (per 500 m of ventilation channel—1492 Pa in the routine mode and 667 Pa in the emergency mode), which is difficult to compensate using the mode fans currently in operation in subway tunnels. The presented parameters are characteristic of mine fans. In view of the fact that the required air flow rate in the routine mode exceeds the air flow rate in the emergency mode (for the tunnels equipped with the ventilation channel), it is expedient to use the longitudinal ventilation (Figure 3).

IOP Conf. Series: Earth and Environmental Science **262** (2019) 012043 doi:10.1088/1755-1315/262/1/012043

Table 1. Geometrical parameters and pressure loss in friction in double-line tunnels and ventilation channels of different design in routine operation.

<table>
<tr><th>Description</th><th>F, m^2</th><th>Π, m</th><th>D_h, m</th><th>v, m/s</th><th>Q, m^3/s</th><th>K_{eqv}, mm</th><th>ΔP, Pa</th></tr>
<tr><td>Double-line tunnel without air duct—longitudinal ventilation (Figure 1a)</td><td>42.3</td><td>26.56</td><td>6.37</td><td>2.8</td><td rowspan="3">117.7</td><td rowspan="3">2</td><td>61</td></tr>
<tr><td>Double-line tunnel witht air duct—longitudinal ventilation (Figures 1c and 1d)</td><td>39.1</td><td>28.16</td><td>5.55</td><td>3.0</td><td>42</td></tr>
<tr><td>Double-line tunnel witht air duct—longitudinal ventilation (Figures 1b and 1e)</td><td>38</td><td>30.4</td><td>5</td><td>3.1</td><td>51</td></tr>
<tr><td>Ventilation channel, concrete—longitudinal–transverse ventilation (Figures 1c–1e)</td><td rowspan="2">9.17</td><td rowspan="2">15.85</td><td rowspan="2">2.31</td><td rowspan="2">6.4</td><td rowspan="3">58.85</td><td rowspan="3">0.1</td><td>569</td></tr>
<tr><td>Ventilation channel, steel—longitudinal–transverse ventilation (Figures 1c–1e)</td><td>269</td></tr>
<tr><td>Ventilation channel, steel—transverse ventilation (Figure 1b)</td><td>4.45</td><td>9.56</td><td>1.86</td><td>13.2</td><td>1492</td></tr>
</table>

Table 2. Geometrical parameters and pressure loss in friction in double-line tunnels and ventilation channels of different design in emergency operation.

<table>
<tr><th>Description</th><th>F, m^2</th><th>Π, m</th><th>D_h, m</th><th>v, m/s</th><th>Q, m^3/s</th><th>K_{eqv}, mm</th><th>ΔP, Pa</th></tr>
<tr><td>Double-line tunnel without air duct—longitudinal ventilation (Figure 1a)</td><td rowspan="2">9.17</td><td rowspan="2">15.85</td><td rowspan="2">2.31</td><td>4.4</td><td>40.45</td><td rowspan="3">0.1</td><td>127</td></tr>
<tr><td>Double-line tunnel witht air duct—longitudinal ventilation (Figures 1c and 1d)</td><td>4.2</td><td>39.35</td><td>120</td></tr>
<tr><td>Double-line tunnel witht air duct—longitudinal ventilation (Figures 1b and 1e)</td><td>4.45</td><td>9.56</td><td>1.86</td><td>8.8</td><td>39.35</td><td>667</td></tr>
<tr><td>Ventilation channel, concrete—longitudinal–transverse ventilation (Figures 1c–1e)</td><td>42.3</td><td>26.56</td><td>6.37</td><td rowspan="3">2.07</td><td>87.6</td><td rowspan="3">2</td><td>33</td></tr>
<tr><td>Ventilation channel, steel—longitudinal–transverse ventilation (Figures 1c–1e)</td><td>39.1</td><td>28.16</td><td>5.55</td><td>80.9</td><td>20</td></tr>
<tr><td>Ventilation channel, steel—transverse ventilation (Figure 1b)</td><td>38</td><td>30.4</td><td>5</td><td>78.7</td><td>23</td></tr>
</table>

Table 3. Air capacity of tunnel ventilations in routine and emergency operation.

Ventilation	Tunnel ventilation capacity, kW	
	Routine	Emergency
Longitudinal	7.2	2.9
Transverse	175.6	52.5
Longitudinal–transverse	18.3	6.5

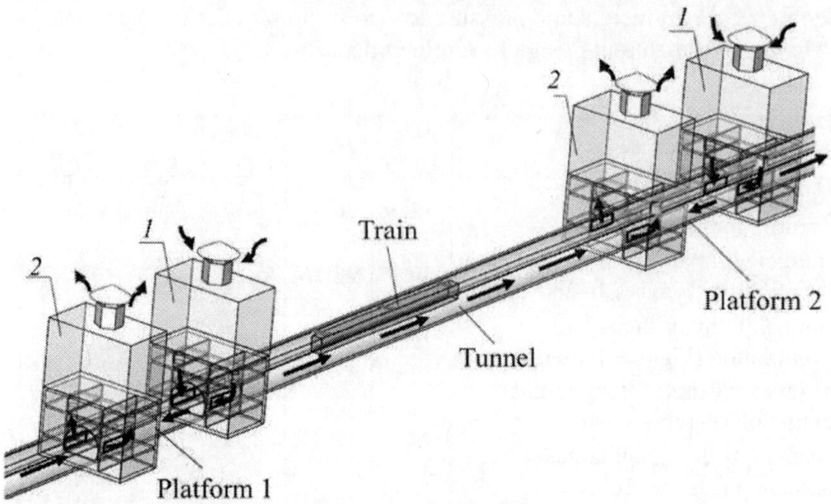

Figure 3. Schematic operation of tunnel ventilation in emergency mode: *1*—supply air chamber; *2*—exhaust air chamber.

3. Conclusions

In the routine mode of tunnel ventilation, air flow rate is the same in the ventilation schemes discussed while presser losses are low and differ insignificantly in different design tunnels. For these reasons, it is economically efficient to use the longitudinal ventilation in routine operation. The transverse ventilation in the routine mode is inefficient due to high drag in the air duct, which requires higher blast pressure and fan capacity (respectively, 5.5 and 24.4 times higher than in the longitudinal ventilation). In the emergency mode, it is advisable to use the longitudinal–transverse ventilation with smoke removed via the continuous air channel and fresh air inflow along the escape ways toward both sides from the fire source in the train.

Acknowledgements

This study was carried out in the framework of the Basic Research Program, Project Registration No AAAA-A17-117091320027-5.

References

[1] SP 120.13330.2012 *Subways* Effective of Jan 01, 2013 Available at: http://docs.cntd.ru/document/1200035937

[2] SP 2.5.1337-03. Sanitary Rules for the Pperation of Subways Effective of May 30, 2003 Available at: http://files.stroyinf.ru/Data1/45/45693

[3] Frolov YuS 2012 Experience and prospects of the development of the transport infrastructure of Spain *Metro i Tunneli* No 3 pp 1–9

[4] Krasyuk AM, Lugin IV, Pavlov SA, Romanov VI and Melnik GA 2015 Russian Federation Patent No. 2556558 Subwaye Ventilation Method *Byull. Izobret.* No. 19

[5] Mawhinney JR and Trelles J 2012 Testing water mist systems against large fires in tunnels: Integrating test data with CFD-simulations *Fire Technology* Vol 48 No 3 pp 565–594

[6] Lou B, Qiu Y and Long X 2017 Maximum smoke temperature in non-smoke model evacuation region for semi-transverse tunnel fire *Journal of Applied Fluid Mechanics* Vol 10 No 4 pp 1139–1151

[7] Pavlov SA 2016 Uniform air distribution in the extended autotransport tunnels *Int. Conf. Proc.: Architecture, Construction and Road Transport—Problems, Prospects, Innovations* pp 52–57 (in Russian)

[8] Krasyuk AM, Lugin IV, Alferova EL and Kiyanitsa LA 2016 Evaluation of ventilation flow charts for double-line subway tunnels without air chambers *Journal of Mining Science* Vol 52 No 4 pp 740–751
[9] Posokhin VN 2008 *Aerodynamics of Ventilation* Moscow: AVOK-PRESS (in Russian)
[10] Tsodikov VYa 1975 *Ventilation and Heat Supply of Subways* Moscow: Nedra (in Russian)
[11] Pavlov NN, Barkalov BV, Amirdzhanov SS et al 1978 *Internal Sanitary Engineering Installations. Part 3: Ventilation and Air Conditioning* Book 2 Moscow: Stroiizdat (in Russian)

Prospects for vortex hydrodedusting in coal mines

VN Makarov[1]* and NV Makarov[1,2]

[1]Ural State Mining University, Yekaterinburg, Russia
[2]Ventilation Ecology Safety, Yekaterinburg, Russia

E-mail: *mnikolay84@mail.ru

Abstract. Efficiency of coal mining and mineral processing is restricted by imperfection of the technology of blasting-generated dust control and mine safety. Using the mechanism of circulating inertia hetero-coagulation in the liquid–solid system, the authors propose the method of vortex high-pressure hydrodedusting. Furthermore, the contact interaction model is adjusted, and the relations are derived for the Stokes and Reynolds criteria, minimized total dust absorption energy, effective wetting angle and minimum diameter of water drops as function of their rotation speed. It is shown that the reduction in the particle size distribution in dispersed dust improves efficiency of dust explosion localization and silicosis control. It is proved that the patented vortex nozzles scale-down both minimum size of absorbed dust by times and water flow rate by 20%.

1. Introduction

Hydrodedusting is one of the most popular tools to prevent explosions of coal dust and create health conditions in mining. As practice shows, intensification of production and introduction of new advanced mining and processing technologies is restricted by imperfection of methods available for localization of coal dust explosions. Low-pressure spraying misses dust fines which are the most explosible and provocative of silicosis [1–3].

A potential for improvement of hydrodedusting efficiency lies in increasing the outlet water pressure as it governs the rate of coagulation and capability of water drops to trap dust particles [2]. At the same time, the increase in the energy consumption by aeration results in lower competitive ability of the eco-technology in mining [3–5].

Regarding micro-size particles showing hydrophobic behavior, the degree of inter-penetration of two phases, i.e. efficiency of coagulation, depends on the nature of surface phenomena in the zone of their contact governed by the relative velocity of water drop and dust particle, their sizes and surface tension at the interface. It I experimentally found that dust particles having diameter less than $5\cdot10^{-6}$ m are hydrophobic [2, 6, 7]. In the meanwhile, dust in coal mines is mainly composed of particles (1.5–250)$\cdot10^{-6}$ m in size [1, 4].

Thus, considerable portion of the most explosive-hazards and harmful dust cloud cannot be wetted effectively as it is hydrophobic due to small size of particles, which greatly weakens performance of high-pressure hydrodynamic dedusting.

The objective of modeling kinematics and dynamics of the water drop–dust particle system in the course of the proposed vortex inertia orthokinetic heterocoagulation is studying the mechanism of kinematic coagulation under the action of bound vortex in the zone of contact induced by a water drop with a rotation speed ω_w [8, 9].

Fixing of particles approaching the drop at a distance of effective adhesion depends on the value of a wetting angle θ. For a water drop to catch hydrophobic particle, it is required to perform work of external inertia forces, which conforms with a kinetic energy W_k of the particle–drop interaction in their contact zone. A water drop traps a dust particle when the kinetic energy W_k equals or exceeds an absorption energy Π_{w-g}, coming up to a sum of an adhesion energy W_{ad} (F_{ad}—force of adhesion), governed by a detachment energy, and a wetting energy W_{w-g} (F_{w-g}—force of surface tension), conditioned by a spreading energy [5, 6].

Based on the known graphical model for the kinetic coagulation of a dust particle by a water drop at $\omega_w = 0$ [5, 6], a model of the vortex kinematic coagulation was constructed such that a water drop rotates at the speed ω_w and induces the bound vortex in the contact zone.

It is seen in the mode of the contact interaction at the moment of collision in the solid–liquid system in Figure 1 that the water drop–dust particle contact are governed by the wetted perimeter diameter d_{wet} directly affects the value of the limiting wetting angle θ. With the smaller curvature of the drop surface in the contact zones, i.e. smaller size drop, the limiting wetting angle θ diminishes, which, consequently, results in the increase in the energy required for the total absorption of a dust particle with a diameter $d_{d\ min}$ by a water drop with a diameter d_w and determined by the surface energy of detachment and spreading.

This study focuses on the mechanism of purposeful control of the limiting wetting angle θ and the kinetic energy W_k of interaction between water drops and dust particles.

2. Mathematical modeling of vortex kinematic coagulation

As the limiting wetting angle θ reduces, the absorption energy lowers, which makes it possible to ensure pre-set efficiency of dedusting at reduced energy input, or to push the range of absorption of smaller size particles, i.e. to enhance dedusting at the preset energy consumption.

Figure 1 shows that when a dust particle collides a water drop with the rotation speed ω_w, the wetted perimeter diameter grows up to $d_{wet\omega}$ as against d_{wet} at $\omega_w = 0$, i.e. under classical heterocoagulation (*1*, *2* in Figure 1).

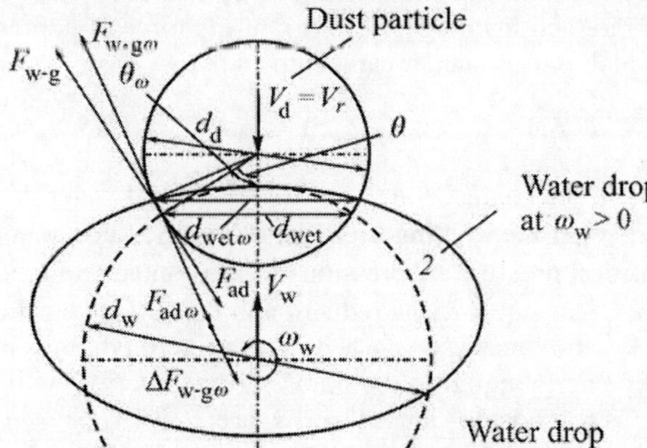

Figure 1. Graphical model of vortex kinematic coagulation of dust particle and water drop: *1*, *2*—respectively, models of classical and vortex orthokinetic heterocoagulation at $\omega_w = 0$ and $\omega_w > 0$.

The Strokes and Reynolds numbers Stk and Re, characteristic of the ratio of inertia and viscosity in the three-phase system of liquid–solid–gas, determine identity properties of kinematic, inertia and aerodynamic phenomena of coagulation [6]. The experimental research in [6] found the critical values of the Stokes number when catching of dust particles becomes impossible.

To push down the energy limit, i.e. energy consumption of high-pressure hydrodynamic dedusting, it is necessary to change the interaction kinematics between a water drop and a dust particle in the contact zone by varying rms value of the Stokes number at the constant forward velocities of the dust particle and water drop, V_d and V_w, respectively. With regard to the aforesaid, this is achievable at the water drop rotation at the speed ω_w around its axis coinciding with the vector of the velocity V_w [7–9].

In the vortex kinematic coagulation, rms values of the Stokes and Reynolds numbers are given by:

$$\mathrm{Stk}_{\omega} = \frac{d_d^2(\rho_d - \rho_g)\sqrt{(V_w - V_d)^2 + 0.25\omega_w^2 d_d^2 \sin\theta}}{18\mu_g d_w}, \qquad \mathrm{Re}_{w\omega} = \frac{d_w \rho_w \sqrt{(V_w - V_g)^2 + 0.25\omega_w^2 d_d^2 \sin\theta}}{\mu_g} \tag{1}$$

where d_w is the water drop diameter, m; ρ_w, ρ_g, ρ_d are the densities of water, gas and dust, respectively, kg/m^3; μ_g is the dynamic viscosity of gas, kg/ms; V_g is the velocity of gas, m/s.

In this fashion, rotation of water drop increases the actual rms numbers Stk_{ω} and $\mathrm{Re}_{æ\omega}$ in the drop–particle contact zone and favors reduction in the surface–adhesion energy barrier and critical level of the aerodynamic energy barrier [6].

When a water drop rotates at a speed ω_w, around it surface and in the contact zone, according to the Helmholtz–Bernoulli condition, a zone of rarefaction by the value of the specific energy ΔW_k of the bound vortex is created; the vortex velocity is determined from the Biot–Savart law from electrodynamics. Accordingly, the bound vortex, governed by the water drop rotation, reduced the static pressure in the contact zone with the dust drop, increases the limiting wetting angle down to θ_ω and contributes to lowering of the hydrodynamic energy barrier [11].

Drawdown in the energy required for the total absorption of dust particle by rotating water drop by the value of the potential rarefaction energy, i.e. work of the force of depression in the contact zone in the dust particle section equal to its diameter can be given by:

$$\Delta\Pi_{w-g\omega} = \Delta F_{w-g\omega} d_d = \frac{1}{2}\rho_w \Gamma_\omega \omega_w d_d S_c S_s^{-1}, \tag{2}$$

where Γ_ω is circulation of gas in the contact zone of dust particle and water drop, m^2/s; S_c is the area of the contact zone conforming with the wetting area, m^2; S_s is the surface area of dust particle, m^2; ΔF_{w-gw} is the force of rarefaction pressure (depression) in the contact zone of dust particle and water drop due to the bound vortex effect, equal to the reduction in the surface tension, N.

The equation of extra kinetic energy to push down the aerodynamic barrier of absorption, or actually to reduce the kinetic energy equal to the bound vortex energy, with regard to (2) as well as equations by Bernoulli and Ostrogradsky–Gauss [10, 12] are:

$$\Delta W_{kw} = \Delta\Pi_{w-g\omega} = \frac{\pi}{8}\rho_w d_d^3 \sin^4\theta\, \omega_w^2 . \tag{3}$$

For the vortex inertia orthokinetic heterocoagulation, the minimum energy of total absorption, considering (3), by analogy with the classical heterocoagulation at $\omega_w = 0$, is written as [5, 13, 14]:

$$\Pi_{w-g\omega} = \Pi_{w-g} - \Pi_{w-g\omega} = 2\delta_{w-g}\cos\theta_\omega , \tag{4}$$

where δ_{w-g} is the surface tension coefficient at the liquid–gas interface, J/m^2.

Taking into account (3) and (4), the limiting wetting angle in the contact zone of the liquid and solid phases at the water drop rotation speed ω_w is given by:

IOP Conf. Series: Earth and Environmental Science **262** (2019) 012044 doi:10.1088/1755-1315/262/1/012044

$$\theta_\omega = \arccos\left(\cos\theta - \frac{\pi\rho_w d_d^3 \sin^4\theta\omega_w^2}{8\delta_{w-g}\cos\theta}\right). \quad (5)$$

Thus, with regard to [6], Eq. (5) and the proposed model for the inertial orthokinetic coagulation in the dust particle–water drop system at the water drop rotation speed ω_w, the minimum diameter $d_{d\,\omega\,\min}$ of a dust particle totally absorbed during capturing and wetting by water drops under the action of surface tension and forward and backward motion inertias is:

$$d_{d\,\omega\,\min} = \frac{\delta_{w-g}\cos\arccos\left(\cos\theta - \frac{\pi\rho_w d_d^3 \sin^4\theta\omega_w^2}{8\delta_{w-g}\cos\theta}\right)}{(\rho_d - \rho_g)(V_w - V_g)^2}. \quad (6)$$

Figure 2 illustrates calculations carried out by the proposed mathematical model of the vortex kinematic coagulation to determine the change in the critical values of the Stokes number Stk_{cr} ($Stk_{\omega cr}$) as function of the rotation speed ω_w of a water drip with the diameter $d_w = 6 \cdot 10^{-6}$ m for the perfectly hydrophobic particles of silica. The isolines prove an essential decrease both in the surface–adhesion energy limit preventing particle adhesion and in the critical level of the aerodynamic energy barrier (line *4* in Figure 2).

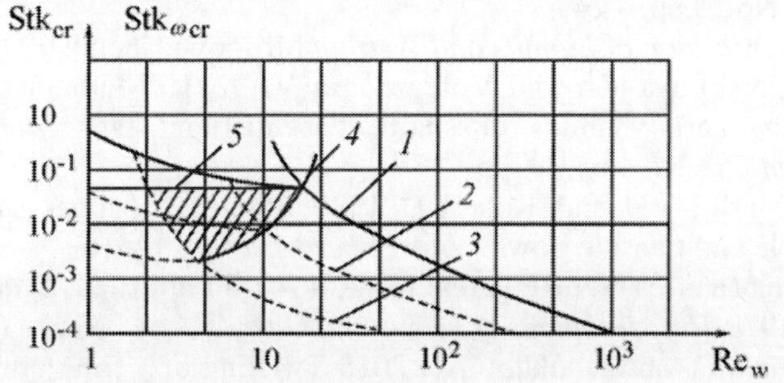

Figure 2. Isolines of water drop rotation speed as function of critical values of the Stokes and Reynolds numbers: *1*—$\omega_w = 0$; $Stk_{cr} = 4.1 \cdot 10^{-2}$; $Re_w = 20$; $d_{d\,\min} = 6 \cdot 10^{-6}$ m; *2*—$\omega_w = 1.5 \cdot 10^2$ s^{-1}; $Stk_{cr} = 9 \cdot 10^{-3}$; $Re_w = 15$; $d_{d\,\min} = 3.5 \cdot 10^{-6}$ m; *3*—$\omega_w = 3 \cdot 10^2$ s^{-1}; $Stk_{cr} = 5 \cdot 10^{-3}$; $Re_w = 6$, $d_{d\,\min} = 1.5 \cdot 10^{-6}$ m; *4*—Stk_{cr} versus $d_{d\,\omega\min}$ at $Stk_\omega = Stk_{cr} = 4.1 \cdot 10^{-2}$; *5*—$Stk_{cr} = 4.1 \cdot 10^{-2}$ versus ω_w and $d_{d\,\omega\min}$.

At the water drop rotation speed $\omega_w = 3 \cdot 10^2$ s^{-1} the critical values of $Stk_{кр}$ and $Re_ж$ drop more than 4 and 3 times, respectively, as against their values in the conditions of the forward movement of water drop, i.e. at $\omega_w = 0$. Moreover, the rms values of the Stokes and Reynolds numbers calculated from (2), in the line *4* in Figure 2, conform with their rms values under total adsorption at $\omega_w = 0$ from the known criterion equations [6].

The reduction in the Reynolds number for water drops under vortex high-pressure hydrodedusting promotes decrease in the water flow rate and required pressure, which improves efficiency of dust control.

The experimental research sufficiently accurately proved the calculated results from the proposed mathematical model and demonstrated high efficiency of vortex inertia orthokinetic heterocoagulation: the water consumption was lower by 20% and the minimum size of absorption of totally hydrophobic dust particles was decreased to $1.5 \cdot 10^{-6}$ m as compared with the classical high-pressure hydraulic dedusting.

3. Conclusions

1. Rotation of water drops under the vortex inertia coagulation raises the rms value of the Stokes number Stk_{ω}, which promotes an increase in the wetting angle and a decrease in the minimum size of absorbable dust particle down to $1.5 \cdot 10^{-6}$ m.
2. The vortex kinematic coagulation allows reduction in water flow rate by 20% as against the classical method of high-pressure hydraulic dedusting.

References

[1] Ksenofontova AI and Burchakov AS 1965 *Theory and practice of Dust Control in Coal Mines* Moscow: Nedra (in Russian)

[2] Saranchuk VI, Kachan VP, Rekun VV et al 1984 *Physicochemical Framewhor for Hydrodedusting and Prevention of Coal Dust Explosions* Kiev: Naukova Dumka (in Russian)

[3] Makarov VN, Potapov VYa, Davydov SY, Makarov NV 2017 A method of additive aerodynamic calculation of the friction gear classification block (SCOPUS) *Refractions and Industrial Ceramics* Vol 38 No 3 pp 288–292

[4] Yanov AP, Vashchenko VS, Gagauz FG et al 1984 *Combating Dust and Harmful Gases in Iron Mines* Moscow: Nedra (in Russian)

[5] Zhuravlev VP, Saranchuk VI, Strakhov IA et al 1990 *Modeling and Designing Hydrodedusting Systems* Kiev: Naukova Dumka (in Russian)

[6] Frolov AV, Telegin VA and Sechkerev YuA 2007 Elements of hydro-dedusting *Bezop. Zhiznedeyat.* No 10 pp 1–24

[7] Zimon AD 1976 *Adhesion of Liquids and Wetting* Moscow: Chemistry (in Russian)

[8] Makarov VN, Gorshkova VA and Volegzhanin VA 2016 Mathematical modeling of active control of the aerodynamics of shaft of centrifugal fans *Mining Informational and Analytical Bulletin* No 4 pp 39–45

[9] Malmyth ND, Marlhi VD and Kole DD 1980 Studies of upper surface blown airfoils in jucompressible and transuic flows *AJAA Paper* No 18 pp 14–16

[10] Makarov NV and Makarov VN RF Patent No 260 14 95 Published: *Byull. Izobret.* 2016 No 31

[11] Loitsyansky IL 1978 *The Mechanics of Fluids and Gas* Moscow: Nauka (in Russian)

[12] Bautin SG, Krutova IY and Obukhov AG 2015 Twisting of a fire vortex subject to gravity and Coriolis forces *High Temperature* Vol 53 No 6 pp 928–930

[13] Bautin SP and Novakovskiy NS 2017 Bautin S. P. Numerical simulation of shock-free strong compression of 1D gas layer's problem subject to conditions on characteristic *Journal of Physics: Conference Series* Vol 894 No 1 Available at: http://iopscience.iop.org/article/10.1088/1742-6596/894/1/012067

[14] Bautin SP, Krutova IYu and Obukhov AG 2017 Mathematical justification of the Earth's revolution effect on tornadoes and tropical cyclones *Vestnik MIFI*Vol 6 No 2 pp 101–107

Additive mathematical modeling and development of high-pressure adaptable local ventilation fans

NV Makarov[1,2]* and VN Makarov[1]

[1]Ural State Mining University, Ekaterinburg, Russia
[2]Ventilation Ecology Safety Ltd, Ekaterinburg, Russia

E-mail: *mnikolay84@mail.ru

Abstract. Intensification of mining operations and application of innovation technologies providing efficient mining and processing of mineral materials is constrained by requirements for aerogasdynamic safety systems. The local-ventilation fans distinguished for deficient adaptability and aerodynamic load are one of energy-intensive components of this system. Modified conformal mapping, Chaplygin singular points, and the theory of associated vortices and residues were used as the basis for construction of an additive mathematical model of a rotating circular grating of aerogasdynamic profiles with vortex sources to control circulation and the boundary layer. The aerodynamic scheme and parametric series of adaptive highly loaded fans of VRVP type were developed for airing of dead-end workings of up to 3000 m in length, The combination of the novel fan and gas-sucking fans increases the aerodynamic barrier up to a load of at least 35 000 t/day. The test results of pilot VRVP-6 justified the increase in adaptability by 70% and the aerodynamic load by more than 20%.

1. Introduction

The energy intensity of conventional ecological technologies reaches 25% at gas-hazardous collieries; about 40 % of energy consumption is low efficient. The provision of high-class mining productivity and energy efficiency often contradicts with energy intensity of supporting processes assuring ecological safety. The deficient aerogasdynamic safety efficiency restricts introduction of novel technologies composing the integrated innovative mineral deposit management. Growth of loads on stoping workings in combination with the aerogasdynamic safety requirements enhances the actuality of the problem dealing with development of the methodology for design and development of nature-like adaptive mine turbomachines, capable to generate adequate, economic, and wanted pressure drawdown fields in terms of optimal subsoil management conception [1].

High potential to raise aerodynamic loading and adaptability is taken into account in active energy methods designed to control circulation and a boundary layer [1–4]. The research publications lack the data on mathematical turbomachine aerodynamics models taking into account the back relation of dependence of energy parameters of vortex system formed nearby profiles on the characteristics of the external network [5–7].

The present paper is aimed at development of theoretical basis and methodology for design of the principal component of ecotechnology in the mining and metallurgical complex in Russia: energy-efficient nature-like turbomachines with controllable circulation, adapted to performance parameters of the basic technological process.

Considering [8], adaptivity principle and hydrodynamic analogy, the aerogasdynamic profile can be represented as a population of local integrated vortexes, imitating combination of a classical profile and a vortex chamber.

2. Problem statement and its solution

According to the general statement of the problem, $(n_i + n_s + n_v + 1) = (n' + 1) + n_v = (n_\Sigma + 1)$-sheet elliptical contour is associated in a plane case with an elliptical circular grid with n_l profiles, n_i, n_s source and runoff flows and n_v local vortexes at every profile. In the first sheet of $(n_\Sigma + 1)$-sheet Riemann surface the physical plane comprises the study circular grid of aerogasdynamic profiles, which vortex chambers have inlet and outlet channels through which air streams are capable to flow in and flow out [2, 7–9].

In k-th sheet ($k = 2, ..., n' + 1$) of Riemann surface the real channel of vortex chamber is patterned by a jet channel with walls extending to an infinite point A_k. In k-th sheet ($k = n', \dots n_\Sigma + 1$) of Riemann surface at location point of local vortexes the intensity of the source and runoff are equal in magnitude and opposite in sign. The research was conducted at two stages on the basis of a graphic model shown in Figure 1 under assumption that in the entire flow range D_z on $(n_\Sigma + 1)$-sheet Riemann surface the flow is stationary and vortex-free, the fluid is ideal, incompressible, weightless, Bernoulli constant is constant [1, 3–6, 7].

(a)

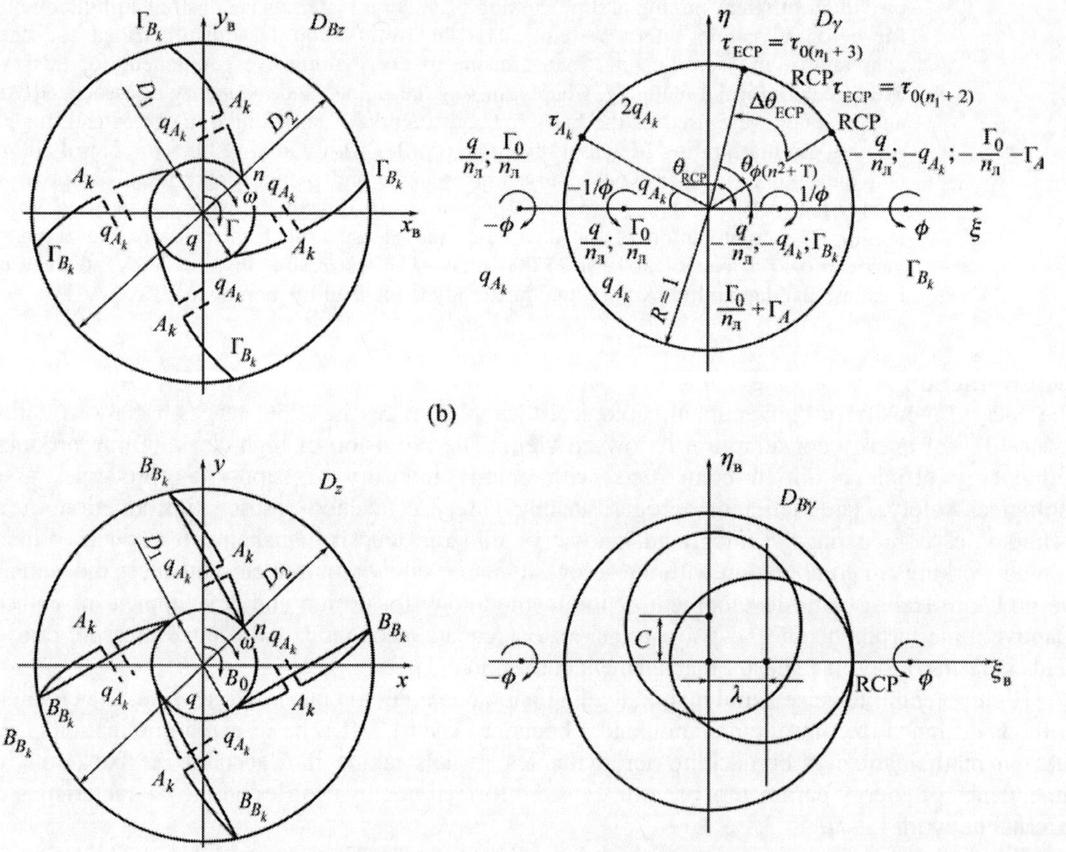

Figure 1. Principal scheme of conformal transformation sequence: (a) transformation of n_{sh}-sheet domain D_γ into n_{sh}-sheet domain $D_{B\gamma}$; (b) transformation of n_{sh}-sheet domain $D_{B\gamma}$ into $(n_\Sigma + 1)$-sheet domain D_z.

Conformal mapping function $Z(\gamma)$ is [5, 6, 7]:

$$n_l \ln z = \ln\frac{\gamma_{-1}+\Phi}{\gamma-\Phi} + e^{2i\beta_l+c}\ln\frac{\gamma-\Phi_1^{-1}e^{i\theta_1}}{\gamma-\Phi_2^{-1}e^{i\theta_2}}, \tag{1}$$

$$z = \left(\frac{\gamma+\Phi}{\gamma-\Phi}\right)^{1/n_l}\left(\frac{\gamma-\Phi_1^{-1}e^{i\theta_1}}{\gamma-\Phi_2^{-1}e^{i\theta_2}}\right)^{(2i\beta_l+c)/n_l}, \tag{2}$$

where $z = re^{iv}$, $\gamma = ce^{i\theta}$ are complex coordinates in domains D_z and D_γ, respectively; r, v are radius and polar angle at plane z respectively; ρ, θ are radius and polar angle at plane γ respectively; Φ is shape-parameter of an equivalent circular grid of profiles in the form of sections of logarithmic spirals; β_l is angle of logarithmic spiral of the equivalent profile grid; $\gamma_1 = (1+\lambda)\Phi_1^{-1}e^{i\theta_1}$, $\gamma_2 = \Phi_2^{-1}e^{i\theta_2}$, $K_\Phi = e^{2i\beta_l+c}$ are complex parameters, constituting a profile shape of the initial circular grid of analytical profiles.

To plot complex potential $F[Z(\gamma)]$ in n_{sh}-sheet Riemann surface of appearance of circular singular-radius domain D_γ we suggest to use additivity principle and Chaplygin singular points method [7, 8].

The general form of complex potential $F[Z(\gamma)]$ of current beyond the single-radius circle in n_{sh}-sheet Riemann domain D_γ can be written as:

$$F[z(\gamma)] = F_0[z(\gamma)] + \sum_{k=1}^{n'} F_{A_k}(\gamma) + \sum_{k=n'+1}^{n\Sigma} F_{B_k}(\gamma), \tag{3}$$

where $F[z(\gamma)]$ is complex potential of the classical conformal mapping problem which expression is described in [5]; $F_{A_k}(\gamma)$ is additional complex potential for a flow, describing the availability of runoff rates q_{A_k} in proper source points:

$$F_{A_k}(\gamma) = \pi^{-1}q_{A_k}\ln(\gamma-\tau_{A_k}) - 0.5\pi^{-1}[q_{A_k}\ln(\gamma^2-\Phi)q_{A_k}\ln(\gamma^2-\Phi^{-2})], \tag{4}$$

$F_{Bk}(\gamma)$ add-on complex flow potential characterizing local vortexes of Γ_{B_k} in intensity at certain points:

$$F_{B_k}(\gamma) = 0.5\pi^{-1}i^{-1}B_{B_k}\ln\frac{\gamma-\Phi^{-1}}{\gamma-\Phi}. \tag{5}$$

The solution is unique by the theorem of solution uniqueness for thr Dirichlet–Neumann problem [6, 7].

According to (3)–(5) we get a formula for a complex flow rate beyond the circle of single radius of n_l-sheet Riemann domain D_γ:

$$\frac{dF}{d\gamma} = 0.5\pi^{-1}n_l^{-1}(q+i\Gamma_0)((\gamma+\Phi)^{-1}-(\gamma-\Phi)^{-1}) - 0.5\pi^{-1}n_l^{-1}(q+i\Gamma_0)(\gamma+\Phi-(\gamma-\Phi)^{-1}) +$$

$$+\frac{(q-n_l q_{\Sigma A}+i\Gamma_0)}{2\pi n_l(\gamma+\Phi)} + \frac{(q-n_l q_{\Sigma A}+i\Gamma_0)}{2\pi n_l(\gamma+\Phi^{-1})} + \frac{(in_l\Gamma_l + in_l\Gamma_{\Sigma B} - n_l q_{\Sigma A} - q - i\Gamma_0)}{2\pi n_l(\gamma+\Phi)} -$$

$$-\frac{(in_l\Gamma_l + in_l\Gamma_{\Sigma B} + n_l q_{\Sigma A} + q - i\Gamma_0)}{2\pi n_l(\gamma+\Phi^{-1})} + \frac{\pi^{-1}\sum_{k=1}^{n'} q_{A_k}}{(\gamma-\tau_{A_k})} + V'_v(\gamma) - u_\tau[z(\gamma)]\frac{ds}{d\gamma}, \tag{6}$$

where V_V, u_τ are tangential components of a plug flow rate and a flow in plane Z; q is coefficient of source consumption; Γ_0, Γ_1 is intensity of vortex in the center of circular grid in plane: D_z: $\sum_{k=1}^{n'} q_{A_k} = q_{\Sigma A}$, $\sum_{k=n'+1}^{n\Sigma} \Gamma_{B_k} = \Gamma_{\Sigma B}$.

Given that vortex source is available at the back critical point of aerogasdynamic profiles, the point of disturbance of conformity mapping of the initial circular grid θ_{RCP} shifts to point of zero potential ambient velocity θ_{ECP}, by magnitude $\Delta\theta_{RCP}$ determined by displacement of a vortex source being adaptive to the external network characteristics [10].

With regard to conjugacy property, Zhukovsky–Chaplygin–Kutta postulate, and displacement of the rear critical point (RCP) to the effective critical point (ECP) under hydrodynamic peculiarities, specific for these locations, the formula to calculate circulation Γ_1 is:

$$\Gamma_{dkl} + \Gamma_{vkl} - 4q[1 + n_l q'(\Phi^2 - 1)^2(\Phi^2 + 2\cos\theta_{0(n'+2)}) + 1] - \frac{\Phi(\Phi^2+1)\sin\theta_{0(n'+2)}}{n_l(\Phi^2-1)^2(\Phi^2 + 2\Phi\cos\theta_{0(n'+2)} + 1)} -$$

$$\frac{(2\pi V'_{v(n'+2)})(\Phi^2 - 2\Phi\cos\theta_{0(n'+2)} + 1)}{(\Phi^2 - 1)} - \frac{4\Gamma_0 \Phi\cos\theta_{0(n'+2)}}{n_l(\Phi^2 + 2\Phi\cos\theta_{0(n'+2)} + 1)} + \frac{\Phi\sin\theta_{0(n'+2)}\sum_{k=1}^{n'} q_{A_k}}{1 - \cos(\theta_{A_k} - \theta_{0(n'+2)})} +$$

$$+ \frac{q_{0(n_\Sigma+2)} \sin_{0(n_\Sigma+2)} - \theta_{0(n_\Sigma+1)}}{1 - \cos\theta_{0(n_\Sigma+2)} - \theta_{0(n_\Sigma+1)}} + \frac{\Phi}{\Phi_2 - 1}\sin\theta_{0(n_\Sigma+2)} - \theta_{0(n_\Sigma+3)}. \quad (7)$$

The resultant mathematical model justifies the above fundamental conclusion on principal specific features of a flow mode in circular grids of aerogasdynamic profiles with vortex chambers and makes it possible to realize the nature-likeness principle for turbomachines.

Figure 2 demonstrates kinematic parameters of a circular grid of profiles adaptive to the external network characteristics [11].

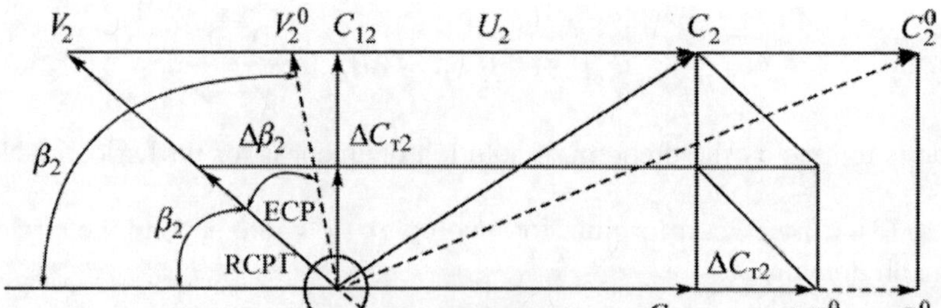

Figure 2. Kinematic flow parameters at outlet from the circular grid of profiles: — parameters of the classical grid; – — parameters of a circular grid with aerogasdynamic profiles; C_{u2}^{a}, β_2 — peripheral speed coefficient, angle of flow exit from the working wheel considering the effect of adaptive vortex source, respectively; $\Delta\beta_a$ — angle of ECP turn relative to RCP under the influence of adaptive vortex source; C_{u2}, C_{r2}, $\beta_{2л}$ — coefficients of peripheral and radial flow velocity and angle of exit of circular grid profile, respectively.

Hydrodynamic analog $\Delta\beta_a$, being a conformal mapping of angle $\Delta\theta_{ЗКТ}$ on the single radius circle of Riemann domain D_γ for domain D_z with account for (1), (7) can be presented in the form:

$$\Delta\theta_{RCP} = \theta_o(n_\Sigma+2) - \theta_o(n_\Sigma+3) = \arcsin\frac{\Gamma_{vkl}(\Phi^2-1)}{\Phi}. \quad (8)$$

Thus, $\Delta\beta_a$ is angle of ECP turn relative to RCP is a function of circulation coefficient Γ_{vkl} of a vortex source, which intensity tends to vary adaptively to parameters of ventilation network. The statement is justified by the scheme of kinematic parameter variations in Figure 2.

Using the new-proposed additive mathematical model of circular grids of aerogasdynamic profiles, the researchers developed and experimentally tested the radial aerodynamic scheme Ts149-20 and proposed the parametric line of local-ventilation fans of VRVP (Table 1).

Table 1. Parameters of type-size line of local-ventilation fans as compared to the latest fans of opponents.

Parameters	Fan type					Parameter variation index
	VMEV-8	VRVP-8	VME-2-10	MS760-22 (Germany)	VRVP-10	
Rated capacity, Q, m^3/s	13.5	14	16	17	19	1.13
Complete pressure, P, Pa	800	890	380	420	510	1.23
Depth of economic adjustment, N, kW	0.44	0/78	0.45	0.55	0.78	1.42
Specific energy consumption	1.67	1.22	1.64	1.41	1.22	1.22
Working length, $L_{в}$, m	2000	3500	1500	1700	2000	1.75

The pilot prototype of VRVP-8 passed tests in the certified laboratory SMK-Center. The commercial tests of the device are held at INTEKS. The design parameters calculated based on the novel mathematical model and the full-scale test results on VRVP-8 differ by 7–10%, thus justifying the acceptable reliability of the new-proposed model.

3. Conclusions

1. The novel parametric line of local-ventilation fans improves the fan adaptability by more than 70% and boosts the developed pressure by 20% to provide efficient blowing of dead-end workings of more than 1500 m in extension.
2. The application of the proposed parametric fan line in combination with gas-suction fans enables to higher the aerogasdynamic barrier of the stoping working load up to at least 35 000 t/day.

References

[1] Makarov VN, Patrakeeva IY and Makarov NV 2016 Genesis of perfection of fans of local airing *GIAB* No 4 pp 54–61
[2] Pat. 2543638 RF 2015 A way of increase of pressure and profitability of bladed turbomachines of radial type, NP Kosarev, NV Makarov, VN Makarov *Byull. Izobret.* No 7
[3] Li H, Song L, Li Y, Feng, Z 2011 2D viscous aerodynamic shape design optimization for turbine blades based on adjoint method *Journal of Turbomachinery* Vol 133 No 3 Article ID 031014
[4] Rao RS, Kuma CG, Prakasham RS, Hobbs, PJ 2008 The Taguchi methodology as a statistical tool for biotechnological applications: a critical appraisal *Biotechnology Journal* Vol 3 No 4 pp 10–523
[5] Loitsyan D 1978 *Mechanics of Liquid and Gas* Moscow: Nauka (in Russian)
[6] Smirnov VI 1974 *Course of Higher Mathematics* Moscow: Nauka Vol 3.4.2 (in Russian)

[7] Amaral S, Verstraete T, Van den Braembussche R and Arts T 2010 Design and optimization of the internal cooling channels of a high pressure turbine blade. Part I: Methodology. *Journal of Turbomachinery* Vol 132 No 2 Article ID 021013
[8] Gostelow DZh 1987 *Aerodynamics of Lattices of Turbomachines* Moscow: Mir (in Russian)
[9] Shi W, Zhou L, Lu, W, Pei B and Lang T 2013 Numerical prediction and performance experiment in a deep-well centrifugal pump with different impeller outlet width *Chinese Journal of Mechanical Engineering* Vol 26 No 1 pp 46–52
[10] Shojaeefard MH, Tahani M, Ehghaghi MB, and Fallahian MA Beglari 2012 Numerical study of the effects of some geometric characteristics of a centrifugal pump impeller that pumps a viscous fluid *Computers and Fluids* Vol 60 pp 61–70
[11] Ivanov OP and Manchenko VO 1986 *Aerodynamics and Fans* Leningrad: Mashinostroenie (in Russian)

IOP Conf. Series: Earth and Environmental Science **262** (2019) 012046 doi:10.1088/1755-1315/262/1/012046

Application and improvement of technologies for underground mining of placer deposits in Yakutia

VS Markov[1]*, VN Labutin[2] and AS Kurilko[3]*****

[1]Mining Institute, Ammosov North-Eastern Federal University, Yakutsk, Russia
[2]Chinakal Institute of Mining, Siberian Branch, Russian Academy of Sciences, Novosibirsk, Russia
[3]Chersky Institute of Mining of the North, Siberian Branch, Russian Academy of Sciences, Yakutsk, Russia

E-mail: *marko-valeri@mail.ru, **Labvn@yandex.ru, ***a.s.kurilko@igds.ysn.ru

Abstract. The historical background is presented, and the experience gained in application of equipment and technologies of underground placer mining in Yakutia is considered. The variant of chamber mining with application of bolter miners and backfilling self-propelling machinery is proposed. The variant allows increase in productivity of development heading and actual extraction by 4–5 and 2.5 times, respectively, and reduces mineral loss and dilution.

1. Introduction

One of the key branches of the mining industry in Yakutia is gold mining. It is connected with discovery of the Aldan gold placer in 1923. In 1925 Aldanzoloto Union Trust was founded. If we take overall level of underground placer mining as the basis, the period from on the start of the industrial development and up to now can be conditionally be divided into stages.

Stage I is 1928–1945. Together with dredging and hydraulic mining of placers, the underground method was widely used since 1928. In 1932 in the Aldan mines, belt conveyors were for the first time tested and introduced for sand haulage, including elevation to ground surface. Some mines employed scrapers and self-dumping skips. Stoping was carried out manually. Efficiency of labor was from 1.12 to 2.3 m^3/manshift. Aldan placer mines in 1932–1938 were the most advanced and best equipped gold mines in the country [1]. In 1931 Aldanzoloto was converted into Yakutzoloto Trust, which enabled geological exploration all over the territory of Yakutia Autonomous Republic. In 1936, on the basis of discovered and explored gold placers, within Yakutzoloto, Allakh-Yun Gold-Mine Management was established and in 1939 transformed into independent Dzhugdzhurzoloto Trust. The Trust carried out some underground operations with the help of placer miners.

Stage II began after 1945 with discovery of deep-buried placers near Allakh-Yun settlement. At the same time, the decision was made on the formation of Indigirka Mine Management. Underground operations were mostly concentrated the placer mines Marshal and Nelkan. Mining was either seasonal or year-round using various methods. Machinery and equipment were selected for every process of underground mining, energy supply systems of the mines were expanded, electric power plants were constructed, a series of all-metal gold washing machines of various designs, drilling rigs, high capacity scraper winches, etc. were engineered. In view of high consumption of timber, difficulties of face-entry drivage and mine ventilation, the Institutes of Permafrostology and

Irgiredmet in 1965–1970 proposed breast-and-pillar systems of mining with roof supported by pillars and timber props, applied until now, mostly, in the regions of Indigirka.

Stage III should be related with gold discovery in the area of the Kular Ridge westward of the Yana River mouth.

Commercial development of the Kular gold field started in 1962. Since 1963 Kular Mine developed gold placers by the opencast and underground method. It was transformed to Kularzoloto Mining and Processing Plant in 1976. Complex geological conditions of placer mining in the Polar Region (repeated-wedge ice in mine roof, expensive imported timber) conditioned application of shortwall system proposed by VNII-1 Institute in 1965. The underground placer mining enjoyed the widest polarity in 1960–1985 when the specific weight of the underground gold mining in the Republic reached 25–35% of the overall gold production. Kularzoloto MPP ensured almost 70–90% of the total gold production. At that time, Yakutzoloto Production Association developed 8 gold fields via annual operation of 30–40 mines with underground gold production of 1.2–2.0 Mm^3/year. The largest fully mechanized projects were the mines in the Arctic zone of Yakutia (Kularzoloto MPP). The company was the first to carry out heading and stoping operations in underground permafrost placer mines tunneling machines GPK in inclined shaft sinking as well as self-propelling machines (drill jumbos and diesel-drive load–haul–dumpers) and powered roof support T-13K in actual excavation. Annual mine productivity totaled 100–150 thousand m^3, and the mines were the largest placer mines in the world. Such success was thanks to introduction of R&D projects of the Institute of Mining of the North, SB RAS, and VNII-1. Economic performance data of the applied mining systems are compiled in Table 1 [1].

In the considered system, given the place width of 100 m, the length the mine fields was 400–500 m for the year-round mines and 250–300 m for the seasonal mines; in case of placers 50–200 m wide, the mine field length was 250–350 m. The maximum length of the mine field reached 1000–1400 m at the placer 40–60 m wide.

Table 1. Economic performance data of mining systems.

Description	Mining system		
	Longwall	Room-and-pillar	Shortwall
Specific weight of development and temporary drives, %	5.0–7.5	11.8	15–25
Sand loss in pillars, %	3.0–4.5	5.5–7.0	10–15
Timber consumption per 1000 m^3, m^3	50	18	—
Underground miner productivity, m^3/shift	4.2	7–9	10–12
Mine productivity, m^3/day	100–170	300	350–500

In the regions of Indigirka, underground mining of of permafrost placers was mostly seasonal (winter). Mining period was 8–10 months, including 30–40% of time spent to get access to mineral and to prepare mine fields of extraction. Subject to current underground mine planning and design, access roadways were driven in October–November, development entries and temporary roadways—in November–December, and actual production was carried out in January–May. The Arctic mine operated year-round, as a rule, and access roadways were driven in summer as well.

Underground mining of Tirekhtyakh tin deposit was initiated in 1997–1998 at the mining depth of 40 m; the available reserves at the given mining rate ensured the mine life for 25 years. The mining technology used drilling-and-blasting with portable and import self-propelling equipment (drill jumbos and diesel-drive load–haul–dumpers). The system of mining was shortwalling with pillars 3 m wide between the shortwalls; the annual mine productivity was 80 thousand m^3.

The recommended self-propelling machines for underground mining of permafrost placers are listed in Table 2 [2].

IOP Conf. Series: Earth and Environmental Science **262** (2019) 012046 doi:10.1088/1755-1315/262/1/012046

Table 2. Self-propelling machines for underground permafrost placer mining.

Operation	Face height 2.0–3.5 m		Face height 3.5–5.0 m		Face height above 5.0 m	
Hole drilling	UBSH-406	Boomer Minimatic	UBSH-406	Minimatic	UBSH-406	Maximatic
Hole charging	PMZSH-2	—	PMZSH-2	—	PMZSH-2	—
Loading and haulage to haulage drift	PD-5A, PD-8B	ST-5A, Toro-200, LK-1, Wagner, GKHKH-4.1	PD-8B	Wagner, Toro-300, Toro-350	PD-8B	Toro-400, Toro-500
Haulage to ground surface	MoAZ 7405, MoAZ 7429	Blaw-Knox AG	MoAZ 7405, MoAZ 7429	Blaw-Knox AG	MoAZ 7405, MoAZ 7429	Blaw-Knox AG ”
Auxiliary operations	Buses PA-40, 1VlG, AP-1, crane PK-5M Haulage machines 1-BOM-1, Fuel filling—GSM, PSA-5					
Inspection and safety of roof	SP-8A MOK-10	Uniscaler T-3	SP-8A MOK-10	Uniscaler T-3	SP-8A MOK-10	Uniscaler T-3
Cleaning of mined-out area	PD-5A, PD-8B	ST-5A, Toro-200, LK-1, Wagner, GKHKH-4.1	PD-8B	Wagner, Toro-300, Toro-350	PD-8B	Toro-400, Toro-500

2. Room-and-pillar system for underground mining of permafrost placers

As an illustration of underground mining of permafrost placers, it is proposed to use the room-and-pillar system with employment of heading machines, self-propelling equipment, and with backfilling (Figure 1).

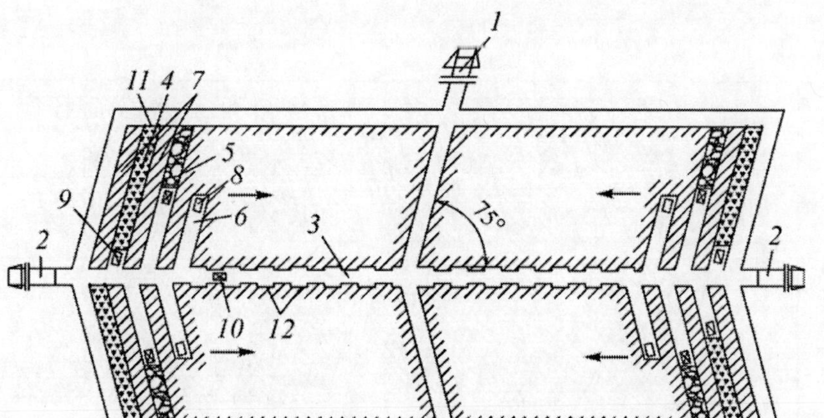

Figure 1. Initial mining stage in extraction panel: *1*—main shaft; *2*—auxiliary shafts; *3*—haulage drift; *4*—mined-out room to be backfilled; *5*—chamber prepared for bedrock cleaning; *6*—stope; *7*—rib pillars (secondary chambers); *8*—cutter–loader; *9*—backfilling machine; *10*—load–haul–dumpers; *11*—bulkhead; *12*—cuts.

At this stage of mining, the equipment pack includes: cutter–loader AM-75, loaders PNB-3K and other, dump trucks MoAZ (or self-propelling cars), load–haul–dumpers LHD for backfilling and backfilling machine UMZK [3].

Access to the extraction panel is got via two inclined shafts (Figure 1). The entry shaft *1* and exit shaft *2* are connected by the haulage drift *3* and perimeter drifts used later on for ventilation or second exits from stopes. The inclined shafts (motot ramps) are driven at an angle of 8–10° by middle-class machines (AM-75, P-220, KSP-42 and others) using LHD at the cross-section of 12–18 m^2.

Development drives are headed by cutter–loaders. Directions and inclinations of the drives are pre-set and periodically controlled by a surveyor. Haulage drift is cut in the productive seam and has a cross-section of 20 m^2; perimeter drive (ventilation drives) have cross-section of 16 m^2. During haulage drift drivage, cuts *12* are made to a depth to 1.5 to serve later on as niches to shelter miners during movement of LHD. Chambers are arranged at an angle to the haulage drift axis *3* for convenient maneuvering of self-propelling machines. Before preparation of secondary chambers, intersections of the primary chambers and haulage drift are reinforced with rock bolts and steel meshing. Extraction both in primary and secondary chambers is carried out after cross connection of the access and perimeter drives. Actual sand production uses cutter–loaders. The project cross-section of the chambers is 20 m^2, the width is governed by the stability of unsupported roof spans.

When the productive seam width is sufficient for operation of self-propelling machines, the seam is extracted full width by cutter–loaders. In thinner seams 0.2–1.0 m thick, extraction is separated: above the same an opening is drive with a width equal to the chamber width and the height sufficient for the self-propelling equipment operation, then the productive seam is extracted. The engineering solutions set the cross-sections of $5 \times 4 = 20$ m^2 both for primary and secondary chambers.

Sand haulage and piling on the ground surface involves underground dump trucks, loading to dump trucks is carried out using loading machines PNB-3K (shoveling). After cross-connection of the chamber and the perimeter drive, cleaning of the mined-out area (bedrocks) by LHD buckets; then, the chamber is assume prepared for backfilling.

After extraction and backfilling of the primary chambers, the pillars between them are developed in the same sequence. The secondary mining (extraction of secondary chambers) is depicted in Figure 2.

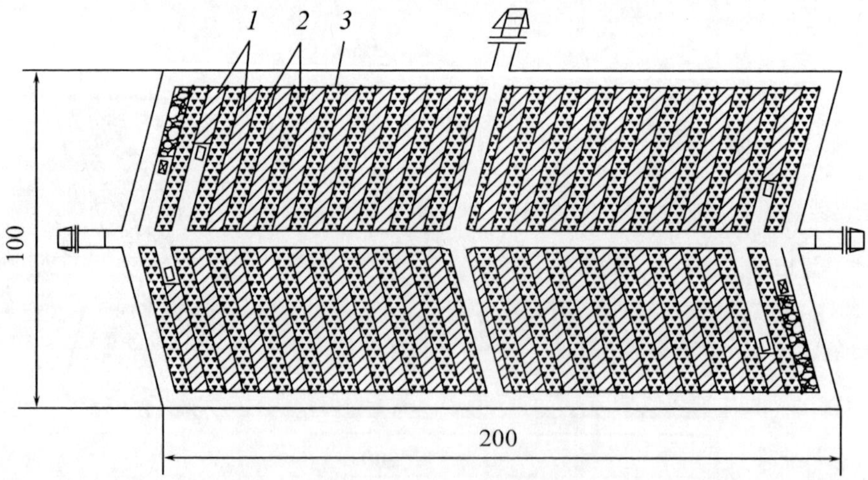

Figure 2. Extraction of secondary chamber: *1*—interchamber pillars; *2*—rock pillars; *3*—bulkhead.

Ground control in this mining system combines rigid interchamber pillars (when primary chambers are extracted) and smooth subsidence of roof on yielding pillars made of backfill (when secondary pillars are extracted). It is recommended to prepare backfill gangue after drivage of chambers and shafts, as well as overburden of operating opencast mine. Retreat backfilling is executed from the mine field boundary to the haulage drift. Haulage of backfill material from shafting and other drivage is performed by LHD, and overburden from the surface is transported via specially drilled large-diameter holes and by dump trucks.

Technical-and-economic comparison of the shortwall mining and room-and-pillar mining shows advantages of the latter. Efficiency of road heading and actual extraction exceeds by 4–5 times and 2.5–3 times, respectively. The volume of road heading makes 30% in shortwalliong and 10.8% in rock-and-pillar. The mineral loss and dilution reduces in the latter case from 19.1 to 2% ad from 45 to 10%, respectively.

3. Conclusions

Improvement of technology and equipment for underground permafrost placer mining in Yakutia should include such basic trends as:

—upgrading of the current technology and machinery for road heading and actual mineral extraction at the increase in mineable thickness to 3 m and more;

—creation of a new cyclic-and-continuous technology providing mechanical (cutting and percussion-drag) destruction of coarsely clastic permafrost rocks as well as modern heading machines.

References

[1] Sherstov VA, Skuba VN, Lubiy KI and Kostromitinov KN 1981 *Underground Placer Mining in Yakutia* Yakutsk (in Russian)

[2] Sherstov VA and Markov VS 2003 Improvement of mechanization of production processes in underground mining of permafrost placers *GIAB* No 1 pp 196–199

[3] Markov VS, Labutin VN and Nikolaeva AA 2014 Chamber system for underground mining of diamond-bearing placers Solur *GIAB* No 9 pp 365–370

IOP Conf. Series: Earth and Environmental Science **262** (2019) 012047 doi:10.1088/1755-1315/262/1/012047

Natural stress estimation by solving inverse problem using acoustic sounding data

NA Miroshnichenko*, AV Panov and LA Nazarov*****

Chinakal Institute of Mining, Siberian Branch, Russian Academy of Sciences, Novosibirsk, Russia

E-mail: *mna@misd.ru, **anton-700@yandex.ru, ***naz@misd.ru

Abstract. Under consideration is the approach allowing quantitative estimation of natural stress field components in the vicinity of underground objects based on solution of an inverse boundary-value problem using the data on acoustic sounding in an extended mine working. To ensure solvability of an inverse problem, the optimal sonic transmitter–receivers positioning and the limiting nominal error of measurements are determined. The objective function is set as a root-mean-square difference between the theoretical and actual head P-wave arrival times at the receivers; unimodelity of the objective function is illustrated using synthetic input data.

1. Introduction

The quantitative information on stress-fields in the vicinity of study objects is of prime importance to solve theoretical and practical problems dealing with geomechanics, geophysics or mining engineering, viz., evaluation of rock mass stability in large-scale construction or mineral deposit development projects, protection of production wells in raw hydrocarbon production, selection of optimal mineral deposit development schemes.

The approach based on definition and solution of inverse problems is known as one of the most efficient methods to evaluate stresses. Its application in geomechanics to diagnose a state and mechanical properties of rocks is described in publications [1, 2]. Applicability of inverse problems to study geotechnical and natural objects is mainly restricted because of quasi-stationary character of most processes running in a rock mass, as well as high cost of in-situ measurements and, as a consequence, scanty volume of input data. Geophysical methods to receive input information are famous for relatively low costs, but interpretation of the obtained data requires special laboratory experiments to plot empirical relationships as “medium state equations” [3–5]. In particular, to realize the acoustic method implies availability of experimental relationships of velocities of elastic wave propagation in rocks versus stresses.

2. Estimation of vertical and horizontal components of the natural stress field

Let consider the approach based on inversion of acoustic sounding data in order to identify vertical and horizontal components of the natural stress field; the idea of the given approach is set forth in [6]. In the publication the input data were results on laboratory tests of argillite specimens under hydrostatical compression within pressure range $0 \le \sigma \le 60$ MPa [7].

Figure 1 demonstrates experimental relationships of elastic waves V_p and V_s velocities vs. hydrostatic stress with approximation with function:

$$V_m(\sigma) = A_m - B_m \exp(-\alpha_m \sigma / \sigma_0), \qquad m = p, s, \tag{1}$$

where $\sigma_0 = 5$ MPa.

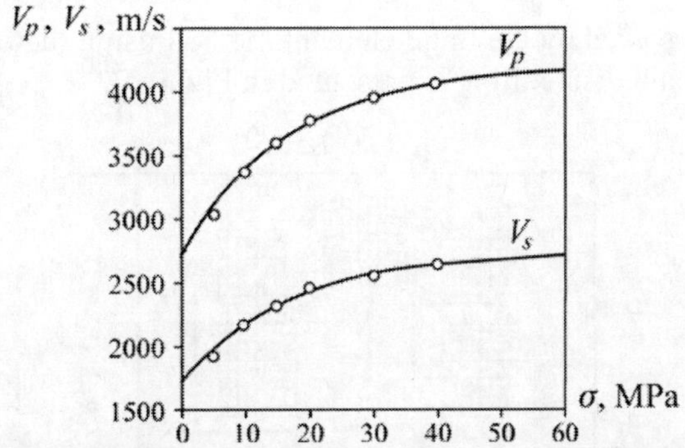

Figure 1. Empirical relationships of elastic wave propagation velocities, obtained from the laboratory experimental data [7].

In the Table 1, the coefficients (1) are calculated by the least squares method, and approximation error is ω.

Table 1. Coefficients of approximation function and relative error.

m	A_m, m/s	B_m, m/s	α_m	ω
p	4212	1408	0.296	0.008
s	2740	1008	0.290	0.012

The study object is a horizontal arched mine working, which extension length is substantially greater than its cross-section dimensions. The study working is in a rock mass, modeled by a homogeneous isotropic elastic medium. The operative direction of the maximum horizontal component σ_H of the natural stress field coincides with orientation of the study working; the minimum component σ_h (cross line) is characterized by lateral factor *q*. This is in compliance with terms of the plane strain state [8].

Calculation domain *R* is a vertical section across extension of the working (Figure 2). The following boundary terms are formulated on ∂R:

$$\begin{gathered}\sigma_{zz}(x,0) = S, \quad \sigma_{xz}(x,0) = 0;. \\ u_z(x,Z) = 0, \quad \sigma_{xz}(x,Z) = 0; \\ u_x(0,z) = 0, \quad \sigma_{xz}(0,z) = 0; \\ \sigma_{xx}(X,z) = q(S + \rho g z) \quad \text{c},\end{gathered} \tag{2}$$

where σ_{ij} are components of stress tensor; u_i are displacements; *S* is lithostatic stress, ρ is medium density; (*x*, *y*, *z*) are Cartesian coordinates, axis *y* is directed along extension of the working; *X* and *Z* are horizontal and vertical dimensions of the domain.

Deformation of the medium in its vicinity is described by equation set of the elasticity theory including equilibrium equation (3), Hook's law (4), and Cauchy relation (5):

$$\sigma_{ij,j} + \rho g \delta_{iz} = 0; \tag{3}$$

$$\sigma_{ij} = \lambda\varepsilon\delta_{ij} + 2\mu\varepsilon_{ij}\,; \tag{4}$$

$$\varepsilon_{ij} = 0.5(u_{i,j} + u_{j,i})\,, \tag{5}$$

Direct problem (2)–(5) was solved by the finite element method using the original program code [9]. A fragment of calculation domain discrediting is presented in Figure 2.

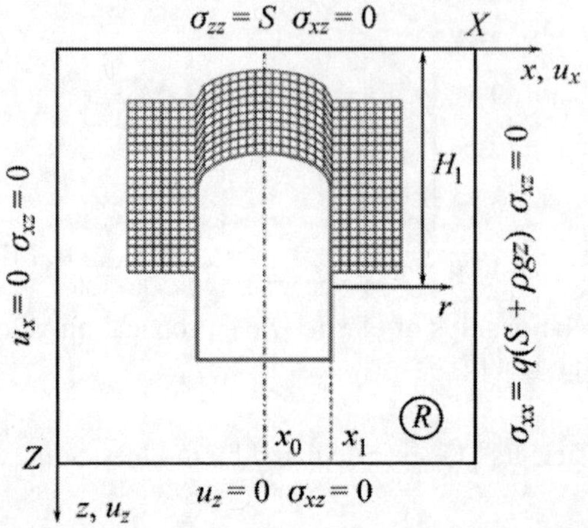

Figure 2. Plan of calculation domain, a fragment of finite element grid, and boundary conditions.

In Figure 3 for X = 92 m, Z = 46 m, ρ = 2000 kg/m^3, λ = μ = 8 GPa, S=10 MPa (stress corresponds to depth H=500 m), q = 1.5 lines of the average stress level $\sigma = (\sigma_{xx} + \sigma_{zz})/2$ and velocity $V_p(\sigma)$, calculated from (1) are shown.

It is established based on the calculation results that the maximum variation in V_p is gained in the working wall at $q > 1$, and in its roof and floor at $q < 1$. It is imperative to consider direction of the sharpest variations in acoustic properties of the medium when it is necessary to select a measurement circuit in order to realize the proposed stress-evaluation process.

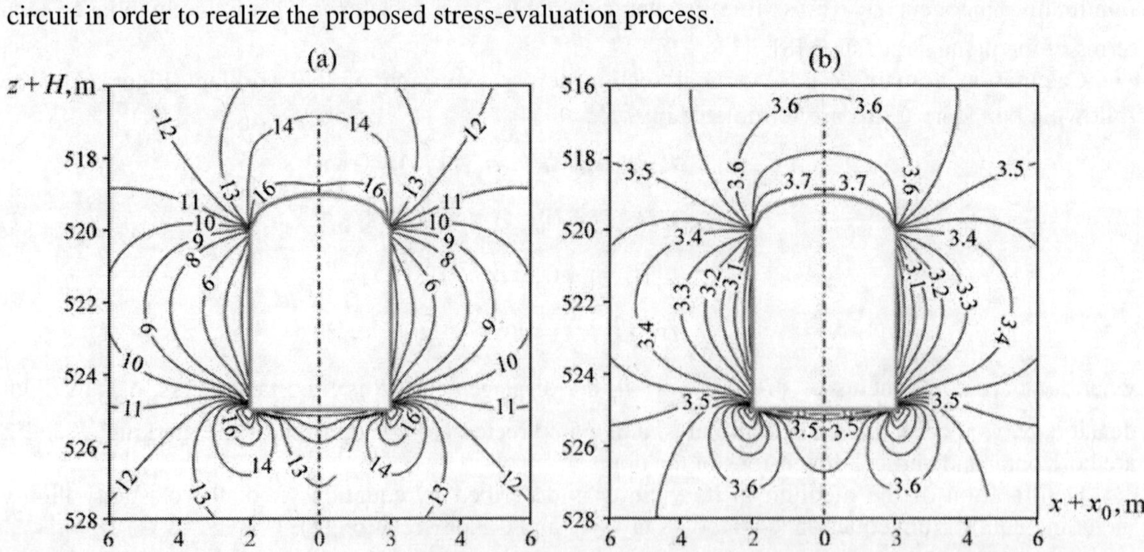

Figure 3. Isolines: (a) average stress σ (MPa); (b) P-wave velocity V_p (km/s).

Let us assume for certainty $q > 1$. We locate a sounding signal transmitter at depth H_1 from the upper boundary of R area in one of working walls and receivers N with interval Δy between them longitudinally in the working. Let t_n be arrival time of the head wave at a receiver of coordinate $y_n = n\Delta y$. As t_n depends only on S and q (medium properties A_m, B_m, α_m, and ρ being known), we introduce the objective function as a mean-square deviation between real t_n^0 and "calculated" t_n time values

$$\Psi(S,q) = \sqrt{\frac{1}{N}\sum_{n=1}^{N}\left[t_n(S,q) - t_n^0\right]^2}. \qquad (6)$$

Values t_n can be determined by using the radial method for the gradient model of the medium in cross-section $z = H_1$, and the spatial distribution of wave velocities can be found from (1), considering (2)–(5). The head P-wave arrival-time curve is computed from relations [10]:

$$y_n(r) = 2\int_0^r \frac{V_p(\xi)d\xi}{\sqrt{V^2(r) - V_p^2(\xi)}}, \quad t_n(r) = 2\int_0^r \frac{V(r)d\xi}{V_p(\xi)\sqrt{V^2(r) - V_p^2(\xi)}}, \qquad (7)$$

where $V(r) = \max\limits_{0\le\xi\le r} V_p(\xi)$, $r = x - x_1$, x_1 — abscissa of working wall (Figure 2).

The minimum of the objective function Ψ is the solution of the boundary inverse problem to determine horizontal and vertical components of the natural stress field using the acoustic sounding data.

To investigate the solubility of the study problem, the researchers implement synthesizing of input data $t_n^0 = (1+\beta_n)t_n(S_0, q_0)$, where β_n is a random magnitude, uniformly distributed on [−γ, γ], γ is a relative noise amplitude, $S = S_0$ and $q = q_0$ is a precise solution.

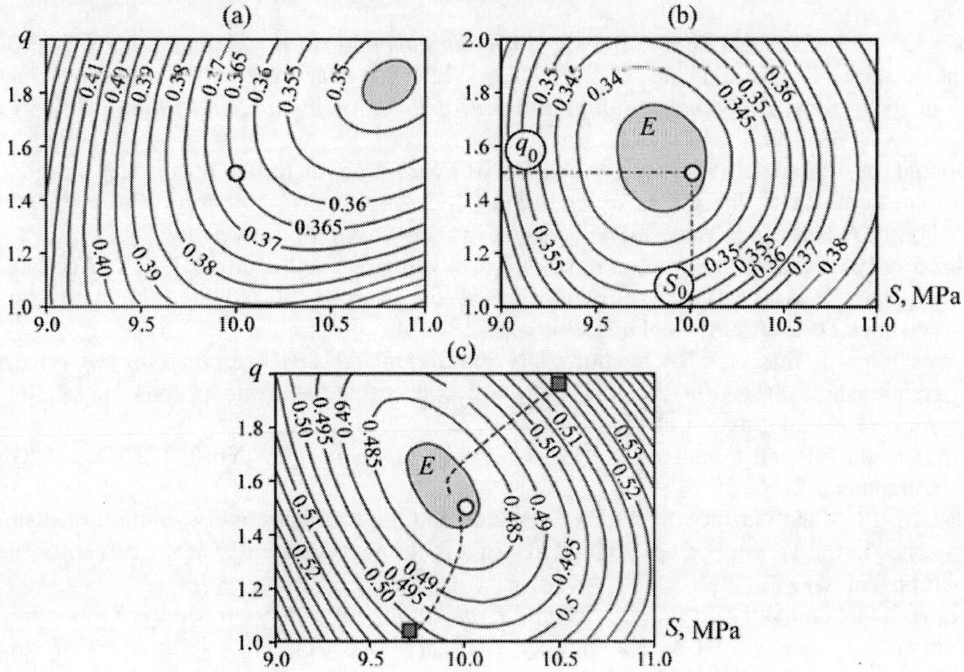

Figure 4. Isolines of target function Ψ (ms) at γ = 0.05 for different number of receivers: (a) N = 25; (b) N = 30; (c) N = 35.

In Figure 4 isolines Ψ are for different number of receivers N at S_0 = 10 MPa, $q_0 = 1.5$, $\gamma = 0.05$, $H_1 = 22$ m, $x_1 = 48$ m, $r = 10$ m, $\Delta y = 1.5$ m. Equivalence domain E, where a relative variation in Ψ does not exceed 2 %, is shaded. Squares mark initial approximations, circles are for precise solution, dashed lines are computation paths. Modified gradient descent method is applied to search for minimum Ψ [11]. The objective function is unimodal at $\gamma \le 0.05$, and its equivalence domain tends to reduce with increase in number of receivers. According to experimental numerical approximation results the extreme point of iteration process locates within domain E at any initial approximation, however, not at any number and/or spacing of receivers (Figure 4a):

$$\mathrm{E}\ (S_0, q_0) \in E. \qquad (8)$$

For S and q the variation range accepted in the present research and prescribed acoustic properties of rocks (Table) the optimal parameters of the recording system are $N = 30$–35 and $\Delta y = 1.5$ m (Figure 4b, c). Increase in relative noise amplitude γ in input data starting from 0.1, spacing variation Δy, number of receivers N, depth H_1 are used to break unimodality conditions Ψ and/or (8).

3. Conclusions

To conclude, the approach set forth in the present research enables to make quantitative estimation of vertical and horizontal components of the natural stress field in the vicinity of a geomechanical object. The approach rests on solution of a boundary inverse problem using acoustic sounding data, recorded in an extended underground working, and empirical relationships of elastic wave velocities versus hydrostatic stress which were obtained based on laboratory experiments on rock specimens. It is established that the single-valued solubility of the inverse problem is feasible, provided that the noise level in input data should be no more than 10 %.

References

[1] Tarantola A 2005 Inverse problems theory and methods for model parameter estimation *SIAM*
[2] Vatulyan AO 2007 Inverse problems in deformable solid mechanics Moscow: Fizmatlit (in Russian)
[3] Jaeger J, Cook NG and Zimmerman R 2007 *Fundamentals of Rock Mechanics* Wiley-Blackwell
[4] Takahashi T, Takeuchi T and Sassa K 2006 ISRM Suggested methods for borehole geophysics in rock engineering *International Journal of Rock Mechanics and Mining Sciences* Vol 43 No 3 pp 337–368
[5] Donald JA, Prioul R, Lei T and Sinha B 2013 Stress characterization in deep boreholes using acoustoelasticity *Rock Characterisation Modelling and Engineering Design Methods* X-T Feng, JA Hudson F Tans (Eds) Taylor & Francis Group pp 309–314
[6] Nazarov LA, Nazarova LA, Romensky EI, Cheverda VA and Epov MI 2016 Acoustic method for rock mass sress determination based on solution of inverse kinematic problem of seismics *Doklady RAN* Vol 466 No 6 pp 718–721
[7] Pervukhina M, Gurevich B, Dewhurst DN and Siggins AF 2010 Applicability of velocity–stress relationships based on the dual porosity concept to isotropic porous rocks *Geophysical Journal International* Vol 181 No 3 pp 1473–1479
[8] Timoshenko SP and Goodier JN 1951 *Theory of Elasticity* New-York–Toronto: McGRaw Hill Company
[9] Nazarova LA and Nazarov LA 2009 Dilatancy and the formation and evolution of disintegration zones in the vicinity of heterogeneities in a rock mass *Journal of Mining Science* Vol 45 No 5 pp 411–419
[10] Boganik GN and Gurvich II 2006 *Seismic Exploration* Tver: AIS (in Russian)
[11] Nazarov LA, Nazarova LA, Karchevsky AL and Panov AV 2012 Estimation of stresses and deformation properties in rock mass based on inverse problem solution using measurement data of free boundary displacement *Sib. Zh. Industr. Matem.* Vol 15 No 4 pp 102–109

IOP Conf. Series: Earth and Environmental Science **262** (2019) 012048 doi:10.1088/1755-1315/262/1/012048

Test stand equipped with a polyharmonic vibration source to study compaction of dispersed material in an enclosed volume

AV Morozov*, LI Gendlina and VM Usoltsev

Chinakal Institute of Mining, Siberian Branch, Russian Academy of Sciences, Novosibirsk, Russia

E-mail: *alex02@ngs.ru

Abstract. The vibration effect on materials to be compacted was analyzed. The design scheme is substantiated for the test stand, intended to study consolidation of a granular material under the polyharmonic oscillations of a working member.

1. Introduction

Compaction of granular materials is an important operation in a good deal of technological processes in different industries. Compaction quality influences durability and reliability of a final product and, as a consequence, the production cost efficiency being of specific value in view of ever-increasing energy resource costs.

Machinery designed to provide consolidation is regularly up-dated with more and more complication of its influence modes on materials under treatment [1–4]. However, the attempts to produce a maximum-density package at relatively low energy consumption, as a rule, failed in spite of a variety of methods and facilities, available and practiced. The appreciable improvement of the compaction process performance requires new comprehensive knowledge about regularities of the process, its usage in combination with experience background in view to develop principally novel technologies.

The researchers working for Institute of Mining, SB RAS, developed the process for vibration compaction of dispersed and powder materials in an enclosed volume and received RF Patent [5] Experimental tests of the process on the finely dispersed material compaction yielded high results. The rational vibroeffect parameters: amplitude-frequency specification and vibration angle were determined.

From Russian and foreign experience in the vibration effect utilization to compact granular materials it is obvious that there is the optimal frequency range for different vibration intensity and particle sizes of a material to be compacted [6–8]. Some researchers recommend to handle a material containing particles of less size with vibration of a higher frequency with the purpose to compact it efficiently. The idea of multi-frequency vibration arose in relation to compaction of concrete mixture containing particles of a wide size range from few micron to few score of millimeters. As the researchers set forth, the presence of two-frequencies in resultant oscillations leads to increase in a velocity of particle motion, hence, higher efficiency of the process.

One of commonly known ways to realize poly-harmonic oscillations of the working member is employment of two or more centrifugal vibroexciters, mounted on the working member, mechanically connected with one another and rotating at different angular velocities. However, the mechanical transmission really complicates a vibration device structure and aggravates its reliability.

The objective of the present research work is to design and manufacture a test stand to generate a polyharmonic high-frequency oscillation mode of the working component operation in a vibration device intended to compact a finely-dispersed material in an enclosed volume.

2. Compaction research on test stand equipped with a polyharmonic vibration source

The idea of a superharmonic vibration drive (as a polyharmonic version) is proposed in publication [9] and rests on practical use of rotation irregularity of eccentric mass in a centrifugal vibroexciter. The reasons for such irregularities can be different, including variable gravitational moment of d eccentric mass relative to rotation axis, given that it is not vertical; variation in resistance to eccentric mass rotation because of design or exploitation factors. Thereto, oscillations of the executive member are characterized with higher harmonics; amplitude of one of them is appreciably enhanced. The resultant polyharmonic oscillations contain a comparable-in-amplitude high-frequency harmonics along with the principal frequency.

The application of low-frequency centrifugal vibroexciter to generate high-frequency oscillations has certain advantages as compared to a high-speed vibrodrive: less power loss in eccentric mass bearings, lower operational noise, and high reliability of vibrodevices.

Figure 1a demonstrates an oscillogram of the working member oscillations in a vibrodevice generated by centrifugal vibroexciter; Figure 1b shows a respective frequency spectrum, thus indicating the presence of higher harmonics in frequency spectrum, but their amplitude is substantially lower than that of basic frequency oscillations. Theoretically it is feasible to enforce any of harmonics, but in practice there is a technical opportunity to do this only with the second or third harmonics.

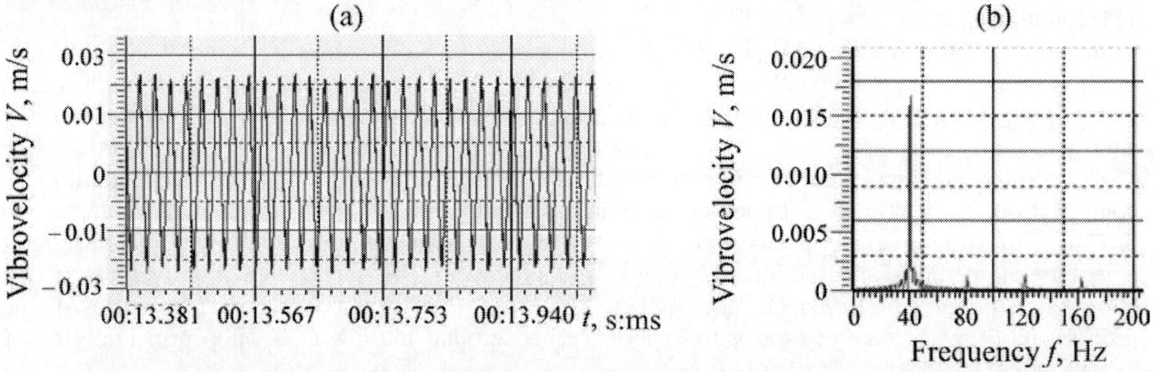

Figure 1. (a) Oscillogram fragment and (b) respective frequency spectrum of vibrovelocity of the working member of the device with centrifugal vibroexciter.

A scheme to realize superharmonic oscillations, presented in Figure 2, is taken as the basis to design a test stand [9]. The presented system with one degree of liberty has support frame *1* connected through buffer damper *2* and spring *3* to the working member *4*. When eccentric masses *5* of a centrifugal vibroexciter rotate, the ideal constraints *6* provide only forward movement of the working member. Tuning of this system to generate resonance in vicinity of the third harmonics leads to increase in its amplitude.

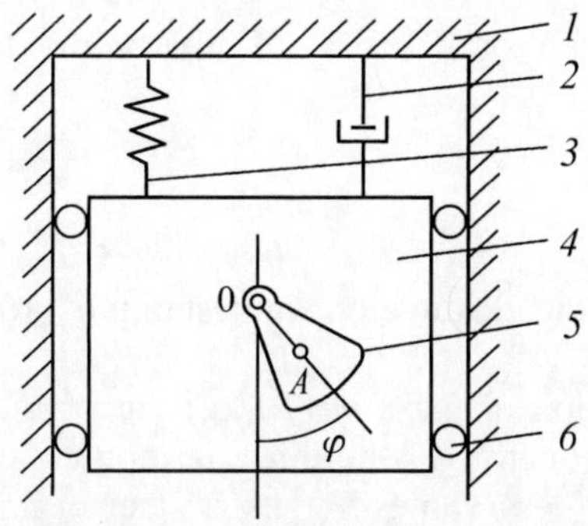

Figure 2. Vibrodrive scheme: *1*—support frame; *2*—buffer damper; *3*—spring; *4*—working member, *5*—debalances of a vibroexciter; *6*—ideal constraints.

The test stand (Figure. 3a) consists of a vessel *1* of rectangular cross-section (500×500 mm), filled with a granular material, and vibration device *2* designed to compact it. Their internal surfaces are made of an elastic material in order to diminish the vibration effect of vessel walls on high-density compaction.

Vibration device includes sealing metallic plate *3* (Figure 3b) and centrifugal directed-action vibroexciter *4*, mounted on base *5*, connected to a sealing plate through elastic elements *6*. The stiffness of the elastic elements is selected based on design calculation in view to provide resonance in the vicinity of the third harmonics of the basic frequency. The strictly vertical motion of the vibrodevice in the operation process is provided due to rigid connection of sealing plate *3* to guiding-frame *7*, where mounted rollers *8* are moving along guide *9* creating one degree of liberty to the vibrosystem.

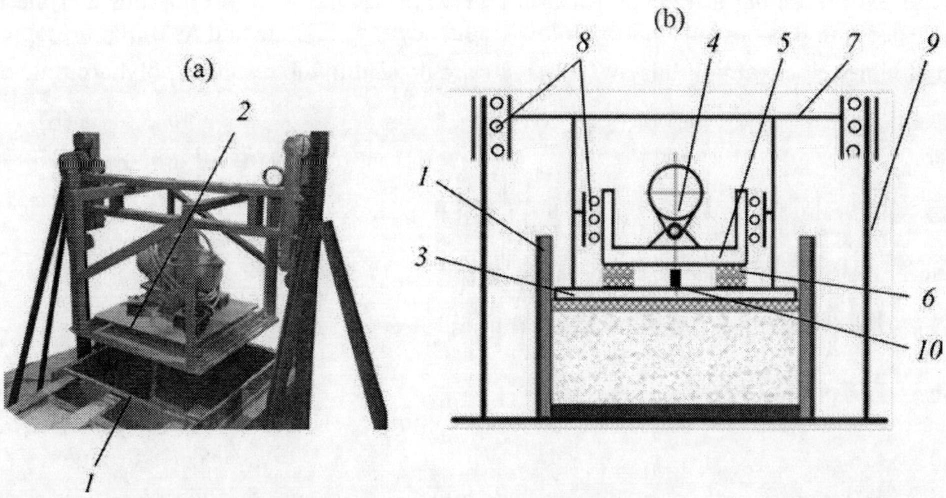

Figure 3. Test stand to study the process for compaction of a dispersed material in an enclosed volume: *1*—vessel; *2*—vibration device; *3*—sealing plate; *4*—vibroexciter; *5*—base; *6*—elastic element; *7*—guiding frame; *8*—rollers; *9*—guide; *10*—sensor.

Basic (low) frequency is selected in the range of 30–50 Hz. At this frequency range, based on earlier research evidence, the efficiency of the dispersed material compaction is the best [10]. Three frequencies of vibroexciter rotation of 30, 40, and 50 Hz (the third harmonics corresponding to them having frequencies of 90, 120, and 150 Hz) are suggested for experimental evaluation of the influence of the complicated superharmonic vibroeffect on the material to be compacted.

In the oscillation system with one degree of liberty the eigen oscillations frequency ω_0 is determined by formula [1]:

$$\omega_0 = \sqrt{\frac{c}{m}},$$

where *c*—cumulative stiffness of the elastic element system; *m*—total oscillating mass of the vibration device.

After preliminary determination of mass *m* (70 kg) the stiffness of elastic elements, responsible for generation of resonance at frequency of the third harmonics, were calculated for each of selected basic frequencies. Numerical stiffness values versus frequency are: 30 Hz — 5.61 MN/m; 40 Hz—9.97 MN/m; 50 Hz—15.6 MN/m.

Three sets of such elements were made of rubber technical plate of 20 mm thickness and Young's modulus of 8.14 MPa. As it is rather complicated technically to generate a resonance mode exclusively by changing parameters of elastic elements, the design of the test stand allows a fine tuning of the system in terms of resonance by varying the basic vibroexciter frequency within a narrow range using a frequency transducer.

The performance of the test stand was verified under the following conditions: the used set of elastic elements had the total stiffness of 10 MN/m, the frequency of a vibrosource, viz., a centrifugal vibroexciter with directed inducing force Pendulum Vibrator type A 200/600, KNAUER ENGINEERING Co. was 40 Hz, expectant frequency of enhanced third harmonics was 120 Hz.

Vibrovelocity of the executive organ – a vibrosealing plate was measured by using piezoelectric accelerometer; its signal passes through charge amplifier and analogue-digital transducer to a personal computer. Information storage and data processing was performed be means of ACTest©" software system.

In Figure 4 presenting a fragment of oscillogram of sealing plate oscillations and the respective frequency spectrum it is obvious that both low frequency $f_1 \approx 42.5$ Hz and its third harmonics $f_3 \approx 128$ Hz, ratio of vibrovelocity amplitude $v_3 / v_1 = 1.26$ are present in the resultant polyharmonic oscilation.

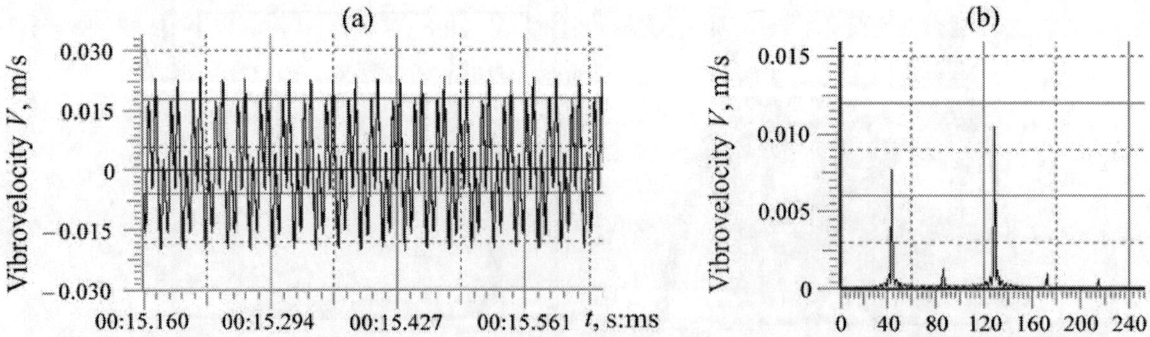

Figure 4. (a) Fragment of vibrovelocity oscillogram and (b) respective frequency spectrum of the sealing plate.

Deviation of frequency values from the calculated values are explained by rough manufacture of the elastic elements.

The follow-on research stage is experimental evaluation of influence of parameters of a superharmonic vibroeffect mode on performance of compaction process.

3. Conclusions

Realization of the polyharmonic mode of the working member operation enabled to generate its oscillations characterized with frequency threefold exceeding frequency of eccentric-mass rotation at comparable amplitude. This makes it feasible to utilize serial commercial vibroexciters with less frequency of eccentric shaft rotation and as a consequence to perform compaction of finely dispersed materials at less energy consumption and mechanical loads on the mechanism, thus appreciably improving the reliability of vibrodrive operation. The new-proposed mode of compaction of finely dispersed materials in an enclosed volume is expected to improve qualitative and quantitative parameters of the process.

References

[1] Lavendela EE 1981 *Vibrations in Technology: Reference* Book Vol 4: *Vibration Processes and Machines* Moscow: Mashinostroenie (in Russian)

[2] Blekhman II 2013 Theory of *Vibration Processes and Devices. Vibration Mechanics and Vibration Technology* Saint-Petersburg: Ruda Metally (in Russian)

[3] Sukhin NV, Bukin SL, Korchevsky AN and Reshevsky AP 1997 Vibration machines with anti-resonance behavior—A new direction of improving technology for processing mineral resources *Progressive Technologies of Mechanical Engineering and Modernity: Int. Conf. Proc.* Donetsk pp. 240–241 (in Russian)

[4] Zakharenko AV 2005 Theoretical and experimental studies of compaction processes of soil rollers and asphalt mixes *Dr. Tech. Sci. Thesis* Omsk: SGADA (in Russian)

[5] Pat. 2553145 RF 2015 Method for fettling electrolyzer cathode assembly with unmolded materials and device for its implementation AV Proshkin, SYa Levenson, VV Pingin and AV Morozov Issued 2015, *Bull. Izobret.* No 16

[6] Garkovenko EE, Nazimko EI, Bukin SL et al 2011 Application of vibration technique with biharmonic oscillations mode in coal benefication *Ugol Ukrainy* pp 41–44

[7] Bukin SL, Kondrakhin VP, Belovodsky VN, Khomenko VN 2014 Excitation of polyharmonic vibrations in single-body vibration machine with inertia drive and elastic clutch *Journal of Mining Science* No 1 pp 101–107

[8] Shatalova IG, Humpbacks NS, Likhtman VI 1965 *Physical and Chemical Fundamentals of Vibration Compaction of Powder Materials* Moscow: Nauka (in Russian)

[9] Bykhovsky II 1968 *Fundamentals of Vibration Technology Theory* Moscow: Mashinostroenie (in Russian)

[10] Gendlina SY. Levenson SY, Eremenko YuI, Vidanov VV 2011 Research results of particulates compaction processes by vibrating method *GIAB* No 8 pp 255–259

New resource-saving technologies for gold recovery from rebellious and hard-to-process material of clayey deposits and mining waste

VP Myazin[1,3]*, DM Shesternev[2] and LV Shumilova[3]*****

[1]Chita Branch Institute of Mining, Siberian Branch, Russian Academy of Sciences, Chita, Russia
[2]Institute of Permafrost, Siberian Branch, Russian Academy of Sciences, Yakutsk, Republic of Sakha (Yakutia), Russia
[3]Transbaikal State University, Chita, Russia

E-mail: *myazinvpchita@mail.ru, **shesdm@mail.ru, ***shumilovalv@mail.ru

Abstract. It is proposed to make leaching piles by layering of crushed fractions by size grades from the bottom layer upward at certain orientation of the pile relative cardinal points and with spraying under waterproof and solution-impermeable light transmissive film. The combination gravity–electrochemical processing technology is proposed for gold extraction from dredge and sluice tailings, and the complete processing line is designed for all-year-round operation, including low-temperature conditions, based on pretreatment of polycomponent chlorine leaching solutions by electrochemical and photochemical synthesis. The magnetic flocculation add-on device is developed for fitting sluices and meant to reduce fine gold loss. The paper presents new engineering solutions that enable higher efficiency and intensity of gold extraction from clayey materials and mining waste, as well as integrated use of natural resources.

1. Introduction

Nonferrous metallurgy in Russia is poorly supplied with raw materials, especially, uranium–copper–gold–manganese–aluminum–nickel-bearing ore [1–3]. At the present time, in view of the lack of large investment in the resource sector of Russia, for production of critical metals, it is urgent to find new technologies for processing uneconomic ore and accumulated old waste of earlier mining and processing operations, as well as clayey gold-containing sand with high content of useful component. Mining companies nowadays tend to reduce financial expenses and risks, they mostly place money in projects with modest capital investment.

The analysis of advancement in technology and equipment for gold recovery from waste and clayey gold-bearing sand in Russia and abroad shows that the top-priority objectives in this regard are reduction in loss of valuable component and improvement of recovery efficiency of fine and very find gold particles. To meet these objectives, it is required to use in a big way innovative engineering solutions [4–8].

2. Intensification of gold recovery

Being proposed for intensification of gold recovery from uneconomic ore and to improve profitability of mining companies, some engineering solutions have increased valuable component extraction by 5–

10% in heap leaching owing to elimination of pore space mudding due to segregation of ore lumps at the pile bottom and by suffusion (Figure 1) [9].

Uneconomic ore is separated on screening-and-crushing plant into size grades of – 5.0, + 5...– 10, + 10...– 15 and + 15... – 20 mm, as a rule. Piling of crushed fractionated ore is carried out by layers, with size grades decreasing from bottom upward, i.e. stock pile layer *1* is composed of size grade + 15 ÷ – 20 mm, the overlying layer is made of smaller size grade +10 ÷ – 15 mm and so on up to the upper layer *2* filled with the finest particles – 5.0 mm in size.

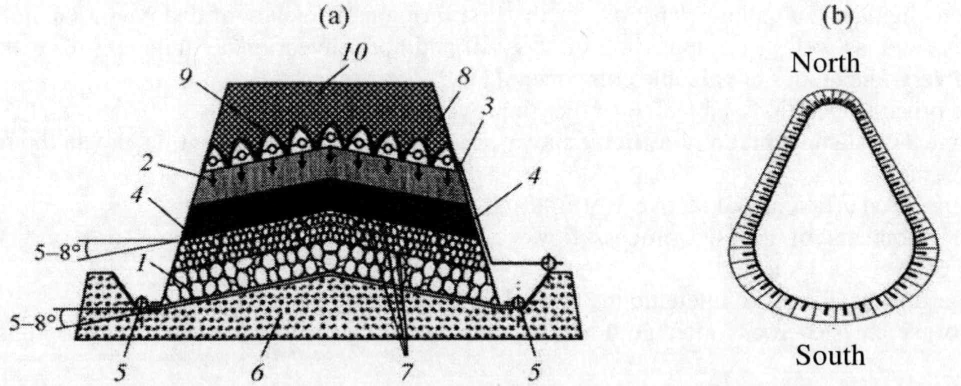

Figure 1. Heap leaching stock pile: (a) elevation view (*1*—bottom layer particles + 15 ... – 20 mm; *2*—upper layer fraction – 5 mm; *3*—spraying and saturation system; *4*—side surfaces of stock pile; *5*—process solution collectors; *6*—stock pile understructure; *7*—perforated polymeric film; *8*—support arcs; *9*—waterproof and solution-impermeable light-transmissive film; *10*—thermal insulator; (b) top view.

Sloped layering of fractionated ore interlayered with perforated polymeric film prevents outflow (washing-out) of clayey and slime particles. Fine particles flow along a an inclined layer toward the side surface of the pile, which crates favorable conditions for uniform penetration of solution to the lower-lying layers, and leaching efficiency and completeness of metal extraction to pregnant solution is increased by 10% as a result.

Stockpiling with its wider part oriented to the south and spraying under waterproof and solution-impermeable light transmissive film make it possible to increase the solution temperature by means of absorption of solar radiation and to totally eliminate energy loss in evaporation and to party reduce energy loss in convective exchange with outer air. The temperature increase is governed by the season, time of day, solar radiation intensity and climatic factors, and is from 5 to 15°C.

Analyzing zoning map of permafrost area of Russia and arrangement of metallogenic belts, it is seen that the key resources of gold mining locate in the Arctic and Subarctic zones, and nearby their southern periphery.

The limiting factors of efficient application of innovations in heap gold leaching in Transbaikalia are: occurrence of most small gold deposits in the areas of frozen and permafrost rocks; complex polycomponent composition of valuable elements and harmful admixtures, as well as elevated content of toxic and potentially toxic components; specific development of temperature inversions in deep ravines and depressions of the Baikal type; little snow failing in winter with clear trend of snow cover redistribution in some regions of the permafrost zone in East Transbaikalia [10].

Much gold is also concentrated in dredge and sluice tailings, intrinsically, fines and flakes, floured gold, or super fine impregnations in ferrous oxides and placer sulphides. The problem of processing of such material consists in the presence of gold within particulate range of 0.07 + 0.05 mm.

The processing technology for dredge and sluice tailings at gold placers in Transbaikalia was tested on semi-commercial scale. The testing included gravity separation on deep filling gateways with concentrate production (magnetite fraction, β^{Au} = 0.75 g/m^3, iron oxides $FeO{\cdot}Fe_2O_3$); first-stage dish leaching by highly active hypochloric acid obtained in electrochemical and photochemical synthesis; second-stage heal leaching by cyanides. The absolute recovery of gold from waste was 95%. The complete processing line ensuring all-year-round operation, including under low temperature, was also developed [9].

Production and processing of gold-bearing clayey sand uses, as a rule, digging machines connected with washing–haulage–transportation systems; therefore the continuity and close interrelationship of all processes predefines improvement of the processing circuit. The loss of valuable component in tailings depends, for the first turn, on efficiency of disintegration and washing of clayey sand, as well as on upgrading of classical and nonconventional equipment for extraction of fine and very fine grades of valuable component [11].

The process operations subject to further improvement include:

—intensive disintegration of difficult clayey sand and removal of cementing clay in the beginning of processing;

—improved efficiency of recovery of fine and very fine gold in separators;

—development of detailed process flow charts for recovery of rebellious fine, very fine and floating gold.

Regarding enclosing nonmetallic material of gold sand, it is recommended:

—to use chiesley rocks after gold-bearing sand washing in other industries (road construction, etc.);

—directional sedimentation of silt-loam particles for making pellets for heap gold leaching and application in other industries (construction).

Within recent years, the team of young researchers of the Chita Division of the Institute of Mining, SB RAS, and Transbaikalia State University has developed and patented complete clayey sand processing lines capable to reduce free gold grains—plate fines, small and large grains with cementing clay. This provides integrated use of mineral raw materials, including enclosing nometallic component (silt–loam settlings and chiesley gangue) in various branches of industry, for instance, in construction and mining.

In the complete processing lines for gold-bearing sand, the scrubbing and sorting machines ensure intensive disintegration of hard-to-wash materials at the head end of the process and to put out the floating gold-containing flow via a spiral classifier discharge and an earth trench for subsequent extraction of floating gold in a special physicochemical module. Then, fine gold is extracted from the pre-washed sluicing tailings on sluices connected with magnetic flocculation add-on device equipped along the length of the chute with alternating polarity magnets made of neodymium–iron–boron alloy. This engineering solution allows holding fine gold in the mass flow by “magnetic fur” formed by floccules of magnetic and strong magnetic particles. The magnetic flocculation add-on device is fitted with a system of frequent rinsing of concentrate, which increase s productive fraction yield and reduces labor-to-output ratio of the process. In the sedimentation structure, by direction physicochemical flocculation, suspended particles in process water are settled and subsequently used as construction materials or anti-seepage filler in the bodies of retaining dams. Furthermore, this clayey product can be used as a binder in agglomeration of mineral raw material in heap leaching. Clarified top layer is returned to the processing circuit, and tailings of concentrate finishing are forwarded to primary concentrate processing plants for separation and utilization later on [11].

3. Conclusions

Thus, new technologies of gold recovery from rebellious and hard-to-process clayey materials and waste will reduce process loss of gold and make it possible to utilize mineral reserves for multiple purposes [9–11].

References

[1] Arens VZh and Chernyak AS 2008 Future of mining is connected with success in chemistry nd microbiology *Vestn. Zabaikal. Tsentra RAEN* No 1 pp 13–18

[2] Arens VZh, Gridin OM, Kreinin EV, Nebera VP, Fazlullin MI, Khrulev A and Khcheyan GKh 2010 *Physochemical Geotechnology* Moscow: MGGU (in Russian)
[3] Fazlullin MI, Saltykov AS, Avdonin GI and Kolpakov GA 2013 Innovative physicochemical geotechnologies for mineral mining *Vestnik RAEN* No 7 pp 4–11
[4] Kulikov VI, Baiborodin BA and Yastrebov KL 2003 Experience of introduction of modern processing technologies for gold-bearing sand in the Irkutsk Region *Vestn. IrGTU* No 2(14) pp 63–67
[5] Zamyatin OV 1975 *Processing of Gold-Bearing Sand and Conglomerates* Moscow: Nedra (in Russian)
[6] Zamyatin OV 2012 Gold-bearing sand sluicing. Basic regularities and process capabilities *Zolotodobycha* No 169
[7] Kowalski P 2004 The method of calculations of the sedimentation efficiency in tanks with Lamella packets *Archives of Hydro-Engineering and Environmental Mechanics* Vol 51 no 4 pp 371–385
[8] Lopez JS, Burgos AJ and, Rodriguez PU 2013 Lamella Settling (FS-PRI-004), Universidade da Coruna
[9] Shumilova LV 2014 *Scientific Reasons for Innovative Technology of Gold Extraction: Monograph* Germany: Palmarium Academic Publishing
[10] Shesternyov DM, Myazin VP and Bayanov AE 2015 Heap leaching in the permafrost zone of Russia *Gornyi Zhurnal* No 1 pp 49–53
[11] Myazin VP, Litvintseva OV and Zakieva NI 2006 *Processing Technology for Gold-Bearing Sand* Chita: ChitSTU (in Russian)

IOP Conf. Series: Earth and Environmental Science **262** (2019) 012050 doi:10.1088/1755-1315/262/1/012050

Mineralogy and process properties of Kolvitsky titanomagnetite ore

YuN Neradovsky[1], SA Alekseeva[2]* and EV Chernousenko[2]

[1]Geological Institute, Kola Science Center, Russian Academy of Sciences, Apatity, Russia
[2]Mining Institute, Kola Science Center, Russian Academy of Sciences, Apatity, Russia

E-mail: *alekseeva@goi.kolasc.net.ru

Abstract. The authors study material composition, texture and structure, as well as process properties of titanomagnetite or from Kolvitsky deposit on the Kola Peninsula. It is found that the ore posses high potential as a unique combination of iron, titanium, vanadium, ilmenite, copper–nickel sulphides and platinum. Pre-treatment of lump material by magnetic separation allows obtaining iron-free nonmagnetic fraction after intermediate crushing. The methods of magnetic, flotation and gravity separation are discussed. The article demonstrates feasibility of integrated processing of this type with production of titanomagnetite, sulphide and ilmenite concentrates for the ferrous and nonferrous metallurgy.

1. Introduction

Global consumption of metals increases, and it is required to enlarge supply of metallurgy with raw materials. Alongside with conventional metal ore, complex ore can become an important resource component for metallurgy in the world and in Russia. Titanomagnetite ore widely spread in the world belong to complex ore. This ore contains iron and titanium, and their commercial value is raised by the presence of vanadium, which is one of the main components of this ore. Moreover, some complex ore deposits contain recoverable copper, cobalt, nickel, gold, platinum, etc. [1, 2].

Shining examples of titanomagnetite ore production and commercial use abroad are deposits of the Bushveld Complex in South Africa, Lac Tio in Canada, Panzhihua in China, etc. Russia holds considerable reserves of this type ore, in amount of round 50% of all global reserves. All in all in the territory of Russia, more than 40 titanomagnetite deposits are explored and appraised (Ural, Siberia, Far East, Karelia, Kola Peninsula). The Kola Peninsula, solely, accommodates more than 10 deposits of titanomagnetite ore [3–8].

Commercial development of titanomagnetite ore is restricted by high mass percent of titanium dioxide in produced iron–vanadium concentrate. Currently, Russian metallurgy uses only low-titanium ore with TiO_2 content not higher than 4% (Gusevogosrsk deposit, Ural). Titanomagnetite with TiO_2 content higher than 5–8% can only be smelted in electrical furnaces, which produce refractory titanian slug needing higher temperature of treatment. Regarding processing technologies for high-titanium titanomagnetite, for instance, in South Africa, raw material (up to 18% of TiO_2) is pre-roasted in a rotating tilted furnace and then is subjected to electrical smelting [9, 10].

Among the Kola deposits, only three fields are low-titanium ore and occur in the hard-to-reach Keivy area, although they can be extracted as infrastructure is developed in the north-east of the Kola Peninsula.

IOP Conf. Series: Earth and Environmental Science **262** (2019) 012050 doi:10.1088/1755-1315/262/1/012050

Other Kola deposits are medium-titanium ore or magnetic titanium–vanadium ore: Gremyakha-Vyrmes, Tsaginsky, Kolvitsky, Tsentralny and other. These deposits are better explored and occure in developed areas with mature communication network, or the development is in process. For medium-titanium ore, the Kola Science Center has developed a technology for processing titanomagnetite concentrate by pyrometallurgical method with production of powdered iron, sodium titanate, titanium vanadate and titanium nitride; moreover, chemical composition of feed stock for this method is of no importance, and lab tests of Kolvitsky, Gremyakha-Vyrmes and Khibiny titanomagnetite produce positive results [11, 12].

Kolvitsky ore is a promising source of titanomagnetite on the Kola Peninsula. The deposit occurs in the south of the Murmansk Region, 50 km eastward of Kandalaksha. The ore possesses high complex potential as a unique combination of magnetic titanium–vanadium–ilmenite ore with contet of copper–nickel sulphides and low-suplhide platinum. Probable reserves of this ore make 100 Mt at the average content of Fe_2O_3 40%, TiO_2 7% and V_2O_5 0.2% [12].

2. Research objects and methods

The tests were carried out on a titanomagnetite sample 450 kg in weight, taken in the detail exploration site of Kolvitsky deposit. The sample is rich in titanomagnetite, and is represented by cores and lumps of densely disseminated and massive ore. Mineral composition of ore, %: titanomagnetite—75–93, ilmenite—to 20, spinel—to 5; sulphides—1–5. Chemical composition (main components), %: Fe_{tot}—51, FeO—32.2, Fe2O_3—37.2, TiO_2—11.1, Cu—0.124, Ni—0.07, S—0.25, V_2O_5—0.5, noble metals—0.806 g/t, platinum group elements—0.012 g/t.

Mineralogy and process properties of the ore was analyzed using a package of methods, including, microscopic, spectral, chemical, grain-size and magnetometric analyses.

The technological studies involved magnetic, flotation and gravity separation. Ore milling used rod and ball mills 30 and 15 l in capacity, respectively. Magnetic separation was implemented on wet magnetic separator 120-T and dry and electric magnetic separator 138 T. Flotation tests were run on lab machines 237 FL and 135-D-FL. Gravity separation was carried out on enlarged lab pant with spiral separators VSR-500 and concentration tables SKO-0.5.

3. Material constitution of ore

Titanomagnetite ore occurs in clinopyroxene and wehrlite, has laminated structure and massive dissemination. By the titanomagnetice content, there is division into low-grade (30%), medium grade (30–50%), high-grade disseminated (50–80%) and massive (> 80%) ore (see Figure 1). The basic minerals are titanomagnetite—42–77%, ilmenite—4–10% and spinel—3–12% [13].

(a) (b) (c)

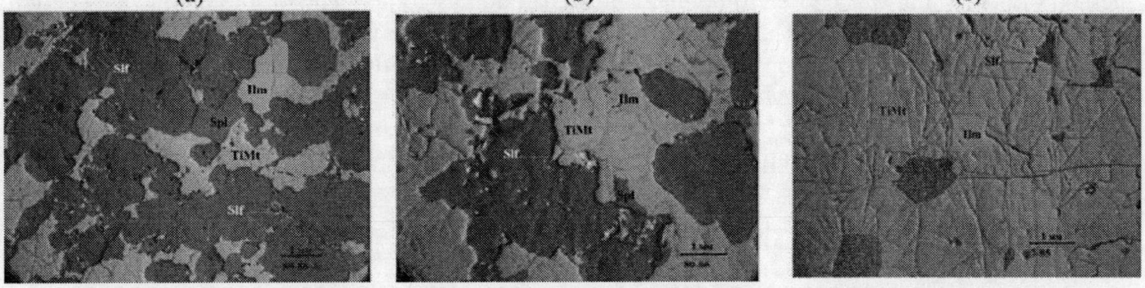

Figure 1. Kolvitsky titanomagnetite ore structure: (a) medium-grade ore; (b) high-grade disseminated ore; (c) massive ore. Reflected light image (TiMt—titanomagnetite; Ilm—ilmenite; Spl—spinel; Slf—sulphides).

Titanomagnetite is represented by a complex four-component system: matrix, ulvospinel, spinel and ilmenite, formed as a result of double disintegration of solid solution. The disintegration products make two persistent paragenesis forms: spinel–ulvospinel and spine–ulmenite. In the first stage, as a consequence of temperature decrease, spinel–ulvospinel (Spl + Usp) evolves with the content of 31.9% Spl and 68.1% Usp. This paragenesis form makes 35.5% of titanomagnetite volume. In the second stage, under influence of tectonic deformation, spinel–ilmenite (Spl + Ilm) evolves with the content of 37% Spl and 63% Ilm. This paragenesis form makes 18.5% in the volume of titanomagnetite. During disintegration of solid solution, vanadium and ulvospinel migrate. Later on, at the stage of ilmenite separation, vanadium remains in the matrix of titanomagnetite. As a result, two phases rich with vanadium form: titanomagnetite matrix and ulvospinel.

Thus, processing of ore can produce magnetic titanium–vanadium (magnetite–ulvospinel) and magnetic titanium(ilmenite) products [14].

Copper–nickel sulphides are present in roks and titanomagnetite ore in the form of dissemination at 5–10%. The dissemination mainly concentrates beyond titanomagnetite and, in connection with this, the content of Cu, Ni and S is 0.232, 0.098 and 0.35% in enclosing rocks and 0.168, 0.075 and 0.36% in titanomagnetite ore, respectively. Sulphides are separated from titanomagnetite in ore, more developed in the layers of disseminate ore and less developed in massive ore (see Figure 1). There are 13 mineral phases sulphides, with dominant copper phase. The base minerals are troilite, chalcopyrite, cubanite and pentlandite; the secondary minerals are bornite, chalcosine, covelline, makinavite, violarite, etc.

Platinum group elements are Pd, Pt nd Rh at the content from 0.020 to 0.814 g/t in rocks and from 0.004 to 0.036 g/t in titanomagnetite ore. The content of Au and AG is high. For instance, the silver content ranges from 0.56 to 1.32 g/t in rocks and from 0.52 to 1.03 g/t in titanomagnetite ore. The total content of platinum group elements, Au and Ag reaches 1.7 g/t in rocks with the sulphide dissemination. It is observed that the contents of platinum group elements, Au and Ag correlate with the content of Cu.

Grains of noble metals vary from 0.2 to 20 μm in size. It is found that noble metals are mostly represented by tellurium–bismuthite of palladium and silver. From the analysis of mineral associations, 78% of noble metal grains are included in the primary sulphide minerals, chiefly in chalcopyrite and cubanite, while 22% of grains occur in streaks intersecting rock-forming silicates [13].

Thus, Kolvitsky titanomagnetite is assumed as the complex Fe–Ti–V–Ni–Cu ore.

4. Process properties of titanomagnetite ore

Processing of titanomagnetite ore uses most often combination of circuits [15–19]. The first stage is production of magnetic concentrate of dry or wet magnetic separation, the second stage is production of ilmenite, sulphide and other concentrates from the nonmagnetic fraction.

For the ore sample under analysis, feasibly of pretreatment of lump material was studied [2]. The method of processing was chosen to be magnetic separation as components of the ore have essentially different magnetic properties. The tests were carried out on lump material – 60 + 30 mm in size using magnetometric method based on measurement of magnetic sensitivity (χ, SI-system units). To this effect, portable susceptimeter SM-30 was used.

Table 1. Magnetometric separation data of the sample, %.

χ, SI units	Yield	Content						Recovery					
		Fe_{tot}	V_2O_5	TiO_2	Cu	Ni	S	Fe_{tot}	V_2O_5	TiO_2	Cu	Ni	S
500–1000	51.99	55.21	0.61	12.12	0.116	0.073	0.25	56.26	52.91	56.98	46.96	54.39	51.39
200–500	33.31	51.61	0.64	11.20	0.132	0.067	0.27	33.70	35.38	33.77	34.10	31.80	35.88
0–200	6.52	14.94	0.34	1.53	0.244	0.068	0.24	1.91	3.66	0.91	12.34	6.32	6.24
Total	91.82	51.04	0.60	11.03	0.131	0.071	0.26	91.86	91.95	91.65	93.40	92.51	93.51
Size – 30 mm	8.18	50.73	0.59	11.27	0.104	0.064	0.20	8.14	8.05	8.35	6.60	7.49	6.49
Initial product	100	51.02	0.60	11.05	0.129	0.070	0.25	100	100	100	100	100	100

By the results of magnetometric separation, three main fractions are distinguished in the sample (Table 1). The product of lumps with magnetc susceptibility higher than $500{\cdot}10^{-4}$ SI units mostly accumulate massive ore containing 75–95% of titanomagnetite and 0–20% of ilmenite. This product yield was 52%. By the content of Fe_{tot}, this product is not finished and needs processing.

The fraction of lumps with the magnetic susceptibility of $200–500{\cdot}10^{-4}$ SI units is middlings; it combines low-grade and high-grade disseminated ore containing 35–75% of titanomagnetite and 0–20% of ilmenite. The yield of this product made 33%.

The fraction of the lumps having magnetic susceptibility of $200{\cdot}10^{-4}$ SI units holds enclosing rocks represented by pyroxinite, olivinite and crystal plagioclase–garnet–pyroxene shale. By the content of titanomagnetite and ilmenite (0–5%), this product can be assumed as tailings but the sulphide mineralization leaves the product in processing. Moreover, this is the smallest fraction, its yield was merely 6.5%, which was typical of the originally high-grade ore. Decision on advisability of the lump separation can be made after testing the commercial-type ore sample.

Thus, magnetic separation of lumps allows removal of fraction that is free from titanomagnetite and ilmenite. This fraction can be dumped, ore dressed separately if its yield is not less than 20–30%.

The lump size was chosen based on the studied kinetics of grinding and dissociation of base minerals. It is found that dissociation kinetics of ilmenite and titanomagnetite has some features. At the early stages of grinding, dissociation of limonite lags behind titanomagnetite. Free grains of titanomagnetite total 90% at grain size of 1.4 mm, while ilmenite reaches such percent when ground to 0.6 mm. The obtained results concord with the size of dissemination: dominating dissemination of titanomagnetite is 0.7–1.5 mm (can reach to 10 mm) while dissemination of ilmenite is 0.5–0.7 mm (to 3 mm). Stating from ground size 0.4 mm, titanomagnetite and ilmenite have almost the similarly high rate of dissociation, corresponding to 95–96% of their free grains. Assessment of dissociation of sulphides shows that dissociation of sulphides is not higher than 70–80% at the ground grain size of –0.4 mm and nearly 90% in the material with the grain size of –0.2 mm.

With regard to the dissociation results, the tests on dressing of size grades –0.4 and –0.2 mm were compared.

The wet magnetic separation data at the magnetic field strength of 900 E were alike. In both cases, titanomagnetite concentrate with the Fe_{tot} content of 61–62% and recovery of 88–89% was produced. Larger size grinding to –0.6 mm failed to result in production of concentrate quality higher than 60% of Fe_{tot}. In this manner, the magnetic separation provides two products:

1. Highly magnetic magnetite–ulvospinel with the content of Fe_{tot} 61–62%, TiO_2 9.6% and V_2O_5 0.54. This product holds around 75% of ore mass and is a complex source for production of iron, titanium and vanadium.

2. Nonmagnetic sulphide–ilmanite with the content of TiO_2 12.5–13.5%, Ni 0.1%, Cu—0.28%, S 0.5–0.6%, noble metals 0.72 g/t and platinum group elements 0.084 g/t. This product holds around 25% of ore mass and is a complex source for production of nickel, copper, cobalt, titanium and noble metals.

The next stage of preparation was processing of nonmagnetic product, production of sulphide andilmenite concentrates is possible both with flotation and gravity separation. The research show that gravity methods are inefficient early in the nonmagnetic fraction processing circuit as they are unable to concentrate sulphides in one of the products, or to obtain products free of sulphides—they evenly spread out in all fractions. For this reason, the nonmagnetic fraction was immediately sent to sulphide flotation.

The flotation wad carried out in an alkaline medium created by caustic ash. The collectors were butyl xanthate and Aeroflot, the frother was agent T-80 and the depressor of rock-forming minerals was carboxymethyl cellulose. The flotation modes were analyzed at the initial grain sizes of –0.4 and –0.2 mm. The size grades were selected with regard to sufficient dissociation of sulphides and producibility of ilmenite concentrate from gravity separation product. The best results were obtained with the size grade of –0.2 mm. The recovery of Cu, Ni and S in the rougher sulphide concentrate was 78.9, 61.1 and 79.6%, respectively, at the content of 1.9, 0.6 and 4.4%. The loss of slupfides in the middlings is mainly owing to sporadic aggregates with silicates. After a cycle of recleaner flotation circuits, the content of Cu in concentrate raised to 4%, Ni—to 1.5% and S—to 11%, while noble metals and platinum group elements increased their cumulative content to 6.4 g/t. In addition, the concentration of ferrous metals (Cu, Ni) in froth product grew relative to initial feed (nonmagnetic fraction) 14 times. The concentration of noble metals increased, too: Ag—8 times; Au—6 times; Pt—3 times; Pd—6 times, Ru—1.5 times; Rh—2.5 times. Silver, gold and palladium are especially considerably concentrated.

For estimation of nonferrous metal recoverability in concentrate, the closed-circuit flotation tests were carried out with middlings to be returned in the circuit. In the closed circuit, at particle size of −0.2 mm, extraction of nonferrous metals in sulphide concentrate made 72% Cu and 71.4% Ni at contents of 4.1 and 1.7%, respectively.

A higher quality concentrate without reduction in recovery can be produced at the smaller size grade of grinding. This was confirmed by the mineralogical assessment of froth product obtained at the size grade of −0.2 mm: the concentrate contained aggregates of sulphides and silicates. In this case, with a view to producing ilmenite concentrate from tailings of sulphide flotation, dissociation of aggregates can be achieved with re-grinding of rougher sulphide concentrate or middlings.

Producibility of a higher quality concentrate was tested. At the ground size of −0.1 mm, the sulphide concentrate was obtained with the contents of Cu 8.6–11.5% and Ni 2.8–3.2%.

Ilmenite concentrate was produced using gravity methods, spiral separators and concentration tables. The gravity separation feed was the flotation midlines subjected to desliming before spiral separation. The spiral separator let out three products: concentrate, middlings and tailings. The tailings with the content of TiO_2 5.6% were withdrawn from processing and dumped, while the concentrate ad middlings were separately sent to recleaning on concentration tables. Table 2 reports final data of the process.

Table 2. Gravity separation of flotation middlings, %.

Product	Yield	Content of TiO_2	Recovery of TiO_2	Processing efficiency
Gravity concentrate	8.55	39.12	22.83	16.7
Mioddlings	18.36	22.80	28.59	12.0
Tailings	42.52	6.77	19.65	−26.8
Slime	30.57	13.86	28.93	−1.9
Feed	100.00	14.65	100.00	0.0

Recovery in the gravity concentrate made 22.8% of the feed at the content of TiO_2 39.01%. The attempts to increase the yield had no success even after more recleaner flotation circuits. After every next recleaner, the yield and quality of the concentrated decreased. This is connected with the fact that in the middlings zone, particle with similar gravity properties accumulate: the product is represented by pyroxene by 60–70%, while pyroxene has similar gravity as ilmenite. Recovery of titanium dioxide in middlings exceeded 25%; accordingly, by handling the problem of dissociation of this product, it would be possible to improve essentially gravity separation efficiency. Aiming to refine gravity concentrate, electromagnetic separation can be used as it provides a product with the content of TiO_2 42–45% at the magnetic field strength of 1800–3500 E.

At this stage, the research demonstrates producibility of ilmenite product. The studies should be continued on a commercial ore sample with a view to analyzing potential of gravity separation efficiency improvement and performance of other mineral processing methods, in particular, flotation.

5. Conclusions

Material constitution of titanomagnetite ore has been studied. Enclosing rocks are mostly pyroxene. The base minerals are titanomagnetite, ilmenite and spinel. Copper–nickel sulphides are presente both in rocks and titanomagetite ore as disseminations at the content of 5–10%. It is found that there 13 mineral phases of sulphides, including copper phases. Platinum group elements are Pb, Pt and Rh, Au and Ag show high contents, and total content of noble metals and platinum group elements reach 1.7 g/t. It is observed that the contents of platinum group elements, Au, Ag and Cu correlate.

It is shown that preliminary magnetic separation of lump ore materials allows obtaining nonmagnetic fraction after intermediate crushing.

As a result of mineralogical and technological research, titanomagnetite, sulphide and ilmenite concentrates have been produced.

The wet magnetic separation yields the titanomagnetiote concentrate with the content of 61–62% Fe_{tot}, 9.6% TiO_2 and 0.54% V_2O_5. Recovery of Fe_{tot} makes 88–89%.

Different regimes of nomagnetic product flotation are analyzed. Achievable recovery of ferrous metals in the concentrate in a closed circuit with the return of middlings back in the flotation scheme is assessed. At the size grade of– 0.2 mm, the recovery of nonferrous metals in the sulphide concentrate made 72% Cu and 71.4% Ni (at the contents of 4.1 and 1.7%, respectively), while the content of noble metals and platinum group elements jointly grew to 6.4 g/t.

Producibility of a higher quality sulphide concentrate with the content of 8.6–11.5% of copper and 2.8–3.2% of nickel was examined at the reduced size grade to – 0.1 mm.

Concentration ability of ilmenite by gravity methods is demonstrated. Using the spiral separation and concentration tables, a rougher concentrate with the content of 39.1% TiO_2 has been obtained. The electromagnetic separation can improve the concentrate quality up to to the content of 42–45% TiO_2.

Thus, the Kolvitsky titanomagnetite ore possesses high complex potential for production of suitable concentrates for the ferrous and nonferrous metallurgy.

References

[1] Vyshegorodskii D 2006 Titanium-magnetite ores—A promising raw material for metallurgy *Ural. Rynok Metallov* No 11 pp 49–53

[2] Borisenko LF, Delitsyn LM, Polubabkin VA and Uskov ED 1997 *Intergated Use of Titanomagnetite Ore* (*Laboratory and Process Research*). *Review* Moscow: Geoinformmark (in Russian)

[3] Rohrmann B 1995 *Supply Demand and Application of Vanadium and Titanium* Johannesburg: Highveld Steel and Vanadium Corporation

[4] Jena BC, Dresier W and Reilly IG 1995 Extraction of titanium, vanadium and iron from titanomagnetite deposits at pipestone Lake, Manitoba, Canada *Miner. Eng.* No 1-2 pp 159–168

[5] Yurong J 1992 Study on beneficiation of titan Panzhihua mine with cone separator by a comprehensive flowsheet *Proc. 1st. Jnf. Conf. Mod. Process. Miner. and Miner. Process* Beyjing pp 329–333

[6] Hиkkaneп E and Welden H 1985 The production of vanadium and steel from titanomagnetites *Int. J. Miner. Process* Vol 815 No 1-2 pp 82–102

[7] Bykhovskii LZ, Pakhomov FP and Turlova MA 2008 Mineral and raw materials base and perspectives of complex use of titanium–magnetite and ilmenite magmatogenic deposits in Russia *GIAB* No 1 pp 209–215

[8] Bartenev NS and Osokin AS 1985 Geological structure and main types of complex ores of the Gremyakhinskoe deposit *Structural Control of Mineralization in the Kola Peninsula Ogneous and Metamorphic Massifs* Apatity pp 31–41

[9] Reznichenko VA, Karyazin IA, Morozov AA and Sadikhov GB 2000 Complex use of titanium–magnetite at a new production development stage *Metally* No 6 pp 3–8

[10] Marsuversky BA, Melamud SG and Avtsinov AF 1995 Smelting of pig iron with the use of nonfluxed pellets of the Kachkanarsky GOK *Metallurg* No 10 pp 18–19

[11] Grishin NN, Neradovskii YuN, Kasikov AG, Voitekhovskii YuL and Rakitina EYu 2015 New powder materials based on titanium–magnetites *Proceedings of the 2nd All-Russian Conference with International Participation: Research and Development in the Field of Chemistry and Technology of Functional Materials* Apatity: IKHTREMS (in Russian)

[12] Grishin NN, Kasikov AG, Rakitina EYu and Neradovskii YuN 2010 RF Patent No 2385962 *Otkryt. Izobret.* No 10

[13] Voitekhovskii YuL, Neradovskii YuN, Grishin NN, Rakitina EYu and Kasikov AG 2014 The Kolvitskoe deposit (geology, material composition of ores) *Vestnik MGTU* Vol 17 No 2 pp 271–278

[14] Voitekhovskii YuL, Neradovskii YuN, Borozdina SV, Groshev NYu, Mokrushin AV, Savchenko EE and Malygina AV 2016 Complex titanium–magnetite ores of the Kolvitskoe deposit (Kola Peninsula) *Proceedings of the 13th All-Russian (with International Participation) Fersman Scientific Session Dedicated to the 50th Anniversary of the Geologist's Day* Apatity pp 67–69

[15] Neradovskii YuN 2014 Investigation of the phase composition of titanium–magnetite (on the example of the Kolvitskoe deposit, Kola Peninsula) *Rational Subsoil Use: Collected Papers* SE Gavrishev (Ed) Magnitogorsk: MGTU pp 158–167 (in Russian)

[16] Neradovskii YuN, Voitekhovskii YuL, Grishin NN et al 2014 Sulphide mineralization of titnaium–magnetite ore of the Kolvitskoe dposit (Kola Peninsula) *Science and Education–2014: Int. Conf. Proc.* Murmansk pp 858–863 (in Russian)

[17] Chizhevskii VB, Shavakuleva OP and Gmyzina NV 2012 Processing of titanium–magnetite ores of the Southern Urals *Vestnik MGTU* No 2 pp 5–7

[18] *Problems of Integrated Processing of Titanium–Magnetites in the South Urals: Int. Conf. Proc.* Magnitogorsk: MDP 2001 (in Russian)

[19] Trofimov NN, Golubev AI, Phillipov NB et al 2002 Results of technological studies of gold–platinoid titanium–magnetite ores of the Pudozhgorskoe deposit *Rudy Metally* No 1 pp 47–55

[20] KushparenkoYuS, Tigunov LP, Ivankov SI et al 2002 A new concept of processing technology for complex titanium–magnetite–ilmenite ore *Resopurce-Saving, Low-Waste and Eco-Friendly Technologies for Subsoil Management: I Int. Conf. Proc.* Moscow: RUDN (in Russian)

[21] KushparenkoYuS, Litvintsev EG, Maslovskii PA, Tigunov LP and Tyutyunnik ND 2005 Rational processing technology for ilmenite-titanium-magnetite ores by the example of the Kuranakhskoe deposit *Razvedak Okhrana Nedr* No 4 pp 61–63

[22] Tereshchenko SV, Veselova EG, Alekseeva SA, Chernousenko EV and Neradovskii YuN 2016 Study of the magnetic properties of the types of ores and rocks of the Kolvitskoe deposit for assessing the possibility of lump concentration *Vestnik Kolsk. Nauch. Tsentra RAN* No 3 pp 53–56

IOP Conf. Series: Earth and Environmental Science **262** (2019) 012051 doi:10.1088/1755-1315/262/1/012051

Cut-and-fill stoping by vertical blocks in Internatsionalny mine

AA Neverov[1]*, SA Neverov [1], AM Nikolsky[1], MV Tishkov[2] and SA Shchukin[1]

[1]Chinakal Institute of Mining, Siberian Branch Russian, Academy of Sciences, Novosibirsk, Russia
[2]Yakutniproalmaz Institute, Republic of Sakha (Yakutia), Russia

E-mail: *nnn_aa@mail.ru

Abstract. The paper describes design and geomechanical assessment of cut-and-fill system of stoping by vertical blocks under protection of artificial roof, with breakage and cemented backfill in the conditions of limited area of Internatsionalnaya kimberlite pipe. It has been found that safe stoping is ensured if the pillar between vertical blocks has the double width of a block.

1. Introduction

Deeper level mining features considerable natural, technological and economic complications. Modern mines use technologies based on the use of self-propelling machines and automated control over basic and auxiliary processes. In this connection, industrial highly mechanized geotechnologies enjoy increasingly extensive development and improvement.

For mining conditions of Internatsionalnya kimberlite pipe currently developed at a depth below 1300 m, it is proposed to use the system of cut-and-fill stoping with drilling-and-blasting in high vertical blocks—CFVB (Figure 1). The application of this technology is conditioned by the extremely low rate of ore production and by the huge amount of development and face-entry drivage in the framework of the actually used top-down slicing with shearing and backfilling.

2. Features of the applied mining system

The feature of CFVB as against the conventional cut-and-fill stoping is considerable reduction of horizontal dimensions of stoping blocks (10×10, 12×12, 15×15 m) and, consequently, smaller exposed areas both in ore and backfill. The latter, particularly at great depths and under highly variable quality of rock mass, contributes to higher structural stability and safety of mining.

The system of mining with CFVB is as follows. Within the limits of a sublevel, vertical blocks VB are cut by drilling cross-cuts in the roof and haulage cuts (18–25 м) in the floor; the blocks have horizontal dimension 5×15 m and height of 20 m. Drilling cross-cuts and haulage cuts are made from the footwall drifts along the boundary of the ore body VBs are caved. In order to increase productivity and improve safety of mining, between vertical blocks under caving and backfill, ore pillars having sizes equal to one or two VB are left. Depending on efficiency of caving and backfilling, as well as on geomechanical behavior of rock mass, the sequence of stoping by VB may be such as illustrated in Figure 2. The stoping sequence can be varied with much experience gained in practicing this system of mining.

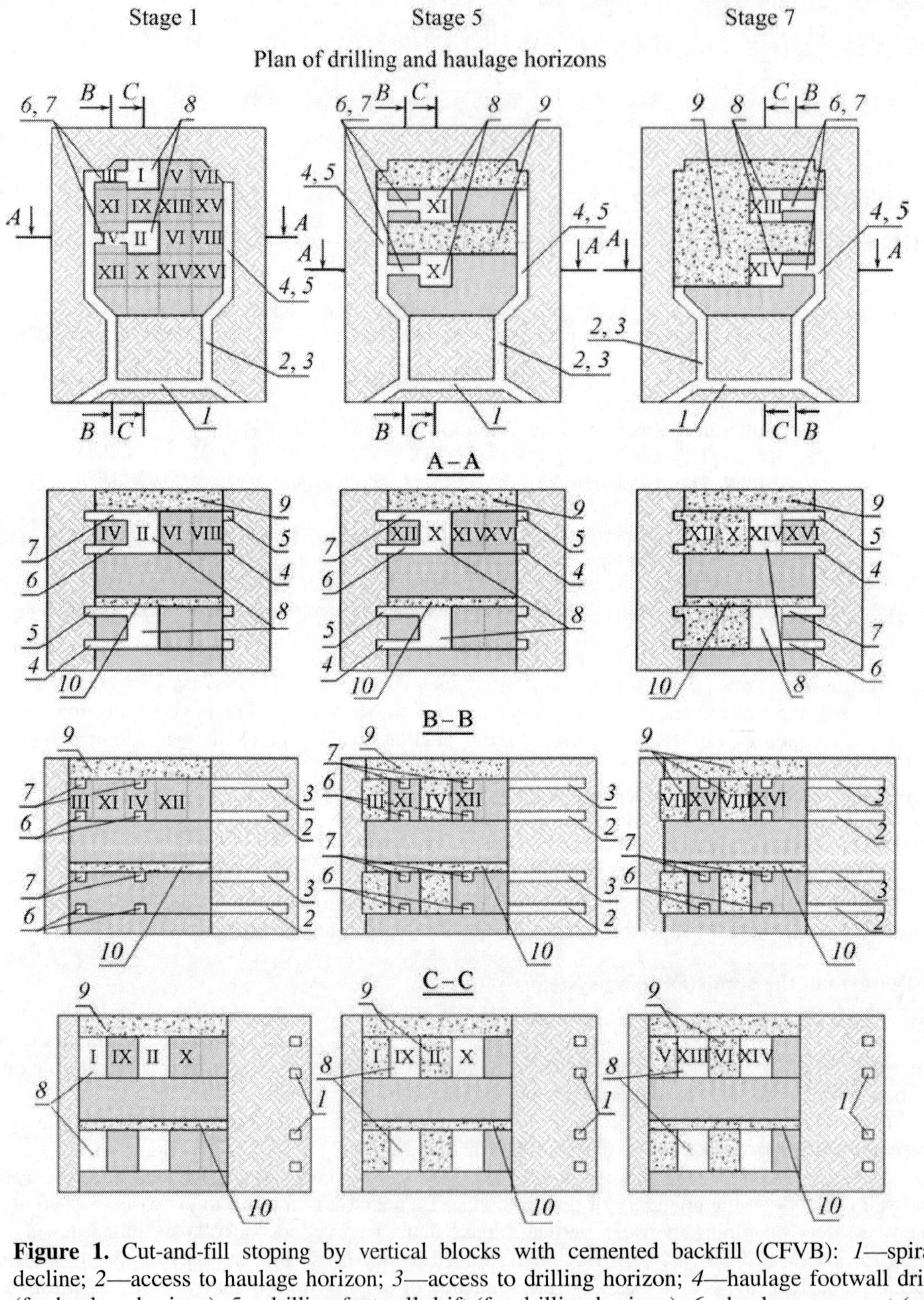

Figure 1. Cut-and-fill stoping by vertical blocks with cemented backfill (CFVB): *1*—spiral decline; *2*—access to haulage horizon; *3*—access to drilling horizon; *4*—haulage footwall drift (for haulage horizon); *5*—drilling footwall drift (for drilling horizon); *6*—haulage cross cut (cut 18–25 m); *7*—drilling cross-cut; *8*—extracted vertical blocks; *9*—filled blocks (backfill); *10*—initial cut layer; I, II, III etc.—sequence of stoping in vertical blocks.

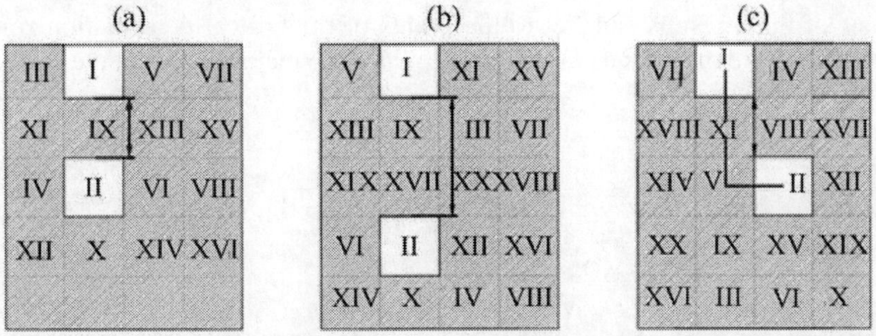

Figure 2. Sequence of stoping by vertical blocks in the system CFVB: (a) single pillar between VB in the same row; (b) double pillar between VB in the same row (in the next rows VB are arranged so that to have only corner-to-corner contact); (c) arrangement of VB to exclude formation of a through pillar (i.e. VB to be soon caved and backfilled): I, II, III, etc.—sequence of stoping by VB.

In this technology, an extraction panel 40 m high is separated from the above-adjacent rock mass by a split layer (SL) to serve as an artificial roof for lower lying stoping. The extraction panel 40 m high is divided into sublevels. The height of a sublevel is set subject to quality of rock mass and kimberlite. In accord with the geological information on the quality of ore and enclosing rocks, the height of sublevels is advices to be no more than 20 m.

Sublevels are prepared by driving spiral annular declines to the drilling and haulage horizons. Then, from the declines, along the boundary of the kimberlite pipe, haulage and drilling footwall drifts are made to ensure accesses to mining machines to the haulage and drilling horizons, respectively, as well as for ventilation of stopes. Drilling cross-cuts under protection of the split layer (the roof of a drilling cross-cut is the artificial fill mass) can be cut to the fill width (length) of the pipe, or as VB are being extracted. Before breakage, in each VB, sot raises are made and enlarged into a cutoff.

Stoping is carried out by layered (or bulk) breakage of ore in downward holes drilled from drilling cross-cuts. Broken ore is drawn from a haulage cut driven on the block floor. As breakage is completed, the floor is cleaned up using a remote-control load–haul–dumper. The machine operates immediately on the floor of the mined-out stope where people are prohibited to enter.

After cleaning-up of the floor, apron walls are constructed and the area between them is backfilled. Stoping in the next VB is only allowed when the backfill reaches the rated strength such that is exposure by a required size stope is safe. For sustainable high output of kimberlite, no less than two extraction panels should be operated.

Sequence of VB breakage is selected with regard to required strength of backfill for stoping in the neighbor blocks and with a view to eliminating equipment stoppage in wait for the backfill to reach the required strength.

The main advantages of this mining system are:

—extraction of diamond ore at permissible loss;

—minimum amount of development and face-entry drivage as against downward slicing;

—high productivity;

—lower labor content as compared with slicing;

—required safety of surface and underground structures as well as the nature from impact of underground mining operations;

—safe labor conditions.

Safe parameters of structural elements in the geotechnology CFVB are determined using 3D FEM-based modeling of stress state of rocks [1–4]. The estimation criterion is the Mohr–Coulomb strength [5–7].

Figure 3 depicts potential damage zones in ore body, rock mass and backfill (colored black) at various stages of stoping (Fig. 1) with single pillar between VB in the same row (Figure 2a). It follows from the analysis of these results, subject to the points of the critical deformation zones in enclosing rocks, ore and backfill, the main loads casing rock failure are major compressive and shearing stresses [8, 9].

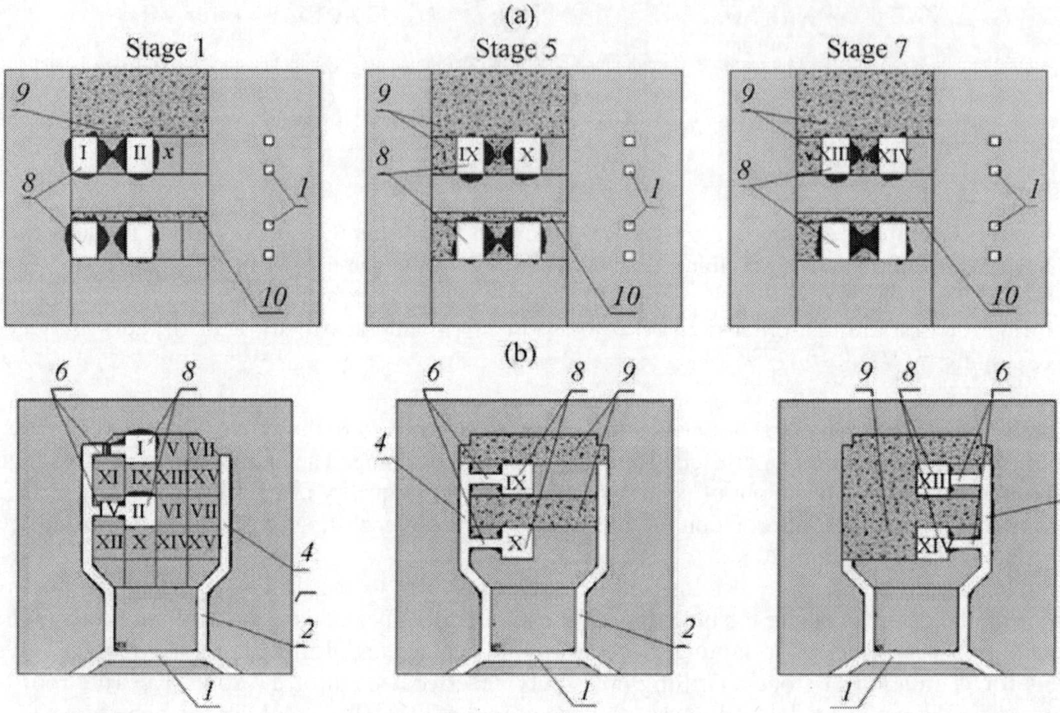

Figure 3. Probable damage zones by the Morh–Coulomb criterion in the structural elements of CFVB technology with single pillars between VB: (a) section C–C in Figure 1; (b) in plan of the haulage horizon in the upper extraction panel (Figure 1). The legend is as in Figure 1.

The structural elements of the mining system, where instability of rocks, kimberlite and backfill is anticipated include [10–13]:

—VB–drive corner zones composed both ore and backfill;

—sidewalls, roof and floor of footwall drilling and haulage drifts. Probable failure zones are local and no more than 1.0 m in size. The situation is the same around the accesses to the haulage and drilling horizons I the area of 10-m damage zone of kimberlite and rocks contact. With distance from the accesses, conditions get much better (almost no damage zones are observed);

—sidewalls, roof and floor of the haulage cross-cut. Probable failure zones vary from 0.5 to 1.5 m in size. The same situation is observed in the sidewalls and floor of the drilling cross-cut. However, in the artificial fill roof of the latter, the probable failure zones have sizes from 0.5 to 2.0 m, which implies that the strength of the cemented backfill should be increased;

—intersections. Damage zones at intersections of various drives are: 2.0–3.0 m in the sidewalls and 1.5–2.0 m in the roof;

—sidewalls, roof and floor of mined-out VB. If the sidewalls are composed of kimberlite and adjacent rocks, at the top and bottom of VB, probable failure zones vary from 1 to 2 m in size, while in the center of VB they may range from 5 m up to a plastic hinge (damage zones of neighbor VB almost touch and even intercross). The situation is the same in VB with sidewalls composed of backfill. Sidewalls beyond mutual influence of VB are in more favorable conditions (VB sidewalls opposite to kimberlite pillars and fills), which is of of the methods for safe arrangement of vertical blocks across the area of kimberlite pipe;

—in the roof and floor of VB, zones of instability of backfill and kimberlite, respectively, are local and vary from 1.5 to 2.0 m in size.

An alternative of safe CFVB is the change of stoping sequence (safe arrangement of VB across the area of the pipe) so that the width of pillars between blocks equals two widths of a block (Figure 2b).

Figure 4 shows instability zones in structural elements of CFVB technology with double pillar between VB in the same row. From generalizations of these results in comparison with the single width pillar between blocks, some conclusions have been drawn.

On the whole, the locations and sizes of concentrations zones of major compressive and shear stress are the same. A slight difference is observed in the stress distribution between vertical blocks under backfill and under breakage. The situation is analogues when studying probable failure zones in rocks, kimberlite and backfill: the sizes and locations are identical. A significant difference of the variant with the double pillar between blocks is the absence of wide (intercross) zones of probable failure in kimberlite and backfill between inter-influencing VB at the stages of backfill and mining, which is an evidence of higher safety of this variant. The local areas of backfill damage in the roof of drives and VB is reflective of insufficient strength of backfill at the given depth (1300 m).

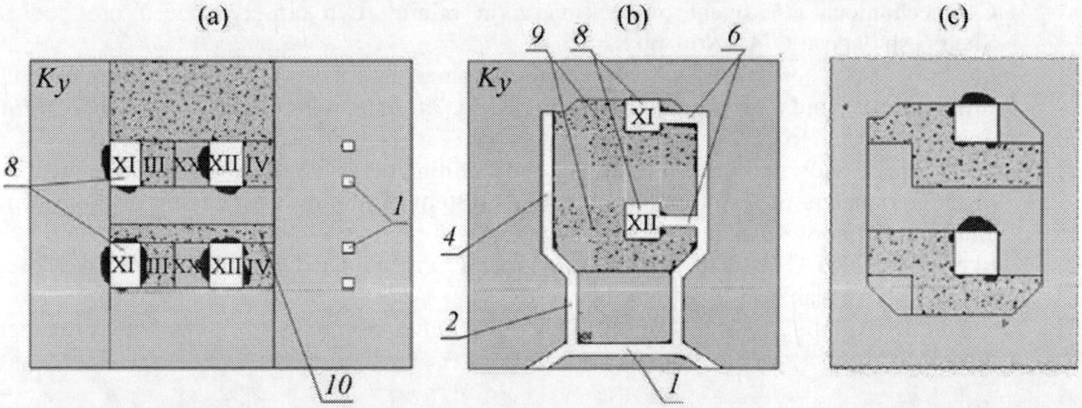

Figure 4. Probable zones of failure by the Morh–Coulomb criterion in the structural elements of CFVB technology with the double width pillars between blocks (similarly to Stage 7 in Figure 3): (a) section C–C in Figure 1; (b) in plan of the haulage horizon of the upper extraction panel; (c) in plan of the center area of the upper extraction panel.

Thus, the main safety constraints of CFVB technology are: relative small horizontal dimension of the kimberlite pipe, which is the critical limitation for safer arrangement of VB in the ore body and, consequently, safer sequence of stoping; poor quality of kimberlite ore and insufficient strength of cemented backfill.

3. Conclusions

According to the performed calculations, the variant of CFVB with the double-width pillar between the vertical blocks in one row ensures the safest mining conditions are the given structural parameters and sequence of stoping. The large zones of instability in rocks and backfill in case of the single pillars impose constraint on the application of this technology at great depths. Therefore, safe application of cut-and-fill stoping by vertical blocks requires obligatory approval by full-scale trial and the increase in the strength of cemented backfill up to 5–6 MPa.

References

[1] Zienkiewich O 1971 *Finite Element Method in Engineering Science* New York: Wiley
[2] Potvin Y and Giles G 2008 The development of a new high-energy absorption mesh

Australasian Institute of Mining and Metallurgy Publication Series pp 89–94
[3] Reiter K and Heidbach O 2014 3-D geomechanical-numerical model of the contemporary crustal stress state in the Alberta Basin (Canada) *Solid Earth* 5 (2) pp 1123–1149
[4] Fadeev AB 1987 The *Finite Element Method in Geomechanics* Moscow: Nedra (in Russian)
[5] Neverov SA and Neverov AA 2013 Geomechanical assessment of ore drawpoint stability in mining with caving *Journal of Mining Science* Vol 49 No 2 pp 265–279
[6] Freidin AM, Neverov SA, Neverov AA and Konurin AI 2016 Geomechanical assessment of geotechnology at a project stage of underground ore mining *Gornyi Zhurnal* No 2 pp 39–45
[7] Bashkov VI, Shtirts VA, Eremenko AA and Konurin AI 2015 Assessment of geomechanical conditions of rockburst-hazardous rock mass under large-scale blasting in Gornaya Shoria *J. Fundament. Appl. Min. Sci.* Vol 2 pp 205–213 (in Russian)
[8] Neverov AA, Konurin AI, Shaposhnik YuN, Neverov SA and Shaposhnik SN 2016 Geomechanical substantiation of sublevel-chamber system of developing with consolidating stowing *16th International Multidisciplinary Scientific Geoconference, SGEM 2016: Science and Technologies in Geology, Exploration and Mining* Vol II Albena, Bulgaria pp 443–450
[9] Eremenko AA, Gakhova LN, Konurin AI, Koltyshev VN, Prib VV and Uzun EE 2018 Geomechanical assessment of rock mass in mining two closely spaced ore bodies at Sheregesh deposit *GIAB* No 1 pp 67–75
[10] Shaposhnik YuN, Neverov AA, Neverov SA and Nikolsky AM 2017 Assessment of influence of voids on phase II mining safety at Artemievs deposit *Journal of Mining Science* Vol 53 No 3 524–532
[11] Boltengagen IL, Korenkov EN, Popov SN and Freidin AM 1997 Geomechanical subtantiation of the parameters of a continuous chamber suystem of mining with caving of the roof rock *Journal of Mining Science* Vol 33 No 1 pp 56–63
[12] Kazikaev DM 2005 Geomechanics of Underground Ore Mining: University Textbook Moscow: MGGU (in Russian)
[13] Turchaninov IA, Iofis MA and Kasparyan EV 1989 *Fundamentals of Rock Mechanics* Leningrad: Nedra

Functional schemes of automated and robotic control of equipment in longwall top coal caving

MS Nikitenko[1]*, SA Kizilov[1] and VK Kuleshov[2]

[1]Federal Research Center of Coal and Coal Chemistry, Siberian Branch, Russian Academy of Sciences, Kemerovo, Russia

[2]Tomsk Polytechnic University, Institute of Non-Destructive Testing, Tomsk, Russia

E-mail:* schum24@rambler.ru

Abstract. The article proposes functional schemes of automated and robotic control of equipment in longwall top coal caving method (LTCC) with coal discharge to face conveyor. The application of the proposed schemes lessens the influence of the human factor, prevents coal dilution, eliminates conveyor overloading, decreases workload of shearer operator in face area and reduces the probability of making errors by operators.

1. Introduction

The effective operation of the longwall complex depends on the coordinated work of powered roof support, face conveyor and shearer. Therefore, as usual, integrated automation systems are used that provide electro-hydraulic control of the support from the control posts located on the support units and remote from the control panel installed on the drift, with reference to the conveyor and shearer position [1].

Thick coal seams development technology with controlled coal discharge to face conveyor involves applying support units with special devices. The features of such supports are the availability of an outlet window, a feeder, which are intended for withdrawal and discharged coal loading to face conveyor [2–4]. Для управления процессом выпуска угля задействованы следующие технические устройства:

—backstop, providing overlapping withdrawal flow gate (outlet window) of support unit;

—feeder initiating a motion process of drawing caved rock mass;

—feeder tray, providing the function of loading on the face conveyor with simultaneous operation of the feeder [1].

It is possible to implement a feeder and feeder tray in the form of a single device, depending on the type of feeder.

2. Methodology

Development of the functional schemes of control is the critical stage of engineering new mining equipment, especially when the equipment is meant for self-operation and is the basis for design of robotic mining complexes for manless mines.

Functional schemes of automated and robotic control of technical devices for top coal caving are worked out on the basis of the developed model of the mechanized top coal caving [1, 5]. Top coal caving process is decomposed into technological operations, and then system-functional approach in the form of IDEF0 methodology is applied [6, 7]. The model aimed at solving a problem of feeder,

IOP Conf. Series: Earth and Environmental Science **262** (2019) 012052 doi:10.1088/1755-1315/262/1/012052

feeder tray and backstop control with taking into account changing parameters of the technological environment.

3. Results and discussion

3.1. Functional scheme of automated control

Functional scheme of automated control of technical devices for top coal caving is composed of necessary and sufficient elements, mechanical and hydraulic connections, data transmission lines as shown in Figure 1. The elements are arranged in thee independent control circuits: slide valve position, feeder inclination and feeder mode of operation.

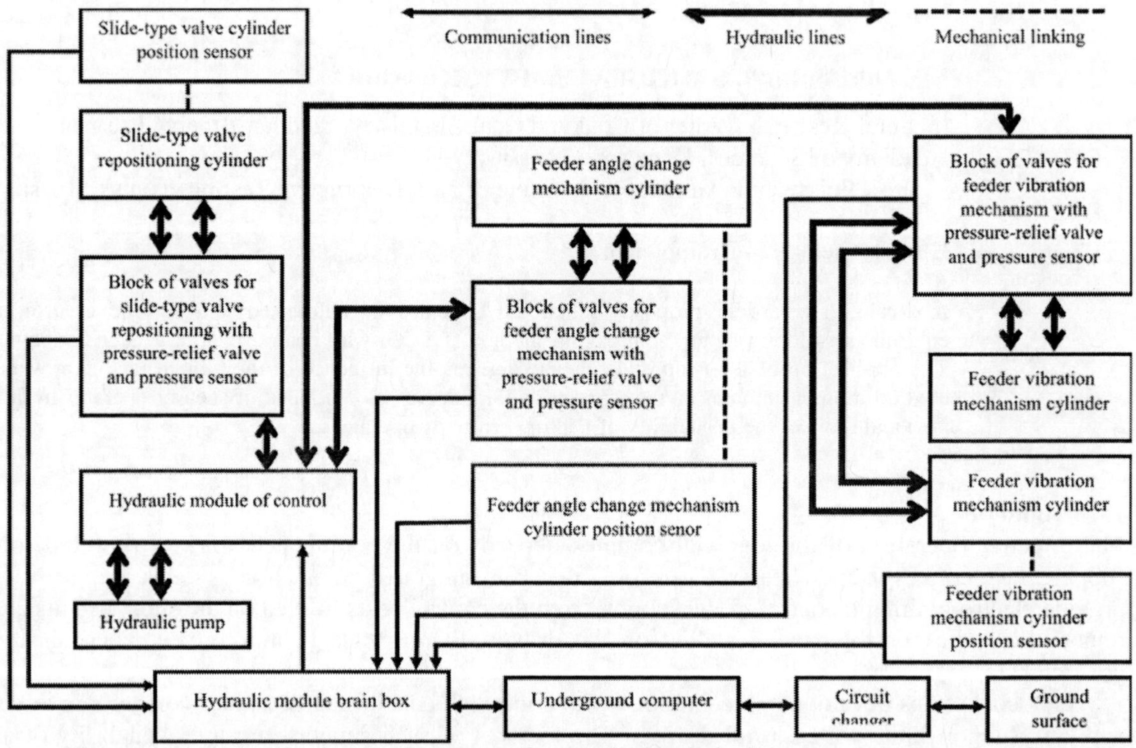

Figure 1. Functional scheme of automated control of technical devices for top coal caving.

As seen in Figure 1, all circuits ensure adjustability of parameters using the hydraulic module brain box arranged on the powered roof support unit. It is provided that double control of mining equipment from the single dispatch office and by underground computer.

Each circuit has independent system of valves to control oil flow to and from the hydraulic cylinder pressure-relief valve of the hydraulic system and oil pressure sensor in the hydraulic circuit. The blocks of valves are connected to the hydraulic block of the roof support control, which allows integrating auxiliary blocks in to common hydraulic system of the LTCC equipment. The data from the oil pressure sensors are sent to the hydraulic module brain box and then to the general control database on the underground computer and to the single dispatch office. Position of the cylinders is controlled by the position sensors which send data to the hydraulic module brain box, which allows an operator to know current condition of any technical device engaged in top coal caving and discharge.

In order to ensure back-and-froth motion of the feeder, the hydraulic cylinder position sensor is used by the control system as the position indicator of the feeder pan for the prompt change in the direction of movement. The signal of automatic change in the direction of movement of the cylinder is set by the underground computer. Since the position sensor continuously informs of the cylinder position, the frequency of the to-and-fro movement is smoothed and physical shocks of the cylinder at the end points are eliminated.

3.2. Functional scheme of robotic control
Functional scheme of robotic control of technical devices for top coal caving is shown in Figure 2 and represents an extended version of the automatic control scheme. It is added with the computer vision and the human–machine interface based on the brain–computer interface.

According to robotic application scheme, the powered roof support is equipped with the computer vision, including lidar to determine discharge rock volume and ash content meters—ah meter and ultrasonic sensors. This set of equipment is to perform key robotic functions when the powered roof support is in self-operation mode. These functions are prevention of rock outlet through discharge opening, determination and overcoming of bridging or arching of coal flow above the powered roof support sections.

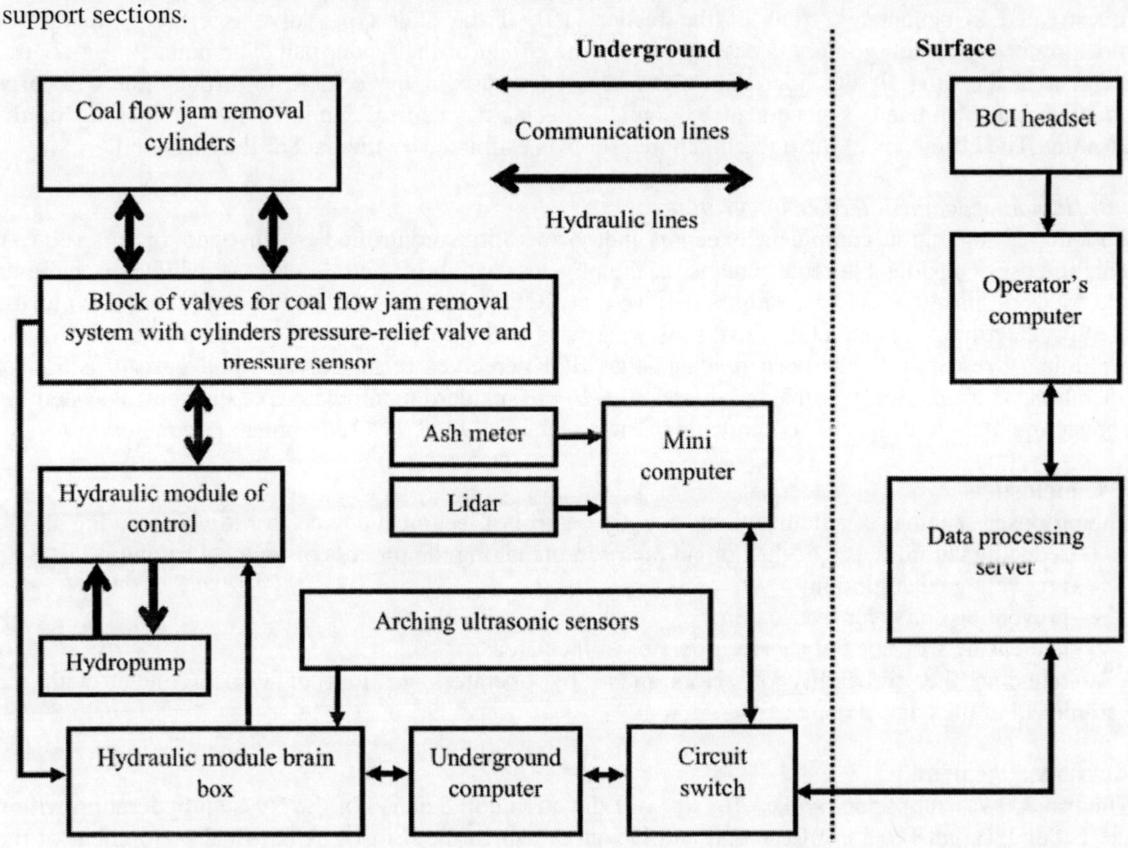

Figure 2. Functional scheme of robotic control of technical devices for top coal caving.

Aimed to prevent rock outlet in the discharge opening during or after coal discharge, discharge rock compositions is inspected by two ash meters arranged one above the other. The first ash meter is in charge of the flow upward from the the inlet opening, and the second ash meter surveys the inlet opening zone. The data are transferred to a minicomputer mounted on the powered roof support section and, then, to the data processing server on ground surface. In case that considerable concentration of ricks is recorded in the zone of the first ash meter, the system switches to ready condition to stop rock flow to the feeder pan. Namely, the velocity of coal flow on the feeder to the conveyor is changed by adjusting the feeder pan inclination. As son as the second ash meter recorded higher rock concentration too, the data processing serves generates the command to shut the inlet opening by the sliding-type valve and to stop and system of the feeder pan excursions after its total unloading. Load of the pan is in this case recorded by the lidar.

Bridging of coal, which prevents rock flow to the feeder pan, is detected by the ultrasonic sensors. The ultrasonic range-finding sensors oriented at angle upward detect voids in the flow. The sensors operate continuously and trace distances from the protection enclosure to the objects in the work zone. The sensors are directly connected with the hydraulic module brain box and transfer data to the underground computer and surface data processing server. The mechanism of bridging removal is developed automatically or with an operator involved. The measures of bridging removal are the change in discharge mode in the other sections of the powered roof support, or employment of special mechanic devices to re-initiate the discharge.

Cessation of coal flow through the discharge opening due to blocking or arching is the off-normal situation which was simulated on the scaled longwall model [8, 9] and will be additionally investigated in benchmark trials of the feeder [10]. If the slide-type valve is closed, the fact of blocking of the discharge flow is determined by the filling of the feeder pan. The main filling control element is the lidar. If the lidar data show clear decrease in the coal flow through the discharge opening, the command is generated to actuate special mechanical facilities for unblocking of the opening. The efficiency of the undertaken measures is estimated by the data of lidar, as well.

3.3. Human–machine interface integration

Robotic self-operation equipment executes many operations controlled by an operator (dispatcher). This imposes considerable requirements on the physical health of the operator, as well as on the level and focus of his attention. The authors believe that BCI application will enable continuous monitoring of physical fitness of operators, levels of weariness and attention [5, 11]. Furthermore, during the preliminary research, it has been revealed that BCI perceives facial gestures and gaze direction of operators. This makes it possible to add gestures to the standard manipulators of the control system by connecting them with specific commands in the interface of the LTCC equipment parameters.

4. Conclusions

The proposed schemes of automated and robotic control of technical devices for top coal caving allow:

—reducing the influence of the human factor in managing the process of top coal caving;

—preventing coal dilution;

—preventing conveyor overloading;

—decreasing workload of shearer operator in face area;

—reducing the probability of errors made by operators, or loss of visual contact with the parameters of the control object.

Acknowledgements

The project was supported by the Ministry of Education and Science of the Russian Federation within the Federal Target Program: Research and Development in the Priority Areas of Development of the Russian Scientific and Technological Complex for 2014–2020, Agreement No 14.604.21.0173 as of Sep 26, 2017, ID No RFMEFI60417X0173.

References

[1] Nikitenko MS, Kizilov SA, Nikolaev PI and Kuznetsov IS 2018 *IOP Conf. Series: Materials Science and Engineering* 354 012014 DOI:10.1088/1757-899X/354/1/012014

[2] Klishin VI, Anferov BA and Kuznetsova LV 2017 Directions for improving top coal caving in thick seams *Innovations in Fuel-and-Energy and Machine Enegineering: Int. Conf. Proc.* pp 57–63 (in Russian)

[3] Klishin VI and Klishin SV 2010 Coal extraction from thick flat and steep beds Journal of Mining Science Vol 46 No 2 pp 149–159

[4] Klishin V, Nikitenko S and Opruk G 2018 *IOP Conf. Series: Materials Science and Engineering* 354 012015 DOI:10.1088/1757-899X/354/1/012015

[5] Kizilov SA, Neogi B, Nikitenko MS, Nikolaev PI and Kuznetsov IS 2017 *Zh. Gorn. Prom.* 6 (136) pp 76–79

[6] Shirobokova SN 2014 *Use of Tools to Support Business Process Reengineering* Novocherkassk: NPI (in Russiian)

[7] Nikolaev PI and Zinoviev VV 2016 Method of substantiation of underground robotic geotechnologies without permanent presence of people in the faces *Vestn. KuzGTU* No 4 pp 26–33

[8] Klishin VI, Varfolomeev EL, Borisov IL and Klishin SV 2018 Investigation of coal disacharge modes on brassboards of powered roof support units *J. Fundament. Appl. Min. Sci.* Vol 5 No 1 pp 66–71 (in Russian)

[9] Klishin VI, Varfolomeev EL, Borisov IL, Kokoulin DI and Khudyntsev EA 2018 Design of brassboard powered roof support and lab-scale test installation for controllable coal discharge *Naukoemk. Tekhnol. Razrab. Uspolz. Miner. Resurs.* No 4 pp 199–203

[10] Kizilov SA, Nikitenko MS and Khudintsev AA 2018 *High-Tech Technologies for Development and Use of Mineral Resources* **4** pp 313–316

[11] Kizilov SA, Nikitenko MS and Neogi B 2018 *IOP Conf. Series: Materials Science and Engineering* 354 012016 DOI:10.1088/1757-899X/354/1/012016

Ozonized petrochemical products used to intensify coal flotation

YuF Patrakov[1]*, SA Semenova[1] and MS Klein[2]*****

[1]Institute of Coal, Siberian Branch, Russian Academy of Sciences, Kemerovo, Russia
[2]Gorbachev Kuzbass State Technical University, Kemerovo, Russia

E-mail: *yupat@icc.kemsc.ru; **semlight@mail.ru; ***m_klein@mail.ru

Abstract. The ozonization method was used to improve the efficiency of oil-based flotation agents for coal dressing. Ozonation of petroleum products contributes to transformation of their hydrocarbon composition with the formation of oxygen-containing compounds of different functionality. The resulting reagents acquire the properties of complex reagents such as collector + foaming agent. The use of ozonized petroleum products leads to an increase in the efficiency of flotation of coking coals: a decrease in reagent consumption, an increase in the yield of concentrate and ash content of waste. Analytical processing of petroleum products is more effective for crude light oils, medium-temperature oil fractions and mixtures of low-boiling and high-boiling fractions.

1. Introduction

Flotation process is intensified by special reagent modes enabling production of low-ash coal concentrates at high recovery ratio. The efficient flotation reagents improve selectivity of the process, lower reagent consumption rates and residual reagent concentration in a liquid phase, viz., they enable to diminish costs of reagents and industrial water treatment. In view of the above the search for novel efficient components and ways to modify routine reagents in order to improve their specific flotation properties acquires specific significance.

The products and wastes of oil refining and petrochemistry are widely used in coal flotation. Apolar reagents: kerosene, thermogasol, diesel fuel [1, 2] and heavy oil fractures in combination with light ones [3] are selected as collectors. Frothers are reagents with polar functional groups and hydrocarbon radicals, for example, residual oil in butyl production (RBA agent) or polyprolene glycol (PROH agent) [1–3].

Flotation modes can be modified by applying a combination of apolar and heteropolar reagents: petrohydrocarbons (gasol) and technical mixtures of alcohols and esters (RBA, PROH, etc) [3–5] olefin oxides [6]. Compounds with functional O-groups, contained in them are capable to participate in both donor and acceptor interaction with active centers of coal particles and to promote higher flotation efficiency [6, 7].

Apart from surfactant additions the coal flotation performance can be improved by oxidative modifications of apolar oil fractions [8]. Thereto, catalytic oxidation of kerosene fractions leads to their higher flotation activity, 3–4-fold less reagent consumption rate as compared to the conventional reagent mixtures [9]. Of interest is also electrochemical oxidation of apolar collectors where hydrocarbons can be oxidized to respective acids [6].

The other promising, in authors' opinion, method of oxidative modification is ozonization. The objective of the present research is investigation into ozonization effect on variations in chemical composition of apolar petrochemicals in order to improve their flotation properties in coal processing.

2. Results of the laboratory experiments

The study objects are marketable specimens of West-Siberian low-viscosity oil, kerosene TS (TU 10227-86), Thermogasol (TU-38.301-18-31-19) and mazut M-100 (GOST 10585-99). Specifications of hydrocarbon reagents are reported in Table 1.

Table 1. Specifications of flotoreagents used.

Reagent	Specimen	Density, kg/m^3	Kinematic viscosity, mm^2/s	Amount of adsorbed ozone, g/kg	Acid number AN, mg/100 g	K_0
Kerosene	Initial	820	1.25	—	0.5	0.12
	Ozonized	—*	—	14.3	4.7	0.47
Oil	Initial	860	8.0	—	1.0	0.23
	Ozonized	—	—	36.7	19.2	1.32
Thermogasol	Initial	910	15.0	—	0.5	0.04
	Ozonized	—	—	34.5	19.2	0.77
Mazut	Initial	990	115.0	—	5.0	0.24
	Ozonized	—	—	25.6	9.3	0.32

*These parameters were not determined for ozonized specimens

Ozonation of petroleum oil was performed at room temperature and atmospheric pressure in a jet-type reactor in a stream of ozone-oxygen mixture (ozone concentration of about 100 mg/l). Ozone concentration was determined by UV-gas analyzer (adsorption by ozone in ultra-violent spectrum brand at wave length of approximately 250 nm).

Infrared spectra were recorded at Fourie spectrometer Infralum-FT-801. The oxidation level K_o was calculated relative to ratio of summary intensity (sum of integral optic densities D) of brands for adsorption of oxygen-bearing hydroxyl (3400 cm^{-1}) and carboxyl (1730 cm^{-1}) groups to D CH$_n$-aliphatic (1460 cm^{-1}) groups:

$$K_0 = (D_{3400} + D_{1730}) / D_{1460}.$$

Flotation activity of petroleum products was evaluated on processing of slimes of meager-lean coking coal with ash content $A_{ic}^d = 10-12\,\%$. The tests were executed at a laboratory flotation machine at reagent consumption of 1.5–5.8 kg/t. Test results were estimated in terms of the yield of concentrate γ_c and wastes γ_w, ash content in concentrate A_c^d and wastes A_w^d, recovery of combustible mass into concentrate E_c, solid content in concentrate, height of froth column h_f and selectivity of the process K_{sel}:

$$K_{sel} = \frac{A_w^d - A_c^d}{100 - A_{ic}^d},$$

where A_w^d, A_c^d, A_{ic}^d is ash content in wastes, concentrate, initial coal, respectively.

The hydrocarbon reagents (HCR) applied in the present research differ notably in component composition and physical properties (Table 1), this factor explains different solubility of ozone and velocity of its interaction with reactive groups of organic compounds. Components of oil and thermogasol (middle-distillate oil fraction T_{boil} = 300–350°C) exhibit the highest reaction activity; this effect is demonstrated in relatively large amount of adsorbed ozone per an equal time interval of processing (Table 1). Ozonation of kerosene (low-boiling petroleum fraction, T_{boil} = 200–300°C) and mazut (high-boiling petroleum fraction, T_{boil} > 350°C) results in the yield of products with acid number (AN) and oxidation ratio K_0 of 2–4 times lower as compared to the rest study reagents (HCR). The process is complicated by notable gumming in the case with kerosene and by highered viscosity in the case with mazut.

The interaction of HCR with ozone results in formation of oxygen-bearing groups of different functionality. In IR-spectrum it is notable the increase in intensity of adsorption bands for OH-alcohols and carbonic acids (3400 cm^{-1}), C=O lactones and anhydrides (1780 cm^{-1}), C=O alifatc (1730 cm^{-1}) and aromatic (1710 cm^{-1}) acids, aromatic cetones (1650 cm^{-1}), C–O cyclic esters, furanes, lactones (1260, 970–1000 cm^{-1}), S=O sufoxides (1300, 1150, 1050 cm^{-1}) (Figure 1).

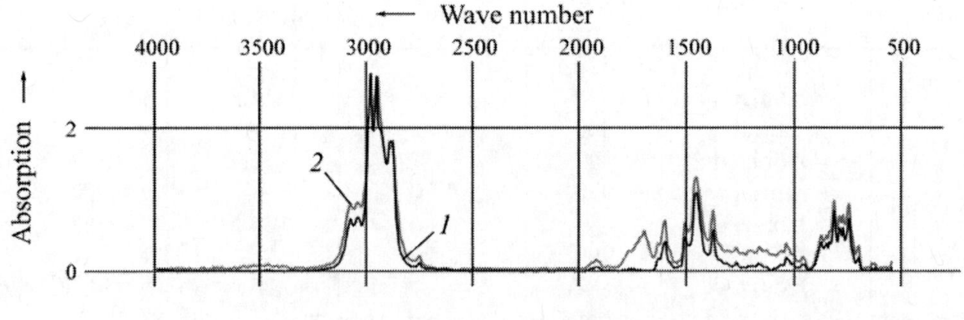

Figure 1. IR-spectra of thermogasol specimens: *1*—initial; *2*—ozonated.

Efficiency of flotation with apolar HCR is evaluated in terms of the peculiarities of component composition and rheological properties of petroleum product [3, 4]. Position of the study reagents in the row of growing viscosity (Figure 2) indicated essential aggravation of their flotation properties despite of increased HCR consumption rate (Table 2). Low flotation properties of mazut as a mono-reagent is due to the complexity in formation of high-dispersed water-oil emulsion, attachment and flowing of a mazut drop over surface of a coal particle [3, 4]. The combination of more or less viscous HCR is often used to eliminate these restrictions. As a case in point, the equal-proportion combination of mazut and kerosene soundly boosted the coal flotation performance at much less reagent consumption rate (Table 2).

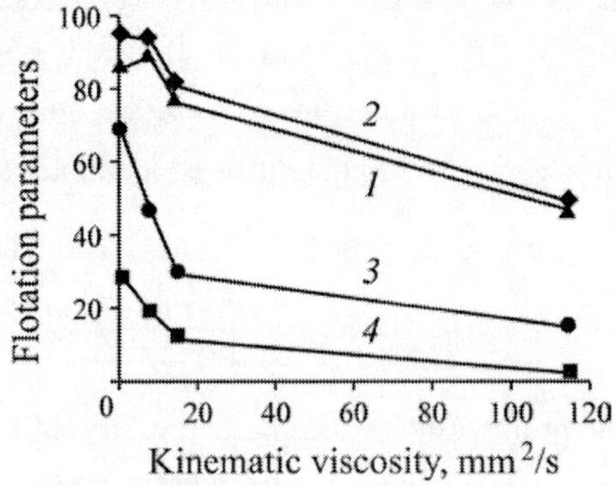

Figure 2. Parameters of coal flotation with petroleum products depending on their viscosity (Table 1): *1*—concentrate yield γ_c, %; *2*—combustible mass yield to concentrate E_c, %; *3*—ash content in wastes A_w^d, %; *4*—froth height h_f, mm.

Table 2. Flotation of meager-lean coking coal with petroleum hydrocarbon reagents.

Reagent	Consumption, kg/t	Concentrate					A_w^d, %	K_{sel}
		γ_c, %	$A_К^d$, %	E_c, %	C_T, g/l	h_f, mm		
Kerosene	2.1	86.2* 87.7	8.7 8.6	94.7 96.4	160 165	28 33	68.0 75.6	0.71 0.81
Oil	2.4	89.0 94.4	6.4 6.4	93.4 98.7	136 166	19 40	46.7 79.6	0.45 0.82
Thermogasol	1.5	76.2 91.6	7.2 6.7	80.8 97.3	135 187	12 30	29.7 71.4	0.26 0.74
Mazut	5.8	46.9 48.0	7.2 6.9	48.9 50.2	86 93	2 4	14.4 14.7	0.08 0.09
Mazut 50% + kerosene 50%	1.8	71.1 87.7	6.5 6.2	74.6 92.2	114 140	12 30	21.6 43.2	0.17 0.41

*Top value corresponds to the initial specimen; lower one corresponds to ozonized specimen

Ozonizing modification of apolar HCR contributes differently to higher coal flotation performance: high yield of concentrate, higher solid content, higher recovery of combustable mass into concentrate ash content in wastes (Table 2). Thereto, selectivity ratio K_{sel} of flotation used to grow 1.1 to 2.8 times for mazut and thermogasol, respectively. High parameters of flotation activity (increase 1.8 times) are exhibited by ozonized light crude oil. Lower ozonation effect on flotation properties of kerosene and mazut is conditioned by initially high adsorption capacity of the first reagent to hydrophobic surface medium-metamorphized coal and a poor dispersability in the pulp of the second reagent. Efficiency of ozonized mazut appreciably grows as HCR in combination with light petroleum fraction (Table 2). The concentrate yield reaches the level of flotation with ozonized kerosene, but ash content in coal descends to the minimum test results (Table 2).

All the ozonized HCR specimens have the increased height of froth column h_f (Table 2). Arising frothing properties can be explained by additional amount of heteropolar compounds in the composition of petroleum fractions, this is justified by higher values of AN and K_0 (Table 1). As a result, HCR start to exhibit complex properties of a collector and a frother.

3. Conclusions

1. It is established that it is feasible to produce the complex-type reagents with properties of a collector and a frother in one in the coal flotation based on ozonized petroleum products.
2. Ozonation of petroleum products contributes to transformation of their hydrocarbon composition with formation of heteropolar oxygen-containing compounds of different functionality: aldehydes, acids, linear and cyclic esters.
3. Preliminary ozonation of petroleum and its fractions improves coke flotation performance , lowers the reagent consumption rate increases concentrate yield and ash content in wastes.
4. Ozonolytic treatment of petroleum products designed to apply them as flotoreagents is more effective for light crude oils, medium-temperature petroleum fractions and mixtures of low- and high-boiling fractures.

Acknowledgements

The authors express gratitude to researchers Kharlampenkova YuA, Shilyaev AB and Malysheva VYu, Federal Research Center of Coal and Coal Chemistry, Siberian Branch, Russian Academy of Sciences, Kemerovo, for their contribution to the research and physicochemical analysis of HCR.

References

[1] Khan GA, Gabrielova LI, and Vlasova NS 1986 Flotation reagents and their application Moscow Nedra pp 271
[2] Abramov AA 2012 Collected works: Flotation. Reagents are collectors Moscow Gornaya Kniga Vol 7 656 pp
[3] Vlasova NS, Klassen VI, and Plaksin IN 1962 The study of the action of reagents in the flotation of bituminous coals Moscow Publishing house of the USSR Academy of Sciences 172 pp
[4] Klassen VI 1963 Flotation of coals, Moscow: State scientific-technical publishing house in mining 380 pp
[5] Melo F, Laskowski JS 2006 Fundamental properties of flotation frothers and their effect on flotation *Miner Eng* Vol 19 pp 766–773
[6] Tyurnikova VI and Naumov VE 1980 Increasing the efficiency of flotation Moscow Nedra
[7] Grishin IA, Knyazbaev JS 2015 The Basic directions of perfection of agent modes in flotation of coal *Successes of modern natural science* No 12 pp 87–90
[8] Khoshdast H, Sam A 2011 Flotation Frothers: Review of Their Classifications Properties and Preparation *Open Miner Process J* Vol 4 pp 25–44

IOP Conf. Series: Earth and Environmental Science **262** (2019) 012054 doi:10.1088/1755-1315/262/1/012054

Method of in-seam gas drainage borehole testing

AV Patutin*, LA Rybalkin

Chinakal Institute of Mining, Siberian Branch, Russian Academy of Sciences, Novosibirsk, Russia

E-mail: *andrey.patutin@gmail.com

Abstract. The paper describes the method of in-seam gas dynamic logging in long drill holes. The method uses a robotic tool based on the straddle packer assembly. The choice of a measurement scheme depends on the time planned for testing, required accuracy and quality of borehole walls. By the testing data, the intervals with low gas influx can be identified for further hydraulic fracturing to stimulate methane drainage, or it is possible to undertake prompt precautions in outburst-hazardous areas.

1. Introduction

Most published gas dynamic research findings are connected with drilling holes from surface down to coal seams. There is hardly any information on interval measurements of gas recovery in in-seam drainage boreholes. This is explained, firstly, by the generally accepted technology of gas drainage when a borehole is immediately connected to a gas drainage system and is not used for studying properties of rock mass, and, secondly, by high cost of work, among other things, due to inapplicability of traditional oil and gas technologies.

Filtration characteristics of a reservoir are most commonly determined using pressure–flow rate relationship, build-up and draw-down curves of gas pressure in a sealed interval of drainage boreholes. These methods are based on features of elastic behavior of oil or gas reservoir when cones of depression form around holes put in operation and expand at a certain rate in the reservoir [1, 2].

The method of gas dynamic logging described in this paper is based on profile measurement of gas influx rate in long drainage boreholes. Recently the mine industry in developed coal mining countries extensively uses directional drill rigs to make boreholes with a preset trajectory up to one kilometer long [3, 4]. It provides much larger gas drainage zone in a coal seam and much higher methane flow rate [5]. On the other hand, there seems to be no research into properties of coal seams using these holes toward optimization of methane recovery. The interval profile measurements of gas influx rate along drill holes at considerable distances from an excavation will reveal features of methane distribution and filtration in coal and rock mass, and will help select most favorable borehole intervals for hydraulic fracturing.

2. Schemes of in-seam gas dynamic drill hole logging

Based on the analysis of experience gained in borehole, including developed in-seam, testing and modern downhole tools, we put forward several approaches to in-seam measurements:

- For drainage boreholes with minor variable cross-section, e.g. in coal seam roof or in hard coal—robotic two-packer tool self-moving along a borehole without help of drill rig;

- For drainage boreholes with greatly variable cross-section due to sloughing and damage of walls, or strong deviation of the borehole diameter from the drill bit design size, which is most often observed in coal seams—single-packer tool pushed into the targeted interval of the hole using the string.

Let us discuss the gas dynamic logging using the robotic two-packer tool moved in hole using a special transport module. The tool has three independent lines of pressure feed, and it is well suited for research of standard and long gas drainage boreholes drilling in coals seam roof or in hard coal [6, 7]. The profile logging procedure is described below (Figure 1).

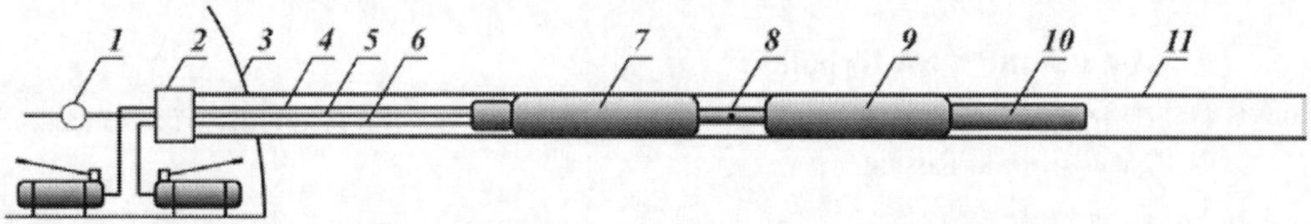

Figure 1. Scheme of gasdynamic logging using two-packer tool: *1*—flow rate meter; *2*—lubricator; *3*—underground excavation; *4*—pressure feed line to closer lying packer; *5*—pressure feed line to farther lying packer; *6*—line for pressure feed/degassing of interval between packers; *7*—closer lying packer; *8*—inter-packer interval; *9*—father packer; *10*—transport module; *11*—hole.

The two-packer tool with the transport module *10* is moved along the hole *11* to its bottom and pulls the pressure feed lines *4*–*6* meant for pumping the packers *7* and *9*, as well as pressure feeding/degassing of the inter-packer interval *8*. The hole is equipped with the lubricator *2* preventing gas outflow from the hole to the excavation *3* during the tests. As the tool is set at the hole bottom, pressure is fed to the packers to seal the inter-packer interval which is the testing interval in this case. Gas released from coal to the test interval is fed using the line *5* and the flow rate meter *1* to the mine drainage system. The air-and-methane flow rate is recorded. Then, the packers are blown off, the tool is moved to a certain distance, and the packers are inflated again. The measurements are taken. In this manner, by moving the tool from the mouth to the bottom of the hole and taking measurements, the gas influx–hole length curve is plotted.

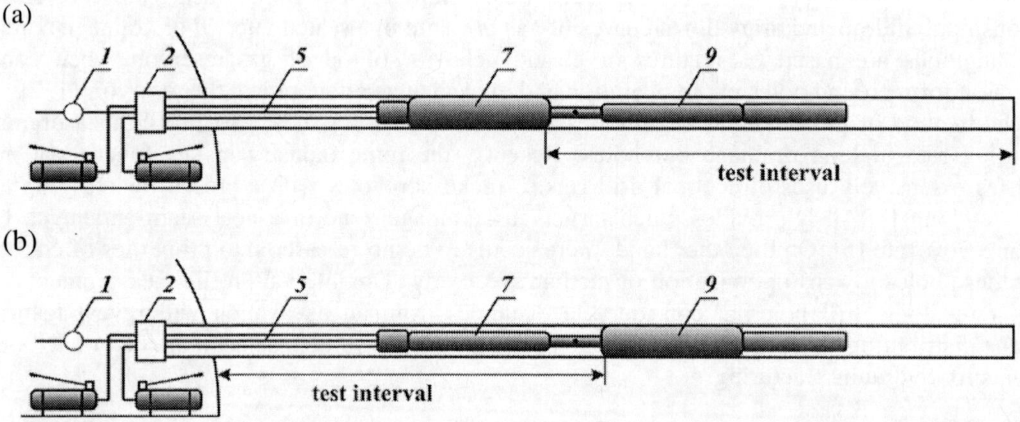

Figure 2. Scheme of fast gasdynamic logging with two-packer tool: (a) measurement interval is between the packer and hole bottom; (b) measurement interval is between the packer and hole mouth; *1*—flow rate meter; *2*—lubricator; *5*—pressure feed line to father lying packer; *7*—closer lying packer; *9*—farther lying packer.

The second gas logging scheme is used when rapid estimate of gas release along the hole is required and featured shorter operating time (Figure 2). In this case, the testing interval is longer due to independent operation of inflated packers within the tool structure; the number of measurement points can be reduced without significant loss of the test information content.

Figure 2a depicts the tool in the hole, the measurement interval is between the closer lying packer *7* and the hole bottom. Gas discharge in the interval is measured by the flow rate meter *1* and recorded. Then, the packer *7* is blown off and the packer *9* is inflated. In this case, the measurement interval is between the packer *9* and the hole mouth (Figure 2b). Gas discharge in this interval is also measured and recorded. Thus, gas influx is measured in two different intervals at the same position of the tool. After the tool is moved closer to the hole bottom, measurements are repeated. The number of measurement points is governed by the wanted precision and time limit of work. This scheme is a quick method to reveal zones of different gas influx along drainage boreholes. Even with two measurements at the same point at the middle of a hole, it is possible to roughly assess which half of the hole has higher gas recovery, and, if necessary, to plan and perform more comprehensive testing.

Figure 3 shows some theoretical curves obtained in modeling gas influx in a 150 m long hole. Measurements are taken each 10 m; at each measurement point two values of gas recovery are recorded for the intervals: closer packer–bottom (Figure 2a), mouth–farther packer (Figure 2b).

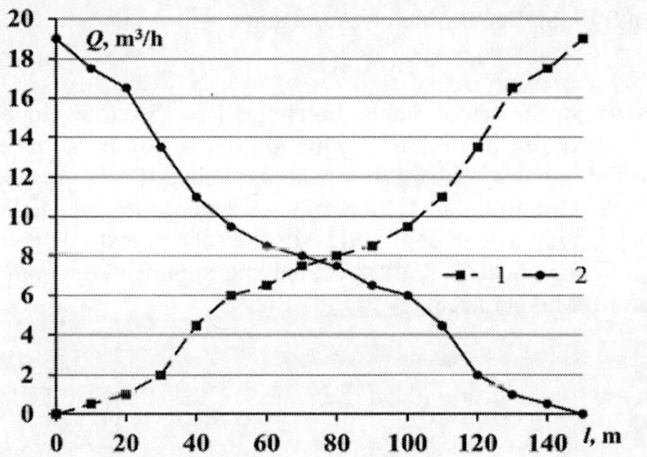

Figure 3. Theoretical curves of modeling gas influx in hole for different measurement intervals: *1*— mouth–farther packer; *2*—closer packer–bottom.

Analysis of the curves allows us to distinguish sections with different gas release, and the curves slope between two measurement points yields gas influx in this section. Plotting two curves creates the opportunity to improve the measurement accuracy by means of averaging test results in long drainage boreholes.

When it is impossible to operate the robotic equipment due to bad condition of the hole walls, the scheme with one elongated packer can be used (Figure 4).

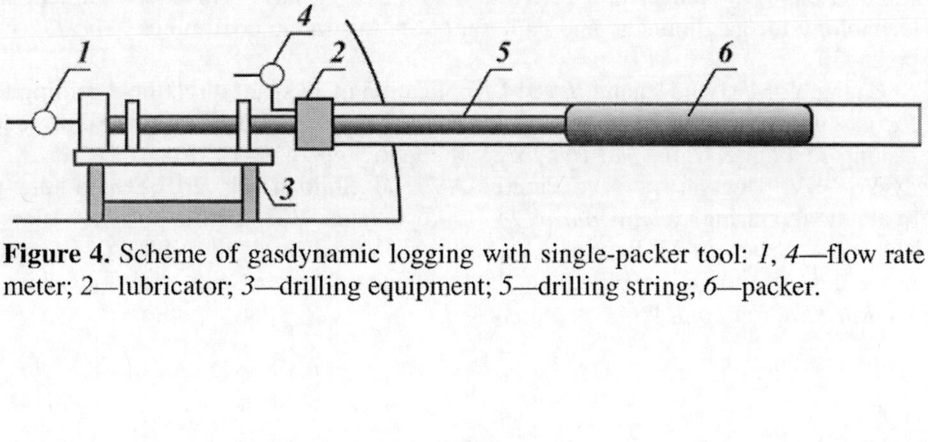

Figure 4. Scheme of gasdynamic logging with single-packer tool: *1*, *4*—flow rate meter; 2—lubricator; *3*—drilling equipment; *5*—drilling string; *6*—packer.

Using the drilling rig *3* or a special system to advance the pipe *5* in the hole, the packer *6* is pushed to the bottom and inflated. An additional rigid tube is placed in each drill pipe and set in special holders at both ends of the pipe. It provides two independent channels for packer inflation and gas extraction. The packer has a through channel for gas flow to the flow rate meter *1*. The flow rate meter *1* records gas influx from the measurement interval between the packer and hole bottom. The flow rate meter *4* mounted on the lubricator *2* shows gas influx from the measurement interval between the packer and hole mouth. Similarly to the express-logging scheme with the two-packer tool two values of gas influx are obtained at one measurement point, which improves the method reliability. After the measurements are taken, the packer is moved toward the mouth, and the process is repeated. The test curves are similar to the curves in Figure 3.

Thus, the efficient gas dynamic logging algorithm is as follows. Firstly, condition of the gas drainage borehole walls is estimated, its real diameter compliance with the standard size drill bit is checked, and gaging is performed. If free displacement of the string is obstructed, the hole is washed and cleaned using auxiliary downhole equipment, e.g. air hammer. Underground mining experience shows that such preparation can be required in holes drilled in soft coal or in seams under high stresses and high jointing. In such cases, the gas dynamic logging is to be carried out using the scheme with the elongated packer mounted on the drill string. In a hole drilled in the roof or in hard coal, it is possible to use robotic system to transport the logging tool in the present hole interval without assistance of the drill rig and to perform full-package profile measurements.

3. Conclusions

The schemes of in-seam gas dynamic logging in long drainage boreholes have been developed. The choice of a scheme is governed by the limited time of work and required accuracy of measurements, as well as by the hole quality. In holes drilled in the coal seam roof or in hard coal, it is suggested using the two-packer tool, which provides the opportunity to take longer measurement intervals and shorter time of testing. For logging in weak rocks in holes with damaged walls the scheme with the elongated packer is proposed, which provides the high quality sealing of the measurement interval. In all schemes it is necessary to use a lubricator for higher accuracy of the results.

Acknowledgements

The work was funded by the Ministry of Education and Science of the Russian Federation, Project No RFMEFI60417X0172.

References

[1] Bourdet D 2002 *Well Test Analysis: The Use of Advanced Interpretation Models* Amsterdam: Elsevier

[2] Thakur P, Schatzel S and Aminian K (Eds) 2014 *Coal Bed Methane: From Prospect to Pipeline* San Diego: Elsevier

[3] Hungerford F, Ren T and Aziz N 2013 Evolution and application of in-seam drilling for gas drainage *Int J Min Sci Technol* Vol 23 No 4 pp 543–553

[4] Wang F, Ren TX, Hungerford F, Tu S and Aziz N 2011 Advanced directional drilling technology for gas drainage and exploration in Australian coal mines *Procedia Eng* Vol 26 pp 25–36

[5] Lu S, Cheng Y, Ma J and Zhang Y 2014 Application of in-seam directional drilling technology for gas drainage with benefits to gas outburst control and greenhouse gas reductions in Daning coal mine, China *Nat Hazards* Vol 73 No 3 pp 1419–1437

[6] Serdyukov SV, Degtyareva NV, Patutin AV and Shilova TV 2016 Open-hole multistage hydraulic fracturing system *Journal of Mining Science* Vol 52 No 6 pp 1210–1215

[7] Rybalkin LA, Azarov AV, Patutin AV, Shilova TV and Serdyukov SV 2015 Geomechanical properties determination based on data obtained from in-seam boreholes logging *ISRM VietRock International Workshop* pp 86–91

Geological aspects of processing zeolite-containing rocks in Eastern Transbaikalia

KK Razmakhnin[1,2]*, AN Khatkova[2] and YaYu Blinovskaya[3]*****

[1]Chita Division of the Chinakal Institute of Mining, Siberian Branch, Russian Academy of Sciences , Chita, Russia
[2]Transbaikal State University, Chita, Russia
[3]Far Eastern Federal University, Vladivostok, Russia

E-mail: *constantin-const@mail.ru, **alisa1965.65@mail.ru;
*** blinovskaya@hotmail.com

Abstract. The resource-potential of zeolite-containing raw materials of Eastern Transbaikalia is estimated. The prospects of mining and processing zeolite-bearing rocks in the Priargunsky District of the Transbaikal Region are studied. The geological aspects of mining and processing of zeolite-bearing rocks in Eastern Transbaikalia are considered. The problems of integrated processing of zeolite-containing rocks are discussed. The influence of climate on the stability of natural objects under man-made impact is estimated. The ways of reducing environmental impact of zeolite production are identified.

1. Introduction

Russia holds considerable reserves of natural zeolite (more than 70 deposits and mineralization zones). The ultimate and probable reserves of zeolite-bearing rocks round 25 billion tons at the average zeolite content of 25–35% [1].

Russia's largest deposits of zeolite ore are: Pashenka, Sakhapta and Voznesenskoe in the Krasnoyarks Territory, Badarma in the Irkutsk Region, Pegas in the Kemerovo Region, Shivyrtui, Kholin, Badin, Talan-Gozagor in Transbaikalia and Buryatia, Serdechnoe and Chugyi in Primorye, Lyutoga and Chekhov in the Sakhalin, as well as Khonguruu, Soros and Chuchuba in Yakutia [1].

The region of Transbaikal possesses the highest potential to create and maintain the largest supply of zeolite-bearing raw materials. At the present time, there are four large-size and twenty medium-size objects of zeolite rocks in the region. All objects occur geographically comfortable, nearby transportation routes and economically mature areas. The most commercially attractive deposits are Shivyrtui, Kholin, Badin and Talan-Gozagor which hold jointly 17 billion tons of zeolite-bearing rocks [2].

Natural zeolite deposits in Transbaikal are mostly composed of clinoptilolite, mordenite and chabasite capable to enjoy wide application in various industries owing to their properties. At the present time in Transbaikal, the Kholin deposit is only under development; minerals from this deposit have found high demand in the medical industry, agriculture and in technologies dealing with sorption. The Shivyrtui deposit is abandoned though the Zhivyrtui zeolite products were rather highly marketable recently. The Badin deposit is closed because of carcinogenic activity of mordenite present in an essential amount in the deposit. The Takan-Gozagor deposit (or mineralization field) holds the most valuable kind of zeolite—chabasite; in the meanwhile commercial development of this project was never started.

This study aims to identify geoecological factors of zeolite-bearing rock mining and processing in Transbaikal, including environmental impact and its mitigation.

2. Zeolite mining and processing in Transbaikal

Much hope with the development of Priargunsky zeolites in Transbaikal is for the first turn connected with the approval of the status of advanced economic and social development territory for the town of Krasnokamensk the Governmental Commission in partnership with the Foundation for Development of Monotowns. In the nearest future, in accord with the Transbaikal Zeolites program, an application will be filed for mining the Shivyrtui and Talan-Gozagor zeolite-holding deposits in the framework of the monotowon management plant. Being supported, this mining project will provide the region with the quality raw material for production of commodities for medicine, ecology, building, agriculture, etc. based on the available processing industry for zeolite-bearing rocks in Krasnokamensk (suspended to date), it would be possible to create a processing and concentration works to obtain marketable products from zeolites of the Shivyrtui and Talan-Gozagor deposits.

Zeolite mining and processing in the area of Krasnokamensk will contribute to the advancement in industry, infrastructure and social sphere. At the same time, it will yield extra environmental pressure on the adjacent territories. The resultant geoecological after-effects will include disfunction of ecosystems, air and soil pollution, surface and underground water contamination, as well as disfigurement of landscape. The negative impacts of mining are the large-scale environmental intervention which is mostly expropriation, partial or total destruction (deforestation, terrain disturbance, etc.) and violation of fauna [3].

It is known that the most environmental impact of mining is limited by an area of direct influence which can be minimized by efficient nature conservation. At the same time, severe exposure of a mining-adjacent territory not only calls for environmental measures but also introduction of resource-saving technologies in all process circuits, starting from mining and finishing with marketable production. The integrate use of zeolite-bearing rocks assumes application of target-oriented methods of energy deposition and physico-mechanical action, often chemical treatment, for instance, acid or alkaline methods in aluminum production, which result in severe ecological problems.

An important point is the integration in the use of waste from processing of zeolite-bearing rocks. At the present time, with regard to geography, industry and infrastructure in the region under discussion, zeolite production waste can be used in construction, in preparation backfill for surface and and underground mines, landfill operations in highway engineering, etc. A system approach to mining waste management will enable total elimination of waste storage in special lots and, thereby, will reduce pressure on the backbone components of the environment, which is of critical concern in the the light of high man's impact in the region. Furthermore, the integrated approach to zeolite-bearing rock processing will allow forgoing overburden dumping and tailing ponds, which essentially disturb ecology, moreover, expenses connected with waste disposal, ore protection from erosion, oxidation, etc. will no more be required. The state-of-the-art analysis of the integration in the use of mineral resources shows that mining, processing and marketable production operations should be looped into a single technology cycle, which is only possible when waste from one production becomes a basic material for another. Thus, construction of process flow sheets for zeolite-bearing rock processing should conform with the wasteless production philosophy with transition from accumulation to separation of zealite-bearing raw materials into structural constituents. Such approach should be pursued at all stages of mining of zeolite deposits, with the selective extraction of process types and grades each of which is to find individual application. Some types and grades of zeolite rocks are usable without pretreatment (in case of zeolite content at the level of 95–98%). However, the listed above deposits are not rich with zeolite, which calls for technologies to be developed for zeolite concentration and removal of admixtures. Such technologies must ensure efficient beneficiation of zeolite rocks, be resource-saving and have minimum environmental impact.

On the whole, the region where the Shivyrtui and Talan-Gozagor deposits are located, is a large industrial cluster. The ecological safety requires a comprehensive system analysis of the environmental impact due to zeolite production activities, prediction of the environmental changes and development of nature-conservative measures in accordance with the anthropogenic effect and the response of nature. The conservation measures should attend regional criteria of weak environmental stability under the human intervention, as well as prevent formation of an unstable ecological system with the infringed structural and functional parameters.

It is worthy of mentioning that the weak environmental stability is partly governed by the severe climate in the Priagrunsky region. The climate is distinctly continental, the average annual temperature is −2.2°C, the average annual wind velocity is 1.3 m/s, summer is dry while winter is low-snowy and very frosty. The relief in the region of the expected production is flat, which allows efficient spread of industrial emission and their higher concentration in the ground surface layer in large residential areas. All these factors, including the distinctly continental climate and deep seasonal freezing of soil in the Priargunsky region, conditioned the non-sustainable natural systems in the region, which can be damaged as a result of the human activities and heavily degrade the environment.

3. Conclusions

Thus, the estimate of the geoecological aspects of zeolite mining and processing in Eastern Transbaikalia has demonstrated the demand of a system approach to zeolite production in the region with regard to compete utilization of the mineral raw material, ecological properties of resource-saving technologies at the overall minimization of the environmental pressure.

Acknowledgements

This study was supported in the framework of the Fundamental Research Program, Project State Registration Number AAAA-A17-117092750073-6.

References

[1] Yusupov TS 1985 Methods of concentration and recovery of zeolites from rocks *Methods of Diagnostics and Quantitative Determination of Zeolite Content of Rocks* Novosibirsk: IGiG SO AN SSSR pp 161–168 (in Russian)
[2] Pavlenko YuV 2000 *Zeolite Deposits of Eastern Transbaikalia* Chita: ChitGU (in Russian)
[3] Faleychik LM 2014 Geoinformation models in the assessment of damage to natural systems as a result of economic activity *Vestnik ZabGU* No 08 (111)

Role of mineralogical and technological evaluation in development of processing technologies for zeolite-containing rocks

KK Razmakhnin[1,2]* and AN Khatkova[2]**

[1]Chita Division of the Chinakal Institute of Mining, Siberian Branch, Russian Academy of Sciences , Chita, Russia
[2]Transbaikal State University, Chita, Russia

E-mail: *constantin-const@mail.ru, **alisa1965.65@mail.ru

Abstract. The article presents a rational package of mineralogical and analytical research methods including a high-resolution optical microscopy, X-ray, morph metric, granulometric, thermal gravimetric and elemental analyses, magnetometry, IR spectroscopy and methods of toxicological, radioactive and hygienic assessing, which provides reliable mineralogical and technological evaluation of complex composition and structure of zeolite-containing minerals in Eastern Transbaikalia. The evaluation consists in defining chemical and mineral compositions of rocks, sizes and micro-concretion of grains, nature of micro-inclusionds and distribution of elements.

1. Introduction

Natural zeolite is a new, unconventional and very promising kind of nonmetallic minerals. The use of zeolites in the industry and agriculture was started in the 1960s after discovery of deposits formed due to conversion of volcanic glass in USA, Japan and Italy. Currently more than 2000 deposits discovered in many countries of the world hold several tens of billions of tons of zeolite reserves, out of which USA, Japan and the CIS countries possess the reserves of 10–20 billion tons. By 2002 Russia had 14 deposits of zeolite-bearing rocks with the total in-place reserves of 649460 thousand tons (Category A+B+C1) and 798631 ths t (Category C2) located mostly (~ 74%) in the Transbaikal Region. These deposits occur in favorable geographic and economic conditions, and offer unbounded capabilities for mining and use of the inexpensive raw material in the industry, agriculture and ecology [1–3].

A series constraint for the wider range application of natural zeolites is insufficient and often nonuniform content in rocks (10–60%) because of genesis conditions. In the meanwhile, some technologies (dewatering of gases and nonaqueous fluids, refinery processing, catalysis and medicine), as the feedstock quality standards, the content of economic zeolite minerals (clinoptilolite, mordenite and chabasite) should be not less than 75–95%.

In this connection, it is of the current concern to develop processing technologies for zeolite-bearing rocks with regard to features of composition, texture and structure, morphology as well as physical, mechanical and chemical properties aiming to enhance potential of applicability of zeolites in science-driven sectors and to replace synthetic analogs [3, 4].

A framework of the uniform theoretical approach to zeolite-bearing rock processing is technological mineralogy with a package of modern tools for the integrated mineralogical and technological evaluation helpful in selecting the most efficient ore pretreatment and concentration methods, and in prediction of zeolite behavior in technological processes using rational flow sheets.

2. Mineralogical and technological evaluation of zeolite-containing rocks

The proposed efficient package of mineralogical analysis methods includes high-resolution optical microscopy; X-ray, morpho metric, granulometric, thermal gravimetric and elemental analysis; magnetometry; IR spectroscopy; electron microscopy; Mossbauer effect study; toxicological, radioactive and hygienic assessing. This package of methods provided the reliable mineralogical and technological evaluation of very complex compositionally and structurally zeolite-bearing rocks of Eastern Transbaikalia, resulting in determination of chemical and mineralogical composition, grain size distribution, micro-structural concretion of grains, as well as the nature and distribution of micro-inclusions.

The tasks of the mineralogical support in creation of highly efficient circuits for zeolite-bearing rock processing are: determination of phase (mineral) composition of source feedstock and its structural characteristics (size of mineral grains and the natures of their concretion with other minerals); description of texture (mutual orientation of mineral aggregates); study of distribution of recoverable elements in rocks; estimation of mineral properties which have influence on the processing efficiency; mineralogical control of processes by means of express phase analysis of the processing products.

The analysis of such complex mineral objects as zeolite-bearing rocks needs modern mineralogical techniques. The development of processing technologies for low-grade zeolite-bearing rocks used a set of the mineralogical analysis methods that enable determination of all mineral phases, including highly dispersed, implementing diagnostics, defining distribution and interaction of mineral phases and estimating process properties of minerals.

The first-stage characterization of the texture and structure of zeolite-bearing rocks uses optical microscopy. The successive mineralogical studies aimed to determine all mineral components and phase composition, contents and properties of rock-forming minerals, nature of dissociation, types of mineral admixtures are carried out with the pre-milled and blended material sorted and fractionated with respect to density and magnetic properties.

Experience shows that conventional mineralogical research (optical–mineral and optical–petrographic) fail to ensure complete determination of mineral composition of rocks, with the diagnostics of all mineral phases and quantification of all contents. The reason is complex poly-mineral structure of rocks and the presence of finely dispersed aggregates. Accordingly, an optimal way out of the situation in integration of the optical-and-mineralogical methods with the X-ray, chemical, physicochemical and instrumental techniques of the phase composition determination, which allow reliable identification of all mineral phases and, then, analysis of variation in their quantitative ratios in the products of processing circuits.

It is impossible to study zeolite-bearing rocks without the electron microscopy analysis which detects surface fines by the direct observation of micro-relief, determines chemical compositions of minerals and informs on the surface distribution of elements.

For the express technological evaluation of zeolite-bearing rocks, examination of mineral distribution in rocks and dressing products, as well as the morphometry of mineral phases, the optical geometric image analysis with Video Master system was for the first time carried out.

The morphometric characteristics of zeolite-bearing rocks (phase/mineral compositions, texture and structure, dissemination sizes) were calculated in terms of geometric parameters of grains: area A (μm^2); length L (μm); width B (μ), perimeter P (μm), elongation $U = L/D$; shape factor $C = 4pA/P2$.

Using the obtained optical geometric information, processability of zeolite-bearing rocks was estimated with the determination of: optimal pretreatment mode to ensure sufficient dissociation of mineral phases, ration process flow sheet and unavoidable process loss, which reduces amount of technological investigations while improving their efficiency.

Inasmuch as zeolite-bearing rocks contain iron either in the form of isomorphic admixture or mechanical micro-spots, Mossbauer effect spectroscopy was used to reveal features of rock composition and explain their magnetic properties.

In order to understand processes of deep conversion of zeolite-bearing rocks, it is important to know physical, physicochemical, mechanical, thermal and adsorption properties which mostly can can be determined by the mineralogical techniques in accord with the regulatory documents of the Scientific Council for Mineralogical Research Methods. In particular, such physical characteristics as density, microhardnessm specific magnetic susceptibility, ionic conductance, bulk weight, moisture content and strength were determined. .

Within the adsorption properties analysis, the textural geometric parameters of zeolite-bearing rocks were measured (volumes and surface areas of pores per sizes grades).

The quality assessment of zeolite-bearing rocks involved the toxicological, radioactive and hygienic assessing. The efficiency of complexing of the mineralogical analysis methods is demonstrated in terms of zeolite deposits in Eastern Transbaikalia (Shivyrtui, Kholin, Badin and Talan-Gozagor).

Based on the accomplished mineralogical and technological research of zeolite-bearing rocks in Eastern Transbaikalia, is found that the main enclosing admixtures, including mineral phases of iron and silica, occur in concretions of zeolite and rock-forming minerals or as inclusions up to 5 μm in size. It is difficult to dissociate zeolite mineral because of their fine intergrowth and to recover selectively zeolites, and it is necessary to apply methods of directed dissociation (disintegration of minerals along the concretion boundaries. The similar physical properties of zeolites and basic associate minerals (montmorillonite, quartz, calcite, opal, calcedony, cristoballite, feldspar, etc.) call for the dissociation of minerals based on improvement of contrast characteristics of their process properties as well as using a set of purposeful physical and physicochemical effects.

3. Conclusions

1. The effective package of methods has been proposed for the reliable mineralogical and technological evaluation of very complex structurally and compositionally zeolite-bearing rocks in Eastern Transbaikali, including chemical and mineral composition analysis, grain size distribution, microstructure of concretions of grains, nature of micro-inclusions and distribution of elements.
2. Based on the accomplished mineralogical and technological evaluation, the key characteristics of zerolite-bearing rocks of Eastern Transbaikalia, which govern selection of processing and concentration technologies toward maximum dissociation of minerals and removal of contamination admixtures (to 99%) as well as for the high-quality marketable production, have been determined.

Acknowledgements

This study was supported in the framework of the Fundamental Research Program, Project State Registration Number AAAA-A17-117092750073-6.

References

[1] Yusupov TS 1985 Methods of concentration and recovery of zeolites from rocks *Methods of Diagnostics and Quantitative Determination of Zeolite Content of Rocks* Novosibirsk: IGiG SO AN SSSR pp 161–168 (in Russian)

[2] Pavlenko YuV 2000 *Zeolite Deposits of Eastern Transbaikalia* Chita: ChitGU (in Russian)

[3] Khatkova AN and Razmakhnin KK 2014 Evaluation of the possibility of complex processing of zeolite-containing raw materials *Razvedka i okrana nedr* Issue 6 pp 48–49

[4] Shushkov DA, Kotova OB, Captains VM and Ignatiev AN 2006 Analcime-bearing rocks of Timan as a promising type of minerals *Scientific Recommendations for the National Economy* No 123 Syktyvkar: Komi Nauch. Tsentr UrO RAN (in Russian)

IOP Conf. Series: Earth and Environmental Science **262** (2019) 012057 doi:10.1088/1755-1315/262/1/012057

Automated device to study the properties of rocks during drilling blast holes in open–pit mines

AS Regotunov* and RI Sukhov

Institute of Mining, Ural Branch, Russian Academy of Sciences, Yekaterinburg, Russia

E-mail:*333vista@mail.ru

Abstract. The paper presents an automated device designed to provide live information on energy intensity of rock drilling, which offers an indirect characteristic of rock strength. The test results of the device are reported.

1. Introduction
Advanced information system and facilities can greatly increase efficiency of production. The Institute if Mining, UB RAS, is developing an automated device to be installed on a roller cutter drilling rig to proved live data on strength of rocks during blast hole drilling (Figure 1).

Figure 1. Automated devices installed in SBSH250-MNA-32 operator's cabin.

The automated device is a set of sensors connected to a controller (Figure 2) with wireless data transmission to an external server. The dedicated data processing program interprets data on properties and structure of rocks, and on condition of drilling machinery. This method of data communication on rock mass strength characteristics was for the first time proposed in [1]. Later on, this research area was developed by Russian [2–6] and foreign scientists [7–11].

2. The measuring method and results
Blast holes are drilled in different strength rocks, which changes the path and performance of drill bit. Sensors trace variations in operating characteristics of motors and remotely send this information to the external server.

This instrumentation has been trialed in open pit mines of Uralbest Company during roller cutter drilling by the rig model SBSH-250MNA-32 in rocks of different drillability. In the course of measurement of drilling energy intensity, huge data amount on rocks of various petrographic composition was obtained.

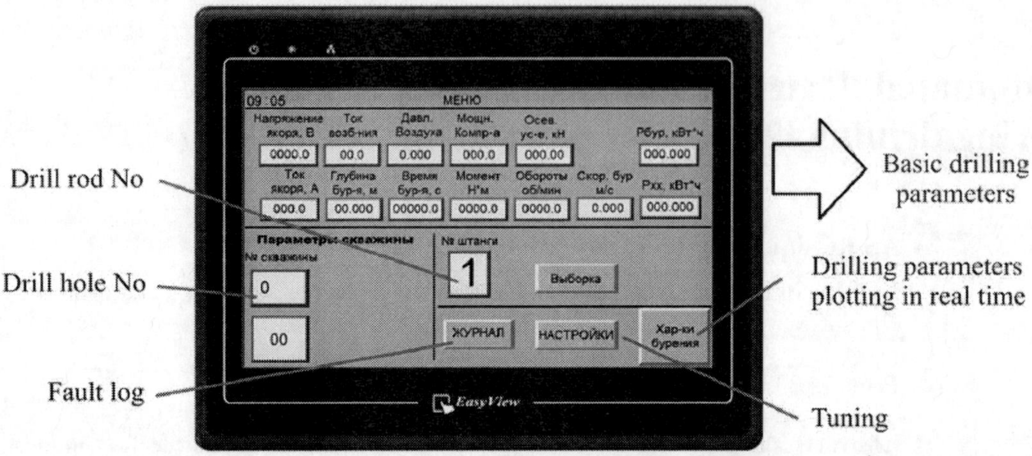

Figure 2. Automated device controller display.

The data processing using a dedicated program allowed converting the data on energy intensity of drilling to the drilling difficulty index P_d by Rzhevsky. After that, by krigging [12], two-dimensional digital models were created. Figure 3 shows one of such models. The models visualized distribution of rocks by the index P_d vertically and along the strike of a block prepared for blasting per each row of blast holes.

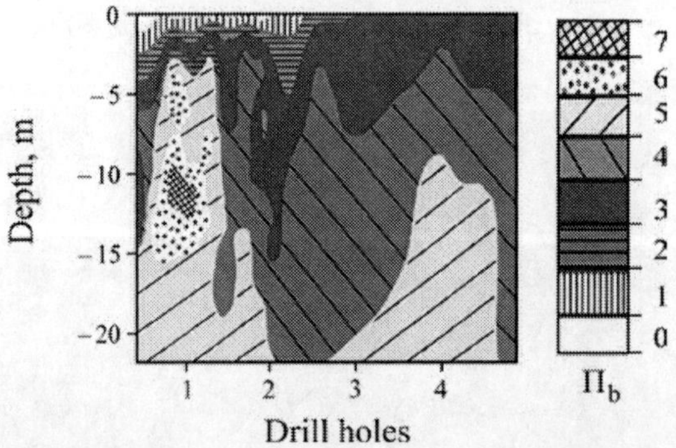

Figure 3. Model of rock distribution by the drilling difficulty index P_d.

Figure 4. Test block with a diorite pocket unbroken after explosion.

The analysis of the models reveals that among readily drillable rocks, there is a pocket of hard-to-drill rocks (chrysotile serpentine recrystallized into antigorite and amorphous serpophite). The earlier prediction provided no densening of blasting pattern or placement of more explosives in holes near by the pocket; eventually, explosion left unbroken a diorite block in the vicinity of rocks of increased strength as per modeling (Figure 4). Thus, the blasting results confirmed the forecast by the modeling.

The studies showed that in the same-type rock mass, energy intensity of drilling varied in wide ranges. The key factors governing such wide variation ranges of energy intensity of rock drilling were nonuniformity, alternation and influence of fault zones. It was also found that SBSH-250MNA-32 rig consumed power of 134 kWh per 1 m drilling of a hole 18 m long, including 20 kWh spent for auxiliary operations. The energy intensity of blast hole drilling (by rotator capacity, excluding compressor power consumed to blow off chips from the hole) was on average 1.4 kWh/m.

3. Conclusions

1. The pilot version of the automated device and program has been developed for getting and interpreting data on strength characteristics of rock in the processing of blast hole drilling.
2. Efficiency of the device has been proved in mining conditions of Ural's large mining and processing plant (Uralasbest Company).
3. The test results might be interesting for many mines operating in hard rock mass, as well as in tunneling and rock construction.
4. The application of the automated device will cut down drilling and blasting cost and will improve quality of rock fragmentation by blasting.

Acknowledgements
This study was supported in the framework of State Contract No 007-00293-18-00 (Topic No. 0405-2018-0015), as well as by extra-budgetary resources from a commercial contract and Competitive Project No. 0405-2018-0001 (Project No. 18-5-5-10).

References

[1] Tangaev IA 1978 *Rock Drilling and Blastability* Moscow: Nedra (in Russian)
[2] Rogalsky BS, Shtogrin EA and Kushmir IS 1985 Evaluation of rock drilling capacity in open-pit mining *Gornyi Zhurnal* No 7 pp 41–44
[3] Zharikov SN 2010 Determination of rock strength by roller drilling parameters in open pit mines *Gornyi Zhurnal* No 7 pp 50–51
[4] Zharikov SN and Shemenev VG 2014 Express study of rock mass properties during preparation for drilling and blasting *Vzryv. Delo* No 111/68 pp 155–164
[5] Sukhov RI, Shakhmatov DS, Pankov IV, Ivanovskaya SV and Veresa AF 1991 Energy intensity of destruction of rocks in blast hole drilling *Gornyi Zhurnal* No 3 pp 26–28
[6] Regotunov AS and Sukhov RI 2016 Results of analysis of rock strength data from drilling using hardware/software package *Problm. Nedropolz.* No 4 pp 121–129
[7] Turo K 1997 Prediction of drillability in hard rock tunneling by drilling and blasting *Proceedings of the World Tunnel Congress'97* Vienna, Austria Vol 1 pp 103–108
[8] Bilgin N and Kahraman S 2003 Drillability prediction in rotary blast hole drilling *1st International Mining Congress and Exhibition of Turkey-IMCET* pp 177–182
[9] Luis E. Izquierdo and Luciano E. Chiang 2004 A methodology for estimation of the specific rock energy index using corrected down-the-hole drill monitoring data *Mining Technology* (Trans. Inst. Min. Metall. A) Vol 113 pp A225–A236 DOI: 10.1179/037178404225006218
[10] Kelessidis VC 2011 Rock drillability prediction from in situ determined unconfined compressive strength of rock *Journal of the Southern African Institute of Mining and Metallurgy* Vol 111 pp 429–436

[11] Yarali O and Soyer E 2013 Assessment of relationships between drilling rate index and mechanical properties of rocks *Tunneling and Underground Space Technology* Vol 33 pp 46–53 DOI: 10.1016/j. tust.2012.08.010, vol. 33, 2013

[12] Saveliev AA, Mukharamova SS, Pilyugin AG and Chizhikova NA 2012 *Geostatistical Data Analysis in Ecology and Nature Science (Using R Package): Text Edition* Kazan: Kazakn. Univer. (in Russian)

IOP Conf. Series: Earth and Environmental Science **262** (2019) 012058 doi:10.1088/1755-1315/262/1/012058

Role of radiation modification in ore pretreatment

VI Rostovtsev

Chinakal Institute of Mining, Siberian Branch, Russian Academy of Sciences, Novosibirsk, Russia

E-mail: benevikt@misd.ru

Abstract. It is found that the increased radiation exposure of rocks treated by accelerated electrons changes the rock deformation and strength characteristics: ultimate compression strength as well as moduli of deformation and elasticity. The uniaxial compression strength of granite is 68.33 MPa without treatment and 35.08 MPa at the absorbed radiation dose of 10 kGy. The established possibility of weakening of rocks by accelerated electrons can be used to reduce consumption of electricity during ore pretreatment and to minimize loss of useful components in processing.

1. Introduction

The present-day development in the mineral mining and processing industry is connected with treatment of rebellious ore and mine waste. This trends conditions also high loss of commercial components at the stages of mining and beneficiation, as well as considerable operating and capital cost of final output.

Up to 35–40% of loss of valuable components in pre-treatment is connected with mineral aggregates and 30–35%—with fines smaller than 10 μm in size [1]. In order to reduce such losses in processing finely disseminated ore without formation of aggregates and over-grinding, conventional crushing and milling in jaw and cone crushers and ball mills should be replaced by selective disintegration. The process of disintegration is the main operation in ore pretreatment. Ore is assumed prepared for dressing when ground to a certain size. As per the coarseness standard adopted in Russia and in the CIS countries, processing of nonferrous and ferrous metal ore accepts size grades of – 71 and – 44 μm. Some foreign processing plants admit size – 38 μm. In recent years, low-grade and uneconomic ore treatment as well as recovery of finely dispersed minerals and pure metals is carried out using equipment of fine (smaller than 20 μm) and super-fine (smaller than 7 μm) milling [2]. Energy inputs are comparatively low at the stages of crushing (to 1.2 kW·h/t) and high at the stages of milling (50 kW·h/t).

The problem has been complicated of later decade as processing increasingly involves rebellious finely disseminated and low-grade ore, and target components are over-ground and lost with slurry in this case. Mineral crystals of many studied and extracted nonferrous metal ore and tin, rare ad rare earth elements are within the size range of 0.015–0.050 mm, and the task of ore pretreatment consists in provision of the above-mentioned coarseness of particles with preservation of the crystal structure of minerals.

2. Different approaches to mineral processing

The commonly used ball mills are low-efficient with the identified ore types due to increased sliming and destruction of useful minerals, which impairs selectivity. In longer period milling, amount of mineral association of higher mechanical strength grows; nondecomposed aggregates in such associations are

30–40 μm in size, and their disintegration requires higher impact effects. The analysis of many studies in the sphere of geomaterial grinding makes it possible to state that fine and super-fine milling requires equipment with much higher energy of crushing than in ball mills using acceleration of gravity [3, 4].

The process properties depend on the constitution and genesis of minerals, which governsproperty contrast of intergrown minerals, their changeability under various effects. The maximum information on composition and structural parameters of minerals under processing can be obtained from the data of process mineralogy. The interactions of minerals and their grain size distribution condition coarseness of disintegration and dissociation ability of minerals under treatment.

In order to reduce mineral loss and energy consumption, it seems promising to treat rebellious minerals by nonmechanical energy deposition that enables overcoming physical intractability and allows selective disintegration without over-grinding [1, 7–10]. High efficiency of energy treatment techniques in mineral dressing is proved by researchers of foreign scientist from Canada, Germany, Japan, South Africa, etc.

The distinguished Russian scientists, Corresponding member of the USSR Academy of Sciences I. N. Plaksin assumed one of the critical trends in mineral processing is such energy deposition that improves completeness and multi-purposiveness of mineral use. He was one of the pioneer researches to justify efficiency of external actions in mineral dressing processes [11].

In this fashion, the problem connected with maximum dissociation of minerals at the minimum over-grinding can be solved through engineering innovative machines [1, 2, 4, 12], or by intensification of available procedure by means of using energy deposition [1, 7, 13]. The review of the studies accomplished by some institutions, including the Chinakal Institute of Mining [8, 9] reveals possibility of improving dissociation of rebellious minerals and enhancing processing efficiency by means of preliminary treatment by accelerated electron flow. According to the analysis of mineral samples from different deposits, they are as a rule polymineral and represented by rock-forming minerals and rocks, for example, limestone, hornfel and granite. The aim of this study was to examine the change in strength of core specimens of the listed rock types as a result of treatment by high-energy electrons.

Limestone is one of the widest spread sedimentary rocks. Limestone deposits make 19–22% of all sediments. Regarding occurrence in the earth crust, limestone is the third after quartz and feldspar. Large quantity of calcite deposits from hot water in hydrothermal veins with sulphides and in effusive rock amygdales with zeolite, chalcedony, quartz and barite.

Hornfel strata enclose copper–molybdenum formations containing copper, molybdenum, gold, silver, lead, zinc, cobalt, bismuth and mercury (Endybai deposit in Yakutia); antimony (Terek in Kirgizia, Xikuangshan in China); tungsten (Tigrinoe in Primorye) and other metals and minerals. Granite is the most important rock is the earth crust. Granite is tightly connected with the deposits of such minerals as Sn, W, Mo, Li, Be, B, Rb, Bi, Ta and Au.

The physical and mechanical properties of minerals and rocks were determined using core samples of regular cylindrical shape in accordance with the effective standards and procedures, on equipment passed metrological certification. The samples had a diameter d of 30 mm and height h of 60 mm; the geometry factor h / d was 2 upon average for the tested samples, which complied with the Russia's State Standard 21153.2-84. Deformability testing procedure agreed with the requirements [14–16]. The prepared core samples were exposed to beam of accelerated electrons on industrial electron accelerator ILU-6 at the Institute of Nuclear Physics, SB RAS, under electron energy of 2 MeV. The key research findings on the effect of electron treatment on the mechanical and acoustic properties of ore samples are presented in Tables 1 and 2 and in the figure.

It follows from the data in Table 1 that treatment by the beam of accelerated electrons changes mechanical properties of limestone, hornfel and granite core samples. The increase in the absorbed dose results in the decrease in compression strength, as well as in deformation and elasticity moduli. Limestone has the compression strength of 49.11 MPa before treatment and 35.24 MPa after absorbing a dose of 15 kGy. Poisson's ratios grow with the absorbed dose: 0.17 before treatment and 0.38 after absorbing the dose of 15 kGy. The same laws are observed in hornfel and granite.

Table 1. Results of mechanical and dynamic testing of limestone, hornfel and granite core samples after accelerated electron treatment.

Sample No.	Absorbed dose, kGy	Sample size $d \times h$, mm	Mechanical properties		
			Compression, strength σ, MPa	Static elastic characteristics	
			Uniaxial compression	Deformation modulus *E*, GPa	Poisson's ratio *ν*
Limestone (ρ = 2.68 g/cm^3)					
0	0	29.5×60.2	49.11	13.20	0.17
5	5	29.5×60.5	76.61	13.32	0.22
10	10	29.5×60.5	38.24	7.44	0.26
15	15	29.5×60.4	35.24	10.43	0.38
Hornfel (ρ = 2.66 g/cm^3)					
0	0	29.5×60.3	123.15	29.70	0.175
5	5	29.4×60.3	115.20	22.72	0.18
10	10	29.4×60.3	78.97	9.71	0.18
15	15	29.5×60.2	135.89	28.64	0.19
Granite (ρ = 2.45 g/cm^3)					
2	0	29.5×61.3	68.33	13.19	0.19
5	5	29.6×56.9	67.98	9.34	0.19
8	10	29.6×60.0	35.08	7.04	0.16
9	15	29.6×60.0	57.53	10.34	0.20

Table 2. Acoustic properties of limestone, hornfel and granite core samples after accelerated electron treatment.

Sample No.	Absorbed dose, kGy	Sample size $d \times h$, mm	Atmospheric conditions			
			Wave velocity, m/s		Deformation modulus *E*, GPa	Poisson's ratio *ν*
			P-wave	S-wave		
Limestone (ρ = 2.68 g/cm^3)						
0	0	29.5×60.2	5021	2657	49.30	0.31
5	5	29.5×60.5	4605	1992	29.50	0.38
10	10	29.5×60.5	4386	1992	29.21	0.37
15	15	29.5×60.4	4485	1982	28.62	0.38
Hornfel (ρ = 2.66 g/cm^3)						
0	0	29.5×60.3	5601.3	3030.7	63.62	0.29
5	5	29.4×60.3	5339.7	2972.7	60.62	0.28
10	10	29.4×60.3	5345.7	3012.0	61.81	0.27
15	15	29.5×60.2	5233.3	2971.3	59.52	0.26
Granite (ρ = 2.45 g/cm^3)						
2	0	29.5×61.0	3747	2287	32.01	0.20
5	5	29.6×56.9	3543	2142	30.29	0.27
8	10	29.6×60.0	4054	2376	37.98	0.26
9	15	29.6×60.0	3905	2339	37.49	0.26

The tests show the change in acoustic properties of limestone, hornfel and granite after radiation exposure (Table 2). In an initial core sample of limestone, velocities of P- and S-waves are, 5021 and 5021 and 2657 m/s, respectively, these parameters after absorption of the does of 15 kGy are 4485 and 1982 m/s. The P- to S-wave ratio is on average 2.14 under atmospheric conditions. This ratio in limestone core sample is 2.2–2.3 after treatment by accelerated electrons and 1.89 before treatment. In case of hornfel tests, it is observed that the ratio of P- and S-waves reduces when the dose of radiation is increased.

Regarding granite (Table 2), it is seen that with an increase in the absorption dose to 15 kGy, velocities of P- and S-waves first decrease and then increase. P- and S-wave velocities are 37478 and 2287 m/s in the initial core sample of granite, 3543 and 2142 m/s after absorption of the dose of 5 kGy and 3905 and 2339 m/s after exposure to the radiation dose of 15 kGy.

It follows from the data in Tables 1 and 2 that the change in the deformation and strength characteristics of granite is connected with the modification of its structure after treatment by accelerated electrons. The same inference is drawn after examining specimens in the figure. Uniaxial compression of granite specimen treated by accelerated electrons results in more particles as compared with the untreated specimen. In case of limestone, two particles are obtained in failure of untreated specimen and from 6 to 10 particles in disintegration of the treated specimen. Thus, the integrated analysis of deformability and strength characteristics of core specimens of different rocks show that treatment of the specimens by the high-energy electrons changes mechanical, dynamic and acoustic characteristics of rocks: uniaxial compression strength, elasticity modulus, P- and S-wave velocities.

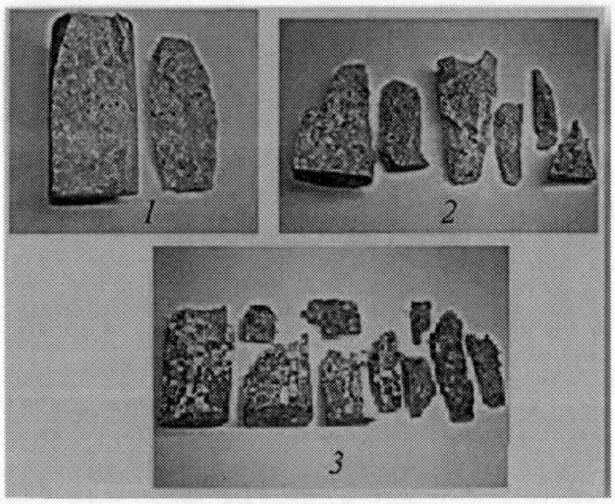

Figure 1. Core specimens of granite after destruction: *1*—original specimen; 2 and *3*—specimens treated by accelerated electrons under doses of 5 and 15 kGy, respectively.

The behavior of limestone, hornfel and granite in failure after treatment by accelerated electrons is connected with the specific nature of interaction between electrons and solids [8, 17]. A beam of electrons generates point defects in minerals in short time of $\sim 10^{-11}$ s: equal number of vacancies and dislocated atoms appear during this period. The equilibrium is achieved by formation of constant clusters or disappear of unstable damages under diffusion transfer. Then, the vacancies reach peak concentrations and merge into caverns that become places of discontinuities in solid material. As a result, displacement of dislocations is simplified and failure of minerals is intensified.

3. Conclusions

Finally, mechanical properties of minerals are changed under treatment of accelerated electrons. Caverns, electric breakdown and accumulation of dislocation along the boundaries result in the micro-cracking and mechanical instability. The revealed weakening of different rock types under influence of accelerated electrons can be used to reduce power consumption in ore pre-treatment and to minimize loss of commercial components in the subsequent processing circuits.

Acknowledgements
This study was accomplished in the framework of R&D No. AAAA-A17-117092750073-6.

References

[1] Chanturia VA and Kozlov AP 2017 Contemporary issues of comprehensive processing of refractory ores and technogenic materials *Plaksin's Readings–2017: International Conference Proceedings* Krasnoyarsk: SFU pp 3–6 (in Russian)
[2] Chanturia VA and Malyarov PV 2012 A review of worldwide achievements and development prospects of the disintegration technology in mineral processing *Plaksin's Readings–2012: International Conference Proceedings* Petrozavodsk: KNTs RAN pp 3–10 (in Russian)
[3] Revnivtsev VI, Gaponov GV, Zarogatsky LP et al 1988 *Selectove Disintergation of Minerals* Moscow: Nedra (in Russian)
[4] Vaisberg LA and Zarogatskii LP 2003 Foundations of optimal mineral disintegration Journal of Mining Science Vol 39 No 1 pp 87–93
[5] Kotova OB, Ozhogina EG and Rogozhin AA 2017 Applications of mineralogy in the refractory mineral processing technology *Plaksin's Readings–2017: International Conference Proceedings* Krasnoyarsk: SFU pp 10–13 (in Russian)
[6] Pirogov BI 2017 Nature and evolution of the process properties of minerals *Plaksin's Readings–2017: International Conference Proceedings* Krasnoyarsk: SFU pp 28–31 (in Russian)
[7] Chanturia VA and Bunin IZh 2007 Non-traditional high-energy processes for disintegration and exposure of finely disseminated mineral complexes *Journal of Mining Science* Vol 43 No 3 pp 311–330
[8] Bochkarev GR et al 1997 Prospects of electron accelerators used for realizing effective low-cost technologies of mineral processing *Proceedinds XX International Mineral Processing Congress* Aachen: Clausthal-Zellerfeld Vol 1 pp 231–243
[9] Kondrat'ev SA et al 2014 Justification and development of innovative technologies for integrated processing of complex ore and mine waste *Journal of Mining Science* Vol 50 No 5 pp 959–973
[10] Chanturia VA, Trubetskoy KN, Viktorov SD and Bunin IZh 2006 *Nanoparticles in Processes of Disintegration and Disclosure of Earth Materials* Moscow: IPKON RAN (in Russian)
[11] Plaksin IN., Shafeev RS, Chanturia VA and Yakushkin VP 1970 On the effect of ionizing radiation on the flotation properties of minerals *Concentration of Minerals: Selected Works* IN Plaksin (Ed)Moscow: Nauka pp 292–300 (in Russian)
[12] Revnivtsev VI 1977 Research and development problems in the improvement of the ore preparation technology *Obog. Rud* No 6
[13] Chanturia VA 2008 Sustainable development prospects of the mineral processing industry in Russia *Advanced Technology for Comprehensive Mineral Processing* VA Chanturia (Ed) Moscow: Ruda Metally (in Russian)
[14] RF State Standard GOST 28985-91 *Rocks. Methods for Determining Strain Properties under Uniaxial Compression*
[15] RF State Standard GOST 21152.2-84 *Rocks. Uniaxial Compressive Strength Measurement Methods*
[16] RF State Standard GOST 21153.7-75 *Rocks. Measurement Method for Elastic P- and S-Wave Propagation Velocity*
[17] Kovalev AT 1999 Possibility of applying radioactive electrization for electrical separation of pulverized mineral mixture *Journal of Mining Science* Vol 35 No 2 pp 199–203

Effect of compressive stresses on permeability of coal: Experimental and modeling estimates

LA Rybalkin, TV Shilova* and AN Drobchik

Chinakal Institute of Mining, Siberian Branch, Russian Academy of Sciences, Novosibirsk, Russia

E-mail:*shilovatanya@yandex.ru

Abstract. Research findings on permeability of fractured coking fat coal from Tikhov Mine are presented. It is shown that rock pressure has considerable effect on coal permeability under various gas pressure gradients. In particular it is found that an increase in the triaxial compression from 1 to 8 MPa reduces coal permeability by 8–1 times if the gas pressure gradient is 0.67–1.34 MP/m and by 6.1–7 times if the higher gradient is equal to 2–3.3 MPa/m. Parameters of cleats and sizes of joints (blocks) in coal matrix are studied before and after loading of coal specimens. It is obtained that the size of the coal matrix blocks in the unloaded state is 530 μm at the width (opening) of cleats of 12.7 μm, which decreases by 1.8 times under the triaxial compression equal to 8 MPa. The measured parameters of cleats and coal matrix blocks are used to assess permeability of coal using theoretical models relating coal permeability with porosity or effective stress (compression) of coal. The comparison of the theoretical estimates and experimental studies of fractured coal shows that the measured permeability of coal under compression from 1 to 5 MPa agrees with the estimates by the permeability–porosity model. If the compression is higher than 5 MPa the estimates by different models coincide. It is found that the highest disagreement between the measured and calculated permeabilities takes place at the low gradient of gas pressure which is probably connected with the Kinkenberg effect.

1. Introduction

Permeability of coal is an important characteristic to be taken into account in underground mine planning, design and operation, in particular, in gas drainage. According to general views, methane flows in natural cleavage system [1–3]. The rate of gas flow depends on properties of cleats, namely, their number, width (opening), interconnectivity, length in the flow direction, etc. [4, 5]. The known theoretical models for coal permeability assessment are based on dependence of gas flow rate on either porosity or effective compression of rocks [6, 7].

This paper presents lab test results on permeability and parameters of cleats in fat coal in Tikhov Mine and analyzes the measured and calculated permeabilities using the known theoretical models [8–10].

2. Equipment

Gas permeability of coal was determined using a test facility designed in the Laboratory of Physical Methods of Impact on the Rock Mass at the Institute of Mining, SB RAS [11]. The facility provides the opportunity to study filtration of gas in cylindrical specimens of low-permeable rocks under separately adjustable axial and lateral pressures equal to 30 MPa. The facility set involves a measurement system for automatic long-term testing by a preset program.

Coal jointing analysis was carried out on instrumentation and software system Mineral S7, including optical microscope OLYMPUS BX51, video camera SIMAGIS 2P-3C, personal computer and specialized software. Additionally, ultraviolet radiation equipment composed of an upright frame platform with fixed LED UV emitters [12].

3. Preparation of specimens

The tests were carried out on coking fat coal sample at 200 m level in Tikhov Mine (Kuzbass). Permeability was tested in cylindrical specimens with diameter D = 30 mm and height L = 30–60 mm. For the cleavage analysis polished sections were made in a way so that their surfaces disclosed microstructure of coal in cross-section of the specimens. The polished sections were finished using abrasive fine diamond powder with particles less than 0.05 μm [13]. Before the tests the prepared specimens were saturated with luminophore in low vacuum. Application of such adsorptive substance on specimens provides the opportunity to find defects of internal volume. The capacity of luminophore EpoDye powder to dissolve at a ratio of 1:40 was used. Then, the specimens were drying at a room temperature in a dark place for 24 h. The cleavage tests involved 5 specimens. A detailed description of the specimen preparation procedure is given in [12].

4. Experimental procedure

The tests on coal permeability were carried out using nitrogen filtration in axial direction of cylindrical specimen under continuous pressure difference (ΔP) at the specimen faces. The pressure difference varied in the tests from 0.01 to 0.1 MPa at an increment of 0.01–0.02 MPa. Triaxial compression P was varied from 1 to 8 MP at a step of 2 MPa. A series of tests was performed for each value of P with different ΔP. The permeability was calculated using the procedure from [14].

Cleavage of coal was examined in a reflected light. After each loading cycle by the pressure P, coal microstructure was analyzed, including determination of the number and width (opening) of cleats, size of blocks in coal matrix and angle between the systems of cleats. The tests were carried out using 5X and 10X lens systems. For each cleat and coal matrix block not less than 100 measurements of width were taken along a cleat. Not less than 5 measurements of angles of cleats were made. The obtained data was processed statistically with determination of average values for the model analysis.

5. Discussion of results

The experimental relationship of the gas permeability coefficient and pressure difference ΔP under different compression P of coal specimens is demonstrated in Figure 1. Review of literature sources shows that the obtained values are typical of heavily jointed coal [15].

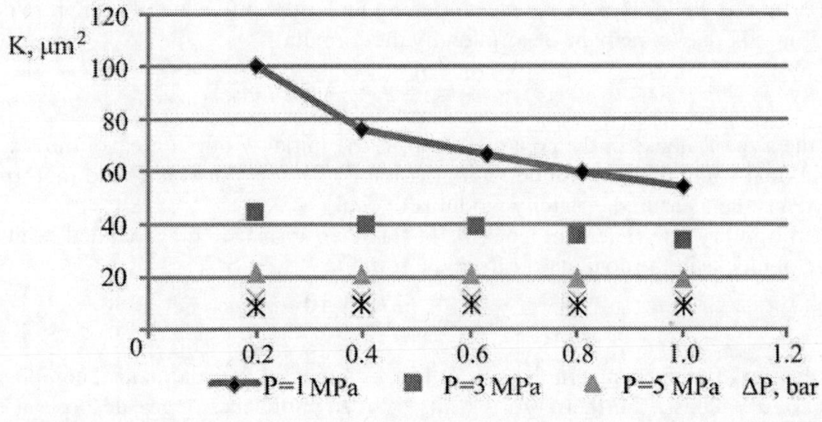

Figure 1. Gas permeability coefficient K_g, versus pressure difference ΔP of nitrogen in coal specimens under triaxial compression by P = 1, 3, 5, 7 and 8 MPa.

According to [16–18], bituminous coal structure is described by a system of cleats, including face cleats and butt cleats which can be perpendicular to each other and to the bedding plane (Figure 2). The similar structure was revealed in the specimens of fractured fat coal in the tests (Figure 3).

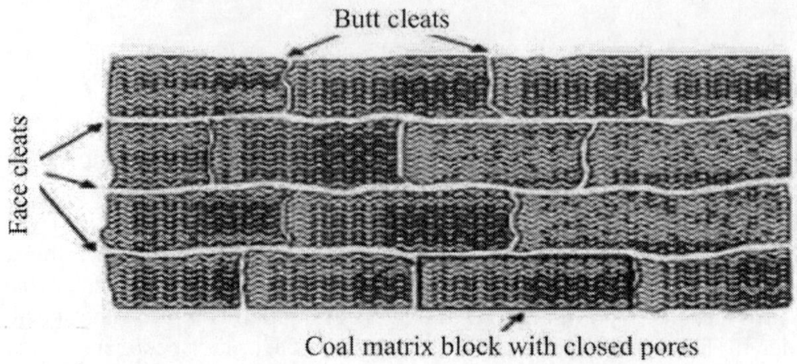

Figure 2. Sketch of typical structure of crevassed coal [19].

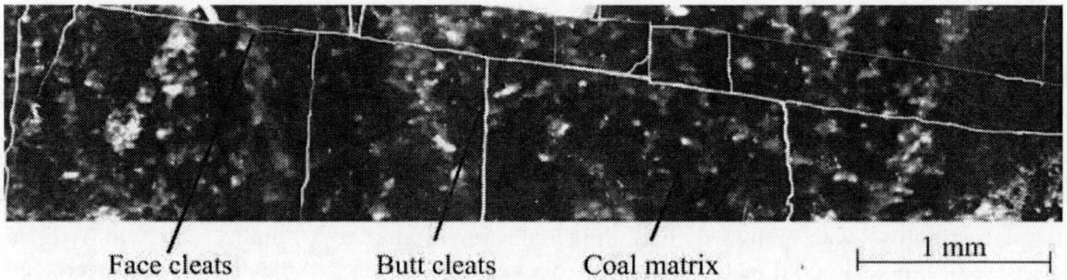

Figure 3. Cleavage in fat coal (level 200 m, Tikhov Mine): cleats are colored white.

Using the instrumentation and software system Mineral S7, the microscopic studies into parameters of cleats and coal matrix blocks were performed under atmospheric pressure (without load) before and after loading of a specimen in a test cell. The average size of blocks in the coal matrix was 530 μm, and the width (opening) of cleats was 12.7 μm. Average opening of cleats reduces by 1.8 times (from 12 to 6.5 μm) as the compression P is increased by 8 times (from 1 to 8 MPa). The average angle between both type cleats was 101°.

The obtained characteristics were used to calculate coal permeability by the Levine model [8]. The model estimates the change in the cleat opening under the influence of compression of rocks and shrinkage of matrix due to methane desorption by the formula

$$\frac{b_2}{a} = \frac{b_1}{a} + \frac{1-2\nu}{E}\left(P - P_0\right), \tag{1}$$

where a is the size of blocks in the coal matrix; b_1 is the initial width of cleats (under compression by P_0); b_2 is the final width of cleats (under compression by P); ν is Poisson's ratio (approx 0.3 for coal); E = 2500 MPa is the measured elasticity modulus of coal.

The permeability coefficient of coal, k, is found in terms of the obtained width of cleat and average size of blocks in the coal matrix from the formula

$$k = \frac{1.013 \cdot 10^9 \cdot b_2^3}{12 \cdot a}. \tag{2}$$

According to [8], the change in the size of blocks owing to the coal matrix compressibility can be neglected (i.e. a = const). Furthermore, we disregarded influence of gas desorption on opening of cleats, which is insignificant in nitrogen.

In the other known model [9, 10], the effect of compression on permeability of coal is assessed by the formula

$$k = k_0 \cdot e^{-3C_f (P - P_0)}, \tag{3}$$

where k_0 is the permeability coefficient under initial compression P_0, μm^2; C_f is the compressibility of cleats, MPa^{-1}; P is the compression of coal, MPa. Compressibility of cleats was assessed using he procedure described in [20]. The measured value of the initial coal permeability k_0 at the atmospheric pressure made 69.9 μm^2 (mD).

The relationship of the measured permeability coefficients and triaxial compression P = 1–8 MPa in the specimens of fat coal is depicted in Figure 4. It is found that gas permeability lowers with increasing P, and the value of the reduction is higher for the smaller gradients of the gas pressure ΔP/L. For instance, at the pressure difference ΔP = 0.02–0.04 MPa (ΔP/L = 0.67–1.34 MPa/m), under the increase in P from 1 to 8 MPa, the permeability drops by 8–11 times, while at the pressure difference ΔP = 0.06–0.1 MPPa (gas pressure gradient 2–3.3 MPa/m), the drop in the permeability makes 6.1–7 times. Alongside the experimental data, Figure 4 shows the permeability coefficients calculated by the formulas (1), (2) of the model [8] (dashed line) and formula (3) of the model [10] (solid line). The statistical processing of the measurements and comparison with the calculated results yields that under the compression P = 1–5 MPa, the Levine model estimate is closer to the measurements [8]. Under the pressure P more than 5 MPa, the calculated results of both models are identical. The most essential disagreement between the experiment and calculation is observed at small gradient of gas pressure (at ΔP = 0.2 MPa). We assume that this is connected with the known phenomenon of gas slippage at the grain interfaces (Klinkenberg effect) promoting an increase in gas permeability of rocks under low reservoir pressure [21].

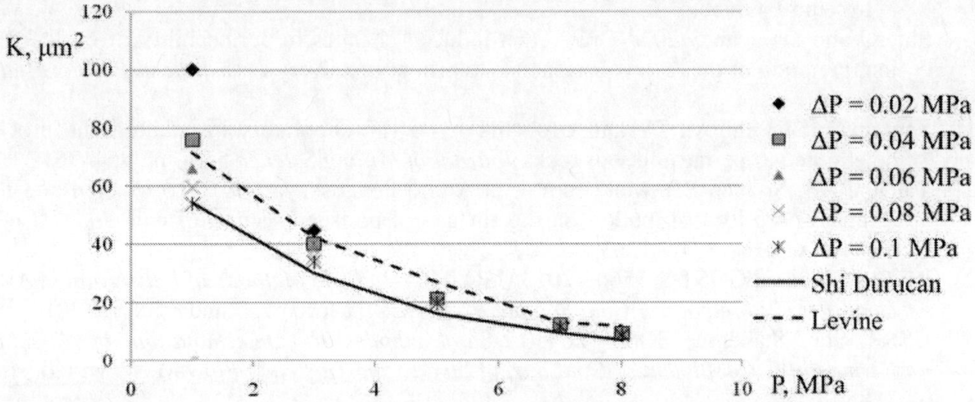

Figure 4. Relationship of gas permeability coefficient K_g and pressure P of coal: points—experiment; solid line—model [9, 10]; dashed line—model [8].

6. Conclusion

Permeability of fractured fat coking coal under increasing triaxial compression to 8 MPa drops from 69.9 μm^2 by a factor of 6–11. This drop depends on the pressure gradient of gas and intensifies as it lowers. For instance, permeability drops by 8–11 times if the gas pressure gradient is 0.67–1.34 MPa/m and by6.1–7 times at the gas pressure gradient is 2–3.3 MPa/m.

The average opening of cleats in the test unloaded fat coal specimens is 12.7 μm, the size of blocks in the coal matrix is 530 μm. These parameters can be used to assess permeability of actual coal seam by the Levine model [8] under rock pressure of 2–3 MPa and more. Under lesser compression and smaller gas pressure gradients (0.67 MPa/m and lower) as well as the average reservoir pressure, the theoretical calculations yield underestimated values of gas permeability in fractured coal (by 30% and more).

The obtained results can be used in prediction of gas permeability in fractured coal under compressive stresses conformable with reservoir conditions.

References

[1] Purl R, Evanoff JC and Brugler ML 1999 Measurement of coal cleat porosity and relative permeability characteristics *SPE Gas Technology Symposium* Society of Petroleum Engineers

[2] Pan Z, Connell LD, Camilleri M and Connelly L 2010 Effects of matrix moisture on gas diffusion and flow in coal *Fuel* 89(11) pp 3207–3217

[3] Lu M and Connell LD 2007 A dual-porosity model for gas reservoir flow incorporating adsorption behavior. Part I: Theoretical development and asymptotic analyses *Transport in Porous Media* 68(2) pp 153–173

[4] Somerton WH, Söylemezolu IM and Dudley RC 1975 Effect of stress on permeability of coal *International Journal of Rock Mechanics and Mining Sciences &Geomechanics Abstracts* Vol 12 No 5-6 pp 129–145

[5] Laubach SE, Marrett RA, Olson JE and Scott AR 1998 Characteristics and origins of coal cleat: A review *International Journal of Coal Geology* 35(1-4) pp 175–207

[6] Pan Z and Connell LD 2012 Modeling permeability for coal reservoirs: A review of analytical models and testing data *International Journal of Coal Geology* 92 pp 1–44

[7] Karmanskii AT, Ilinov MD and Gizatulina IN 2013 Laboratory tests of permeability of porosity of coal with regard to phase–physical and stress state *Zapi. Gorn. Inst.* Vol 205

[8] Levine JR 1996 Model study of the influence of matrix shrinkage on absolute permeability of coal bed reservoirs Geological Society London *Special Publications* 109(1) pp 197–212

[9] Seidle JP, Jeansonne MW and Erickson DJ 1992 Application of matchstick geometry to stress dependent permeability in coals *SPE Rocky Mountain Regional Meeting* Society of Petroleum Engineers

[10] Shi JQ and Durucan S 2004 Drawdown induced changes in permeability of coalbeds: A new interpretation of the reservoir response to primary recovery *Transport in Porous Media* 56(1) pp 1–16

[11] Serdyukov SV, Shilova TV and Drobchik AN 2017 Laboratory installation and procedure to determine gas permeability of rocks *Journal of Mining Science* 53(5) pp 954–961

[12] Tanaino AS, Sivolap BB, Maksimovsky EA and Persidskaya OA 2016 Method and means to estimate porosity distribution on the surface of polished section of coal *Journal of Mining Science* 52(6) pp 1216–1223

[13] RF Stat Standard GOST R 55663-2013 (ISO 7404-2:2009) *Methods of Petrographic Analysis of Coal. Part 2: Methods of Coal Sample Preparation* Moscow: Standartinform 2014

[14] USSR Stat Standard GOST 26450.2-85 *Methods for Determination of Absolute Gas Permeability Coefficient of Coal during Steady and Unsteady Flow* Moscow: Izd. standartov 1985

[15] Sander R, Pan Z and Connell LD 2017 Laboratory measurement of low permeability unconventional gas reservoir rocks: A review of experimental methods *Journal of Natural Gas Science and Engineering* 37 pp 248–279

[16] Close JC 1993 *Natural Fractures in Coal* Chapter 5

[17] Nelson CR 2000 Effects of geologic variables on cleat porosity trends in coalbed gas reservoirs *SPE/CERI Gas Technology Symposium* Society of Petroleum Engineers

[18] Pattison CI, Fielding CR, McWatters RH and Hamilton LH 1996 Nature and origin of fractures in Permian coals from the Bowen Basin, Queensland, Australia Geological Society London *Special Publications* 109(1) pp 133–150

[19] Harpalani S 1999 Compressibility of coal and its impact on gas production from coalbed reservoirs *Rock Mechanics for Industry* Amadei, Dranz, Scott, Smeallie (Eds) American Rock Mechanics Association: Vail, Colorado pp 301–308

[20] Pan Z, Connell LD and Camilleri M 2010 Laboratory characterization of coal reservoir permeability for primary and enhanced coalbed methane recovery *International Journal of Coal Geology* 82(3-4) pp 252–261

[21] Klinkenberg LJ 1941 The permeability of porous media to liquids and gases *Drilling and Production Practice* American Petroleum Institute

Investigation into operation dynamics of a downhole eccentric-type vibration source

AV Savchenko*, MN Tsupov, and DS Evstigneev

Chinakal Institute of Mining, Siberian Branch, Russian Academy of Sciences, Novosibirsk, Russia

E-mail: sav@eml.ru

Abstract. Wave-type oscillations generated by a downhole unbalanced seismic source are studied on the laboratory test stand. The mode of source operation with an inner chamber filled with a fluid is investigated at shaft rotation frequency ranged from 3 to 8 Hz. The difference of rotation frequencies of vibration source and magnetic field of asynchronous motor stator is determined. The amplitude-frequency characteristic of the vibration source in the low-frequency range is obtained.

1. Introduction

The vibration– and wave-type processes are used to rectify and to enhance oil yield from production active reservoirs [1–9]. Surface-mounted vibration sources with controllable frequency and effect embracing a whole oil reservoir are efficiently applied to generate elastic oscillations [1–5]. Generated seismic oscillations passing through a rock mass thickness are recorded by downhole geophysical devices mounted at the active reservoir level. When processing a seismic signal, the carrying frequency of a vibrosource and a response of a rock mass to its effect are singled out. After tuning in terms of resonance frequency of an active reservoir the vibration source generates a continuous flow of seismic energy for a long term, up to three months in some cases [1–2] Deviation from the resonance frequency of the active reservoir aggravates the wave effect efficiency with no notable increase in oil production [5, 6]. Tuning of the vibrosource in terms of resonance frequency of the reservoir takes much time and is considered an expensive prior operation, as it implies demounting and pulling out equipment from a number of bore wells.

The downhole immerse vibration sources are used to increase density of the seismic energy flow supplied to the active reservoir. Tuning of carrier frequency of an emitted signal in terms of resonance frequency of the oil bed is not practiced for impulse generators [7 – 9].

The eccentric-type sources are capable to adjust carrier frequency of an emitted signal and allow its compliance with resonance frequency of the production reservoir by increasing an oil yield flow to the productive well. A The tuning of emitted signal of the source is performed by adjusting frequency of rotation of electrodrive shaft which actuates an accentric mass executes. Downhole vibration sources work in a fluid occurring in a hole; such fluids in different deposits can differ in density and viscosity. In high-viscosity oil deposits the downhole accentric vibrosource is subjected to substantial drag force, acting on rotating eccentric mass. Tuning of a vibrosource relative to resonance frequency of an oil bed is realized by a generator drive being an asynchronous electric drive. Increase in load on drive shaft should result in reduction in rotation frequency. In other words, instead of operation at expectant resonance frequency of oil reservoir the vibration source would operate at lower frequency and this fact can negatively affect the oil production rate per this oil well. The present paper aims at determination of discrepancy between rotation frequency of eccentric vibrosource operation in a fluid and the prescribed frequency of a magnetic field of the starter of the asynchronous electric drive. To gain these objective requires to study dynamics of low-frequency operation of the vibrosource.

2. Description of the test stand and experimental procedure

The test stand to study dynamics of a downhole vibration source consists of a metallic case with an eccentric mass mounted on it and actuated by asynchronous electric drive, the water supply and drain system, seismic oscillation recorder (Figure 1).

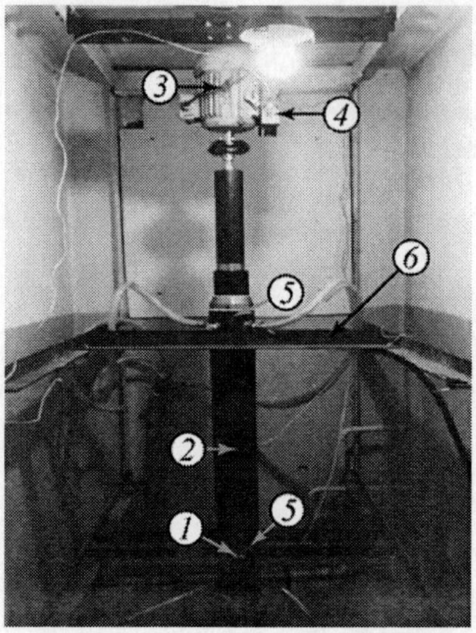

Figure 1. Scheme of eccentril mass location in the test stand: *1*—seismic sensor; *2*—external case of the vibration source; *3*—asynchronous electric drive; *4*—digital temperature gague; *5*—bearing units; *6*—metallic case of the stand.

The eccentric vibrosource is mounted vertically on the test stand. The rotation moment is transmitted from asynchronous electric drive *3* through shaft and flexible coupling. Seismic sensor GMT-12.5 *1* mounted at the level of bearing units *5* records frequency of magnetic field generated by source 2 under GOST ISO 10816-1-97 [10]. The analogue seismic signal is received by measurement transmitter HANDYSCOPE HS4 DIFF and then its digital form passes to a personal computer. The heating of electric drive and bearing blocks is controlled by the monitoring system consisting of digital temperature sensors, microcontrollers and a computer. Before the test the internal chamber of the vibration source is filled with water. A series of measurements is fulfilled at frequencies of magnetic field of electrodrive starter: 3, 4, 5.1, 6.2, 7.2, 8.2 Hz.

3. Procedure for processing of experimental vibrovelocity signal data

Vibrovelocity signal transmitted from analogue-digital converter to a personal computer was digitally processed to single out main harmonics with respective carrier frequency at which an eccentric device operates. Chebyshev's narrow-band filter with the final impulse characteristic was used [11–13]. In Figure 2 the amplitude-frequency spectra of the initial and filtered signals for 5 Hz carrier frequency are reported.

Fields of electric-drive starter are preset by frequency converter ABB IP20 / UL Open type.

After preliminary filtration of signals obtained at different frequencies their amplitude- frequency spectra were plotted on cumulative graph in view to determine a frequency difference Δf_i, corresponding to *i*-th carrier frequency of the vibration source (Figure 3).

IOP Conf. Series: Earth and Environmental Science **262** (2019) 012060 doi:10.1088/1755-1315/262/1/012060

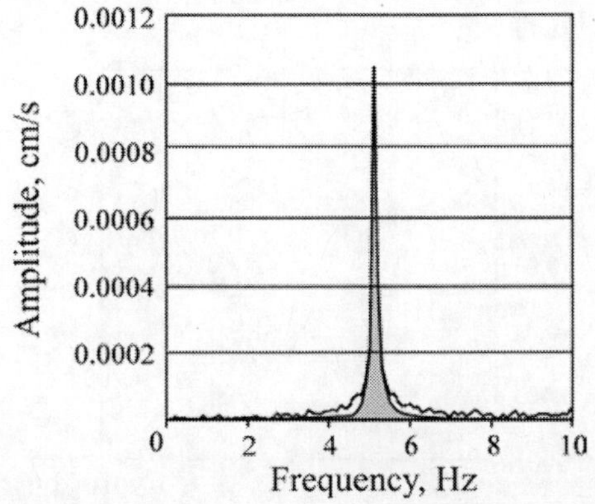

Figure 2. Amplitude and frequency distribution in initial signal of vibrovelocity (unshadowed) and after application of narrow-band filter (shadowed).

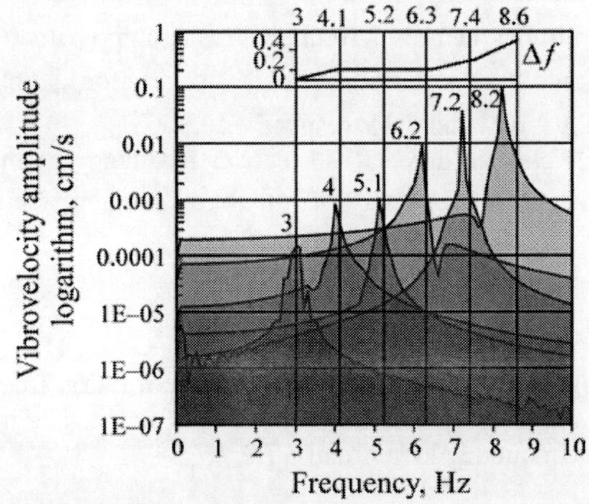

Figure 3. Summary graph of amplitude-frequency spectra of seismic signals (solid lines with filling) and respective rotation frequencies of the magnetic field starter (vertical straight lines).

The wave shape of a filtered signal was used to evaluate average amplitude of vibrovelocity at the preset frequency. In the tests the frequency range was selected within 3–8 Hz with step of 1Hz. The wave form of the filtered vibrovelocity signal with carrier frequency of 5.1 Hz is shown in Figure 4. The measurement data and follow-on processing of the signals were used to obtain amplitude-frequency characteristics of the eccentric source in the low-frequency range (Figure 5).

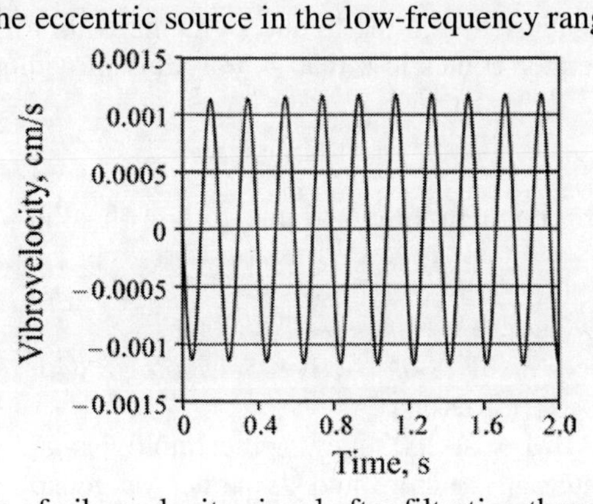

Figure 4. Wave form of vibrovelocity signal after filtration through narrow-band filter.

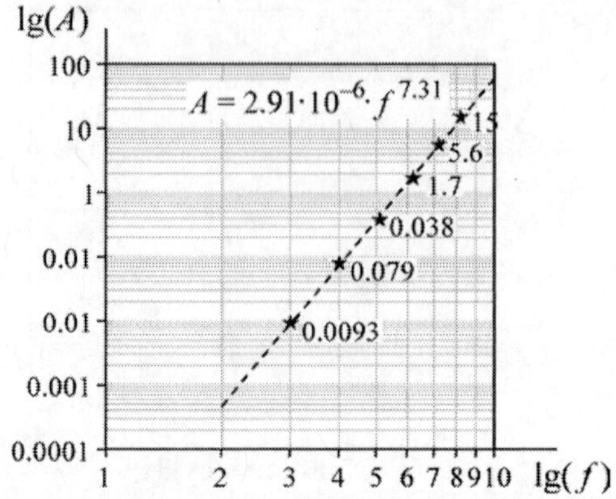

Figure 5. Amplitude-frequency characteristic of the eccentric source in the low-frequency range.

Discrete values of amplitudes at preset frequencies are approximated by power dependence of $A = kf^b$, where A is amplitude of vibrovelocity oscillations, cm/s; f is frequency, Hz; k, b are constants, equal to $2.91 \cdot 10^{-6}$ and 7.31, respectively.

Sliding of asynchronous electric drive at i-th carrier frequency of the source is determined as a ratio of frequency difference Δf_i to frequency of rotationof magnetic field starter f_{st_i}, expressed in [14–16]:

$$s_i = \frac{\Delta f_i}{f_{st_i}} 100\ \% .$$

The sliding is increasing with growth of eccentric source rotation frequency. The maximum value of sliding amounts to 4.7 % and corresponds to frequency of 8.2 Hz. The instrumental determination of sliding could not be made at frequency lower than 3 Hz

4. Conclusions

With growth of rotation frequency of a downhole vibration source the drag force of the fluid acting on an eccentric source and load on the drive tend to increase and, as a consequence, difference of frequencies of magnetic field of the starter and rotor of the asynchronous electric driver is increasing.

Calibration curve is plotted for frequency difference allowing tuning of vibrosource operation in a well filled with a fluid in terms of the resonance frequency of the oil reservoir; analytical dependence of the amplitude-frequency specification was used to compute amplitude of vibrovelocity, generated by the source.

It is established experimentally that in the frequency range within 3–8 Hz the maximum sliding of the electric drive of the generator equals less than 4.7 %, thus providing the reliable operation of the generator drive.

Acknowledgements

This work was carried out in the framework of the Basic Research Program, Project No 0321-2016-0005.

References

[1] Alekseev AS, Geza NI et al 2004 Active *Seismology with Powerful Vibrational Sources* Novosibirsk: SB RAS (in Russian)

[2] Alekseev AS, Glinsky BM et al 2002 New geotechnologies and complex geophysical methods for studying the internal structure and dynamics of geospheres *Vibration Geotechnology* Moscow: Regional public organization of scientists for problems of applied (in Russian)

[3] Vasil'ev VI, Evchatov GP et al 1976 Experimental studies of the process of excitation of seismic waves by a vibrational source *Questions of Excitation of Seismic Waves by a Vibrational Source* Novosibirsk: IGiG SB AS USSR pp 65–86 (in Russian)

[4] Sorokin VN 2004 Distribution of pressures under the radiating plate of a surface seismic source *Omsk. Nauch. Vestn.* No 1(26) pp 86–88

[5] Sorokin VN 2004 On the possibility of vibroprocessing oil deposits at several dominant frequencies *Neft. Prom.* No 11 pp 44–47

[6] Chichinin IS and Jushin VI 1977 Frequency method of vibroseismic research *Problems of Vibrational Transmission of The Earth* AV Nikolaeva and IN Galkina (Eds)Moscow: Nauka pp 14–31 (in Russian)

[7] Dyblenko VP 2008 *Wave Methods of Impact on Oil Reservoirs with Hard-to-Recover Reserves: Review and Classification* Moscow: VNIIOENG (in Russian)

[8] Kravtsov JaI and Marfin EA 2011 Wave influence on reservoir as a universal way of increasing the efficiency of extraction of heavy oils and natural bitumen *Georesursy* No 3(39) pp 17–18

[9] Marfin EA, Kravtsov JaI et al 2014 Field testing of the wave impact on the oil production process at Pervomaiskoye oilfield *Georesursy* No 2(57) pp 14–16

[10] GOST ISO 10816-1-97 2007 *Mechanical Vibration. Evaluation of Machine Vibration by Measurements on Non-Rotating Parts. Part 1: General Guidelines* Moscow: Standartinform (in Russian)

[11] Costa VL, Schettino HV et al 2017 Digital filters for clustered-OFDM-based PLC systems: Design and implementation *Digital Signal Processing* Vol 70 pp 166–177 DOI: doi.org/10.1016/j.dsp.2017.08.004

[12] Saramäki T, Mitra SK anfd Kaiser JF 1993 Finite impulse response filter design *Handbook for Digital Signal Processing* Vol 4 pp 155–277

[13] Mitra SK 2011 *Digital Signal Processing Computer-Based Approach* McGraw-Hill

[14] Boguslawsky, Korovkin N and Hayakawa M Large AC 2016 *Machines: Theory and Investigation Methods of Currents and Losses in Stator and Rotor Meshes Including Operation with Nonlinear Loads* Springer—Japan KK Tokyo DOI: 10.1007/978-4-431-56475-1

[15] Pollefliet J 2018 Power electronics: Drive technology and motion control, Elsevier Academic Press 412 pp DOI: doi.org/10.1016/C2017-0-00733-4

[16] Carravetta A, Houreh ShD, Ramos HM 2018 Pumps as turbines: Fundamentals and applications Springer Int Publ 218 pp DOI: 10.1007/978-3-319-67507-7

IOP Conf. Series: Earth and Environmental Science **262** (2019) 012061 doi:10.1088/1755-1315/262/1/012061

Numerical modeling of eccentric mass rotation in chamber filled with fluid

AV Savchenko*, DS Evstigneev and MN Tsupov

Chinakal Institute of Mining, Siberian Branch Russian, Academy of Sciences, Novosibirsk, Russia

E-mail: *lion_ltd@ngs.ru

Abstract. Numerical modeling of eccentric mass rotation in fluid under different water and oil pressures is performed. The relationship of extra-consumed motor power to overcome fluid resistance in downhole operation of screw pump at the maximum design depth of n is obtained. The power consumption of vibration sources is evaluated in the productive field of low- and high-viscosity oil in Mongolia and Russia.

1. Introduction

Borehole mining method in oil production uses sucker-rod, electric-centrifugal or screw pumps. The application depth of sucker-rod pumps is limited to 1400 m. Deeper level oil recovery involves electric-centrifugal or screw pumps. The highest efficiency in well cleanout and flow rate stimulation is achieved with the method of vibro-treatment of wellbottom zone using downhole vibration sources [1–5]. The unbalance vibration source for downhole operation at depths greater than 1400 m with any available types of pumps is proposed in [6, 7]. The source generates harmonic vibrations due to unbalance mess of load moving relative to the shaft rotation axis and transmits them directly to reservoir. The inner chamber of the vibration source is unsealed, and fluid, entering the chamber from the well, prevents rotation of eccentric mass, which increases the pump motor load. This study aims at finding power consumption of downhole unbalance vibration source in recovery of different-type oil at a depth down to 2500 m.

2. Modeling rotation of eccentric mass in fluid

The downhole vibration source consists of an internal chamber, shaft and eccentric mass, and is placed at the occurrence depth of pay zone. The internal chamber is unsealed and filled with fluid. The eccentric mass is rotated by the shaft; it is made of steel in the form of a semicylinder rotating clockwise at an angular rate $\vec{\omega}$ about the axis Oz (Figure 1). In numerical modeling, the computational domain encompassing the eccentric mass and the internal chamber, $\Omega = \Omega_1 \cup \Omega_2$, is divided into two sub-domains: Ω_1, in which the eccentric mass rotates, and the immobile sub-domain Ω_2. Such approach was for the first time proposed in [8] and later on was named the method of Multiple Reference Frame—MRF [9–13]. Equations of hydrodynamics with steady-state flows are solved using one of numerical methods. The net domain remains static, i.e., the net is unchanged during the calculation. In the sub-domain holding the rotating eccentric mass, the equations of fluid flow and continuity include the angular rate $\vec{\omega}$. In the immobile sub-domain, in the equation of motion, inertia forces are neglected and fluid mass transfer at the boundary of two domains is set as the equality of fluid pressures and normal

IOP Conf. Series: Earth and Environmental Science **262** (2019) 012061 doi:10.1088/1755-1315/262/1/012061

component of absolute velocity. At the boundary of the immobile sub-domain Ω_2 and inface of the chamber, sliding condition is set [14–16], and modeling of flows in the boundary layer at the hard surface of the eccentric mass in the sub-domain Ω_1 uses well functions [17].

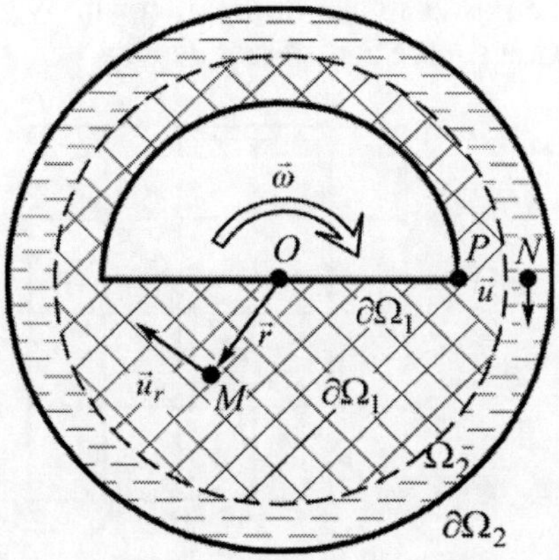

Figure 1. Arrangement of eccentric mass in fluid-filled chamber of vibration source: line [---] marks the boundary between Ω_1 [hatched] in which eccentric mass rotates and Ω_2 [dashed] at the chamber wall.

Rotation axis of the eccentric mass passes through the point *O*, middlelength section of the eccentric mass blade—*OP*; the the points *M* and *N*, fluid flow velocities are directed along the vectors $\vec{u}_r$ and $\vec{u}$, respectively, the points *O* and *M* are connected by the radius vector $\vec{r}$.

3. Modeling results of vibration source operation in water

Let us set water pressure in well in the range from 0.101325 to 25 MPa with an increment of 5 MPa and calculate change in density and dynamic viscosity versus depth. The results are compiled in the table below.

Table 1. Change in density and dynamic viscosity of water with increasing depth.

Pressure, MPa	Depth, m	Density, kg/m^3	Dynamic viscosity, 10^{-6} Pa·s
0.101325	0	998.2	1003
5	509.6	1000.6	1000
10	1016.7	1003.0	998.1
15	1521.4	1005.4	996.1
20	2023.7	1007.8	994.2
25	2523.6	1010.2	992.3

The power required for the motor to overcome the fluid resistance during operation of the unbalance vibration source at different depth is assessed is given by [14, 16]:

$$P = 2\pi f \Delta T_q , \tag{1}$$

where *f* is the rotation frequency of the eccentric mass, Hz; ΔT_q is the difference of the moments of forces; T_q is the moment of force, N, found by integrating pressure distribution on the middlelength section of the eccentric mass blade with a diameter *D* and a length *L* [16]:

$$T_q = \iint_S \vec{r} \times p(x, y) d\vec{\sigma} = L \int_0^{D/2} p(r)\, r dr . \tag{2}$$

The pressure distribution *p*(*p*) on the blade of the eccentric mass is determined from numerical solution of equations of hydrodynamics with rotating sub-domain. The input parameters for the dynamic model are taken from the table above. The rotation frequency of the eccentric mass is assumed as 5 Hz. The calculations in ANSYS Fluid Flow (Fluent) used the Reynolds-averaged Navier–Stokes equations with the standard two-parameter k–ε model of turbulence for steady-state flows [16, 17, 20, 21]. Figure 2 shows the relationship between the power intake of the motor in overcoming of water resistance during operation of the unabalance vibration source at different depths.

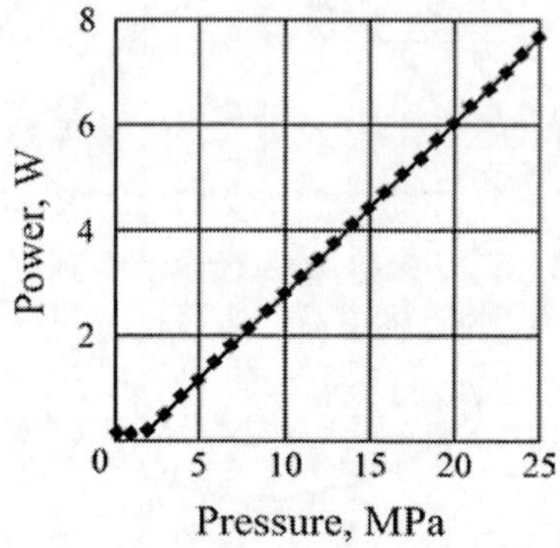

Figure 2. Power consumption as function of pressure in well at various depths (points show the calculated values, solid line is the linear interpolation).

Aiming to analyze influence of oil viscosity and density on motor load in rotation of the eccentric mass, two reservoirs of low-viscosity light and high-viscosity heavy oil are selected.

Verkh-Tarskoe field lies in the north of the Novosibirsk Region, 340 km away from Novosibirsk. The oil-bearing stratum U_1^1 with the average occurrence depth of 2455.6–2530.4 m contains light (density < 850 kg/m^3) low-viscosity ($260\cdot10^{-6}$–$780\cdot10^{-6}$ Pa)·oil with small content of sulfur 0.3% and paraffine 1.81% [22]. The operation of the unbalance vibration source at the occurrence depth of the producing stratum $Ю_1^1$ at the well pressure of 23.9 MPa will take 7.43 W to overcome head resistance of oil on the eccentric mass blade and 7.29 W in case that the well is filled with water.

Tsagaan-Els oilfield is located in the south-east of Mongolia, in the East Gobi basin. This is a sandwich-type reservoir represented by a number of producing blocks with tectonic and lithological partings. The main pay zone is from 1000 to 2500 m deep. Oil is highly paraffinic and contains not less than 11% by mass of solid hydrocarbons nand ot more than 0.2% by mass of sulfur. Well No. 1410 produces viscous ($178040\cdot10^{-6}$ Pa·s) heavy oil (density 921.2 kg/m^3) at a depth of 1170 m [23]. The power to be consumed to overcome head resistance of oil is 8.75 W and 3.24 W in the well filled with water.

4. Discussion

It follows from the relationship between the power consumption and well pressure in Figure 2 that there is almost no increment in the power intake down to the depth of 306 m as the dynamic viscosity of water changes insignificantly. Starting from the pressure of 3 MPa, the power consumption grows by the linear law with a slope of 0.327 and initial ordinate −0.527. At the depth of 2523.6 m, the consumed power is 7.7 W.

The linear power–pressure dependence follows from the functional connection of the power with the density and dynamic viscosity of fluid in which the eccentric mass rotates. In 1934 White and Brenner [24] obtained the expression for the power consumption:

$$P = N_p \rho \omega^3 D^5 , \tag{3}$$

where ρ is the fluid density, kg/m^3; ω is angular velocity, rad/s; D is the eccentric mass diameter, m; N_p is the dimensionless number of power, or the Newton number [14, 16, 20, 21]. The functional dependence between N_p, Reynolds number Re, Froude number Fr and geometrical dimensions of turbin and tank where rotation takes place is given in [25, 26]:

$$N_p = \frac{P}{\rho\omega^3 D^5} = k\left(\frac{\rho\omega D^2}{\mu}\right)^\alpha \left(\frac{\omega^2 D}{g}\right)^\beta f(l,w,T,c), \quad (4)$$

where k is the coefficient of proportionality; μ is the dynamic viscosity of fluid, Pa·s; α, β are the experimentally determined exponents; $f(l,w,T,c)$ is the functional connection between geometrical sizes of turbine and tank; l, w are the length and width of blade; T is the length of the tank; c is the clearance between the blade and the tank bottom.

Inasmuch as the density and dynamic viscosity of fluid changes with increasing pressure, at the same rotation frequency of eccentric mass, the number of power and the density will change by the law: $N_p(p) = 1.35 \cdot 10^{-3}p - 0.00214$, $\rho(p) = 0.48p + 998.2$. Placing $N_p(p)$ and $\rho(p)$ in (3) produces: $P(p) = 0.00016p^2 + 0.323p - 0.513$. Disregarding the quadratic term in the latter expression, we obtain the linear dependence: $P(p) = 0.323p - 0.513$.

In the producing well in the stratum U_1^1 in Verkh-Tarskoe oil field, the difference between the power intakes by the eccentric mass in fluid and water is 0.14 W. Such small difference is due to low viscosity of oil as compared with water and similarity of pressure distributions on the eccentric mass blade (Figure 3a).

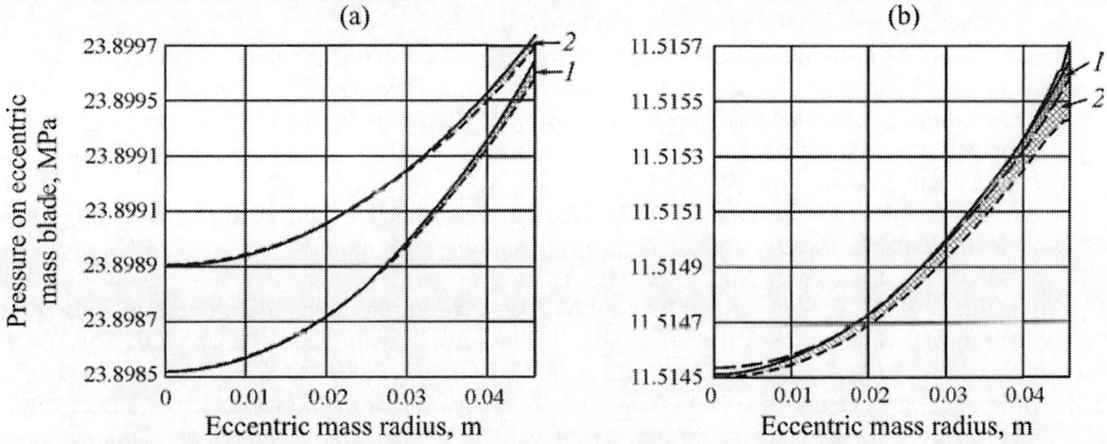

Figure 3. Distribution of pressure exerted by water (*1*) and oil (*2*) on the middlelegth section of eccentric mass blade: (a) Verkh-Tarskoe field oil (Novosibirsk Region, Russia)); (b) Tsagaan-Els field oil (East Gobi, Mongolia). Solid and dashed line show fluid pressure on the front and back sides of the blade, respectively.

For Tsagaan-Els field and well No. 1410, the difference in pressure distributions is higher for oil as against water (Figure 3b). Consequently, the power consumption is 2.7 times higher in oil than in water. The densities in water (1003.7 kg/m^3) and heavy oil (921.2 kg/m^3) differ insignificantly [23], and for this reason, the difference between values of pressure distribution on the eccentric mass blade ($r = 0$ m) is also small—29 Pa.

5. Conclusions

The power consumption required to overcome water resistance in operation of downhole unbalance vibration source grows linearly with depth.

The unbalance source can operate in fields of light, low-viscosity and heavy viscous oil. In terms of Verkh-Tarskoe field of light and low-viscous oil in the stratum U_1^1 at the average depth of occurrence ranging from 2455.6 to 2530.4 m, it is shown that overcoming of oil resistance in operation of downhole unbalance vibration source will consume 7.43 W additionally to 0.9 kW taken by the source. In Tsagaan-Els field of heavy viscous oil in productive stratum at a depth of 1170 m, 8.75 W of the motor net power will be consumed.

Acknowledgements
This work was carried out in the framework of the Basic Research Program, Project No 0321-2016-0005.

References

[1] Dyblenko VP 2008 *Methods of Wave Stimulation of Scavenger Oil Reservoir. Review and Classification* Moscow: VNIIOENMG (in Russian)
[2] Hua G, Falcone G, Teodoriu C and Morrison GL 2012 Comparison of multiphase pumping technologies for subsea and downhole applications *Oil and Gas Facilities* pp 36–46
[3] Snarev AI 2010 *Design of Machines and Equipment for Oil and Gas Production: Educational and Training Aid* Moscow: Infra-Inzheneria (in Russian)
[4] Kostrov S and Wooden W 2005 In seismic stimulation shows promise for revitalizing mature fields *Oil & Gas Journal* Vol 103 No 15 pp 43–49
[5] Westermark RV, Brett JF and Maloney DR 2002 Enhanced oil recovery with downhole vibration stimulation *SPE/DOE Thirteenth Symposium on Improved Oil Recovery* Tulsa, Oklahoma pp 1–8 DOI: doi.org/10.2118/67303-MS
[6] Savchenko AV, Stupin VP et al 2016 Development of borehole unbalance vibrosources and test benches for their research *Proceedings of Interexpo GEO-Siberia-2016: International Scientific Conference* Vol 4 pp 3–7 Novosibirsk: SSGA (in Russian)
[7] Tsupov MN and Savchenko AV 2016 Development of seismic vibration sources for rock mass treatment *Mashinovedenie* Vol 4 pp 62–67
[8] Luo JY, Issa RI and Gosman AD 1994 Prediction of impeller induced flows in mixing vessels using multiple frames of reference *European Conference on Mixing—IChemE Symp Series* UK: Univ. Cambridge No 136 pp 549–556
[9] Tonello N, Eude Y, Meux B and Ferrand M 2017 Frozen rotor and sliding mesh models applied to the 3D Simulation of the Francis-99 Tokke turbine with Code_Saturne IOP *Conf. Series: Journal of Physics: Conf. Series* 782 pp 1–12 DOI: 10.1088/1742-6596/782/1/012009
[10] Giorgi MG, Donateo T et al 2017 Numerical investigation of the performance of contra-rotating propellers for a remotely piloted aerial vehicle *Proc. 72nd Conf. of the Italian Thermal Machines Engineering Association* Lecce, Italy pp 1101–1018
[11] Kim Ch, Park J, Kim D and Baek J 2016 Numerical analysis on non-equilibrium steam condensing flow in rotating machinery *ASME 28th Symposium on Fluid Machinery* Washington pp 1–11 DOI: 10.1115/FEDSM2016-7588
[12] Lopes Jr GB, Gómez LC and Bock EGP 2016 Mesh independency analyses and grid density estimation for ventricular assist devices in Multiple Reference Frames simulations *Technische Mechanik* Vol 36 No 3 pp 190–198
[13] Stein P, Pfoster Ch et al 2015 Computational fluid dynamics modeling of low pressure steam turbine radial diffuser flow by using a novel multiple mixing plane based coupling—Simulation and validation J. *Eng. Gas Turbines Power* Vol 138 No 4 DOI:10.1115/1.4031388
[14] Gülich JF 2010 *Centrifugal Pumps* Springer-Verlag Berlin Heidelberg
[15] Rielly CD and Gimbum J 2019 Computational fluid mixing *Food Mixing. Principles and Applications* Chichester: Wiley Blackwell pp 125–174
[16] Kresta SM, Etchells III et al 2016 *Advances in Industrial Mixing: A Companion to the Handbook of Industrial Mixing* John Wiley & Sons
[17] Schlichting H and Gersten K 2017 *Boundary-Layer Theory* Springer-Verlag Berlin Heidelberg

[18] Charny IA 1951 *Unsteady Motion of Real Liquid in Pipes* Leningrad: Izd. Tekhniko-Tekhnich. Lit. (in Russian)
[19] Aleksandrov AA and Grigor'ev BA 1999 Handbook: Tables of thermophysical Properties of Water and Water Vapor Moscow: MEI (in Russian)
[20] Edward LP, Atiemo-Obeng VA and Kresta SM 2004 Handbook of Industrial Mixing: Science and Practice New Jersey: John Wiley & Sons
[21] Cengel YA and Cimbala JM *Fluid Mechanics* McGraw Hill Education
[22] Antonov DA 2002 Efficiency of the Use of Oil Gas in Verkh-Tarskoe Oil Field (Novosibirsk Region) Bachelor's Thesis (in Russian)
[23] Gehrehlmaa T 2010 Investigation of the processes of biodegradation of viscous oils in Mongolia for creating methods for increasing oil recovery and reclamation of oil-contaminated soils *Theses of Candidate of Chemical Sciences Dissertation* Tomsk (in Russian)
[24] White AM and Brenner E 1934 Studies in Agitation vs the correlation of power studies data *Trans. Ame. Ins. Chem. Eng.* Vol. 30 pp. 585–579
[25] Bates RL, Fondy PL and Corpstein RR 1963 Examination of some geometric parameters of impeller power *Industrial & Engineering Chemistry Process Design and Development* Vol 2 No 4 pp 310–314 DOI: 10.1021/i260008a011
[26] Chudacek MW 1985 Impeller power numbers and impeller flow numbers in profiled bottom tanks *Industrial & Engineering Chemistry Process Design and Development* Vol 24 No 3 pp 858–867 DOI: 10.1021/i200030a056V

IOP Conf. Series: Earth and Environmental Science **262** (2019) 012062 doi:10.1088/1755-1315/262/1/012062

Applicability of in-situ leaching and divided blasting combination

AG Sekisov[1]* and AYu Lavrov[2]**

[1]Mining Institute, Far East Branch, Russian Academy of Sciences, Khabarovsk, Russia
[2]Chita Division of the Chinakal Institute of Mining, Siberian Branch, Russian Academy of Sciences, Chita, Russia

E-mail: *sekisovag@mail.ru; **lavrov_2001@mail.ru

Abstract. The article gives the theoretical and experimental substantiation for application of in-situ leaching combined with divided blasting of ore blocks. Blasting also involves injection of hot steams of chemical solutions with active oxidizers in rocks. The authors discuss the designs of explosive charges with inserted shells of chemical agents in them for the divided blasting and injection as well as the compositions of agents for in-situ leaching of gold, silver, molybdenum and copper leaching from ore pillars and deep-level rebellious ore.

1. Introduction

Depletion of ore reserves at the closing stage of mining under resultant change in stress state of rocks [1] actually requires transiting mining systems with backfill and/or in-situ leaching widely applied in the uranium and copper production industries. In case of ore bearing such valuable components as gold, silver, molybdenum, wolfram, nonferrous metals and rare earths, the specified method, despite its evident advantages over conventional geotechnologies, is only used sporadically in the form of dull-scale experiments. The primary causes are the difficulty of oxidation of most ore types and selective extraction of valuable elements from them, high density as well as structural and textural features of ore bodies, which governs low permeability of broken rock blocks for leaching solutions. It is possible to improve efficiency of in-situ leaching and stimulate recovery of useful components from rebellious ore by increasing temperature of preparation and/or leaching solutions and by raising concentration of highly active oxidizing and complexing agents in them. The technical implementation of this approach involves special structure borehole charges, namely, thin-walled bulbs containing various chemical agents are arranged along the axis of an explosive charge [2]. During blasting, the chemical agents mix with explosion gases, and the provoked chemical reactions form a set of high-active oxygen-bearing compounds, including ion-radical and radical forms (Figure 1). Generated under heating of the solutions by blasting, steams bearing hydrated active ion-radical and radical forms of oxidizes are pushed by residual pressure into opening micro-cracks.

Bottom-up divided blasting by sections, in particular, Vertical Crater Retreat method VCR widely used in Canada [3, 4] is known to produce the most uniform fragmentation with the dominant yield of particle size –20 mm by the limited length charges. By placing thin-wall plastic shells with chemical agents in the center of such explosive charge, it is possible to ensure quasi-spherical propagation of blast waves as well as oxidizing solutions and steams. In such conditions, the ore in the blasted section will be uniformly broken and soaked with condensed steams of chemical agents injected in the blast-generated

IOP Conf. Series: Earth and Environmental Science **262** (2019) 012062 doi:10.1088/1755-1315/262/1/012062

micro-cracks and pores. Such blasting treatment enlarges the ore–solution contact area as well as increases concentration gradient of solutions and dissolved metals in the system of film and pore water–leaching solution flows, which accordingly stimulates diffusion processes of mass exchange.

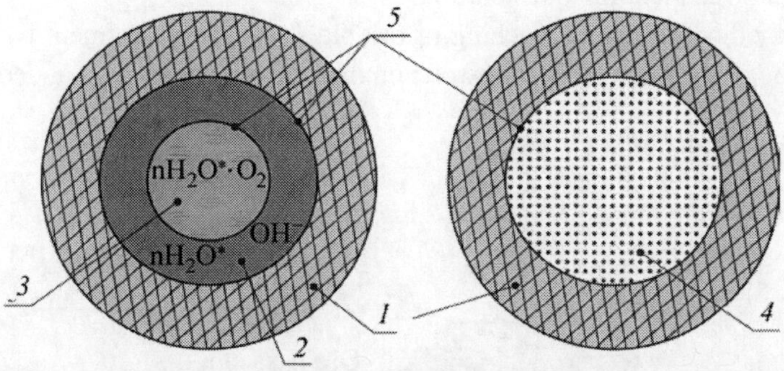

Figure 1. Injection-type charge structure: *1*—explosive; *2*—solution 1 (catolyte); *3*—solution 2 (anolyte); *4*—explosive with ceno-spheres; *5*—plastic shell.

Under divided blasting, a thin layer of ore is broken, or bottomhole zone of long vertical holes drilled in an ore block. The unblasted parts of such holes can be used to feed activation solutions (water-and-gas emulsions) to pretreat rock mass. Such solutions of active hydrated oxidizers and/or complexing agents are prepared using a special photoelectrochemical reactors equipped with a system of electrodes and an ultraviolet source. The hydrated compounds with active oxygen accelerate oxidation of iron, copper and sulfur with the concentration of the resultant sulphates in film water.

Then, the next upper-lying section of the hole is loaded with the same structure explosive charges, and the technology cycle is repeated until the blasting front approaches the block roof. After that, all broken ore soaked with the preparatory solutions is cured, which results in the oxidation of a certain portion of iron and sulfur as well as the primary leaching of copper (in the mode of diffusion). The main stage of in-situ leaching and metal extraction from broken ore by solutions of chemical agents is better to be carried out by steps. First, an ore block is fed with preparatory alkaline–peroxide, sulfuric acid–peroxide, hypochlorite and peroxide–carbonate solutions with active oxidizers (dependent on mineralogical composition of ore and enclosing rocks, as well as on physicochemical properties of commercial-value metals). For instance, the injection-and-blasting treatment of skarn ore with high carbonate modulus can use oxygen-saturated solutions and alkaline photoelectrically activated solutions which will mix with hot explosion gases and form hydrated peroxide compounds. In the injection-and-blasting treatment of gold, copper and complex ore before leaching, for the conversion of sulphide and sulphosalt minerals to sulphate form, or for the oxidation of carbon-bearing and organic shots, the most effective agents are sulfuric acid and group of peroxide compounds of hydrogen and oxygen, including hydric dioxide and residual hydroxyls. Thermobaric activation of a mixture of sulfuric acid and a solution of peroxide–hydroxyl compounds, due to ionization of hydrogen, creates a highly active plasma-like medium containing both hydrated ions and radical-ion water clusters generally represented by the compounds of the type of ($H_2O{\cdot}HO^- (H^+{\cdot}H^-){\cdot}O^*{\cdot}2H_3O^+$), as well as hydrated active, e.g. sulfate, groups:

$$HO^*{\cdot}mH_2O{\cdot}H_2O_2{\cdot}H_2SO_4 = (m - n - 3)H_2O_2{\cdot}HO^- (H^+{\cdot}H^-)O^*{\cdot}2H_3O^*{\cdot}nH_3O{\cdot}SO_4^{2-*}.$$

After the preparatory treatment of ore by active oxidizing solutions before leaching, broken ore is subjected to dynamic penetration–infiltration, infiltration and seepage of working solution containing complexing and oxidizing agents for the valuable ore components. Removal of dissolved and leached elements can be carried out using drainage raises or orepasses. At this step of in-situ leaching, metal recovery from ore uses mostly chloride solutions with active oxygen-bearing compounds synthesized in the photoelectrochemical reactor. In particular, the best fitted method for gold recovery is a chloride–peroxide system (cyanides are applicable in mines after major preparation and expensive environmental protection, thiosulphates and thiocarbamides decompose under intense thermobaric effect). In-situ leaching of molybdenum from quartz–molybdenum ore is advisable to include cyclic treatment of ore by sodium hypochlorite solution and peroxide–carbonate solution resultant from photoelectrochemical treatment of soda solution.

The proposed method on in-situ leaching of ore blocks with pretreatment by the divided injection-and-blasting is applicable in the additional extraction of ore from pillars, and in deep-level mining.

2. An example of in-situ leaching application

An application object for the discussed method of activated leaching may be the Udokan cupriferous sandstone deposit.

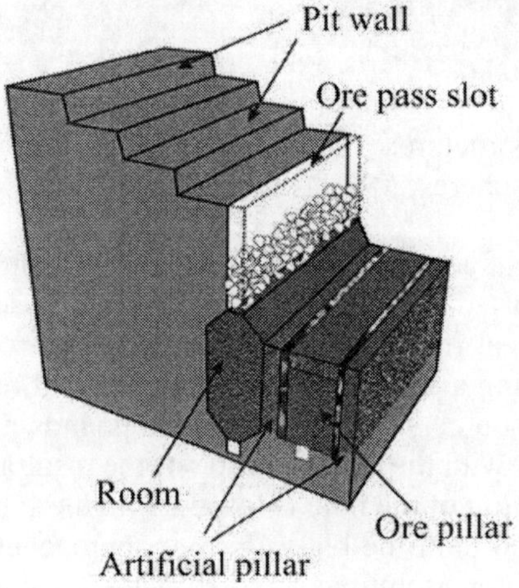

Figure 2. Schematic of the proposed combination in-situ leaching method.

For deep levels of Udokan operations, it is proposed to use a combination of the conventional underground mining system with backfill and in-situ leaching in blocks (Figure 1). Rich ore (copper content of 15–20 %, the portion in the total reserves is 5–7%) is extracted from pillars of the rebellious crude ore mining by cut-and-fill, and metals are later on leached after treatment by the injection-and-blasting method. for safe mining in the seismically active Udokan deposit and to eliminate leakage of leaching solution in faults and fractures, it is expedient to use backfill of two types: cemented paste backfill along the boundary of rooms and "dry" backfill composed of run-of-mine complex ore and sulphide ore from open pit mines in the center of rooms. The main component of the cemented backfill should be mill tailings of rich sulphide ore. The tailings will be sulfating (for 5–10 years), i.e. it will become possible to recover copper and associates from the backfill, which means wasteless mining. Ore pillars in stoping in rich chalcosine–bornite ore are composed of mineralized sandstone with dominating sulphide minerals of finely impregnated bornite or (much more seldom) chalcosine and chalcopyrite. This implies economic inexpediency flotation. For this reason, after copper and silver leaching from disintegrated mineralized "dry" backfill, such ore pillars can be subjected to in-situ leaching and chemical treatment. Copper can be leached from disintegrated ore pillars in in the pulse static mode (as per Professor Lunev's classification), which means alternating cycles of ore soaking with active sulfuric acid–peroxide solution, curing, discharge of pregnant solution and feed of re-activated solutions to the upper part of blocks for the next cycle of treatment until the solutions contain the required concentration of copper. During deeper level mining in the Udokan deposit, leaching of metals can grasp overall ore body rather than backfill and pillars.

The experiments on leaching of copper, silver and gold from the Udokan sulphide ore (bornite–chalcosine paragenesis) by active solutions produced in the lab-scale photoelectrochemical reactor involved coarse samples (average size of 30 cm) 10.5 kg in weight. A sample contained 17.82% of copper, 140.5 g/t of silver and 0.28 g/t of gold. First, crushed ore sample was fed with sulfuric acid solutions with active forms of oxygen synthesized by combining the processes of pre-aeration by bubbles, stage-wise electrolysis and photolysis of the initial aqueous solutions of sulfuric acid. Once a day 250 l batches of sulfuric acid–peroxide solutions were fed in the sample until this modeled in-situ leaching block reached a solid : liquid ratio of 10 : 1, i.e., the total volume of active sulfuric acid solution in the sample reached 1 l. Then, within 6 days, the solution was periodically (once a day) discharged, re-saturated with activated oxygen (bubble aeration after ultraviolet radiation) and again fed in the sample. Upon completing the "sulfuric acid" cycle of leaching, the primary pregnant solution was discharged and the liquid sample was sent for the analysis. According to the analysis outcome, recovery of copper and silver in pregnant solution made 26.7 and 3%, respectively, in 10 days. No gold was observed in the solution. Considerable content of micron native silver and cerargyrite in the cement of cupriferous sandstone, as well as in chalcosine and bornite (alongside with dispersed gold) predetermined the second-stage leaching with active solutions of sulfuric acid and nitric acid and the third-stage leaching with chloride groups. The chloride groups, the basis of which is hydrated meta-stable hypochloric acid HClO, were prepared on the photoelectrochemical reactor and used at the closing stage of dispersed gold and copper leaching. Finally, the recovery of copper was 82.6%, which is comparable with the performance of the conventional circuit of mining–flotation–hydrometallurgical processing of flotation concentrate; silver—61.3% and gold—64.3%. Furthermore, more valuable metals can be additionally extracted after secondary treatment by the injection-and-blasting.

3. Conclusions
In the rockbursting conditions of deep-level mining in the Darasun and Teremki deposits composed of gold–sulphide–quartz loads, efficient modes and parameters were experimentally determined for in-situ leaching of gold with the injection-and-blasting pretreatment of ore blocks. The chemical preparation process was modeled as blasting with injections of peroxide–chloride solutions in the cemented mill tailings using propellant charges. After such preparation, the block was subjected to leaching by active chloride–peroxide solutions in the pulse static mode. The final gold recovery totaled 80% in three days.

References
[1] Oparin VN, Sashurin AD, Kulakov GI et al 2008 *Modern Geodynamics in the Top Lithosphere* Novosibirsk: SO RAN (in Russian)
[2] Sekisov AG, Shevchenko YuS and Lavrov AYu 2012 Explosion injection approach to ore preparation to leaching *Proc All-Russian Conf: Fundamental Problems of Geoenvironment Formation under Industrial Impact* Novosibirsk: IGD SO RAN Vol 1 pp 283–287 (in Russian)
[3] Crocker CS 1979 Vertical crater retreat mining at the Centennial Mine *Mining Congress Journal* No 6
[4] Shevchenko YuS, Manzyrev DV, Popova GYu and Milukina AI Progressive technologies of mineral mass preparation to leaching *GIAB* SI No 10 pp 14–20

Specifics of disseminated gold leaching from oxidizable and non-oxidizable mineral matrices by activated water-and-gas emulsions

AG Sekisov*, AV Rasskazova and TG Konareva*****

Institute of Mining, Far East Branch, Russian Academy of Sciences, Khabarovsk, Russia

E-mail: *sekisovag@mail.ru, **abbot87@mail.ru, ***tat_konareva@mail.ru

Abstract. The article gives theoretical and experimental substantiation of features of finely dispersed gold leaching from oxidizable and non-oxidizable mineral matrices by water-and-gas emulsions which are photo- and electro-chemically activated. Alternative compositions of agents used to prepare rebellious ore for leaching and, specifically, for gold leaching are considered.

1. Introduction

Dispersed and encapsulated gold used to accumulate in ores: either siliceous and alumosilicate matrices or sulfide, sulfoarsenide and sulfosalt mineral matrices [1]. Gold nanoparticles encapsulated in sulfides in mineral ores is an important argument to survey deposits of commercial significance. Nano-size gold is also associated with secondary supergenic concentrated zones, it participates in formation of surface geochemical anomalies and gold-bearing concentrated occurrences. Comprehensive investigation into forms of nano-size gold in an ore makes the ground to revise the rating of some known deposits in view to expand their commercial potential and to update ore processing techniques, as well as to explore new deposits by using the latest prospecting techniques [2].

The recovery of nano-size gold implies, firstly, necessity of preliminary super-fine ore grinding down to 20–30 μm and less [3, 4], secondly, ore processing and oxidation of flotation concentrates in autoclaves or bioreactors [5–7]. Combination of superfine grinding and pre-oxidation also gives good results, but at substantial costs [8]. The pretreatment of rebellious ores is reviewed in details in [9]. The reported techniques provide a rather high gold recovery in follow-on adsorption cyanidation, but high capital and operating expenditures do not allow their application to process poor ore or mining waste.

2. Features of disperse-type gold leaching

The most prospective direction in search for solution to this problem can be the application of compounds containing active forms of oxygen as basic oxidizers in working solutions at an ore-preparation stage before leaching of productive mineral matrices bearing dispersed gold [10, 11]. Such compounds provide high-rate chemical reactions at relatively low temperature of the solution, possibility to oxidize not only respective mineral-forming elements, but dispersed gold included in crystalline lattice of mineral-carriers [12].

The active oxidizers can be produced by radiation of air or chemically pure diatomic oxygen with UV-spectrum light or by electrodischarge treatment with follow-on bubble aeration of the pulp, but the process is accompanied with appreciable gold loss because of recombination of gold particles prior to transition from the gas phase to aqueous one.

The use of integrated electrochemical [13] and photochemical synthesis of oxidizers [14] and chelating complexes directly in concentrated leaching solutions, viz., preparation of reactive-active water-and-gas emulsions to treat the gold-bearing mineral mass at low consumption (L : S = 1 : 10 – 1 : 12) would enable to overcome the available complexities.

It is known that such high-active oxidizers as ozone, atomic oxygen, and hydroxyl-radical can form diatomic oxygen and water vapor under action of hard ultra-violet radiation in the range of electromagnetic wave length of 180–250 nm. In electrolysis of a solution of readily dissociating alkali and oxygen-bearing acids the bubbles of oxygen with water content evolve onto an anode. In the subsequent photochemical reactions water molecules in the volume of evolved bubbles are excited as a result of adsorption of UV radiation and electrolytic diatomic oxygen and decompose into active atoms and radicals or are subjected to ionization. The decomposition products interact with other excited molecules and produce secondary active radicals, ions, ion-radicals and strong molecular oxidizers:

$$O_2{}^{-UV} - 2O^*\ O_2 + O^* = O_3,\ \ H_2O^{-UV} - H_2O^+ + e^-,\ \ H_2\,O_2{}^{-UV} - H^+ + HO_2{}^-,\ \ H_2O^+ + HO_2{}^- = 3OH^* \quad (1)$$

Under coalescence the bubbles of oxygen and hydrogen evolved onto cathode are involved into mutual diffusion, the UV-radiation of the resultant water-and-gas emulsion provokes more intensive yield of active oxygen and hydrogen compounds. As bubbles of electrolytical gases are surrounded with water, then ozone, atomic oxygen, hydroxyl-radical and other active compounds formed in photochemical reactions tend to diffuse into water in the film and can manage to form active hydrate complexes prior of recombination. Thus photoelectrochemical synthesis allows a high yield of such strong oxidizers as $\mathbf{H_2O_2}$, $\mathbf{OH^*}$ along with their hydrated forms in the solution. Using proper initial soluble substances and the integrated electrochemical and photochemical treatment of soda and chloride-sodium solutions, it is possible to synthesize reactive-active oxygen and hydrogen compounds with other elements: carbon and chlore: $\mathbf{C_2O_2^*}$, $\mathbf{C_2O_4^*}$, **HClO***, if required.

Treatment of rather simple and cheap initial reagent solutions by the new technique makes it possible to produce water-and-gas emulsions, containing both high-active oxidizers and chelating agents. Hydroxyl-radicals, synthesized under photoelectrochemical treatment of the solution, are capable to concentrate in water the film and small-cracks and also to get rid of hydrate membrane on the internal boundary of the double electric layer, to diffuse into internal layers of crystalline lattice of ore minerals. In this connection, the chemical oxidation reactions with participation of hydroxyl-radicals can proceed not only in near-surface laters of mineral matrices, but also inside of them. The redox potential of hydroxyl-radicals amounts to 2800 mW [15], this makes it possible to use them to oxidize iron and sulfur, as well as clustered atoms of dispersed gold and platinoids, therefore, transition of them into ionic form and growth of diffusion activity in the volume of the crystalline lattice of minerals. In these processes other active components in solutions resulted from photoelectrochemical treatment under electrodissociation of water or acids can play an important role. They are cations of hydrogen – protons H^+ and metastable hydrogen $\mathbf{(H^-)}$ anions, capable to differentiate periodically under dissociation from hydroxyl-ions $\mathbf{OH^-}$ or hydroxyl-ion-radicals $\mathbf{OH^{*-}}$, $\mathbf{(OH^- = O + H^-,\ OH^{*-} = O^* + H^-)}$. Hydroxyl-ion-radicals are formed in photoactivated hydrolysis of hydroxyl-radicals: $\mathbf{OH^* + nH_2O = H^+ + OH^{*-} + OH(n-1)H_2O}$. As the points of crystalline lattice of mineral-carriers acquires impulses from diffusing hydrogen ions and hydroxyl ions and displaces periodically at distance exceeding the equilibrium level, then there is a high probability for precious metal ions to tunnel through interstitial lattice site.

Hydroxyl radicals are capable to form hydrate metastable complexes of **Au*OH*OSiO** type with atoms of silicium and aluminium associated directly in crystalline mineral lattice, with atoms of dispersed gold, thus gold gets an opportunity to realize intercrystalline migration as a component of compound **AuOH.** “Internal” dissociation $\mathbf{OH^* = (O^{*-} {*} H^+)}$ can proceed periodically in hydroxyl-radical. Hydrogen atom in hydroxyl-radical periodically displaces to a nominal boundary of electron shell and becomes a free proton-particle of a rather high energy. Respectively, its location is not limited exclusively with electron shell of oxygen ion-radical associated to it. So, system ($\mathbf{O^{*-}*H^+}$) can be considered as a meta-stable ion-radical pair $\mathbf{O^{*-}*p^+}$, capable to function on the one hand as proton acceptor and on the other hand as electron acceptor. Initial proton p^+, detached from radical OH*, penetrates into electron shell of the neighboring atom of other element, gold, in particular. This can initiate appearance of mobile metastable **AuH**-type compounds. Ion pairs $\mathbf{H^+ \cdot H^-}$ induce re-dislocation of impurity atoms in crystalline lattice and transit impulses to them with formation new defects in crystalline lattice and form metastable compounds $\mathbf{Au_2H_2}$, thus breaking gold bonds with silicium and oxygen It is well known that even in natural water there are stable gold form $\mathbf{AuOH(H_2O)^0}$ in a wide range Eh – pH. Under certain conditions in the mineral medium – technological solution system the gold declustering by active hydrate complexes is possible with dislocation of atomic (ionic) gold in the composition of compounds AuOH* and/or AuH* from internal layers of crystalline lattice to surface of microcracks and pores of a mineral-concentrator. Later metastable gold compounds react with respective chelating components of the active water-and-gas emulsion with formation of more stable gold complexes directly or through formation of its secondary hydrated clusters. The active reagent medium, containing ion-radical clusters $\mathbf{(H^+ \cdot nH_2O \cdot HO^{*-})}$, associated with hydrated cyanide complexes with base $\mathbf{[(Na^+)OH^-H^+(CN^-)]}$ intensifies leaching of gold accumulated nearby free internal and external surfaces of crystalline lattices:

$$Au_n + 2[H^+ nH_2O^* 2OH^{*-} (Na^+)OH^- H^+(CN^-)]$$

$$Au_{n-2} + [Au_2(CN)_2 (OH)_2]^- 2mH_2O\ H^+ + 2[NaOH(n - m + 1)H_2O]. \quad (2)$$

Thus, water-and-gas emulsions prepared in photoelectrochemical reactors on the basis of alkali and soda solutions can be used to leach ores containing conventionally non-oxidizable mineral-concentrators for dispersed gold.

In leaching the preparation of ores, bearing dispersed gold with high content of sulfide minerals with sulfide sulfur more than 5 % and arsenopyrite as well as 0.1–1 % content of carbonaceous matter cab be based on the application of water-and-gas emulsion, containing containing sulfuric-peroxide as well as chloride compounds along with hydroxyl complexes and percarbonic acids. Sulfide-carbonic gold-bearing ores contain adsorption-active organogenic carbon and residual organic compounds, leading to re-precipitation of gold dissolved in leaching process. These ores are distinguished for really low gold recovery parameters, as a rule, no more than 20 %. Moreover, in this-type ores gold can be present in bielement Au – S, Au – S – C, and Au – C clusters, resisting to cyanidation. The use of water-and-gas emulsions, containing active oxygen, to prepare rebellious ores for leaching makes it possible to apply the percolation-leaching process.

In this case interaction of $(O^{*-}*H^+)$ pair with carbon can result in the process: $C + (O^{*-}*H^+) = CO + H^*$. Oxidation of sulfur and carbon in sulfuric-peroxide medium proceeds at the expense of their interaction with metastable permonosulfuric and persulfuric acids $H_2S_2O_8$, H_2SO_5, which can be obtained by treating the initial solution in electrochemical and photochemical reactors.

In treatment of rebellious ores with water-and-gas emulsions the carbon in ore is oxidized by metastable permonosulfuric acid to monoxide in reaction:

$$H_2SO_5 + C = H_2SO_4 + CO \quad (3)$$

Next carbon monoxide is further oxidized by hydrogen peroxide or hydroxyl-radical:

$$CO + H_2O_2 = CO_2 + H_2O. \quad (4)$$

Dispersed gold with a carbon bond preserved after its oxidation can interact with hydroxyl-radical with formation of metastable complexes of Au – Au (OH, HCO_3) type.

Figure 1. Laboratory bench for percolation tests of ores.

In geotechnological evaluation of mineral materials, bearing dispersed goldmainly associated with sulfide sulfur and carbon the investigation was undertaken on gold-bearing carbonaceous-clayey carbonated ore of Southern Ural origin. The pre-oxidation of the ore was performed with pelletization of 5 mm ground fraction batches with cement, followed with irrigation with photoelectroactivated solutions with sulfuric-peroxide and carbonate-peroxide base. The pelletized batches were mixed and fed to percolation columns (figure). This integrated ore preparation scheme provided an appreciable increase in gold recoverywith subsequent percolation-leaching by active cyanide-hydoxyde solutions at 52% gold recovery into the production solution as compared to the routine gold recovery level of 18%.

3. Conclusions

The investigation into leaching of the study-type ore batches is in progress with active hydroxochloride solutions, subjected to integrated oxidation (direct activated hydrochloration). Hydroxochloride solutions obtained in electrophotochemical treatment of the initial sodium-chloride solutions provided additional oxidation of carbonaceous matter and suppression of its sorption activity as well as gold recovery at 70% level.

References

[1] Starostin NV and Ignatov PA 2006 *Geology of Ore Deposits* Moscow: Akadem. Proekt (in Russian)

[2] Hough RM, Noble RRP and Reich M. 2011 Natural gold nanoparticles *Ore Geology Reviews* Vol 42 No 1 pp 55–61

[3] Birol O and Ucurum M 2016 Investigation of the cyanide leaching optimization for ultrafine grinding gold-silver ore *Particulate Science and Technology* Vol 34 No 5 pp 633–638

[4] Gonzalez-Anaya JA, Nava-Alonso F and Pecina-Trevino ET 2011 Gold recovery optimization of a refractory concentrate by ultrafine grinding—A laboratory study *Minerals & Metallurgical Processing* Vol 28 No 2 pp 94–101

[5] Gertenbach DD 2016 Scaleup of pressure oxidation processes *Minerals & Metallurgical Processing* Vol 33 No 4 pp 178–186

[6] Ubaldini S, Veglio F and Abbruzzese C 2000 Combined bio-hydrometallurgical process for gold recovery from refractory stibnite *Minerals Engineering* Vol 13 No 14–15 pp 1641–1646

[7] Balaz P, Kupka D and Bastl Z 1996 Combined chemical and bacterial leaching of ultrafine ground chalcopyrite *Hydrometallurgy* Vol 42 No 2 pp 237–244

[8] Wu Z, Huang Zh and Ruan R 2013 Effect of temperature on column bioleaching of a refractory gold ore *20th International Biohydrometallurgy Symposium (IBS 2013, Antofagasta Chile* Vol 825 pp 352

[9] Nan XY, Cai X and Kong J 2014 Pretreatment Process on Refractory Gold Ores with As, *ISIJ International* Vol 54 no 3 pp 543–547.

[10] Pat. USA 4.752.412 1987 Method for recovery of precious metals using Ozone Van Antverp Ph Lincoln; 20.07.87.

[11] Pat. 5.942.098 USA. 1996 International Method of treatment of water and method and composition for recovery of precious metal Sekisov Artuor Paronyan Aromais Kouzin Vladimir Lalabekyan Natella 12.04.96.

[12] Sekisov AG, Rubzov YI and Lavrov AY 2016 Activation of heap leaching of low-sulfide ores the invisible gold *Journal of Mining Institute* Vol 217 No 4 pp 96–101

[13] Karlsson RKB 2016 Selectivity between oxygen and chlorine evolution in the chlor-alkali and chlorate processes *Cornell Chemical Reviews* Vol 116 No 5 pp 2982–3028

[14] Rahimi S, Ayati B and Rezaee A 2016 Kinetic modeling and determination role of sono/photo nanocatalyst-generated radical species on degradation of hydroquinone in aqueous solution *Environmental Science and Pollution Research* Vol 23 No 12 pp 12185–12198

[15] Zhelovitskaya AV, Yermolaeva EA and Dresvyannikov AF 2008 Oxidation of organic compounds by means of hydroxide-radikal generated in solutions chemical and electrochemical methods *Vestn. Kazan. Tekh. Univer.* No 6 pp 211–229

Room-and-pillar mining of thick coal seams in the conditions of high gas dynamic hazard in Kuzbass

VV Sementsov[1]* **and SA Prokopenko**[1,2]**

[1]VostNII Science Center, Kemerovo, Russia
[2]Tomsk Polytechnic University—TPU, Tomsk, Russia

E-mail: *v.sementsov@nc-vostnii.ru; **sibgp@mail.ru

Abstract. Room-and-pillar coal mining is effective in the areas unsuitable for fully mechanized longwalling. This mining method has advantages and disadvantages. The application of the room-and-pillar technique to mining of thick coal seams in the conditions of high gas dynamic hazard is face \d with such problems as impossibility of full-height extraction; heading without preliminary prediction of rock bursts and outbursts, or without early prevention of floor fracturing, etc. The ways to solve such problems in terms of coal mines in Kuzbass could be useful for mines in other regions.

1. Introduction

By expert estimates, to 30–40% of coal reserves in Kuzbass occur in subsoil areas unsuitable for fully mechanized longwalling. Such areas of coal seams are limited in size to 400–600 m, have irregular shapes and contain geological discontinuities. It is economically inefficient to use the system of longwall mining in such areas due to huge amount of assembling, low labor productivity and high cost of coal.

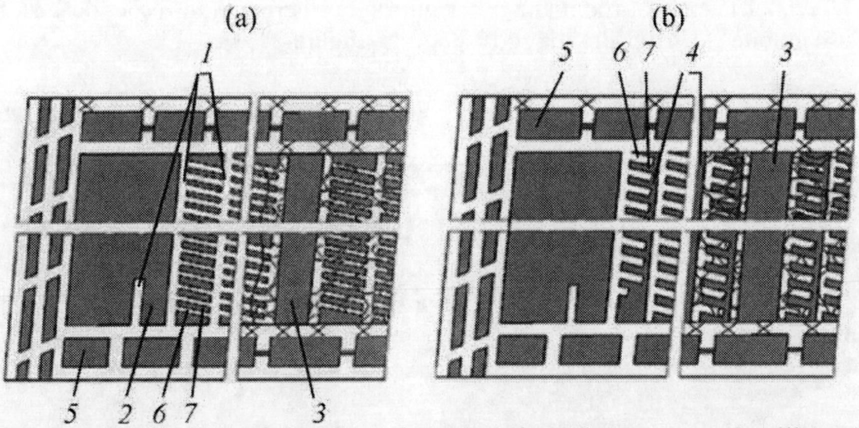

Figure 1. Room and pillar mining flow charts: (a) through entry ways in pillars; (b) blind entry ways in pillars; *1*—room; *2*—extraction panel; *3*—safety pillar; *4*—rib pillars; *5*—protective pillars; *6*—entry way; *7*—pillars protecting miners and machines from roof fall.

In such conditions in flat-dipping and inclined seams, it is possible to employ shortwall systems, in particular, room and pillar mining (RPM) [1]. In this technology, coal is first extracted from rooms separated by pillars. Heading is followed by coal loading to self-propelling cars for hauling to conveying lines. Roof support method is rock bolting. After heading and extraction of two–three rooms, coal is extracted from rib pillars (Figure 1) [2, 3].

Advantages of RPM include:

—mineability of difficult-to-extract reserves, increment in marketable coal output and extension of mine life;

—next-level efficiency of subsoil management;

—capital cost saving due to elimination of purchase of expensive heading equipment.

Disadvantages of RPM are:

—relatively high loss of coal without extraction panels;

—complex control of behavior of pillars and risks of roof falls in rooms;

—purchase of roof support equipment and large amount of roof support activities;

—measure to combat inflammation of coal remaining in mined-out voids.

Application of RPM method in Kunetsk Coal Basin is restrained by the:

—absence of legislation, regulations and procedures for shorwall mine planning and design;

—prohibition of use of RPM in rockburst-hazardous coal mining until recently;

—difficult ground conditions of coal seams (steep dip angles, high gas and water content; high jointing of enclosing rocks; weak roof rocks, structure of coal deposits as strata series);

—unsuitability of available machines for RPM.

2. Room and pillar mining projects in Russia

The countries with the mature coal industry (USA, Australia, South Africa, etc.) widely use room and pillar method of coal mining [4, 5]. This choice is facilitated by the more favorable ground conditions as against Kuzbass and by dynamic support of operations by science and technology. Structurally uniform enclosing rocks of high quality reduce the risk of sudden roof falls in rooms and save cost of roof support. Dedicated machinery opens wide prospects for mechanization of room and pillar mining [6, 7].

Rooms are extracted by heading machines equipped with coal cutter jigs. From a heading machine to a conveyor line, coal is hauled in shuttling self-propelling cars. Roof support is carried out using special high-productive roof bolters [8]. Such equipment is not manufactured in Russia (except for the cars), which affects efficiency of room and pillar mines. Furthermore, in recent decade, foreign mines increasingly use mobile electro-hydraulic roof supports (Figure 2) [9].

(a) (b)

Figure 2. Mobile Roof Support MRS by USA with extended height of (a) 1.6 m and (b) 5 m.

The mobile roof supports eliminate the major disadvantage of RPM—high loss of coal. The mobile roof supports are placed under caving line in a room [10]. After heading machine cuts an entry way in the pillar, the support is advanced behind the machine along the room and protects potential caving zone. Coal reserves are extracted from pillars by sequential mining of entry ways with roof support in the rooms. Such mobile roof supports enable nearly complete extraction of coal from rib pillars. Coal extraction ratio in a panel has risen to 90–95%. This is comparable with the longwall system performance, which improves prospects of shortwalling.

Earlier in Kuzbass, RPM was used in Usinskaya, Sibirginskaya and Raspadskaya coal mines [11]. At the present time, Lenin and Raspadskaya-Koksovaya mines take advantage of this technology. Raspadskaya-Koksovaya mine employs RPM in experimental extraction of coal in gently dipping seam III with a dip angle of 1–12 ° thickness of 10 m. The seam is categorized as prone to rockbusting, gas outbursting and spontaneous combustion [12].

The RPM flow chart is planned and designed with scientific support provided by the VostNII and VNIMI research institutions. Parameters of RPM are determined by numerical modeling of geomechanic behavior of coal and rocks [13–15]. The technology is subjected to commercial-scale trial with heavy scientific work load on evaluation of safety of shortwalling [3, 16, 17].

The application of RPM in Raspadskaya-Koksovaya mine deals with the problems described below.

1. It is required to cut the seam by layers 3.6 m thick stage-wise undercutting of rooms (Figure 3) [18].

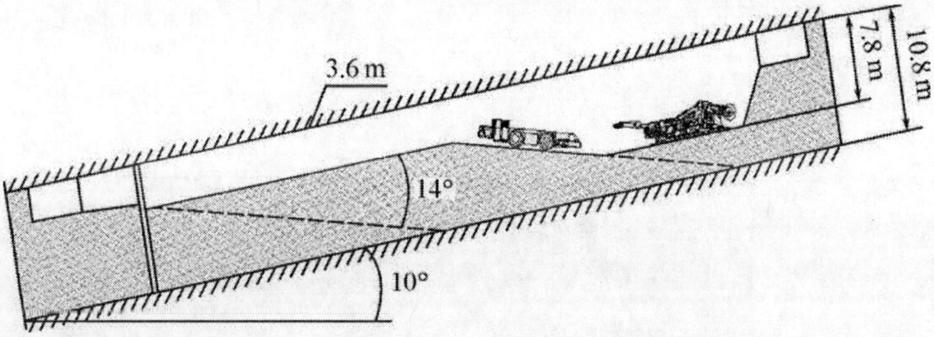

Figure 3. Layer-by-layer cutting of coal in a room in Raspadskaya-Koksovaya mine.

2. It absolutely necessary to predict rock bursts and outbursts. Making rocks rockburst-nonhazardous involves preventive control (Figure 4a) and perimeter drilling of safety pillars (Figure 4b) [18, 19].

3. It is required to drill holes in the room floor to prevent fracturing. In view of the earlier gas dynamic phenomena accompanied by huge methane emissions, it is necessary to drill destressing and drainage holes per each 6 m of heading (Figure 4c) [19].

4. Safety pillars should be left after 5–6 rooms are extracted. These pillars ensure temporal support in case that primary span of the mane roof collapse is exceeded.

5. Configuration of extraction panels should be planned wit regard to cleavages. The undertaken re-arrangement of rooms has resulted in higher stability of roof and sidewalls.

Nonetheless, despite these difficulties, RPM has application prospects in Russia. This is backed up with the experience of the technology application in difficult conditions of Olzherass coal field as well as with the international experience and increasing technical capabilities of efficient coal mining [4, 10, 18–20].

For example, Mezhegeiugol and Denisovskaya mines in Tyva and Yakutia, respectively, operating in stable host rocks and using modern import equipment annually enjoy increment in coal production output. Successful negotiations and future purchase of mobile roof support in USA promise an increase in extraction of coal reserves from pillars.

Russian scientists contribute to expansion of technological capabilities of RPM. The VostNII institute has undertaken R&D aimed to design an advanced rock cutting tool [21] and simpler and cheaper mobile roof support capable to prevent roof falls in rooms during extraction of coal from pillars. These projects will enable substituting the expensive foreign equipment [22]. Manufacture of RPM equipment by Russian machine building plants will expand capabilities and improve efficiency of the technology.

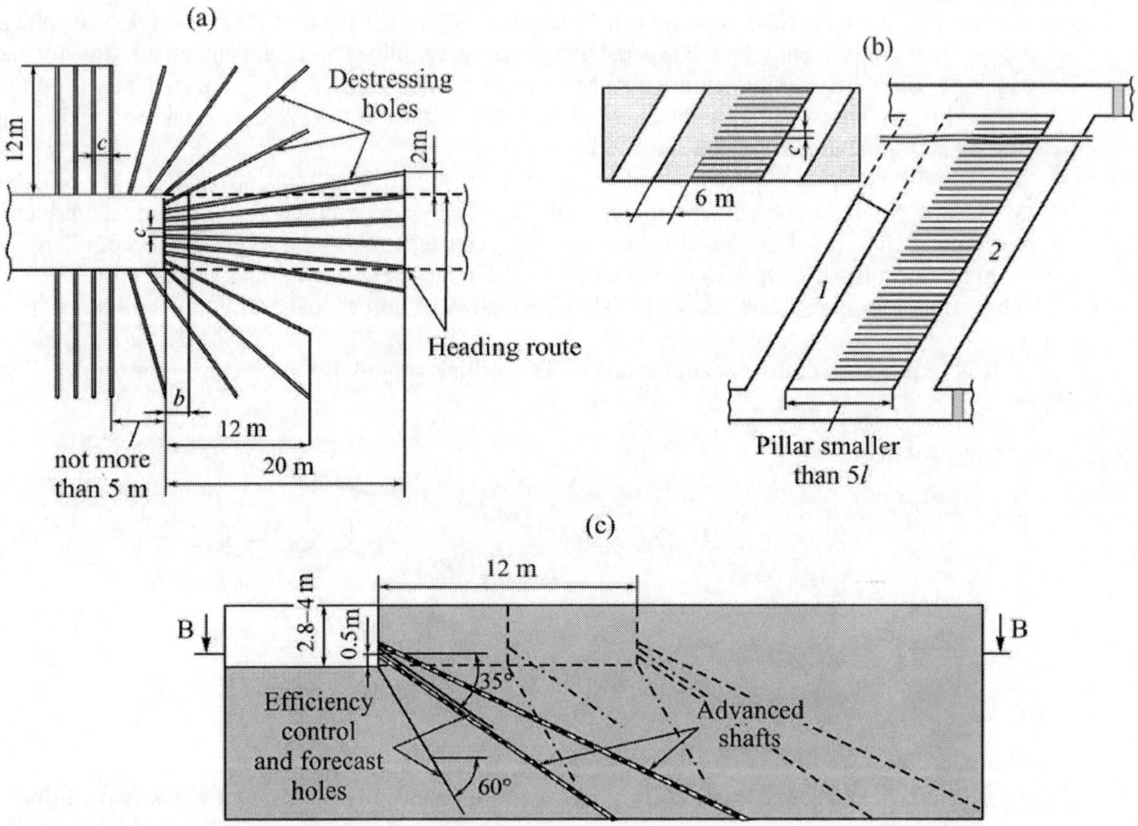

Figure 4. Destressing drilling: (a) heading; (b) treatment of safety pillar; (c) treatment of upper layer floor.

3. Conclusions

1. The method of room and pillar mining has been successfully used for many years in coal mines abroad (USA, Australia, South Africa), which is facilitated by easy ground conditions of coal strata and dynamic support of mining from the side of science and technology.
2. In Kuzbass mines, RPM application requires handling some engineering problems governed by complex nature of coal and rock mass in the Kuznetsk basin, which involves implementation of extra amount of work promoting safety but reducing efficiency of coal production.
3. The problems connected with the proper provision of RPM necessitate manufacture of special-purpose equipment in Russia. The global politics (sanctions and countersanctions) facilitates the execution of the task as foreign equipment leaves the Russian market and the domestic industry is motivated toward innovations.
4. Russian scientists are prepared and can design promising and efficient machines, facilities and technologies to support coal mines and promote expansion of application field of RPM.

References

[1] Yakovlev DV, Baskakov VP, Rosenbaum MA and Kalinin SI 2015 To the question of shortwall mining of rockburst-hazardous coal seams *Ugol* No 7 pp 13–16

[2] Mark C and Zelanko J 2001 Sizing of final stumps for safer pillar extraction *20th International Conference on Ground Control in Mining* Morgantown Virginia USA pp 59–66

[3] Sementsov VV, Dobrovolsky MS, Nivanov EV and Shabalin MP 2017 The application of the shortwall mining in coal seams prone to dynamic phenomena *Vestnik VostNII* No 2 pp 27–31

[4] Lind GH 2002 Key success elements of coal pillar extraction in New South Wales *Journal of the South Institute of Mining and Metallurgy* pp 199–205

[5] Mark C, Gauna M, Cybulski J and Karabin G 2011 Application of ARMPS (Version 6) to practical pillar design problems *30th International Conference on Ground Control in Mining* Morgantown West Virginia University pp 30–38

[6] Maleki H, Owens J and Endicott M 2001 Field evaluation of mobile roof support technologies *20th International Conference on Ground Control in Mining* Morgantown West Virginia University pp 67–77

[7] Maleki H and Owens J 1998 Analysis of the interaction between mobile roof supports and mine strata *Conference on Design and Construction in Mining, Petroleum and Civil Engineering* Escola Politecnica da Universidade de Sao Paulo pp 287–393

[8] McTyer K and Sutherland T 2011 The Duncan method of partial pillar extraction at Tasman Mine *11th Underground Coal Operators' Conference* University of Wollongong pp 8–15

[9] Wilson HG 1991 Mobile roof support for retreat mining The *10th International Conference on Ground Control in Mining Proceedings* S. Peng (Ed.) WV University pp 103–114

[10] Howe L Two Decades of Mobile Roof Support Applications http://www.jhfletcher.com/articles/Two DecadesOfMobile Roof Supports.pdf

[11] *Test Room-and-Pillar Mining in the Upper layer of Coal Seam III in Lening Mine Using Import Equipment JOY: R&D Report* Proropyevsk 2006 (in Russian)

[12] *Procedure and Program of for Semi-Commerial Testing of Shortwall Mining System in Coal Seam III in Mine 2 of Ozherass Coal Field, Raspadskaya-Koksovaya Company* Mezhdurechensk 2015 (in Russian)

[13] Pavlova LD and Fryanov VN 2018 Geomechanical evaluation of deep-level robotic coal mining by the results of numerical modeling *Gornyi Zhurnal* No 2 pp 48–52

[14] Cherepov AA and Pavlova LD 2017 Assessment of agreement between the numerical modeling results and in-situ measurements of stres sstate parameters *Vestnik KuzGTU* No 6 pp 61–68

[15] Eremenko VA, Louchnikov VN, Sandy MP and Shelukhin IS 2015 Evaluation of roof conditions in room-and-pillar mining in permafrost zone *Gornyi Zhurnal* No 3 pp 24–32

[16] Gromov YuA et al 2003 *Guideines for Selecting Geomechanical Parameters for Shortwall Mining Technology for Coal Seams* Saint-PetersburgL VNIMI (in Russian)

[17] Cherepov AA, Shiryaev SN and Kulak VYu 2017 Stress and strain distribution in geomass under room-and-pillar mining of a thick gently dipping coal bed *GIAB* No 9 pp 170–178

[18] *Process Flow Charts and Recommendations for Extraction of Coal Reserves from Seam 6-6a by Room and Pillar Mining Method in Raspadskaya Mine: R&D Report* Prokopyevsk 2006

[19] *Rockburst Hazard Aalysis and Reduction Recommendations. Optimization of Rib Pillar Parameters : R&D Report* Prokopyevsk 2015 (in Russian)

[20] Prokopenko SA, Ludzish VS and Li AA 2017 Recycling possibilities for reducing waste from cutters on combined cutter-loaders and road builders *Waste Management & Research* Vol 35(12) pp 1278–1284 DOI: 10.1177/0734242X17731154

[21] Prokopenko SA, Ludzish VS and Kurzina IA 2016 Improvement of cutting tools to enhance performance of heading machines in rocks *Journal of Mining Science* Vol 52 No 1 pp 153–159 DOI:10.1134/S1062739116010248

[22] Prokopenko SA and Ludzish VS 2014 Problems of innovation course of development of russian mining enterprises *Gornyi Zhurnal* No 1 pp 47–49

IOP Conf. Series: Earth and Environmental Science **262** (2019) 012065 doi:10.1088/1755-1315/262/1/012065

Sealing of hydraulic fracturing interval with polymer bridges in coal seam

SV Serdyukov, AV Azarov*, AV Patutin

Chinakal Institute of Mining, Siberian Branch, Russian Academy of Sciences, Novosibirsk, Russia

E-mail: *antonazv@mail.ru

Abstract. The paper describes in-seam hydraulic fracturing with packing of intervals in uncased holes using solidifiable polymer materials. The idea of this approach includes the usage of two pipe columns nested in one another. The external column is employed for hole cleaning and for delivery of hydrofracturing facilities in the selected interval of the hole. The internal column is composed of five separate hydraulic channels, allowing the polymer bridge components and breakdown fluid to be injected independently. Using the mathematical modeling the authors analyze stress distribution on hole walls in the vicinity of isolated interval depending on elastic properties of polymer bridges.

1. Introduction

With the advance of the directional drilling technologies more elongated holes with a length of more than 400 m are used for pre-mine drainage of coal. A year of degassing increases the gas drainage of the holes to 70% and, thus, improves the safety of coal mining with modern high-productive longwall machines [1, 2]. At the same time, despite the evidential success, current operation of gas drainage boreholes has faced some problems, such as gradual decrease in recoverable coalbed methane due to violation of initial conditions in rock mass by narrow pattern arrangement and mutual effect of drainage boreholes. In this case in-seam hydraulic fracturing creates the opportunity to keep the overall methane flow rate at wider spacing of holes [3]. Additional well tests make it possible to choose efficient intervals for hydraulic fracturing, including estimation of the deformation characteristics in coal and rock mass [4, 5].

One of the problems of hydraulic fracturing application is sealing of a fracture interval in a hole with the cross-section along the borehole axis that can vary greatly in size and shape [6]. In this case regular inflatable packers are of low use. The reason is that their cord-reinforced expendable shells fail to shut down the whole borehole in which the maximum and minimum cross-sections can vary by 3 times. The increase in length of inflatable packers to 5 m and more complicates in-seam operation, and reliable sealing of fracturing intervals is not ensured due to high fluid pressure that can reach several hundred bars.

In present paper authors consider sealing of fracturing intervals with unremovable bridges made of a solidifiable material to be the only method of multi-stage hydraulic fracturing in such conditions. This is a kind of a return to the early hydrofracturing methods in oil production based on the modern engineering solutions and polymers.

IOP Conf. Series: Earth and Environmental Science **262** (2019) 012065 doi:10.1088/1755-1315/262/1/012065

2. Multi-stage hydraulic fracturing

The approach includes solution of such technical problems as:

- Delivery of downhole equipment and polymer materials in the preset interval of a hole prone to breakout;
- Sealing of a fracturing interval in a variable cross-section hole;
- Pressure injection of working fluids (suspensions) with proppant in hole;
- Drilling-out of polymer bridges to recover connection between the fracturing interval and the hole bottom/mouth.

The proposed method of sealing with unremovable bridges includes pre-sealing of a set interval with a formwork made of phenol resins readily spumescent when mixed with catalysts. The unremovable bridges are created by polymerization of the two-component mixture of rigid polyurethane or organomineral resins at high cohesion with rocks [7].

In the developed engineering solution of the downhole pack for coal degassing stimulation by multi-stage hydraulic fracturing two unconnected pipe columns nested one in another are used. The external column equipped with centering skid and ring cutter is employed for cleaning and washing-out of the hole, as well as delivery of hydrofracturing tools in preset intervals.

The internal column with five separate hydraulic lines can have a cartridge design (all lines are pipes, e.g. Figure 1) or mixed type design (breakdown fluid feed line is pipes, the other lines are flexible high-pressure hoses). The internal column is freely rotatable and movable along the external column. A multi-point injector is mounted at the end of the internal column. It feeds formwork and polymer bridge materials in the required intervals of hole or injects breakdown fluid (suspension with propant) under pressure in the fracturing interval.

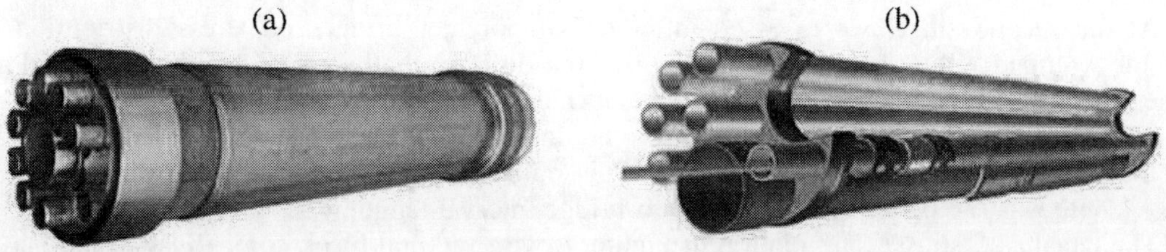

Figure 1. Multi-channel cartridge-design pipe columns (HMPS by SOLEXPERTS, Switzerland): general view of column width central fluid feed line (a); cartridge with off-set fluid feed line (b).

The sequence of multi-stage hydraulic fracturing suing the developed engineering solution is in detail described in [7].

3. Mathematical modeling results

For the estimation of the effective stresses on the borehole walls at the sealed interval the following problem was considered. A hole had a cylindrical form and a radius in a uniform isotropic medium with elasticity modulus $E_1 = 3.6 \cdot 10^9$ Pa and Poisson's ratio $\nu_1 = 0{,}3$; E_2, ν_2 are the elasticity modulus and Poisson's ratio of polymer bridges with a length H installed in the hole. In the interval h, the fluid pressure is P (Figure 2).

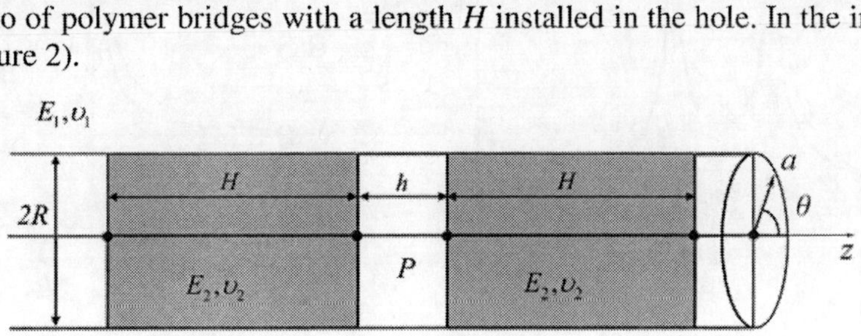

Figure 2. Problem formulation to calculate effective stresses.

The problem was solved by the finite element method (FEM). Firstly, we tested FEM and compared the obtained results and the analytical solution. Figure 3 illustrates calculation of axial and circumferential stresses normalized to the fluid pressure P in the radial loading of the hole interval equal to $2a$. In this case, the analytical solution is obtained using the scheme from [8]. The positive stresses are tensile. The results show that the numerical and analytical solutions are similar, thus, the selected numerical method can be used to solve the set problem.

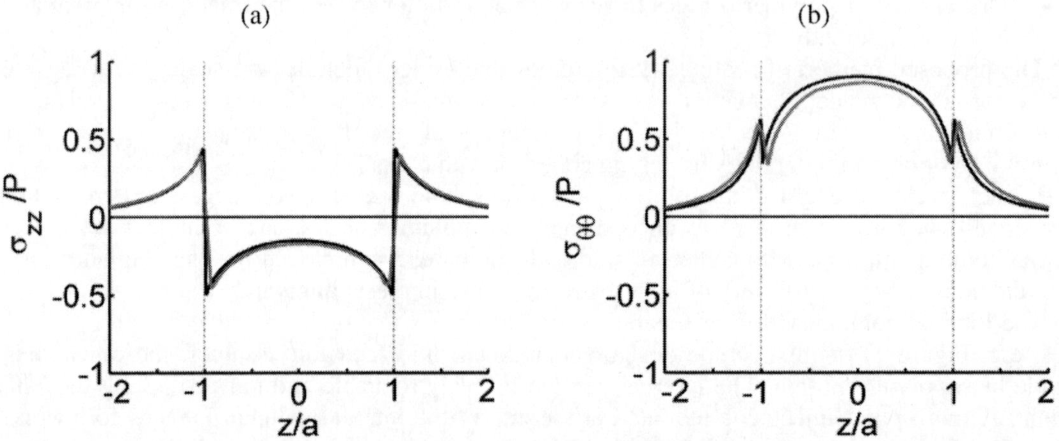

Figure 3. Axial σ_{zz} (a) and circumferential $\sigma_{\theta\theta}$ (b) stresses on the borehole walls, normalized to fluid pressure P: red—numerical solution, black—analytical solution.

At the second stage, we assessed influence of polymer bridges on stress distribution on the borehole wall under fluid feed. It is assumed that the fluid exerts the pressure P on the sealed interval in the hole and on the surface of the polymer bridges from the walls in this interval. The analysis included two cases with the bridge's elastic modulus E_2 higher ($2.5 \cdot 10^{10}$ Pa) and lower ($2.5 \cdot 10^{8}$ Pa) than the enclosing rock elasticity modulus E_1. Poisson's ratio was assumed as 0.3 for all materials, the bridge length was $H = 0.4$ m, and the bridge-to-bridge interval length was $h = 0.01$ m.

The calculated stresses are plotted in Figure 4, with red and black color illustrating data for the low and high elastic moduli, respectively. It is seen in Figure 4 that as the elastic modulus decreases the axial stresses σ_{zz} and the circumferential stresses $\sigma_{\theta\theta}$ slightly grow outside the bridge-to-bridge interval. A change in the elastic modulus has no effect on σ_{zz} stresses within the bridge-to-bridge interval. Also note that σ_{zz} exceeds $\sigma_{\theta\theta}$ in value.

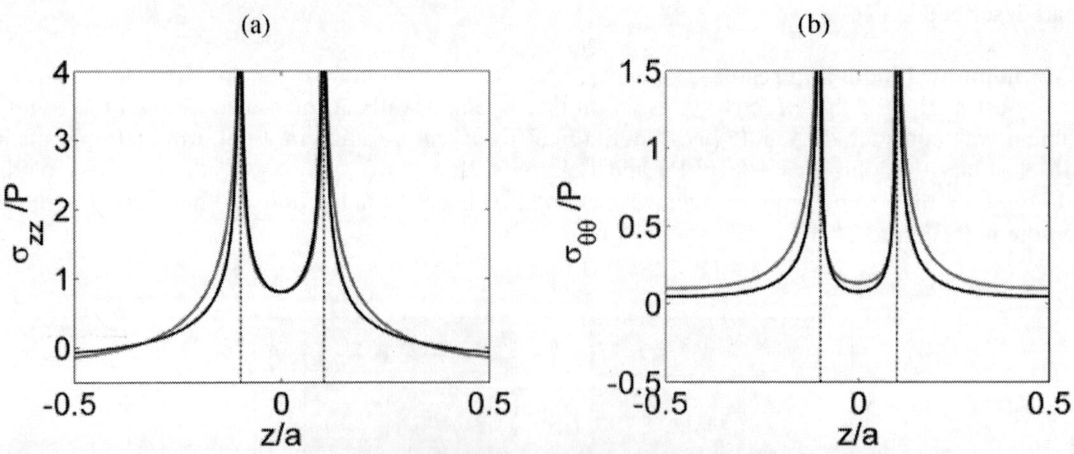

Figure 4. Axial σ_{zz} (a) and circumferential $\sigma_{\theta\theta}$ (b) stresses on the borehole wall normalized to fluid pressure P: red—$E_2 = 2.5 \cdot 10^{8}$ Pa, black—$E_2 = 2.5 \cdot 10^{10}$ Pa.

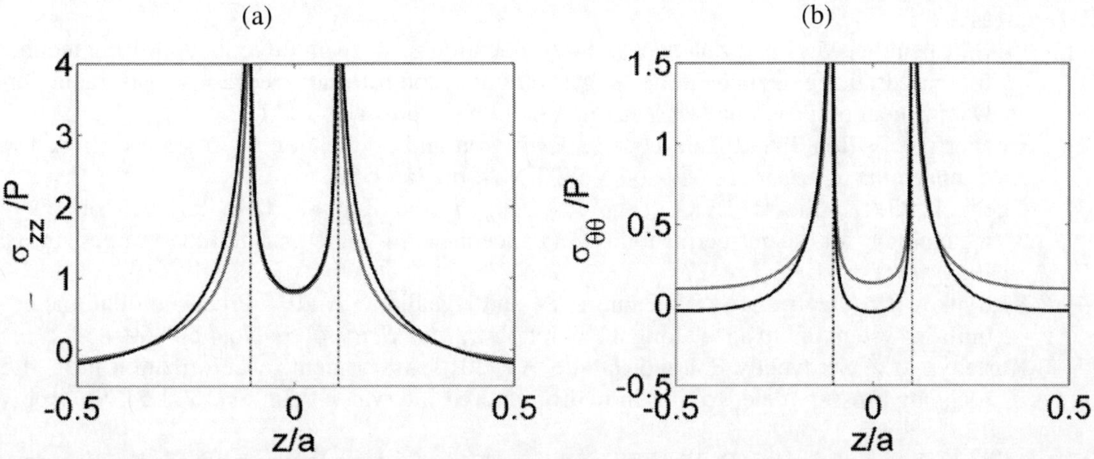

Figure 5. Axial σ_{zz} (a) and circumferential $\sigma_{\theta\theta}$ (b) stresses on the borehole wall normalized to fluid pressure P at $E_2=2.5\cdot10^8$ Pa: red—$\nu_2=0.3$, black—$\nu_2=0.1$.

Figure 5 presents the comparison between stresses for two different ν_2 at $E_2=2.5\cdot10^8$ Pa. It is shown that an increase in Poisson's ratio of the bridges results in higher $\sigma_{\theta\theta}$ in the vicinity of the bridges. On the other hand, the change in Poisons' ratio causes no considerable alteration in the axial stresses σ_{zz}. Additional numerical calculations showed that effect of the Poisson's ratio of polymer bridges decreases with increasing elastic modulus.

The experimental calculation results show that the chemical compounds of the polymer bridges should be carefully selected in order to design and perform hydraulic fracturing based on the proposed engineering solution. Note that the numerical calculations were carried out in the framework of the linear theory of elasticity.

4. Conclusions

The authors have proposed the new method for sealing of hydraulic fracturing intervals in holes of variable cross-section using irremovable bridges made of polymer materials. This method and the developed engineering solutions facilitate in in-seam multi-stage hydrofracturing in gas drainage boreholes.

Using numerical methods effective stresses on borehole walls are modeled mathematically as functions of elastic properties of the polymer bridges. It is shown that using two closely spaced polymer bridges creates favorable conditions on borehole wall for the transverse fracturing. In the framework of the linear theory of elasticity the following results were obtained:

1) As the elastic modulus decreases the axial stresses σ_{zz} and the circumferential stresses $\sigma_{\theta\theta}$ slightly grow outside the bridge-to-bridge interval. A change in the elastic modulus has no effect on σ_{zz} stresses within the bridge-to-bridge interval; but it is found that a decrease of the elastic modulus leads to an increase of the circumferential stresses $\sigma_{\theta\theta}$.

2) An increase in Poisson's ratio of the bridges results in higher $\sigma_{\theta\theta}$ in the vicinity of the bridges. On the other hand, the change in Poisons' ratio causes no considerable alteration in the axial stresses σ_{zz}. Test results showed that effect of the Poisson's ratio of polymer bridges decreases with increasing elastic modulus.

Acknowledgements

The work was supported by the Ministry of Education and Science of the Russian Federation, Project No RFMEFI60417X0172.

IOP Conf. Series: Earth and Environmental Science **262** (2019) 012065 doi:10.1088/1755-1315/262/1/012065

References

[1] Lu S., Cheng Y, Ma J and Zhang Y 2014 Application of in-seam directional drilling technology for gas drainage with benefits to gas outburst control and greenhouse gas reductions in Daning coal mine, China *Nat. Hazard* Vol 73 No 3 pp 1419–1437

[2] Hungerford F, Ren T and Aziz N 2013 Evolution and application of in-seam drilling for gas drainage *Int. J. Min. Sci. Technol.* Vol 23 No 4 pp 543–553

[3] Cheng L, Ge Z, Xia B, Li Q, Tang J, Cheng Y and Zuo S. 2018 Research on hydraulic technology for seam permeability enhancement in underground coal mines in China *Energies* Vol 11 No 2 p 427

[4] Serdyukov SV, Degtyareva NV, Patutin AV and Rybalkin LA 2015 Precision dilatometer with built-in system of advance along the borehole *J. Min. Sci.* Vol 51 No 4 pp 860–864

[5] Kurlenya MV, Serdyukov SV and Patutin AV 2015 Assessment of deformation properties of rocks by pressuremeter testing in hydrofractured interval *J. Min. Sci.* Vol 51 No 4 pp 718–723

[6] Mills K, Jeffrey R, Black D, Meyer T and Carey K 2006 Developing methods for placing sand-propped hydraulic fractures for gas drainage in the Bulli seam *Coal Operators' Conference* pp 190–199

[7] Kurlenya MV, Serdyukov SV, Patutin AV and Shilova TV 2017 Stimulation of underground degassing in coal seams by hydraulic fracturing method *J. Min. Sci.* Vol 53 No 6 pp 975–980

[8] Azarov AV, Kurlenya MV, Patutin AV and Serdyukov SV 2015 Mathematical modeling of stress state of surrounding rocks around the well subjected to shearing and normal load in hydraulic fracturing zone *J. Min. Sci.* Vol 51 No 6 pp 1063–1069

IOP Conf. Series: Earth and Environmental Science **262** (2019) 012066 doi:10.1088/1755-1315/262/1/012066

Geomechanical assessment of processes for staged development and cross-section support setting in extended mine workings

VM Seryakov

Chinakal Institute of Mining, Siberian Branch, Russian Academy of Sciences, Novosibirsk, Russia

E-mail: vser@misd.ru

Abstract. The stress state of a rock mass and support members in the staged advance of two parallel workings is calculated. The feasibility is demonstrated to solve such problems by using the stiffness matrix of the calculation system which remains unchangeable in the development stages. Two versions of mining and construction processes are considered. Specific features of stress-field formation in the support system are discussed. The recommendations on selection of a mining system, best from the geomechanical viewpoint, are worked out and set worth.

1. Introduction

It is impossible to substantiate the applicability of most modern mining processes and respective technological parameters disregarding geomechanical state of a rock mass in mining operations [1]. Configuration of a worked-out space is continuously changing, support members and constructions regularly erected in mine workings carry a partial load arising with mining advance. The mining with consolidating stowing appears the most demonstrative process in this aspect [2]. At every mining stage the workings earlier worked-out are filled with a material which physical and mechanical properties differ from ones of the host rock mass. In this connection a representative description of stress and strain field redistribution in the vicinity of a worked-out space can be gained by considering all the mining stages. The theoretical approach to modeling of the stress-strain state of a rock mass considering a mining sequence stumbles across a number of mathematical and technical obstacles [3]. The researchers working for MI SB RAS developed the methods, applicable to settle all these challenges and to study redistribution of stresses in a rock mass in the course of stoping and stowing operations [4].

2. Different mining sequences and stress state features

In the present paper the novel methods are used for geomechanical studies of advance and support setting in two parallel tunnels. Figure 1 demonstrates one of mining and support setting processes applicable to drive two parallel tunnels [5]. At every mining stage a fresh-driven section of a tunnel is reinforced with temporary or compressible support. The support system and surrounding rocks form a construction apt to deformation when initial stresses release in surrounding rocks. The permanent support is set after the release of initial stresses acting in a rock mass is actually realized, so permanent supports experience insufficient portion of acting load in fact. Actual strain and loading of supports is mainly observed as a consequence of further driving of parallel tunnels. Therefore, the stress field in support members and a surrounding rock mass can be calculated by using a software system designed to compute a stress-strain state of "supports–host rock mass" with account for a mining sequence practiced in this very case [1, 4].

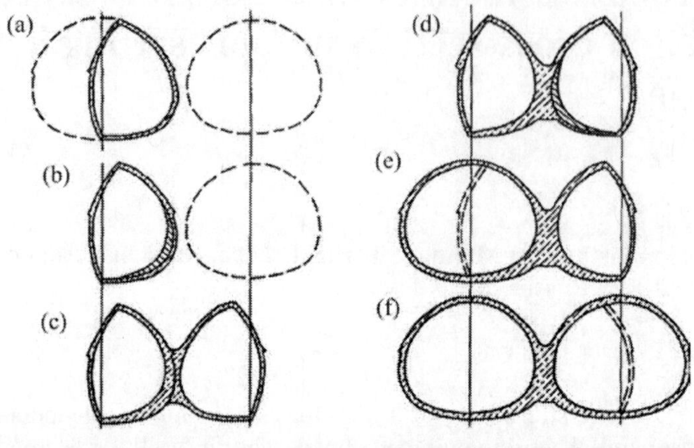

Figure 1. Staged opening of cross-sections of parallel tunnels [5]: a—opening of the right adit in the left tunnel with a temporary apron made of reinforced shotcreting; b—multilayed shotcreting on the right side of the adit; c—opening the left adit of the right tunnel and erection of a temporary apron made of reinforced shotcreting; d— multilayed shotcreting on the left side of the adit; e—complete opening of cross-sections of the left and right tunnels.

The idealized version of the mining sequence under consideration is shown in Figure 2a. Another possible version of a mining sequence and reinforcing of parallel tunnels is demonstrated in Figure 2b. It is impossible to justify a rational mining sequence disregarding regularities in redistribution of stress fields in a host rock mass and support members.

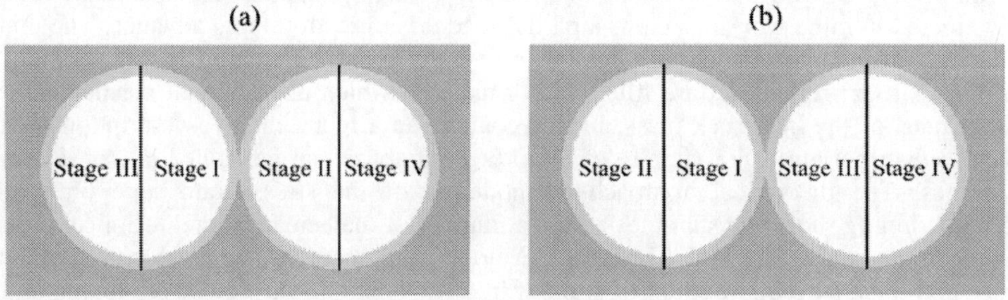

Figure 2. Versions of mining sequence to drive and to reinforce parallel tunnels: a—the first variant; b—the second variant.

Specific features of stress-field formation in supports and surrounding rock mass were earlier considered for the second mining variant [6]. According to the basic calculation conclusions the application of the calculation method for computing a stress-strain state of a rock mass with account for the mining sequence to solve the problem of opening cross-sections of parallel mine workings enables to establish specific features of the formation of concentrated tension and compression stress zones in support members and to identify mining stages where they reach the maximum. To the point, the most failure-hazardous situation in the support system is resulted from the third-stage operations on opening cross-sections of parallel tunnels.

Let compare stress fields, formed in a rock mass and support members in similar stages of the tunneling variants under consideration. Like in [6] we assume that the length of mine workings is

IOP Conf. Series: Earth and Environmental Science **262** (2019) 012066 doi:10.1088/1755-1315/262/1/012066

significant, ant the plane strain terms can be accepted. Boundary conditions on outer contours of the calculation domain are suggested as follows: zero values of the horizontal component of displacement vector u and tangential component of stress tensor τ_{xy}. These terms correspond to the initial stress state of a rock mass with stress tensors: $\sigma_y^0 = \wp H$; $\sigma_x^0 = \nu \wp H / (1-\nu)$; $\tau_{xy}^0 = 0$, known as "gravity" initial stress field. Here σ_x^0, σ_y^0, τ_{xy}^0 are normal and tangential components of stress tensor; $\wp$ is volumetric rock weight; H is distance from the daylight surface to the top boundary of mine workings. Axis Ox is horizontal and Oy is vertical.

The upper horizontal boundary of the calculation domain is free of external load action. The zero vertical component of displacement vector v and tangential component of stress tensor τ_{xy} are suggested on the horizontal bottom boundary. The calculations are made at the following mechanical properties: Young's modulus E= 25000 MPa; ν = 0.25 for host rocks and E = 50000 MPa, ν = 0.2 for the support system. Volumetric rock weight is assumed equal to 0.03 MN/m^3.

The analysis of calculation results by using the final element method [7, 8] for the stress state of a rock mass and support members in the second mining variant made it possible to establish that the maximum concentration of compression stresses in both supports and rocks is reached in the area of connection of parallel workings. In view of the above, let consider variations in the stress state in details in this very area in the course of mining operations. The distribution isolines of the second principal stress after the third mining stage in the mining variants under consideration are shown in Fig. 3.

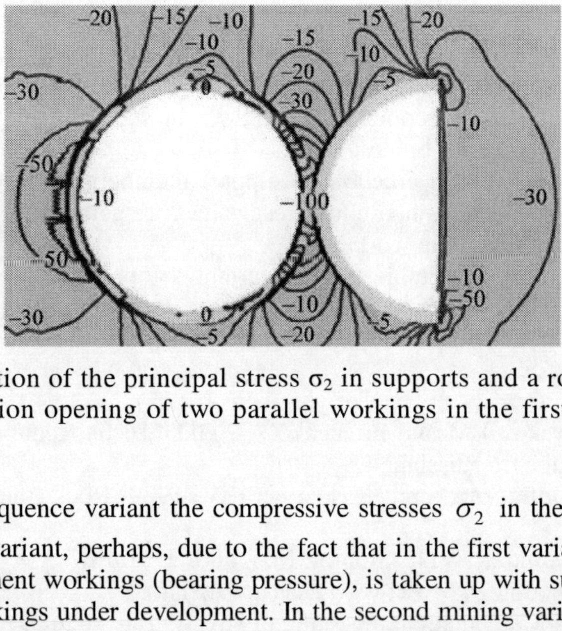

Figure 3. Distribution of the principal stress σ_2 in supports and a rock mass in the third stage of cross-section opening of two parallel workings in the first and second mining sequence variants.

In the first mining sequence variant the compressive stresses σ_2 in the support system are lower than those in the second variant, perhaps, due to the fact that in the first variant the weight of the rock, occurring above development workings (bearing pressure), is taken up with support members erected in both first and second workings under development. In the second mining variant the bearing pressure in the point of connection of workings is taken up by support members erected only in one working.

The final pattern of distribution of the principal compressive stress σ_2 at the point of parallel workings conjugation after fulfillment of all the mining stages is presented in Figure 4. It is obvious that at conjugation point the compression level in the supports is higher in the second variant and exceeds 100 MPa by absolute value.

The analysis of the character of stress redistribution in support members in mining with opening of cross-sections of parallel workings enables to establish some more specific features of their stress state. Thus, stepwise erection of support members in the development workings induce formation of tensile stress areas. Location of such areas and tension level in them vary in the course of mining operations, moreover, the highest stress values are gained at intermediate mining stages. The final distribution picture of the first principal stress, which enables to establish zones of tensile stress formation, reveals a tension level much lower as compared to tension levels at preceding tunneling stages. All the above confirms the must to consider the stress-strain state of the support system and a surrounding rock mass in the staged driving and support setting in mine workings.

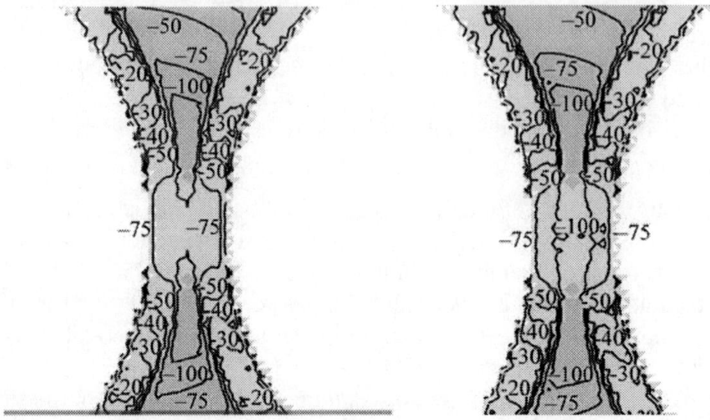

Figure 4. Distribution pattern of stress σ_2 in conjugation area for two parallel workings in the final stage of the first and second tunneling variants.

3. Conclusions

1. Calculation of a stress-strain state of a rock mass and support members in compliance with conditions of the staged advance and support setting in two parallel workings makes it possible to establish specific features of formation of tensile and compressive stresses in support members and to identify actual mining stages at which they reach the highest values.
2. It is established that the stepwise erection of support members in development workings induce formation of tensile stress zones, the maximum stress values are gained at intermediate stages intended to open cross-sections of parallel mine workings.
3. Geomechanically, the most acceptable mining variant to open and to reinforce cross-sections of parallel workings is a mining sequence where formation and reinforcement of sections of parallel workings are provided in areas of their conjugation at first stages.

References

[1] Kurlenya MV, Seryakov VM and Eremenko AA 2005 Technogenic geomechanical stress fields Novosibirsk Nauka (in Russian)
[2] Turchaninov IA, Iofis MA and Kasparyan EV 1989 Rock mechanics Leningrad Nedra (in Russian)
[3] Kuznetsov SV, Odintsev VN, Slonim ME and Trofimov VA 1981 The rock pressure calculation methodology Moscow Nauka (in Russian)
[4] Seryakov VM 2014 Mathematical modeling of stress-strain state of rock mass in the application of technology with the laying of the developed space *Journal of Mining Science* Vol 50 No 5 pp 51–60
[5] Kartozia BA, Fedunets BI, Shuplik MN et al 2003 *Mine and Underground Construction: University Textbook* Moscow: Gornaya Kniga (in Russan)
[6] Seryakov VM 2018 Impact of sequence mining on the stress state of the lining and the surrounding rock for the excavation of two adjacent working Interexpo Geo-Siberia-2018 Conf. Proc. Vol 5 pp 204–210 (in Russian)
[7] Zienkiewicz O 1971 *The Finite Element Method in Engineering Science* McGRaw-Hill
[8] Fadeev AB 1987 *Finite Element Method In Rock Mechanics* Moscow: Nedra (in Russian)

Recommendations on integrated evaluation of preparedness of coal deposits for mining

SV Shaklein and MV Pisarenko*

Federal Research Center of Coal and Coal Chemistry, Siberian Branch, Russian Academy of Sciences, Kemerovo, Russia

E-mail: *mvp@icc.kemsc.ru

Abstract. Considering high importance of geological information quality for mining business, a legal basis for the initiation of technical projects on mineral mining is the state geological survey report on the extent of exploration and preparedness of a mine field for industrial development. To ensure rationality, safety and efficiency of subsoil management, it is proposed to implement such integrated evaluation at the pre-project stage in two directions: technological and geological preparedness.

1. Introduction

Project design choices and current decision-making on mine construction and expansion are not based on the actual data but on perceptions of properties, quality and occurrence of subsoil mineral resources gained from geometrical models based on assaying results. Such models always differ from reality, i.e. have an error. Inaccuracy of geological information inevitably turns into wrong judgments on technology, investment, etc. and sometimes ends with substantial financial loss and abandonment of mining projects; for this reason, it is necessary that geological errors fall within industrial standards.

Considering critical importance of the quality of geological information for mining business [1–5], the Russian Government decrees that the expert report of the state geological survey should provide estimation of "…preparedness of a mineral deposit or its parts for industrial development…" assuming "preparedness" as the extent of exploration such that technical-and-economic performance of a mine project is achieved irrespective of possible difference between actual geological information and values revealed at the mine exploration stage. A positive expert opinion allows a subsoil user to start a mine project. However, there are some omissions.

A legal basis for the initiation of a mine project is the subsoil use licensing. De factor, a mining licensee acquires a legal right to destroy non-renewable mineral resources which belong absolutely to the nation. The key mission of Russia as an owner is efficient use of mineral wealth in the interests of the present and future generations of the country (RF Law on Subsoil, Article 35). Implementation of this function determines evaluation of reserves unrecoverable within a mine project and meant to be lost. Such evaluation is included in the state geological survey.

2. Development of recommendations for the evaluation of preparedness of coal fields

On the ground of the accomplished research findings [6, 7], the recommendations have been developed for the evaluation of preparedness of coal fields for the efficient development at an industrial scale. The evaluation has two directions (see Figure 1) described below.

• *Technological Preparedness.* This is an estimate of technological capacities available in the mining and processing industry for the complete extraction of mineral reserves. It is set by the value of a technological preparedness coefficient which is a ratio of uneconomic reserves Z_{unec} to total proved reserves Z_{tot} inside a mining lease. When the ratio is higher than the permissible value (for coal fields, the standard is 20% unless otherwise is substantiated), a field is assumed technologically unprepared and off operation. Mining to be prospective in such fields should use non-conventional processes and technologies aimed to increase recoverable reserves.

With a view to delineating technologically prepared reserves, a geometrical model is constructed using the technological preparedness coefficient calculated from the structure maps of coal seams within the limits of a mining lease [7].

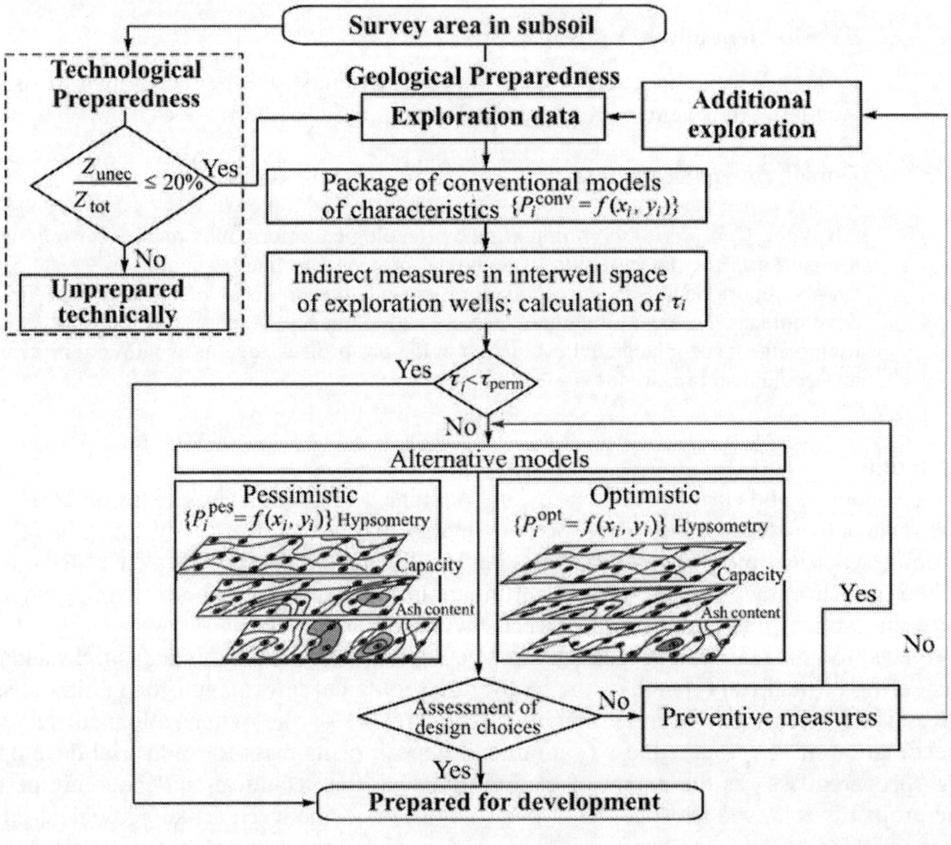

Figure 1. Block-diagram of algorithm for integrated evaluation of mineral reserves preparedness for efficient development at industrial scale: τ_i—preparedness criterion (capacity, ash content, etc.); τ_{perm}—permissible value of τ for characteristic i.

After the technological preparedness of a mine field has been acknowledged, the stage opens for the second evaluation.

• *Geological Preparedness.* This is the quality of conformance between the geological exploration of a deposit and the objectives and claims of the industry. This estimate is governed by individual project designs targeted at mining efficiency and includes:

—determination of indirect measures in interwell space of exploration wells (coal seam hypsometry) and squares of network of exploration wells (other characteristics);

—conventional $P_i^{conv} = f(x, y)$ and alternative ($P_i^{pes} = f(x, y)$ pessimistic and $P_i^{opt} = f(x, y)$) optimistic geometrical modeling using the information on hypsometry and quality of coal seam and rock mass; analysis of efficiency of design choices on schemes of accessing and preparing mineral reserves;

—conventional $P_i^{conv} = f(x, y)$ and alternative ($P_i^{pes} = f(x, y)$ pessimistic and $P_i^{opt} = f(x, y)$) optimistic geometrical modeling of thickness and structure of coal seams, as well as other characteristics; analysis of efficiency of design choices on schemes of mining and time schedules.

Is is worthy of mentioning that the condition of obligatory alternative modeling is the excess of the exploration extent criterion τ_i of a permissible value τ_{perm}.

When the efficiency of design choices in the conventional models preserve in the alternative models of mineral deposits and mining, this is a proof of sufficient geological survey, extent of exploration and preparedness of a deposit for the commercial development. Otherwise, depending on the level of actual mismatch, it is required to carry out additional exploration within the limits of inacceptable discrepancy of actual data and modeling results, or to add the project with preventive measures which would mitigate disamenity due to incomplete knowledge on the mineral object.

This approach suggests including evaluation of coal field preparedness for industrial development in geological survey reports in the section Evaluation of mineral reserves exploration and preparedness for industrial development, or in project documentation in the subsection Geological exploration of underground (surface) mine field. The obligation of these sections in the geological and project documentation is dictated by the effective regulations. However, as practice shows, these section lack arguments required for reasoned decision-making. This omission is explained by complexity of the evaluation of mineral reserves preparedness for development and by the absence of proper instructions and guidelines on the implementation.

3. Conclusions

The integrated evaluation of coal field preparedness for the efficient commercial development, with sub-evaluations of technological and geological preparedness, included in the geological survey and mine project documentation submitted for the state expertise enables reasoned and viable decision-making. For a subsoil user, such evaluation is a proof of the conformity between the extent of geological exploration of a mineral deposit and objectives of individual design choices or sustainability and performance of a mining investment project. For the nation as a subsoil owner, this evaluation is a guarantee that the subsoil user will fulfill obligations on resource conservation as well as on industrial and ecological safety of mineral mining.

This approach in the subsoil management is aimed at decrease of risks, sustainable utilization of mineral wealth and at promotion of unconventional innovative technologies of mining and processing in implementation of mining investment projects.

References

[1] McCarthy P 2003 Managing technical risk for mine feasibility studies *Mining Risk* TheAuslMM ISBN 978-1-920806-00-2

[2] Rogova TB and Shaklein SV 2011 Reliability of coal reserves *Quantitative Assessment and Monitoring* LAP LAMBER

[3] Bogatsky VV 1971 Possibility of a quantitative estimation of reliability of results of prospecting of stratal minerals *Geology of Coal Deposits* Moscow: Nauka Vol 2 (in Russia)

[4] Kazhdan AB 1977 *Prospecting of Mineral Deposits* Moscow: Nedra (in Russia)

[5] Australian Code for Reporting of Exploration Results, Mineral Resources and Ore Reserves (The JORC Code) Prepared by the Joint Ore Reserves Committee of the Australasian Institute of Mining and Metallurgy, Australian Institute of Geoscientists and Minerals Council Of Australia (JORC) 2004.

[6] Pisarenko MV and Shaklein SV 2014 Evaluation of the preparedness of solid mineral deposits to industrial development *Miner. Resursy Rossii* N 6 pp 42–45

[7] Pisarenko MV 2016 Mining and geometric support for evaluation of the technological preparedness of the field for development *Marksheder. Nedropolz.* No 3 pp 40–43

IOP Conf. Series: Earth and Environmental Science **262** (2019) 012068 doi:10.1088/1755-1315/262/1/012068

Experience of application of polyurethane and phenolic resins in roadheading in fractured and heavily disintegrated rock mass

YuN Shaposhnik[1]* and DA Shokarev[2]**

[1]Chinakal Institute of Mining, Siberian Branch, Russian Academy of Sciences, Novosibirsk, Russia
[2]Expert PRO, Ust-Kamenogorsk, Kazakhstan

E-mail: *shaposhnikyury@mail.ru, ** denshok82@mail.ru

Abstract. The problem of mining and ground support in an unstable rock mass under stoping is discussed. The analysis of the experience gained in application of phenolic and polyurethane resins to fill domes-like voids and for preventive consolidation and roadheading in disintegrated rock mass is presented. The feasibility to fill domes formed under roof caving with Blockfil phenolic resin in combination with grouted bolting and the use of Blockpur polyurethane resin for consolidation of disintegrated rocks is investigated. The laboratory test results on strength properties and time of polyurethane resin flowability loss are presented. The procedures are developed for eliminating accidents due to dome formation under roof and sidewalls collapse, as well as for the support of weakened roadways by yielding steel arcs. It is emphasized that geotechnical conditions worsen due to dome formations during recovery works in mines.

1. Introduction

The Orlov deposit of polymetallic sulphide ore features difficult geological and geotechnical conditions. Massive barite–polymetallic and copper–zinc ore as well as massive and disseminated sulphuric copper ore and varieties possess high strength and elasticity. Therefore, mining operations in such conditions should account for risk of dynamic phenomena of rock pressure.

The strength of rocks drops in the zones of weathering, dislocation and hydrothermal alteration from a few meters to 100–150 m thick [1]. Adjacent rocks are very unstable, with numerous differently oriented micro-joints with calcite and pyrite fill, and many slickensides. Under exposure, relaxation of high stress and wetting, such rocks become soft and fall in mine roadways.

Ground support in roadways in unstable rocks (stability category IV) in Orlov Mine includes SVP-22 steel frames and with timber cushioning of voids in back of the support. In difficult ground conditions (in unstable rocks) under higher stress state, steel frames are greatly deformed and roof rock fall in large volumes. As a result big “domes” form in the roofs, which makes roadways prone to accidents and dictates by-pass road heading [2].

For example, in January 2018, loosening of damaged rocks occurred in the roof and sidewalls in haulage no. 1 of level 15 under construction by Vostokshakhtstroi. As a result of collapse of the roof and sidewalls in the roadway, the estimated volume of voids was 140 m^3, including dome volume of 90 m^3, volume in cross section of 15 m^3, volume of voids in back of the metal frame support of 35 m^3.

The experience of mining knows different resins used for filling “domes” or for advanced reinforcement of damaged rock mass, for instance, in mines of the Polar Division of NorNickel, Kaz Minerals and Kuznets Coal Basin [3–6]. Injection reinforcement of rocks enjoys wide application in coal mines in Kuzbass [7–11], in shafts of Udachny Mine of ALROSA [12] and at Don Mining and Processing Plant [13, 14]. Technologies of preliminary cementation of heavily jointed rocks are known as well [15–17]; for instance, in Orlov Mine, heavily damaged and jointed rock mass was reinforced by injection with micro-cement [18].

2. The technology of voids healing by phenol resin: practical results

Aimed to heal voids after rock falls in roof and sidewalls, as well as in order to support the accident span with SVP-22 steel yielding arch support, jointly with DSI Tekhno, the following technology was proposed. It was planned to fill of voids in the rock fall section, over the entire cross-section of the roadway, with BlokFil foaming phenol resin and, then, to undertake the advanced injection of roof rocks with Blokpur polyurethane resin. The characteristics of Blockpur resin are compiled in Table 1.

Table 1. Characteristics of Blokpur resin components.

Description	Component A	Component B
Density at 25°C, g/cm^3	1.12 ± 0.1	1.22 ± 0.02
Color	Transparent to light-yellow	Dark-brown
Flash point, °C	> 200	> 200
Viscosity at 25°C, MPa·s	300 ± 100	200 ± 100

The polyurethane resin had compressive strength of 24.61 MPa by the data of lab-scale experiments at the certification experimentation center of the East-Kazakhstan State Technical University, or 64.51 MPa at 10 % deformation in the age of 2 h by the DSI Tekhno testing results. The flowability loss time at + 25°C was 44 s.

Figure 1 depicts the sequence of heading and re-support in haulage roadway no. 1 (Figure 1). First, voids in the back of 8 frames in the roof and sidewalls are filled with BlokFil foaming resin. Then, in the dome, along the sidewall and roof, two steel feed types with a diameter of 50 mm are laid to guide plastic injection pipes (3 pipes, 2 m long each) to feed the resin. Injection is started in tube 1 installed in the sidewall. After its filling (approximate volume of 20 m^3 ~ 90–1000 kg of BlokFil resin), injection is initiated in tube 2 (approximate volume of 30 m^3 ~ 1400–1500 kg of resin). After that, a light timber apron walls is installed, and filling of residual voids is completed.

When the voids are filled and the phenol resin is cured, the hardened resin is removed within the project cross-section of roadway, by small batches, for the installation of one steel frame. The spacing of the frames in the rock fall section is recommended to be not wider than 0.5 m. After installing and straining of the first frame, the second frame is mounted, etc. In this manner, the whole accident section to a pillar (face) is covered. Before face drilling, in the roof of the roadway, advanced injection bolts DIVI Drill are installed using a self-propelling drill rig or hammer hand drills.

After rock bolting to a depth of 4 m (two 2 m long coupled rock bolts), in the internal hole of each bolt, the reinforcement mixture is injected—two-component polyurethane resin Blokpur. Rockbolting is made in the roof, at an angle of 20–25° to the longitudinal axis of the roadway to a depth of 3.8 m (bolt end of 200 m is left outside to be connected with the injection equipment). The spacing of the rock bolts is 0.5–0.7 m, and there are 6 rock bolts installed in the roof of the roadway 4 m wide.

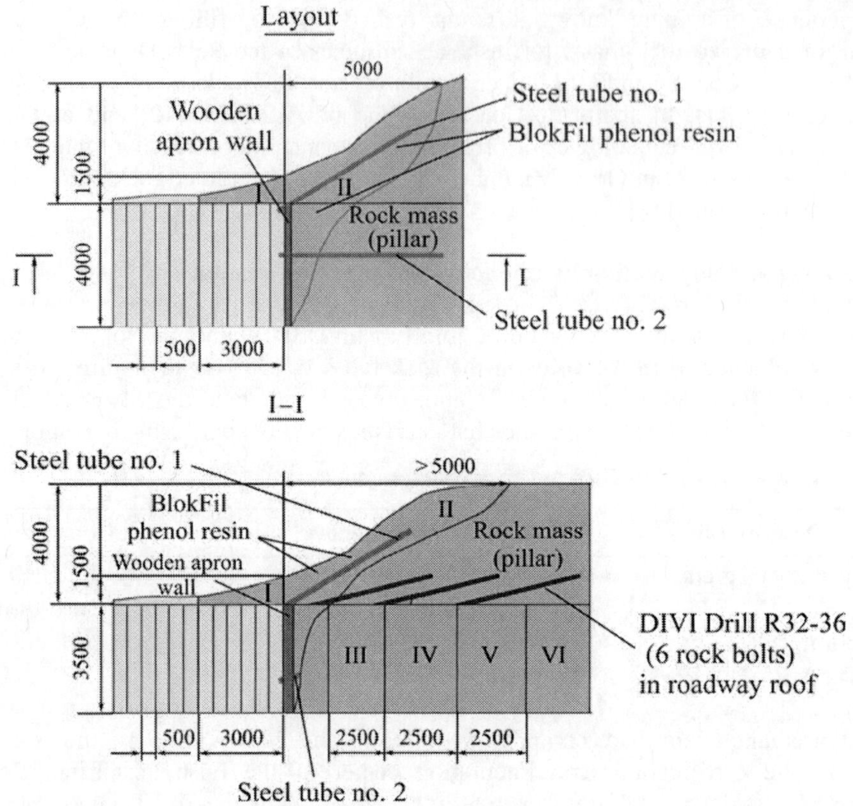

Figure 1. Schematic of "dome" elimination and re-support of the accident span in haulage roadway no. 1 on level 15 in Orlov Mine.

The approximate consumption of Blokpur per one injection DIVI Drll bolt is 30 kg. After completing the rock bolting, drilling is started in the roof for the advanced injection reinforcement. Between the DIVI Drill bolts, three holes with a diameter of 43 mm are drilled to a depth of 3 m in the roof at an angle of 40–45° to the longitudinal axis of the roadway. In the holes, in turn, a sealer with the injection pipes is installed, the injection equipment is connected and Blokpur resin is injected under pressure of 80–100 bar at the rate of 144 kg per injection hole. Then, heading is carried out to a length of 2.5 m, by cuts of 0.5 m and with spacing of the installed steel frames of 0.5 m. After that, the advanced reinforcement and injection cycle is repeated. In case that the cross-section of the roadway enlarges during the heading, the new voids are filled with BlokFil resin. Finally, the actual condition of the roof and sidewalls in the roadway is evaluated with a view to assessing potentiality of the further heading without advanced reinforcement.

In order to heal voids after rock falls in the roof and sidewalls, as well as to reinforce the accident span of the roadway with steel yielding arched support, in haulage roadway no. 1 on level 15 in Orlov Mine, 120 m^3 of BlokFil phenol resin was injected in the gap between between the support and rocks inside the interval of 6 installed frames. Then, the pre-installed apron wall was removed. Prior to continue heading, in the face roof of the roadway, advanced injection DIVI Drill bolts 4 m long were set in two stages at an angle of 20–25°. In the first stage, 0.5 away from the last frame, five bolts were installed to be injected with two-component mix Blokpur, as well as three holes with a diameter of 51 mm were drilled to a depth of 3 m for the injection of Blokpur resin by a sealer with the injection pipes at a pressure of 80–100 bar. In the second stage, 2.6 m away from the first frame, five DIVI Drill bolts were installed and filled with Blokpur, as well as one injection hole was drilled and filled with the resin. Furthermore, other three holes were drilled for the injection of Blokpur to reinforce rocks.

The total volume of injected polyurethane resin Blokpur was 1056 kg. All in all, 15 SVP-22 steel frames were installed with spacing of 0.6 m, which makes a span of 7.2 m in the damaged roof from the entrance of haulage roadway no. 1. In this fashion, the section of doming filled with foaming BlokFil resin was passed. It should be mentioned that at frame 12, geotechnical conditions worsened with rock falls from the roof, left sidewall and face. The looseness parameters were: doming 3. M high above the SVP-22 steel frames and to 1 m in the left sidewall; no loosening of rocks were observed in the right sidewall. The total rock fall volume was 25.7 m^3. It was recommended to fill the voids with BlokFil to avoid further doming. Heading was advised to continue suing jackhammers. Drilling and blasting was only permitted at the bottom of the right sidewall by 2–3 holes 0.5 m long and small explosive charges.

In ventilation cross heading 9 on level 13 in Orlov Mine, loosening of damaged rock mass took place in the roof and sidewalls with the prediction volume of voids of 140 m^3. Aiming to heal domes and recover the mine support in that section, a new approach was proposed. First, it was required to fell gaps in back of the support in the roof and sidewalls within the 40 m-long interval of 8 pre-installed frames with foamed phenol resin BlokFil (Figure 2). The estimated consumption of BlokFil was 1350 kg. Then, strapping made of round wood was made on four frames nearby the rock fall; at the same time, into the dome, along the roof of the heading, two steel feed tubes with a diameter of 50 mm were laid. Using plastic pipes 2 m long, filling of gaps between the support and rocks with BlokFil was completed. Through the steel tubes, plastic injection pipes (6 pipes, 2 m long each) were driven in the dome for the complete filling of the void with phenol resin (approximate consumption of BlokFil was 4950 kg). From the reinforced boundary toward the fallen-in rock mass, DIVI Drill R32-360 injection bolts 3 m long were drilled at an angle of 30–35° to the longitudinal axis of the heading to a depth 2.8 m (the bolt end 200 m long was left outside for the connection of the injection equipment). The spacing of the bolts was 0.5–0.6 m, 18 bolts were installed. After that, the bolts were in turn connected with the injection equipment, and Blokpur resin was fed in the fallen-in rocks under pressure. It was advised to inject resin starting from the lower bolts.

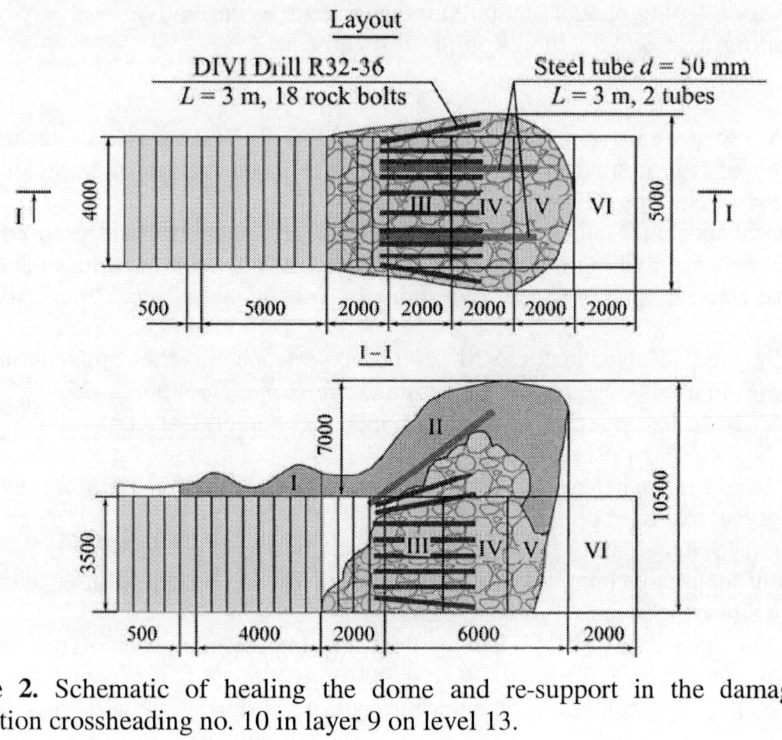

Figure 2. Schematic of healing the dome and re-support in the damaged area of ventilation crossheading no. 10 in layer 9 on level 13.

IOP Conf. Series: Earth and Environmental Science **262** (2019) 012068 doi:10.1088/1755-1315/262/1/012068

The expected consumption of Blokpur resin per one DIVI Drill injection bolt in the conditions of fallen-in rocks is 96 kg. The fallen-in rocks are carefully removed using LHD in such amount that one steel frame can be installed. The frames are set and strapped with wood at spacing of 0.5 m. The gaps in back of the support are filled with phenol resin BlokFil. These operations in the indicated sequence are fulfilled in the span of 2 m of the crossheading, and the cycle is repeated.

Consumption of materials for the void volume of 140 m^3 in haulage roadway no. 1 was:

—phenol resin BlokFil—6300 kg (voids) and 900 kg (in back of the support);

—DIVI Drill—24 bolts (2 m long);

—polyurethane resin Blokpur—1224 kg.

The cost of materials and ancillaries for the breakdown elimination in haulage roadway no. 1 on level 15 made 1917 Rub, including 1375 Rub for phenol resin and 301 thou Rub for polyurethane resin. Furthermore, the rent cost of the two-componet pneumatic pump was 22 thou Rub per moth. Thus, the cost of materials and ancillaries at the void volume of 140 m^3 was 13.6 thou Rub/m^3, or 1917 thou Rub in the accident span 7 m long. Given the cost of development heading of 5 thou Rub/m^3 in Orlov Mine, the cost of driving a by-pass roadway 7 m long makes 490.3 thou Rub, which is much less than the cost of recovery of the heading using phenol and polyurethane resins.

The estimated consumption of materials at the void volume of 140 m^3 in ventilation crossheading no. 10 with drilling and advanced reinforcement of the roof span 8 m long using DIVI Drill bolts is:

—phenol resin BlokFil—4950 kg (voids) and 1350 kg (in back of the support);

—DIVI Drill—72 bolts (3 m long);

—polyurethane resin Blokpur—6912 kg.

3. Conclusions

1. The technology developed and trialed in Orlov Mine in healing of voids by foaming phenol resin with advanced injection of rocks with polyurethane resin and with re-support of roadways by steel frames made it possible to eliminate accident in the conditions of roof rock fall, and to recover safety of mining.

2. The commercial trial of the technology showed that the cost of materials for the recovery of underground excavations using phenol and polyurethane resins is currently essentially higher than the cost of by-pass heading: 13.6 and 5.0 thou Rub/m^3, respectively.

References

[1] Krupnik LA, Shaposhnik YuN and Shaposhnik SN 2017 Experience in introducing the technology of filling “domes” with foaming resins in the mines of Vostoktsvetmet *Bezop. Truda v Prom.* No 5 pp 62–67

[2] Krupnik LA, Shaposhnik YuN and Shaposhnik SN 2016 Technical and economic comparison of the technology of filling the “domes” with foaming resins in the mines of Vostoktsvetmet *Innovative Development of the Mining Industry: Int. Internet-Conf. Proc.* Krivoi Rog: KNU pp 90–93

[3] Klimchuk IV and Malanchenko VM 2007 Experience in the application of polymer technologies in mining enterprises in Russia *Gorn. Prom.* No 4 pp 22–25

[4] Klimchuk IV 2007 Introduction of new polymer technologies at coal mining enterprises of Kuzbass *Glukauf* No 1 (2) pp 88–90

[5] Klimchuk IV and Malanchenko VM 2008 Solution of security problems at mining enterprises of Russia *Glukauf* No 2 (3) pp 95–97

[6] Galaov RB, Nagovitsin YuN, Pliev BZ, Andreev AA and Vilchinsky VB 2014 Improvement of excavation support techniques in mines of the Norilsk Nickel Mining and Metallurgical Company *Gornyi Zhurnal* No 4 pp 25–28

[7] Chubrikov AV and Rib SV 2016 Development and improvement of polymer technologies in coal mines of Kuzbass *Vestnik SibGIU* No 2 (16) pp 3–6

[8] Mayorov AE 2007 Consolidation of adjacent rock mass during support installation in mines

IOP Conf. Series: Earth and Environmental Science **262** (2019) 012068 doi:10.1088/1755-1315/262/1/012068

Vestnik KuzGTU No 1 pp 6–11
[9] Khyamyalyainen VA and Mayorov AE 2010 New methods of cementation reinforcement of rocks *GIAB* No 9 pp 212–217
[10] Khyamyalyainen VA 2015 Development of injection methods for compacting rock masses in Kuzbass *Vestnik KuzGTU* No 5 pp 25–31
[11] Mayorov AE 2010 Deformation and filtration characteristics of solid particles in a suspension or granular material in rock srengthening with cementation and pre-stressed bolting *Journal of Mining Science* Vol 46 No 2 pp 160–166
[12] Matveeva OI, Sokolova EYu, Fedorova GD and Zyryanov IV 2016 Composition of cement grouting of mine shafts in permafrost zone *Gornyi Zhurnal* No 1 pp 82–86
[13] Tretyak AV 2013 Choice of the optimal type of fastening of mine workings on the basis of modeling of the stressed state of an underground structure *Gornyi Zhurnal* No 5 pp 31–34
[14] Til VV, Bekeev MM and Bolikov VE 2013 Geomechanical problems of the construction of shafts at the Donskoy GOK *Gornyui Zhurnal* No 5 pp 22–26
[15] Heinz WF 1988 Pre-cementation of deep shaft International *Journal of Mine Water* December Vol 7 Issue 4 pp 1–12
[16] Qingbin M, Lijun H, Weiguo Q, Dengge L and Jiadong F 2014 Support technology for mine roadways in extreme weakly cemented strata and its application *International Journal of Mining Science and Technology* Vol 24 Issue 2 pp 157–164
[17] Qingbin M, Lijun H, Weiguo Q, Dengge L and Jiadong F 2013 *International Journal of Mining Science and Technology* Vol 24 Issue 2 pp 157–164
[18] Krupnik L, Shaposhnik Yu, Shaposhnik S and Konurin A 2017 Technology of micro-cement injection of destroyed and fractured rock mass at Orlovsky Mine *Mining of Mineral Deposits* Vol 11 (2017) Issue 1 pp 87–92

IOP Conf. Series: Earth and Environmental Science **262** (2019) 012069 doi:10.1088/1755-1315/262/1/012069

Modeling rock destruction under blasting of closely spaced borehole charges

EN Sher

Chinakal Institute of Mining, Siberian Branch, Russian Academy of Sciences, Novosibirsk, Russia

E-mail: ensher@gmail.com

Abstract. The design scheme is developed, and simulation of radial cracks growth under blasting of closely spaced borehole charges is carried out. The case studies of confined explosion of two charges located in neighboring boreholes are considered. Blasting of decoupled charges in neighboring boreholes is investigated. The free surface effect on a shape and size of radial cracks under explosion of two closely spaced boreholes rising to the surface is simulated. The blast pattern to gain the maximum size of the destruction zone is determined.

1. Introduction

The mining blasting operations are mainly realized by coyote blasting technique. According to the design pattern a number of blast holes are preliminarily made in blasting rocks. The blasthole pattern parameters: spaces between holes in a row and between rows are determined in terms of a size of a rock breaking zone made by blasthole charges. Estimates of a size of such zones are reported in publications [1–3]. It is essential to consider interaction of neighboring charges under concurrent blasting in order to make a precise theoretical evaluation of a breaking zone size.

In blasting of elongated pre-charges and blasthole charges in brittle monolithic rocks the destruction is mainly falls on the zone of radial cracks. The software to compute the evolution of uniformly angularly distributed flat radial cracks is developed with 3D problem statement in order to estimate a size and a shape of such cracks induced by blasting of an alongated preset-length charge [4]. In compliance with the zone model of blasting [5–7], the detonation of a charge in-depth of a rock mass is followed by propagation of an elastic compressive wave and later by the destruction-wave front. In the course of wave propagation the elastic wave stress tends to decay, and the destruction wave front is slowed down. The generation and evolution of radial cracks gets feasible within the interval when a velocity of the destruction wave evolution lowers, fracturing gains the maximum velocity of its propagation, and azimuth tension stresses arise [3, 8]. Thereto the destruction-wave front comes to a stop, the radial shear of the elastic medium is fixed at the boundary with broken rocks. The elastic medium expansion gained in the first stage of blasting used to preserve due to the broken rock resistance to radial compression. The broken rock is prone to deformation under Coulomb friction law. Such expansion leads to development of a radial crack system in an elastic zone of a rock mass. As the final size of radial cracks under blasting appears much greater than radius of a breakage zone, the assumption is made in modeling that the fracturing starts in the elastic plane from an initial radial system N of orthogonal cracks which originate at axis of an extended charge, and their dimensions are equal to a charge length and a radial size of the destruction zone r_d. The initial crack edges are supposed to be under constant pressure p_d, stimulating opening of cracks d_0 in pursuance of displacement of boundaries of the elastic zone and the destruction zone resulted from the first-stage blasting effect. To establish a shape of radial cracks in the final stage of their dynamic development, the researchers consider the quasi-static process of the crack development under conditions of a continual increase in opening d_0. At every design stage starting from the initial one the stress state of the elastic medium is estimated nearby the front of cracks in view to establish a probable destruction of a medium and development of a crack.

2. Numerical simulation and results analysis

The paraclase technique is employed to calculate 3D stress state of the medium in an elastic space with a radial system of uniformly angularly distributed flat cracks being under internal pressure [8, 9]. According to this method the crack surface is divided into square elements at pitch a, within these elements the opening and shear of crack edges are assumed constant. Thus, a crack is represented by a set of dislocation elements, described by Burgers vector, which components can be unknown in advance. They are computed in terms of the requirements to fulfill boundary conditions concerning stress in centers of dislocation elements by solving a respective system of linear equations. Coefficients of these systems are coefficients of mutual influence of elements on each other. The influence coefficients were calculated by Pitch–Keller formulas where stress tensor components are presented in a random point in an elastic space in the vicinity of a dislocation discontinuity through contour integrals along its boundary. The same formulas were used to calculate stresses nearby crack edges in order to identify their evolution.

To simplify the study problem on the influence of charge interaction in blasting, the study case of development of a radial crack system, consisting of two flat cracks in a plane of axes of neighboring blastholes was considered.

Estimating calculations of destruction parameters under blasting of closely spaced blasthole charges were performed for a few configurations shown in Figure 1. Scheme in Figure 1a demonstrates two blasthole charges loaded in neighboring blastholes.

As an example of calculations by this scheme, in Figure 2a the first quadrant of plane (x, z) in dimensionless form, referred to a calculation grid space $a = 0.5$, contains the shapes of radial cracks induced by blasting of two charges of 40 in length, loaded in neighboring blastholes with space $2h = 12$ between them, the initial crack openings being different. It is obvious that increase in opening leads to fusing of cracks into one, the front of the resultant crack embraces both charges. Efficiency of blasting of two neighboring blasthole charges can be evaluated by fracturing area S. The relationships of this area versus spacing between blastholes are presented in Figure 2b for different values of initial crack openings d_0. It is evident that there is distance $2h$, at which the crack has the maximum area.

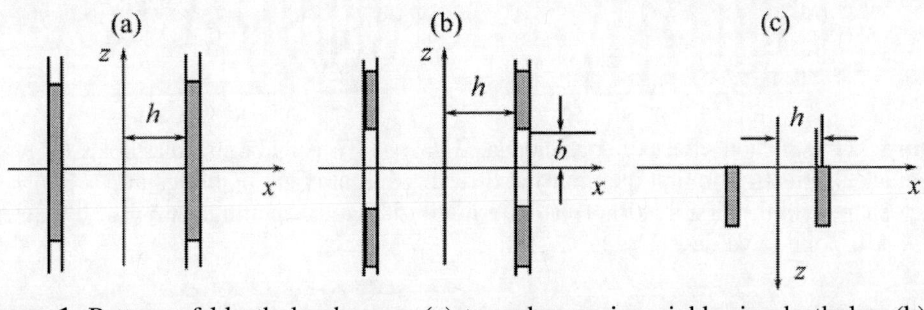

Figure 1. Pattern of blasthole charges: (a) two charges in neighboring lastholes, (b) four charges in two neighboring blastholes, (c) two rising-to-the-surface charges in two neighboring blastholes.

It follows from diagrams in Figure 2b that distance between blastholes, at which the area of cracks is maximum, depends on the initial crack opening, governed by the specific explosive consumption.

According to the calculation data the optimal space is proportional to a transverse asymptotic value of crack size b_m, generated by blasting of a single long charge: $h \approx 1.1 b_m$.

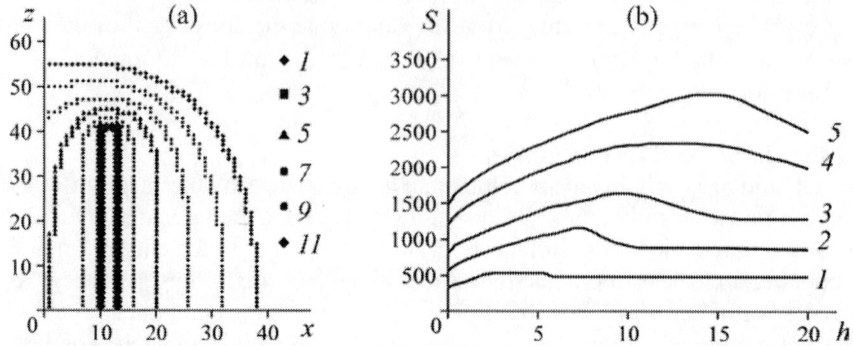

Figure 2. Destruction effect of two long charges in neighboring blastholes (Figure 1a): (a) shape of fractures at their differing initial opening in the first quadrant (curves *1–11* correspond to values: d_0 = 0, 0.2–1); (b) dependences of crack area *S* vs. distance 2*h* between blastholes at the initial openings: d_0 = 0.3, 0.5, 0.7, 1, 1.2 (curves *1–5*).

It can be stated from blasting experience that quality of blasting is higher at the uniform distribution of the charge along blasthole length. The charge is divided into fragments and distributed uniformly in a blasthole with gaps 2*b* relative to each other. Figure 1b illustrates such disposition of four charges. The calculation of crack evolution under simultaneous blasting of four charges loaded at such charge pattern is shown in Figure 3 as an example. Figures 3a and 3b demonstrate in the first quadrant the dimensionless form of cracks shapes induced under blasting of four charges of length 20, loaded by pairs in blastholes with gaps 2*b* = 4; 8 at space between holes 2*h* = 8; 16, respectively. Curves *1–8* are for crack shapes, corresponding to blasting of different-diameter charges; in calculations different diameters are characterized by different values of initial crack openings d_0.

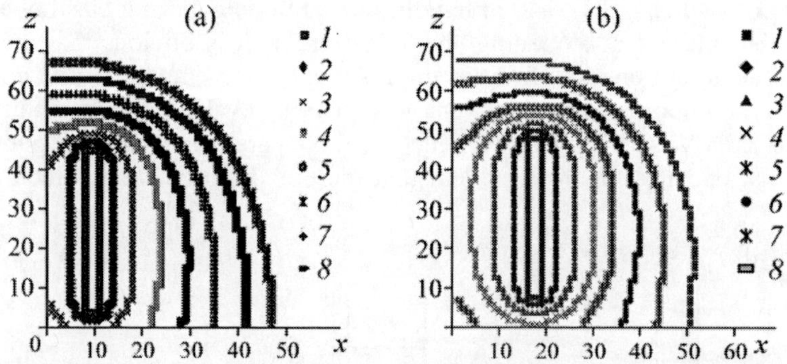

Figure 3. Destruction effect of four elongated charges in neighboring blastholes (Figure 1b): (a) crack shapes (in the first quadrant) at differing initial crack openings at *b* = 2, *h* = 4; (b) crack shapes (in the first quadrant) at differing initial crack openings at *b* = 4, *h* = 8. Curves *1–8* correspond to values: d_0 = 0, 0.2–1.4.

Analysis of calculation results at different values of parameters *b* = 1–6 and *h* = 2–12 revealed that there are intervals of charge dislocations relative to each other, at which the common crack area embracing these four charges is maximal. Approximately such charge disposition is determined by ratios: $h \approx b_m$, $b \approx b_m/2$. Figure 4 demonstrates diagram of dependence of the crack area around four charges on intervals *h* and *b* for a charge, which initial opening amounts to d_0 = 0.9. Column rows 1, 2, 4 are for blasting of charges with parameters $b = 1, 2, 4$. The maximum value $S = 2033$ is gained at inter-charge gap $2b = 6$ in a blasthole and inter-blasthole space $2h = 9$.

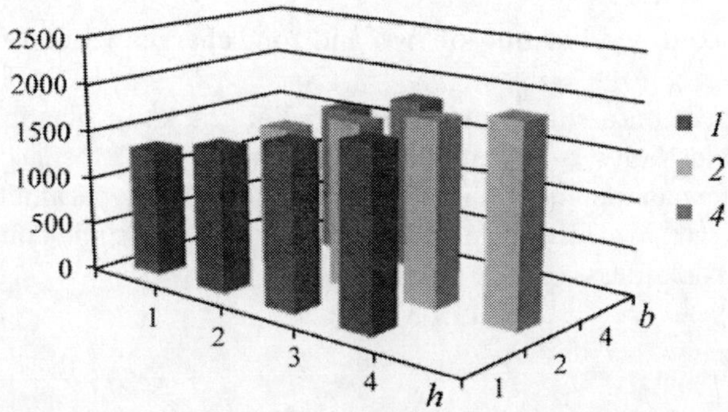

Figure 4. Diagram of dependence of dimensionless area of a crack, generated under blasting of four charges, versus charge dislocation distance h and b.

In the case when an elongated charge starts from the surface the shape and size of the radial crack zone should be evaluated with account for the free surface effect. The software is developed to solve problems where free surface is modeled as a load-free flat crack of a large size as compared to charge length [4].

The calculation is made for the problem of crack evolution under simultaneous blasting of two downhole charges loaded in neighboring blastholes spaced at distance $2h$ from each other (Figure 1c).

As an example Figure 5a demonstrates in the first quadrant the planes (x, z) of radial crack shape induced under blasting of two charges of length $l = 14$ with interval $2h = 12$ between blastholes with different values of the initial crack opening $d_0 = 0$, 0.2 – 1.4 (curves *1*–*8*). From calculations of crack sizes and their areas it can be set forth that there is distance between blastholes, at which the fracturing area is maximal. It is explicit in Figure 5b where relationships of the fracturing area in its dimensionless form versus distance h at different openings of initial cracks d_0 are shown.

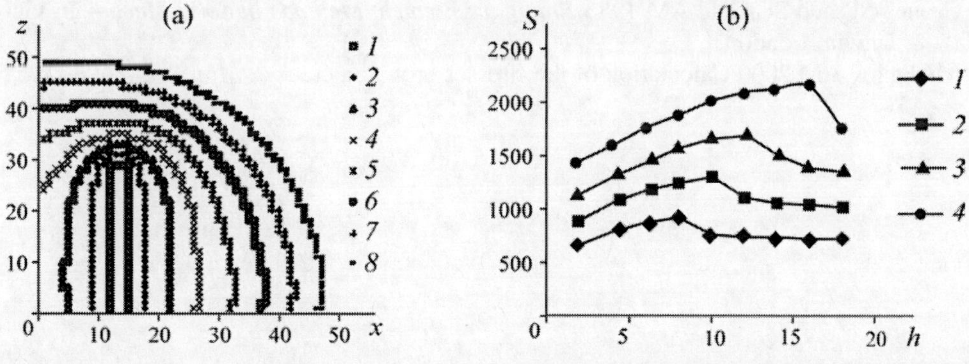

Figure 5. Calculated results on crack evolution under simultaneous blasting of two cord charges rising to the surface: (a) calculated crack shapes in the first quadrant at different values of initial crack openings; (b) relationships of crack area vs. intercharge distance h at some initial crack opening values $d_0 = 0.6, 0.8, 1.0, 1.2$ (curves *1*–*4*).

Comparison of the optimal distance between neighboring blastholes with the size of cracks in axial and lateral directions revealed that the inter-hole space providing the maximum fracturing area is close to a value of growth of a crack initiated by a single charge rising to the surface at lateral direction $h \approx b_m$.

3. Conclusions

The calculation scheme of modeling the crack evolution in blasting closely spaced blasthole charges is worked out.

The study cases of confined blasting of two and four charges loaded in neighboring blastholes are considered.

The free surface effect on a shape and size of radial cracks in blasting two downhole charges loaded in neighboring blastholes is investigated. The parameters of charge disposition at which a size of destruction zone can be maximal are established. Such parameters enable to cut down explosive consumption rate in drilling operations with mass blasting, controlled blasting, blasting with slot formation in a rock mass.

References

[1] Mosinets VN 1976 *Crushing and Seismic Action of an Explosion in Rocks* Moscow: Nedra (in Russian)

[2] Kutuzov BN and Andrievskiy AP 2002 *A New Theory and New Technologies for Destruction of Rocks by Elongated Explosive Charge* Novosibirsk: Nauka (in Russian)

[3] Aleksandrova NI and Sher EN 1999 Effect of dilation on rock breaking by explosion of a cylindrical charge *J. Min. Sci.* Vol 35 No 4 pp 400–408

[4] Sher EN and Chernikov AG 2015 Calculation of the parameters of the radial system of cracks formed during the explosion of an elongated charge in fragile rock *J. Fundament. Appl. Min. Sci.* Vol 2 pp 299–303 (in Russian)

[5] Grigoryan SS 1967 Some questions of the mathematical theory of deformation and destruction of solid rocks *J. Appl. Math. and Mech.* Vol 31 No 4

[6] Rodionov VN, Adushkin VV and Romashev AN 1971 *Mechanical Effect of Underground Explosion* Moscow: Nedra (in Russian)

[7] Chadwick P, Cox., Hopkins HG 1964 *Mechanics of Deep Underground Explosions* Phil. Trans. Roy. Soc. London A 256, 235–300

[8] Crouch SL and Starfield AM 1983 *Boungary Element Methods in Solid Mechanics* George Allen & Unwin, London

[9] Mikhailov AM 2000 Calculation of the stresses around a crack *J. Min. Sci.* Vol 36 No 5 pp. 445–451

Effect of compressive stresses on permeability of coal with fracture filled with lightweight proppant

TV Shilova* and SV Serdyukov

Chinakal Institute of Mining, Siberian Branch, Russian Academy of Sciences, Novosibirsk, Russia

E-mail:*shilovatanya@yandex.ru

Abstract. The paper presents laboratory testing results on effect of hydraulic fracturing on permeability of coal with and without wedging of created fractures by proppant. The tests were carried out with hollow aluminosilicate microspheres ASPM-500 without and with additional polymeric thermoreactive coating, specimens of dense long flame (grade D) coal with permeability of less than $3 \cdot 10^{-3}$ μm^2 and fractured fat coking coal (grade ZH) with permeability over $60 \cdot 10^{-3}$ μm^2. It is found that creation of fractures with and without propping considerably improves gas permeability of dense coal with weakly developed natural jointing. Created fractures are less effective in highly jointed highly permeable coal. It is experimentally proved that filling of created fractures with small opening of 0.4 mm with lightweight proppant under uniform compression of 1–5 MPa improves permeability of dense coal by 7–19.5 times. In jointed coal, application of lightweight proppant under the same conditions improves permeability less considerably—by 1.2–2.6 times. It is revealed that efficiency gas drainage in coal by hydraulic fractures without propping lowers as compressive stresses decrease. Reduction in permeability of coal with propped fracture under growing compression is probably connected with redistribution and compaction of proppant particles in the fracture. The obtained results show that application of lightweight propping materials is a promising way of advance in the technology of hydraulic fracturing for stimulation of coalbed methane drainage. The long-range research, in the authors' opinion, should be focused on the development of lightweight proppants with improved strength characteristics.

1. Introduction

Gas drainage in coal seam under higher confining pressure is difficult due to low permeability of coal—not higher than 10^{-4}—$5 \cdot 10^{-3}$μm^2 as a rule [1]. For methane drainage at admissible rate and recovery, a dense grid of wells has a spacing of 3–10 m. The number of wells is only possible to be reduced by means of increasing the well radius. An efficient approach to this problem's solution is creation of many drainage channels in wells by hydraulic fracturing [2].

Analysis of the hydraulic fracturing experience in coal seams shows that productivity of drainage wells can be improved using arching materials (proppants). Numerous drainage channels filled with sand proppant (30/60 mesh) with opening of 3–4 mm and radius to a few meters, at a spacing of 4–6 m from each other, improve drainage borehole productivity by 5 times in coal seams with permeability of 0.02–0.03 μm and to 180 times in less permeable coal seams for several months [4]. At the same time, traditional proppant materials used for hydraulic fracturing of coal, such as quartz sand, ceramic spheres, early deposit in drainage holes and fractures. Combating this phenomenon with highly viscous fluids and/or high pumping velocity upvalues greatly hydraulic fracturing implementation, which results in preference of density drainage drilling layout [6, 7]. An admissible solution to this problem aimed to cut down meterage drilled is the use of lightweight proppants having similar density as water-based fracturing fluids. Such proppants are readily transferred in holes for a long time and well spread along the created fractures [7, 8]. However, they are soft and can be broken under closure of fracture edges under the action of rock pressure.

This article presents experimental research data on influence exerted by aluminosilicate microspheres ASPM-500 and lightweight proppant based on hollow aluminosilicate microspheres ASPM-500 with additional polymeric thermosetting envelope on permeability of coal under uniform compression. The comparative analysis of measured and calculated permeabilities from the known model of the effect of stresses and sorbing gas flow on evolution of permeability of coal filled mono-layered proppant is given [5].

2. Experimental research

Gas permeability of coal was determined on an installation designed at the Laboratory of Physical Methods of Impact on the Rock Mass at the Institute of Mining, SB RAS [9]. The installation is meant for determining gas-permeability of rocks in a testing cell with adjustable axial and lateral compression of cylindrical samples. The installation includes measurement system for automatic long-term investigation of rock permeability by preset program. Specification of the installation is given in [9].

Examination of the jointed coal structure and proppant grains with thermoreactive coating was carried out on the instrumentation and program system Mineral S7, including optical microscope OLYMPUS BX51, video camera SIMAGIS 2P-3C, personal computer and specialized software.

2.1. Samples

The tests were carried out on grade D coal (long flame) from the Karakan coal deposit (Permyakov open pit mine, Kuznetsk Coal Basin) and grade ZH coking coall (fat) from Tikhov Mine (Kuznetsk Coal Basin).

Two cylindrical specimens of 3 m diameter and 3 and 6 cm lengths were manufactured. Before the filtration tests were carried out, jointing of the specimens had been examined on two specimens of each coal grade. The preparation procedure of specimens is described in detail in [10, 11].

Figure 1. Coking coal grade ZH with a cutout after manufacture of cylindrical specimen (left) and long-flame (grade D) coal specimen with fracture filled with lightweight proppant (right).

In the cylindrical specimens, through axial fractures were made and, then, filled with proppants (Figure 1). Average opening of propped fractures under atmospheric pressure was 0.4 mm. In the tests of coal grade D, the proppant was a hollow aluminosilicate microsphere ASPM-500 (Figure 2a). Later on, the aluminosilicate microsphere were covered with thermoreactive coating and used as the proppant in the filtration tests of coking coal grade ZH (Figure 2b). The thermoreactive coating made of epoxy resin with magnesium particles and phenol-formaldehyde resin improves the strength of proppant, ensures cohesions of the sphere is the drainage channel and prevents their outflow to the drainage borehole [12]. Using the solid microstructure analyzer SIAMS Mineral S7, the size and shape of the main proppant fraction were determined. Average diameter of aluminosilicate microsphere was 140 μm, around 70% of particles had roundness above 0.8. Average diameter of the proppant grains with the polymeric coating was 400 μm, and their roundness was 0.65–0.7.

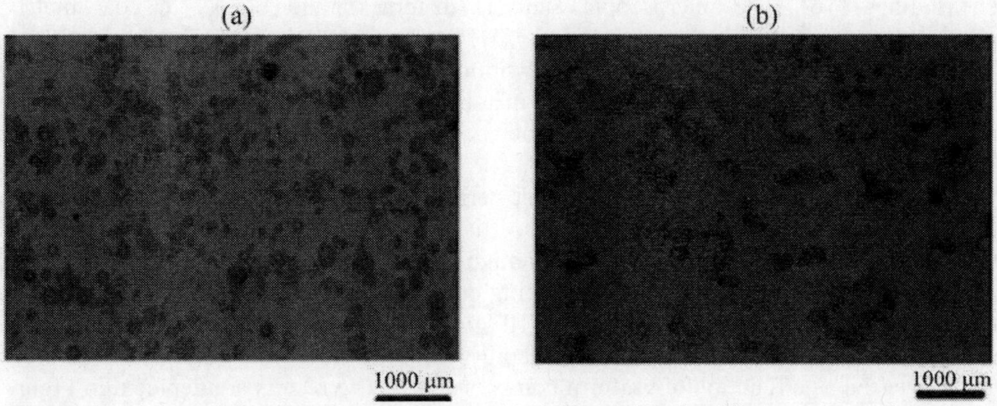

Figure 2. Lightweight proppants for tests: (a) hollow aluminosilicate microspheres ASPM-500; (b) microspheres ASPM-500 with polymeric coating.

3. Experimentation and data processing

Before the filtration tests, jointing of coal was examined in a reflected light. The microstructure analysis included determination of a number and width (opening) of cracks, size of blocks in coal matrix and angle between joint sets. For each crack and coal matrix block, not less than 100 measurements of crack width were taken along the crack strike. Not less than 5 measurements of angles between cracks were taken. The obtained data were statistically processed and averaged.

Coal gas permeability was tested using nitrogen flow in axial direction at constant pressure difference ΔP between the specimen faces and varied compression P.

At the first stage, permeability was tested without simulated hydrofracture. The uniform compression P was varied from 1 to 5 MPa at an increment of 1 MPa for long-flame coal and from 1 to 8 MPa at an increment of 2 MPa for fat coal. At each value of P, a series of tests was carried out at different ΔP from 0.01 to 0.1 MPa at an increment of 0.01–0.02 MPa. At the second stage, permeability of long-flame coal with a through crack without proppant was tested. At the third stage, the tests of permeability of coal with fracture filled with lightweight proppant were performed. In the long-flame coal tests, hollow aluminosilicate miscrosphere ASPM-500 were used; the tests of the fat coal were conducted with the ASPM-500 microsphere covered with polymeric coating. Proppant was uniformly spread on the fracture surface, the specimen was placed in the rubber cut-type seal, and a series of tests was implemented at P and ΔP specified above. For each value of ΔP, at least three experiments were carried out.

Gas permeability factor of specimens before and after proppant addition was calculated from the formula for linear gas flow and stationary filtration process [13]:

$$K_g = \frac{2 \cdot 10^4 V \mu_a P_3}{tS} L \frac{1}{P_1^2 - P_3^2}, \qquad (1)$$

where K_g is the gas permeability factor, 10^{-3} μm^2·(mD); $P_1 = P_3 + \Delta P$ is the pressure at inlet *1* of the test cell (Figure 3), 10^{-1} MPa (bar); P_3 is the pressure at outlet *3* of the test cell (atmospheric),

IOP Conf. Series: Earth and Environmental Science **262** (2019) 012070 doi:10.1088/1755-1315/262/1/012070

10^{-1} MPa (bar); V is the volume of gas flow through the specimen under P_3, cm^3; μ_a is nitrogen viscosity, mPa·s (sP); S is the cross-section area of specimens, cm^2; L is the specimen length, cm; t is the time of gas flow through specimen, s [5]. For ach test series at a certain ΔP, the average gas permeability factor was calculated ($K_g av$).

4. Discussion of the results

According to the common knowledge, coalbed methane flows in natural cracks [14–16]. Thus, coal permeability directly depends on natural crack parameters. This is confirmed by the experimental research findings [16]. The microscopic analysis of cracks and blocks in coal matrix under atmospheric pressure shows that fat coking coal (ZH) has a typical jointing structure described in [14–16]. The average size of blocks is 530 μm, the width (opening) of cracks is 12.7 μm. Average angle between both type joints is 101°. The structure of long-flame coal specimens in the transverse cross-section to the cylindrical specimen axis is characterized by a set of natural joints 47 μm wide at the size of blocks of 4350 μm.

The first series of filtration tests shows that permeability of coal speciemsn D and ZH without crated fracture depends on uniform compression. It is found that permeability of denser D coal under compression increase by 5 times reduces on average by 3.1 times (from $2.8 \cdot 10^{-3}$ μ^2 at P = 1 MPa to $0.9 \cdot 10^{-3}$ μm^2 at P = 5 MPa). Permeability fat coking coal (ZH) under the increase in compression by 8 times reduces on average by 6.7 times (from $60 \cdot 10^{-3}$ μm^2 at P = 1 MPa to $9 \cdot 10^{-3}$ μm^2 at P = 8 MPa).

In the second experimental series, permeability of long-flam coal (D) with fracture without proppant was tested as a function of uniform compression P. The results are depicted in Figure 3. It is found that permeability drops on average by 17.6 times as compression is increased 5 times from 1 to 5 MPa. This implies low efficiency of gas drainage using created fractures without propping.

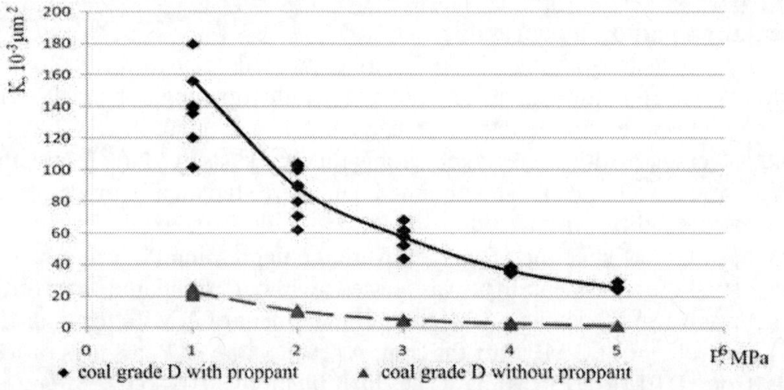

Figure 3. Gas permeability factor K in long-flame coal (grade D) with fracture without and with proppant (hollow aluminosilicate microspheres ASPM-500) versus compression P. Solid curve is plotted by averaged values of K for different pressure differences ΔP in specimen with proppant; dashed curve is plotted by average K for different ΔP in specimen without proppant.

In the third test series, we assessed influence of compression P on permeability of coal grades D and ZH with a through fracture filled with lightweight proppant. In D coal the fracture was filled with hollow aluminosilicate microspheres ASPM-500. Initial opening of the fracture (prior to compression) was 0.4 mm. It is found that permeability of coal grade D lowers on average by 6.2 times as the compression P is increased by 5 times from 1 to 5 MPa (Figure 3). The most probable causes of the permeability decrease are redistribution of proppant grains in the fracture (compaction of proppant), partial destruction and indentation of grains in coal under compression.

In the specimen of fat coking coal grade ZH, the fracture was propped using aluminosilicate microspheres ASPM-500 with polymeric coating. Initial opening of the fracture was 0.4 mm. the obtained results are demonstrated in Figure 4. It is found that coal grade ZH permeability decreases by 1.6–2 times as P is increased from 3 to 8 MPa. Thus, polymeric coating improves operating abilities of hollow aluminosilicate microspheres.

Permeability of coal with fracture filled with proppant in in-seam conditions was assessed using the known theoretical model [5]. The model describes influence of proppant indentation in the created fracture walls under the action of compressive stress, as well as effect of gas flow on opening and permeability of cracks in coal. In the model, fracture is filled with monolayered proppant. In the calculations, we disregard influence of sorption capacity of coal on fracture opening as for nitrogen the sorption effect is insignificant. According to [5] permeability is calculated from the expression:

$$k = k_0\left(1-\frac{R(1-\sqrt{1-\sigma/\pi C}}{b_0}\right)^3, \tag{2}$$

where k is the permeability factor, 10^{-3} μm^2; k_0 is initial permeability, 10^{-3} μm^2; R is the radius of proppant grains, mm; σ is the effective stress, MPa; $b_0 = 2R$ is the initial opening of fracture, mm; C is the cohesive strength of coal, MPa.

The modeling assumed some initial parameters. Initial permeability fits its value under under uniform compression P = 3 MPa and equals 42 10^{-3} μm^2. The proppant grain radius is the average radius of not less than 200 particles after microscopic analysis. The initial fracture opening b_0 is assumed to equal 2R, or 0.4 mm. The cohesive strength value C = 7 MPa is taken from literature [17].

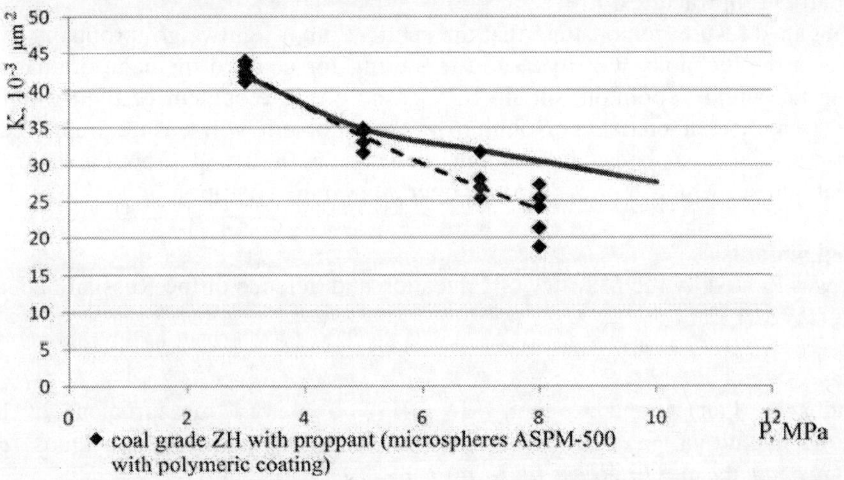

Figure 4. Gas permeability factor K in fat coal (grade ZH) with fracture filled with proppant (hollow aluminosilicate microspheres ASPM-500 with polymeric coating) versus compression P. Dashed curve is plotted by average values of K for various pressure differences ΔP. Solid curve is calculated using the model by Kumar, 2015.

The obtained results are shown in Figure 4. It is found that as the compression is increased, the disagreement of the experimental and calculated data grows. When P < 5 MPa the experiment and calculation curves coincide, while when P > 5 MPa they move apart, and the difference increases with higher compression. Thus, under compression less than 5 MPa, reduction in fracture opening and, consequently in permeability, depends, first of all, on indentation of proppant grains in the fracture walls. Considering strength characteristics of proppant (crushing strength limit 15–20 MPa), it is possible to conclude that the main cause of the disagreement between the experimental and theoretical data under compression higher than 5 MPa is redistribution and compaction of proppant grains.

The implemented research proves that lightweight proppants used to wedge created fractures improves permeability of coal, especially with low initial permeability and weaker jointing. In any case, efficiency of proppants grows under higher compression.

It is found that hydraulic fractures improve permeability of dense coal grade D by 2–8 times under increase in compression by 5 times. Wedging of fractures 0.4 mm initially wide by lightweight proppant improves permeability of coal on average by 7–8.5 times under compression of 1–2 MPa, by 11–14 times under P = 3–4 MPa and by 19.5 under P = 5 MPa. For fractured fat coking coal (grade ZH) under the same conditions, permeability is improved from 1.2 times under P = 3 MPa to 2.6 times under P = 8 MPa. Lower efficiency of hydraulic fracturing in fractured coal is governed by higher initial permeability of coal (more than 60 μm^2) due to developed system of natural fractures.

5. Conclusion

It has been found that hydraulic fracturing with and without propping considerably improves gas permeability of coal ($<3\cdot 10^{-3}$ μm^2) with weakly developed natural jointing. Effect of created fractures on permeability of heavily jointed highly permeable coal ($>60\cdot 10^{-3}$ μm^2) is less than in denser coal. Wedging of 0.4 mm wide fractures by lightweight proppant in denser coal without created fractures improves permeability by 7–8.5 times under compression P = 1–2 MPa, by 11–14 times under P = 3–4 MPa and by 19.5 times under P = 5 MPa. For fractured fat coking coal (grade ZH) permeability is improved by 1.2 times under P = 3 MP and by 2.6 times under P = 8 MPa under the same conditions.

It has been proved that efficiency of gas drainage in coal with hydraulic fractures unfilled with proppant reduces with increasing compressive stress. Reduction in coal permeability with propped fractures under increasing compression is probably connected with redistribution and compaction of proppant particles in fractures.

The obtained results demonstrate that the application of lightweight proppants is a promising way of advance in the technology of hydraulic fracturing for coalbed methane drainage. The long-range research, in the authors' opinion, should be focused on development of light-weight proppants with improved strength characteristics. Solution of this problem will ensure improved permeability of created fractures in coal beds, at pumping of working fluid and propping materials using small-capacity equipment, which is of concern for operation in the condition of a mine.

Acknowledgements

The work was funded by the Ministry of Education and Science of the Russian Federation, Project No RFMEFI60417X0172.

References

[1] Sander R, Pan Z and Connell LD 2017 Laboratory measurement of low permeability unconventional gas reservoir rocks: A review of experimental methods *Journal of Natural Gas Science and Engineering* 37 pp 248–279

[2] Kurlenya, MV, Serdyukov SV, Patutin AV and Shilova TV 2017 Stimulation of underground degassing in coal seams by hydraulic fracturing method *Journal of Mining Science* 53(6) pp 975–980

[3] Serdyukov SV, Degtyareva NV, Patutin AV and Shilova TV 2016 Open-hole multistage hydraulic fracturing system *Journal of Mining Science* 52(6) pp 1210–1215

[4] Jeffrey R and Boucher C 2004 Sand propped hydraulic fracture stimulation of horizontal in-seam gas drainage holes at Dartbrook Coal Mine *Coal Operators' Conference* Wollongong, Australia pp 169–179

[5] Kumar H 2013 Poromechanical Response of Naturally Fractured Sorbing Media

[6] Puri R, King GE and Palmer ID 1991 January damage to coal permeability during hydraulic fracturing *Low Permeability Reservoirs Symposium* Society of Petroleum Engineers

[7] Liang F, Sayed M, Al-Muntasheri GA, Chang FF and Li 2016 A comprehensive review on proppant technologies *Petroleum* 2(1) pp 26–39

[8] Shilova TV, Kurlenya MV and Serdyukov SV 2017 Experimental estimate of fracture permeability with proppant made of alumosilicate microspheres in shallow hydraulic fracturing in oil reservoirs *Journal of Mining Science* 53(6) pp 1085–1089

[9] Serdyukov SV, Shilova TV and Drobchik AN 2017 Laboratory installation and procedure to determine gas permeability of rocks *Journal of Mining Science* 53(5) pp 954–961
[10] RF State Standard GOST R 55663-2013 (ISO 7404-2:2009) *Methods of Petrographic Analysis of Coal. Part 2: Method of Coal Sample Preparation* Moscow: Standartinform 2014
[11] Tanaino AS, Sivolap BB, Maksimovsky EA and Persidskaya OA 2016 Method and means to estimate porosity distribution on the surface of polished section of coal *Journal of Mining Science* 52(6) pp 1216–1223
[12] Serdyukov SV and Shilova TV 2018 A proppant and working fluids for drainage channels formation in a coal seam using hydraulic fracturing *Interexpo GEO-Siberia* 4 pp 27–32
[13] RF State Standard GOST 26450.2-85 *Methods to Determine Absolute Gas Permeability Factor under Stationary and Nonstationary Flow* Moscow: Izd. standartov 1985
[14] Purl R, Evanoff JC and Brugler ML 1991 Measurement of coal cleat porosity and relative permeability characteristics *SPE Gas Technology Symposium* Society of Petroleum Engineers
[15] Pan Z, Connell LD, Camilleri M and Connelly L 2010 Effects of matrix moisture on gas diffusion and flow in coal *Fuel* 89(11) pp 3207–3217
[16] Pan Z and Connell LD 2012 Modeling permeability for coal reservoirs: a review of analytical models and testing data International *Journal of Coal Geology* 92 pp 1–44
[17] Melnikov NV, Rzhevskii VV and Protodyakonov MM 1975 *Reference Book (Cadaster) on Physical Properties of Rocks* Moscow: Nedra (in Russian)

IOP Conf. Series: Earth and Environmental Science **262** (2019) 012071 doi:10.1088/1755-1315/262/1/012071

Effect of cryogenic treatment on disintegration of clay aggregates in washing drum

GV Shirman

Institute of Mining of the North, Siberian Branch, Russian Academy of Sciences, Yakutsk, Russia

E-mail: shirman@inbox.ru

Abstract. The article addresses the problems arising in alluvial clay sand washing in drum-type machines. The author proposes a method to increase disintegration efficiency by cryogenic treatment of initial sands. The results of the experimental studies on formation and destruction of dense clay aggregates in the laboratory model of a washing drum after preliminary freezing–thawing are presented. The investigations have experimentally proved that after 3 freeze–thaw cycles in the temperature range from 243 K to 293 K, a high-clayey lump breaks down at the first minute of disintegration and dissolves in the washing medium, which eliminates subsequent formation of compact clay balls.

1. Introduction

As a result of reduction in number of placers of readily dressible minerals, including diamonds and precious metals, the processing industry is increasingly often faced with materials with high content of clay, which complicates recovery of useful components.

At the present time, clayey sand is disintegrated in drum-type machines which yield satisfactory results in case of readily and moderate-washable sands and offer a drop in efficiency with highly clayey materials (content of silt and clay more than 20%). The key problem of this type washing machines is generation of secondary clayey aggregates—hereinafter, balls, which, depending in size, either go to dredge tailing dumps (40% of balls) or come to tailings after mineral dressing. Owing to excessive adhesion ability, non-decomposed clayey balls can hold some useful minerals which eventually are lost.

As a rule, aimed to improve gold recovery from rebellious materials without modification of processing technologies, it is suggested to use multi-stage washing and concentration circuits which only results in the higher cost of production. One of the methods for improving washing efficiency with highly dispersed clayey sands in the Far North conditions is cryogenic pre-treatment.

The research into the frost effect on clayey rocks, both in Russia [1–5] and abroad [6–9], addressed structural changes in clayey materials in static environment without regard to the process of disintegration. The analyses of cryogenic pre-treatment aimed to weaken sands before washing focused on dispergation in water-and-air medium [10], preparation of rock mass [11], etc. This article describes studies into the effect of freezing on disintegration of sands in dynamic environment, in particular, in drum-type machines.

The earlier research has divided the process of ball formation in a dead-end washing drum into three stages: gain in weight, compaction and disintegration [12]. The aim of the new cycle of investigations is to estimate the effect of cryogenic pre-treatment on the behavior of clay balls in washing drums and to reveal details of change in adhesion ability of highly dispersed geomaterias, as well as to identify significant factors of freezing (time and temperature, number of freeze and thaw cycles) which have influence on the dynamics of formation and disintegration of compact clayey aggregates.

2. Experiment results

The tests were carried out on a lab-scale model of a washing drum *1* (Figure 1) with a seal cap *2* placed on supporting rollers *3* mounted on a frame *4* with a pressure regulator *5* and an electric motor *6*.

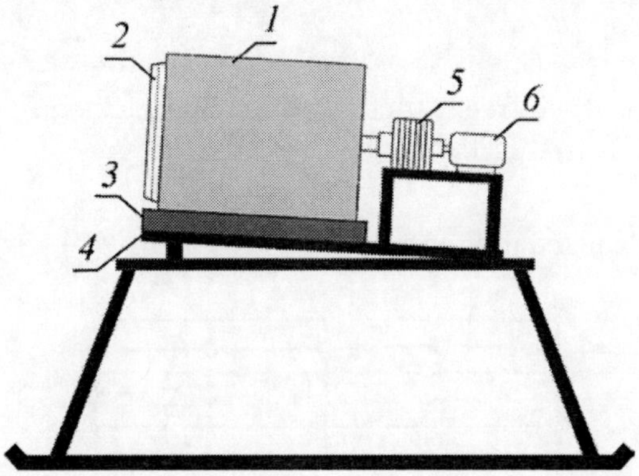

Figure 1. Lab-scale model of dead-end washing drum.

The balls for the tests were made of a dispersed material from a Kuranakh group placer, Aldan Region in the Republic of Sakha (Yakutia). The content of fines –0.05 mm in the initial sand was 45%, the size of +2 mm was pre-separated. The moisture of the balls was 25% and the weight was 120 g. Some balls were air-dried, the other balls were subjected to pre-freezing. In addition, balls with the preset moisture content without cryogenic pre-treatment were also examined. Frosting of specimens was implemented outdoor, in sealed containers, under the temperature of 243 K for not less than 15 h. Thawing was carried out in a lab room, under the temperature of 293 K, in sealed containers for 15 h.

To estimate dynamics of lumping, a clay ball with the certain properties (weight, moisture content, with and without cryogenic pre-treatment) was placed in a dead-end washing drum with water-and-clay mixture (660 g of mixture with the content of fines –0.05 mm of 10 and pebbles –5+2 mm in size of 340 g) with the preset moisture content of 50% to simulate a washing medium. The ball was taken out of the drum and weighted in 1 min.

The tests of clay lumping in the dead-end washing drum demonstrated significant effect of cryogenic pre-treatment on disintegration of compact clayey aggregates. In the tests, the air-dried balls did not swell and fail for 1 h, and showed no appreciable adhesion ability for lumping, while the balls with the moisture content of 25% placed in the drum without cryogenic pre-treatment gained on average 20% in weight in the form of pebbles and clay from the washing medium during the first minutes, swelled by the 10th minute into compressed sphere and completely lost the spherical shape and became clots by the 15th minute of the test.

The samples subjected to washing after one cycle of freezing and thawing showed low adhesion and gained weight insignificantly (round 10%) with the material from the washing medium but became more resistant to disintegration—the balls lost the gained weight by the 20th min and turned into clots by the 40th min.

After two freeze–thaw cycles, the samples showed no adhesion ability, became unstable, swelled and became plastic in 5 minutes. The surface of the balls after three freezing cycles was cracked (Figure 2a), the samples became brittle and some of them decomposed during thawing (Figure 2b). The test proved that after three freeze–thaw cycles, the balls completely dissolved in the drum in 1 minute.

Figure 2. A ball after three freeze–thaw cycles: (a) cracks on the surface of the frozen ball; (b) decomposed ball after thawing.

The testing results are depicted in Figure 3.

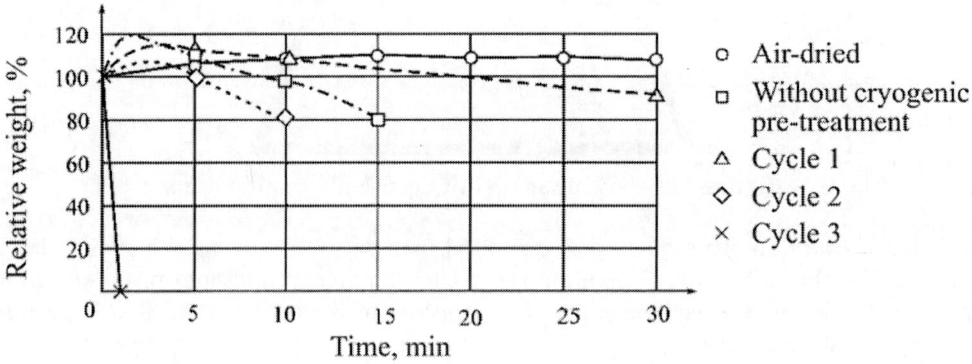

Figure 3. Dynamics of disintegration of clayey balls in washing drum after various techniques of pre-treatment.

3. Conclusions

As a result of the estimate of the cryogenic pre-treatment effect on disintegration of most highly clayey balls in the dead-end washing drum, it has been experimentally found that after three cycles of freezing–thawing in the temperature range of 243–293 K, the highly clayey balls with the moisture content of 25% fails in the first minute of disintegration and disperses in the washing medium, which eliminate subsequent formation of compact lumps.

References

[1] Ershov ED 1979 *Moisture Transport and Cryogenic Textures in Disperse Rocks* Moscow: MGU (in Russian)

[2] Tsytovich NA 2009 *General and Applied* Mechanics *of Frozen Soils: Textbook for Engineering and Construction Universities* 2nd Edition Moscow: Librokom (in Russian)

[3] Balyasnikov GG 1973 Study of frozen soil soaking *Problems of Construction in the North* Krasnoyarsk (in Russian)

[4] Kurilko AS 2004 *Experimental Studies of the Effect of Freezing–Thawing Cycles on the Physical and Mechanical Properties of Rocks* Yakutsk: YaF GU SO RAN (in Russian)

[5] Beylin A Y 1988 *Technology of Disintegration for High-Clayed Sands of Gold, Tin, and Diamond Placers in the Air–Water Environment* Yakutsk (in Russian)

[6] Yıldız M and Soğancı AS 2012 Effect of freezing and thawing on strength and permeability of lime-stabilized clays *Scientia Iranica* Vol 19 Issue 4 pp 1013–1017

[7] Ergulera ZA and Ulusay R 2009 Assessment of physical disintegration characteristics of clay-

bearing rocks: Disintegration index test and a new durability classification chart *Engineering Geology* Vol 105 Issue 1–2 pp 11–19

[8] Ergulera ZA and Shakoorb A 2009 Relative contribution of various climatic processes in disintegration of clay-bearing rocks *Engineering Geology* Vol 108 Issue 1–2 pp 36–42

[9] Zhen-Dong Cui, Peng-Peng He and Wei-Hao Yang 2014 Mechanical properties of a silty clay subjected to freezing–thawing *Cold Regions Science and Technology* Vol 98 pp 26–34

[10] Zakharova SM 1989 Influence of mineral composition and physical structure of high-clay materials on the process of disintegration in water-air environment *Thesis of Cand Tech Sci* Yakutsk: IGDS (in Russian)

[11] Kislyakov VE, Kataev CB, Shakina YV and Korotenko VV 2001 Development of technology for cryogenic and softening of clayey gold-bearing sands *Problems of Geology and Exploitation of Mineral Resources* Tomsk: TPI pp 661–662 (in Russian)

[12] Matveev AI and Shirman GV 2011 Dynamics of formation of clay pellets in disintegration of high-clay sands in washing drums *GIAB* No 10 pp 266–268

Substantiation of process parameters in separation of large-size granite blocks from a rock mass

PN Tambovtsev* and AA Neverov

Chinakal Institute of Mining, Siberian Branch, Russian Academy of Sciences, Novosibirsk, Russia

E-mail: *tambovskiyp@mail.ru

Abstract. On the basis of experimental data, a technological scheme to separate large-size granite blocks from a rock mass by an impact method is proposed, the process parameters of block preparation for separation are verified by physical modeling, and a rock-breaking tool is proposed; the dependence of the main crack length vs. impact parameters is determined.

1. Introduction
In the world practice the drill-and-blast and mechanical rock-breaking processes are used to produce stone blocks [1–4]. The present paper deals with consideration of a process for separating large-size granite blocks from a rock mass by applying the percussion breakage of rocks under the tension strain effect. The general process diagram is shown in Figure 1. A rock mass is stripped in the front-and-flat manner with a pass length multiple of a final block length; thereto a width and height of the pass is equal to width and height of a block. The diagram implies availability of three open surfaces of a stone block, a natural fracturing in a rock bed, end slot dozing, drilling of blasthole rows at vertical rear and end surfaces. After delineation of blocks with blastholes one or few impact tools (special wedges) are intruded into blastholes at rear surface and separate a monolith from a rock mass; next, they are inserted into blastholes at end surfaces; and a monolith is divided into final blocks.

2. Technology of granite blocks separation
The working member of the tool (Figure 1c) is made as a two-step cylindrical bar with a taper flange *12* [5]. When the tool penetrates into a hole the surfaces of taper flange between taper planes *13* act with horizontal resultant force F on hole walls (Figure 1d). Downhole penetration of the tool results in wedging of a rock mass and growth of a crack.

The specific feature of the wedge tool is that it provides an acceptable opening of crack edges δ in the zone nearby a blasthole with a downhole tool. Excessive values of δ promote an increase in bending stress in a crack tip with possible breakout of the main fracture through the front surface of a bench. A direction of crack growth is governed by a set of factors, viz., location of blasthole rows at a preset surface, taper surface of the tool relative to the normal to a crack plane, and admissible opening of a crack $\delta = d_1 - d$ (Figure 1c).

Main parameters of block separation from a rock mass are (Figure 2): H—bench height (a block); B—width of a pass (a block); L—length of a block; L_1—length of a working area; d—blasthole diameter; h—blasthole length; a—pitch of blastholes; b—pitch of tool installation; A_{ed}—impact energy; n—number of impacts.

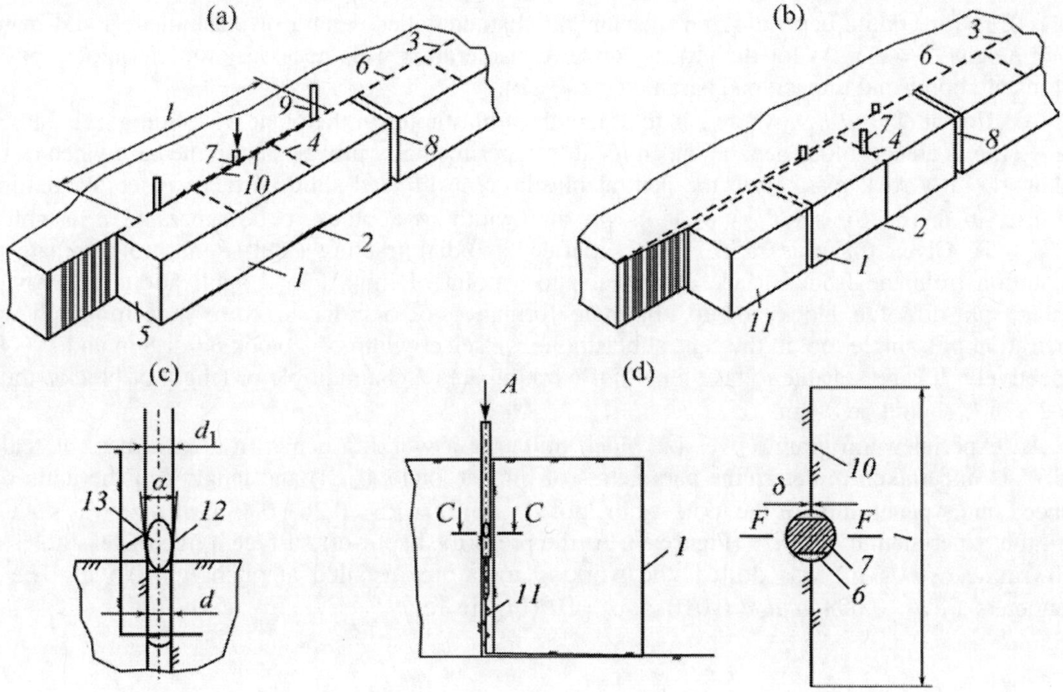

Figure 1. Technological flow sheet to separate stone blocks from a rock mass by drill-and-wedge process with formation of the main fracture at rear surface (a); with separation of final blocks (b); with the tool before its sinking into a blasthole (c); cross-section of a block at the moment of its detachment from a rock mass (d): *1*—front surface; *2*—crack in a bed; *3*—blast hole row at the rear surface; *4*—blasthole row at the end surface, *5*—end surface; *6*—working blasthole; *7*—tool; *8*—end crack; *9*—tool before immersion; *10*—the main crack; *11*—a final block; *12*—taper flange; *13*—flattened surface; *d*, d_1, α—tool parameters.

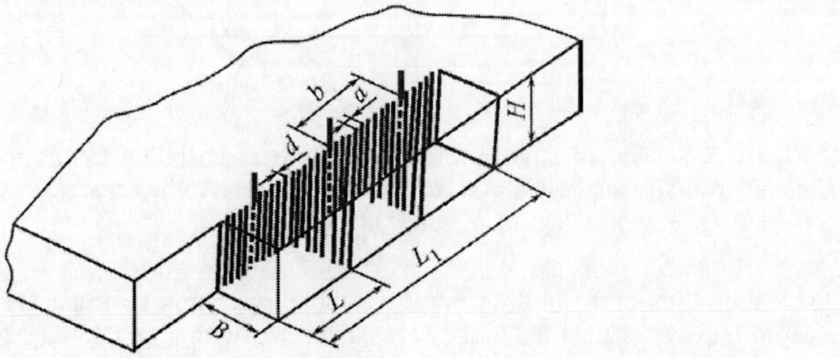

Figure 2. Prime process parameters of block preparation for separation from a rock mass.

When selecting parameters of a production area (H, B, L_1) it is imperative to consider a location of natural fractures (stratal, vertical, longitudinal, and transverse) which split a rock mass into discrete natural joints. The maximum size L of a final block does not usually exceed 2.8 m, its width B and height H are no more than 2 m [4]. Parameters of idle blasthole rows (d, a, h) are determined considering specifications of drill units and rock strength. In production of stone blocks the line drill units are employed to make blast holes of d = 0.032–0.042 m, a recommended spacing of drilling is a

= 0.1–0. m for granite depending on mechanical characteristics, length of blastholes should match a bench height (h = H). As for the process under consideration, it is necessary to determine a pitch of tool installation b and impact load parameters (A_{ed}, n).

Coefficient $Q = zd / S$, where z is total length of blastholes in the plane of splitting; d is blasthole diameter; S is area of block separation. In [6] it is experimentally proved that in the case when a single tool works in a rock mass from the central blasthole, a directed split-off takes place at coefficient $Q \geq 0.12$–0.16. The threshold ratio of length and width of a block to be separated is established ($L/B < 5$). Given that this ratio is not satisfied, a radial fracture (fault) tends to originate with orientation from the front surface of a bench to a point of impact load application. To provide a wanted split-off of a block and to eliminate formation of a radial fracture in a monolith under penetration of a single tool in the central blasthole, we select width of a block B = 1.8 m and $L_1 < 9$ m, respectively. It is reasonable to take length of working area L_1 be multiple of length of blocks, thus, at L = 2.8 m L_1 should be 8.4 m.

An experiment on granite oversize block imitating a working bench in a rock mass at scale of 1 : 5.75 is undertaken to determine parameters of impact load (A_{ed}, n) and length l of the main crack induced under penetration of the tool. A block of $B \times H \times L = 1.5 \times 0.26 \times 0.45$ m in size was separated from the experimental oversize (Figure 3). At the prescribed split-off surface a line of blastholes (d = 0.008 m, a = 0.05 m) was drilled; the working tools are installed at pitch b = 0.5 m. The tool parameters are d = 0.008 m, d_1 = 0.010 m, α = 10°(Figure 1c).

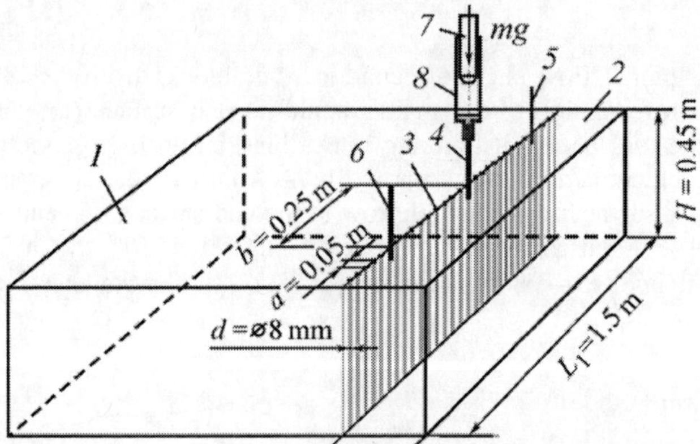

Figure 3. Scheme of physical modeling of block separation from a rock mass: *1*—section of granite oversize suggested as a rock mass, *2*—block to be separated, *3*—line of blastholes, *4*–*6*—tools, *7*—impact device, *8*—guide device.

First, tool *4*, mounted in the central blasthole, was sunk by impacts, then tools *5*, *6* followed in succession. Impacts against the tools were performed with a certain energy by means of throwing an impact device of 20 kg in mass along a guide line from height of 0.2–0.46 m. Penetration x of the tool into a blasthole, an induced fresh fracture, fracture length l along a line of blastholes, and fracture opening δ in the tool zone were measured after every or few impacts.

According to the physical modeling results (Table 1, Figure 4) it is obvious that the impact energy should be increased with growth of an induced crack, thereto, crack formation and its growth are stepwise in character and depend on a magnitude of crack edge opening (Figure 5). To provide penetration of the central tool the required impact energy was A_{ed} =90 J, accumulated energy was $A_{(4)}$ =$A_{ed(4)}$ J; n = 2310; for tools *5* and *6*, J: $A_{ed(5)} = A_{ed(6)}$ =40, $A_{(5)}$=660, $A_{(6)}$ =560.

Table 1. Parameters of physical modeling of block separation from a rock mass.

n	A_{ed}, J	*A*, J	*x*, mm	*l*, m	δ_1, mm	Notes
Penetration of the central tool *4* (Figure 3)						
			0	0		—
1–2	40	80	4	0	—	—
3–5	40	200	10	–“–	—	—
6	40	240	12	0.3	0.1	Initial fracture
7	40	280	14	–“–	0.1	—
8	40	320	16	–“–	0.2	—
9–16	35	600	36	–“–	0.3	Notable withdrawal of the impact device
17–21	90	1050	42	–“–	0.6	—
22	90	1140	44	0,6	0.7	—
23–24	90	1320	45.5	0.6	0,8	—
25	90	1410	46	0.7	0.9	—
26–34	90	2220	49	0.7	1.5	—
35	90	2310	52	0.8	1.7	—
Penetration of tool *5* (Figure 3)						
			0	0		—
1	20	20	3	0.1	0.1	Initial fracture
2–10	20	200	20	–“–	0.2	—
11	20	220	23	0.22	0.2	—
12–14	20	280	31	0.22	0.3	Extension of the fracture to end surface
15	20	300	33	0.5	0.3	Fracture length at the end surface - 300 mm
16–25	40	700	53	–“–	—	–“– 400 mm
26–35	40	660	63	–“–	—	–“– 440 mm
Penetration of tool *6* (Figure 3)						
			0	0		—
1–2	40	80	7	0	—	—
3	40	120	9	0,05	0,1	Initial fracture
4–9	40	360	26	–“–	0.2	—
10	40	400	30	0,5	0.3	Extension of the fracture to end surface
11–14	40	560	40	–“–	0.4	Fracture length at the end surface 440 mm (Figure 4d)
Disintegration	—	—	—	—	—	Mechanical retraction of a block (Figure 4f)

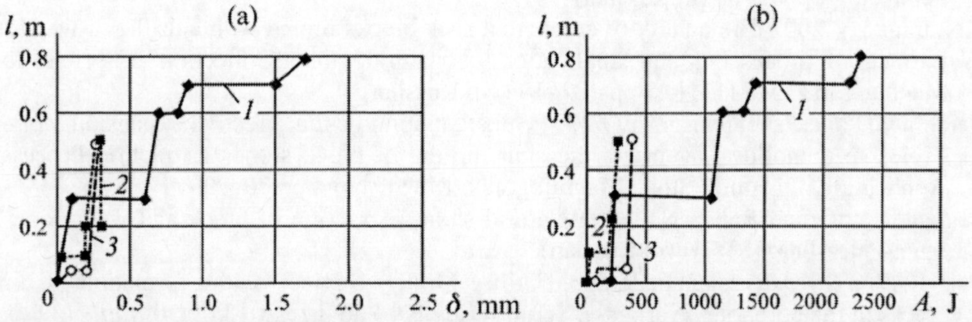

Figure 4. Main crack formation along blasthole line: (a)—dependence of crack length vs. crack edge opening in a blasthole with a tool; (b)—dependence of crack length vs. accumulated energy of impacts (*1*—fracture induced under penetration of the central tool *4*; *2*, *3*—cracks induced under penetration of side tools *5*, *6*).

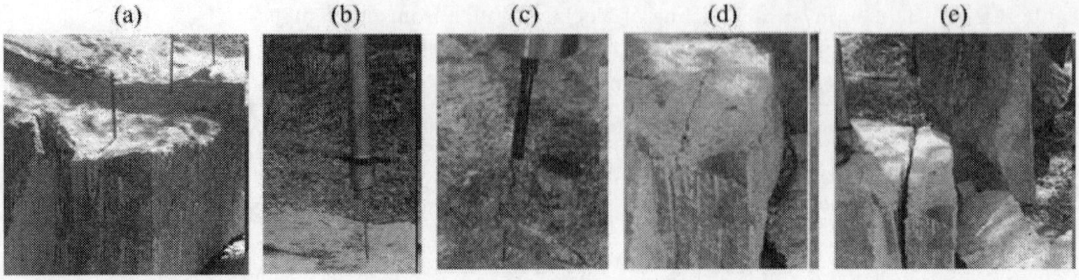

Figure 5. Separation of a granite block: (a)—installation of downhole tools; (b)—mounting of an impact device on the tool *4*; (c)—the initial crack formation induced by the central tool; (d)—crack length of 400 mm at the end surface under penetration of the tool *5*; (e)—resultant block separation.

Considering the principles of geometric and dynamic similarity [7], we determine basic parameter values for a real block separation at an actual open pit section. Geometric scale of modeling is $a = B_{nat} / B_{mod} = 1.5 / 0.26 = 5.77$, where B_{nat}, B_{mod} is width of a block to be separated in the real open pit and in the model. The scale of impact energy is $в = A_{nat} / A_{mod} = m_{nat} / m_{mod} = a^3 = 192.1$, where A_{nat}, A_{mod} are energy of a single impact of a commercial and modeled hammer; m_{nat}, m_{mod} are mass of impact devices of a commercial and modeled hammers. In view of the above, the process parameters of rock-mass preparation to separate blocks are as follows: *H*—2.9 m, *B*—1.5 m, L_1—8.6 m, *L*—2.9 m, *d*—46 mm, *h*—2.6 m, *a*—0.29 m, *b*—2.9 m. Energy of a single impact is $A_{ed(4)}$—17.1 kJ for the central tool, $A_{ed(5,6)}$—7.6 kJ for the side tools, accumulated energy of impacts, kJ: $A_{(4)}$—439.1; $A_{(5)}$—125.5, $A_{(6)}$—106.5.

3. Conclusions

The process diagram is proposed to separate large-size granite blocks from a rock mass by impact punching with wedge tools providing an admissible opening of crack edges in the zone nearby a blasthole and excluding exposure of the main crack at the front surface of a bench.

The physical modeling method is implemented to determine a range of impact energy, applied to the tool; a pitch of the tool installation in the process of monolith separation is determined.

It is demonstrated that the required impact energy should be increased with growth of a crack; formation of a crack and its growth are stepwise in character and depend on a level of crack edge opening and the accumulated impact energy.

References

[1] Tkachuk KK and Grebenyuk TV 2012 Stone block splitting off with static load Problems of subsoil use Proc. International Forum-Competition of Young Scientists Part 1 Saint-Petersburg pp. 82–85 (in Russian)

[2] Kokunina LV 2007 The quality of competent rock blocks mined with a drill-and-wedge method Extraction, processing and application of natural stone Collection of scientific papers Magnitogorsk: MSTU No 7 pp 102–108 (in Russian)

[3] Pershin GD and Pshenichnaya EG 2004 Substantiation of the effective expansion work of non-explosive demolition agents in the spur mining of block stones Extraction Processing and Application of Natural Stone Magnitogorsk: MSTU No 4 pp 70–79

[4] Karasyov YuG and Bakka NT 1997 Natural stone Extraction of block and wall stone Tutorial Saint-Petersburg: SPGU (in Russian)

[5] Pat. 124922 2013 Russian Federation Utility Model, Device for the formation of directional cracks in the boreholes (variants), Tambovtsev PN Pub date 2013/02/20 *Bull.* No 5

[6] Tambovtsev PN 2015 Physical simulation of stone block cutting under impact action on plastic substance in drill hole Journal of Mining Science Vol 51 No 1 pp 73–80

[7] Alabuzhev PM, Geronimus VB, Minkevich LM et al 1968 Theory of similarity and dimensions Modeling Moscow: Vysshaya Shkola (in Russian)

Influence of part-to-part gaps in air distribution system on operation of DTH hammer

VV Timonin*, SE Alekseev, DI Kokoulin, B Kubanychbek and EM Chernienkov

Chinakal Institute of Mining, Siberian Branch, Russian Academy of Sciences, Novosibirsk, Russia

E-mail:*timonin@misd.ru

Abstract. The authors illustrate the role of pneumatic percussive machines in drilling. The article presents the elementary diagram of down-the-hole (DTH) hammer with the marked places of air leakages in the system of air distribution and describes and their influence based on the research data. The variation in the basic power characteristics of the machine is given as function of the size of gaps in the sliding joints. It is shown that the machine design can be improved by means of using elastic valve to allow discharge of air working chambers during the idle stroke of the hammer.

1. Introduction

Mineral mining industry both in Russia and in the world widely uses percussive machines driven by fluid energy source—compressed air, pressurized fluids or mixtures. Such machines operate in drilling hard rocks in underground and surface mines.

In the mid-1900s in Russia, the first prototypes of air-driven percussive machines were designed and manufactured. These machines were to be placed at the bottom of drill hole and named down-the-hole punchers (hammers). Such arrangement ensures best energy transfer to botttomhole and allows higher penetration rate and longer drilling. Compressed air is used as an energy source and a cleaning agent, which makes drilling even more rational. The introduction of that technology made it possible to break ore in long holes.

Drilling with DTH hammers is feasible in rocks of any hardness and jointing. These machines feature high productivity, reliability and simple maintenance. Drilling rate is independent of the hole length. Air flush system is also highly efficient. Down-the-hole equipment provides low noise and reduced deviation of drilling path from straight line. These advantages, as well as simplicity and relatively low cost make these machines highly attractive.

The piston in the pneumatic percussive machines is a part of the air distribution system. The piston is movable part, and its seat areas are made with clearances governed by the diameter. Air inevitably leaks through the clearances. Lack of sufficient knowledge on this point complicates mathematical modeling of the machine.

Analysis of the machine structure shows that influence of air leakage on performance is different per design. For instance in throttled and spay systems, air leakages weakly affect energy efficiency of machines and mainly result in the increased air consumption. In some air-driven hammers, air can leak in-between operative chambers of idle and power strokes. In this case, the influence of air leakage is more appreciable as it worsens energy efficiency of machines. Air leaks from the air feed line to the working chambers are the most hazardous. The number of leak places is of concern.

2. Design features of DTH air hammer

It is most often that clearance leakages take place in critical or supercritical regimes; for this reason, the mass air flow rate Q_{mass}, kg/s, is given by [1]:

$$Q_{mass} = f_{gap}\sqrt{\frac{2K}{K+1}\left(\frac{2}{K+1}\right)^{\frac{2}{K-1}}\frac{P}{\nu}},$$

where f_{gap} is the cross-section area of the gap, m^2; K is the adiabatic exponent (for air, $K = 1.4$); P, ν ate the absolute pressure and specific volume of compressed air in the chamber in front of the clearance, kg/m^3.

It follows from the formula that the highest influence on air flow rate is exerted by the clearance cross-section area. At the same time, it is important where the leak occurs and how it influence the machine performance.

At the Institute of Mining, SB RAS, a basic diagram of air-driven hammer has been developed (Figure 1) to meet to the full extent the requirements imposed on the machines [2–4]. The closed air hammer circuit with air exhaust to the hole bottom prevents inlet of abrasive particles in the hammer and ensures best cleaning of the bottom from drilling chips.

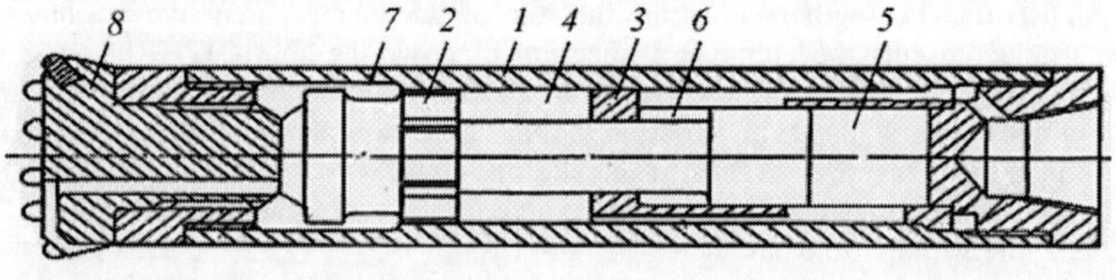

Figure 1. Basic diagram of DTH air hammer: *1*—body; *2*—piston; *3*—air distribution barrel; *4*—ring space of power stroke chamber; *5*—end power stroke chamber; *6*—constant-pressure idle stroke chamber; *7*—counterbore; *8*—drill bit.

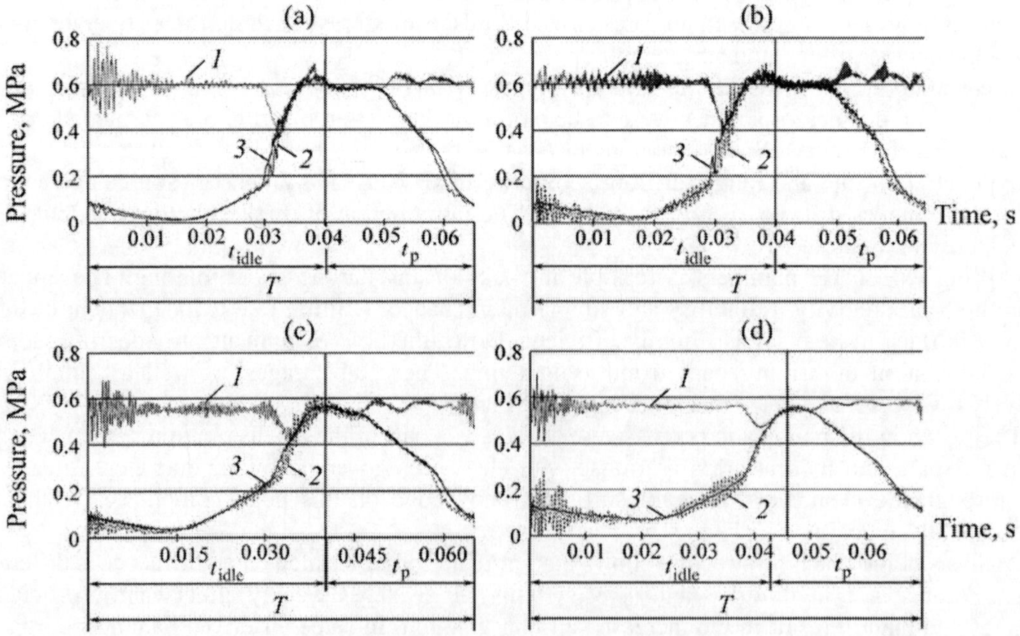

Figure 2. Pressure in the power and idle stroke chambers at (a) S = 0.587 cm^2; (b) S = 0.747 cm^2; (c) S =1.207 cm^2; (d) S = 1.527 cm^2: *1*—pressure p_{idle} in the idle stroke chamber, MPa; *2*—pressure p_{ring} in the ring power stroke chamber, MPa; *3*—pressure p_{end} in the end power stroke chamber, MPa; t_{idle}—idle stroke duration, s; t_p—power stroke duration, s; T—cycle time, s.

In this circuit, it is most hazardous when leakages take placed in gaps of running fits of piston *2* and barrel *3*. During idle stroke of the piston, additional inflow of compressed air in the power stroke chamber obstructs travel of the piston. Since the idle stroke area is comparatively small, leakages can be influential and, thus, their admissible value should be determined. For the more precise estimation of influence exerted by air leaks on the machine energy efficiency, full-scale testing was undertaken.

By the basic diagram above, DTH air hammer PV170 has been designed and manufactured for hole drilling with diameter of 170 mm using energy source with a wide pressure range from 0.6 to 2.0 MPa. The weight of the piston is 17.1 kg.

The tests were carried out on vertical lab bench with recording of pressure charts in the twin chambers of power stroke and in the in-between idle stroke chamber. A blow marker was involved I the tests. The pressure charts were recorded at different adjustments of the machine with the total clearance area *S* of 0.587, 0.747, 1.207 and 1.527 cm^2. The compressed air pressure was 0.6 MPa (Fig. 2).

The pressure charts were processed using the known procedure [5]. The main parmeters of the machine were determined using the pressure charts (Table 1).

Table 1. Main parameters of PV170 DTH hammer at different total areas of clearances

Total clearance area *S*, cm^2	Blow energy *A*, J	Blow frequency *n*, min^{-1}	Recoil factor k_o	Blow velocity *v*, m/s	Piston travel length *L*, mm	t_{idle}/t_{power}
0.587	208.84	922.37	0.21	4.95	82	1.45
0.747	206.95	928.07	0.26	4.92	84	1.47
1.207	160.09	929.04	0.29	4.32	73	1.61
1.527	159.55	857.14	0.29	4.33	71	2.03

The relationship of the energy efficiency characteristics *A* and *n* with the value of *S* is shown in Figure 3. It is seen that the total area of clearances most affects the unit blow energy.

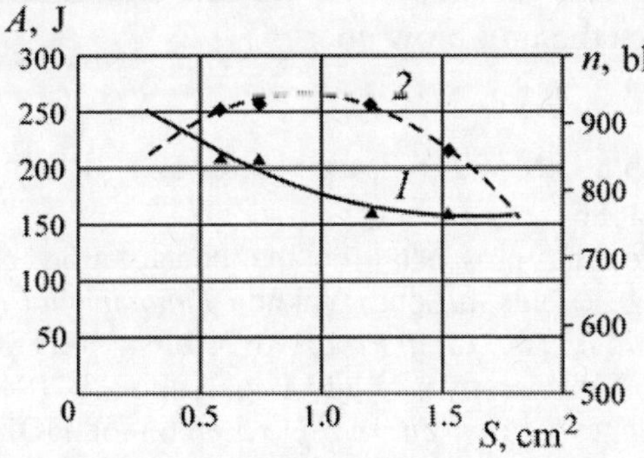

Figure 3. Unit blow energy *A* (*1*) and blow frequency *n* (*2*) versus clearance cross section area *S*.

During the idle stroke, air inflow in the power strike chamber, due to large difference between the values of the effective areas of the piston, impedes and shortens travel of the piston. Due to this constraint, the piston fails to increase blow energy during the power stroke.

Aiming to slacken this effect, the structure is added with a special elastic ring to act as an unloading valve (Figure 4) [6].

The elastic ring *6* in the groove of the front barrel *4* with a gap relative to the inner wall of the body *3* acts as an unloading valve. During the idle stroke travel of the piston *5*, air compressed in the ring *10* and end *11* power stroke chambers after exhaust as well as leaked air inflow in these chambers from the clearances of running fits of the piston can pass to the atmosphere through these clearances and grooves in the front barrel.

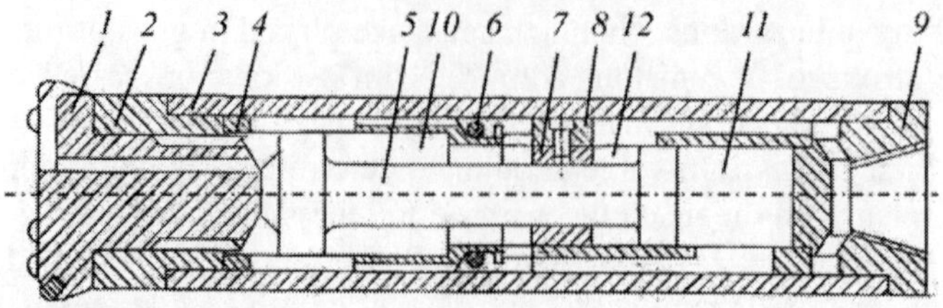

Figure 4. Basic diagram of PNB76R DTH hammer with power stroke discharge: *1*—drill bit; *2*—axle box; *3*—body; *4*—front barrel; *5*—piston; *6*—elastic ring; *7*—separation ring; *8*—back barrel; *9*—adaptor; *10*—ring power stroke chamber; *11*—end power stroke chamber; *12*—idle stroke chamber at constant pressure.

Since the piston area on the side of the power strike chambers in several times larger than on the side of the idle stroke chamber, compressed air in both power stroke chambers decelerates idle travel of the piston and cut its length. Air outlet from the end and ring power stroke chambers after exhaust is shut by the elastic valve makes it possible to reduce counter pressure and, as a consequence, to slacken resistance to the piston movement, and the idle stroke travel of the piston lengthens in the meanwhile. During the power stroke the piston is longer under the action of compressed air pressure from the side of the power stroke chamber due to the longer travel, which elevates energy of unit blow and impact capacity of the machine in whole.

The elastic ring valve conveniently fits in the structure of the air hammer without violating the design simplicity [7]. The further research has shown that the use of such unloading valve allows higher unit blow energy by 20%.

3. Conclusions

The research has shown that the increase in the area of gaps in the mobile joints in the air distribution systems considerably worsens energy efficiency of the machine, mainly, due to reduced energy of unit blow. This drawback can be reduced by using an elastic valve to discharge air from both power stroke chambers during idle stroke of the piston, which lengthens the power stoke travel of the piston and increases the unit blow energy.

References

[1] Timofeev NG and Skryabin RM 2011 Large diameter drilling equipment *Polzunov. Almanakh* No 4/2 pp 84–86

[2] Harlamov YuP 2009 State and prospects for the construction of pile foundations in the development of oil and gas fields in South Yakutia *Fundamental Problems of Geoenviroment Formation under Industrial Impact: Conf. Proc.* Novosibirsk: IGD SO RAN Vol 2 pp 58–61

[3] Kondratenko AS, Timonin VV, Karpov VN and Popelyukh AI 2018 Ways to improve rotary-percussive drilling efficiency *Gornyi Zhurnal* No 5 pp 63–68 DOI: 10.17580/gzh.2018.05.09

[4] Fedulov AI, Arkhipenko AP and Mattis AR 1980 Selection of Size of Gaps in Frictional Pairs of Pneumatic Hammers Novosibirsk: Nauka (in Russian)

[5] Alekseev SE RF Patent No 2090730 *Byull. Izobret.* 1997 No 26

[6] Esin NN 1965 *Procedure for Testing and Perfection of Air-Driven Hammers* Novosibirsk: SO RAN (in Russian)

[7] Timonin VV, Alekseev SE, Kokoulin DI and Zabolockaya NN RF Patent No 2647716 *Byull. Izobret.* 2018 No 8

[8] Petreev AM and Primychkin AYu 2016 Ring-type elastic valve operation in air hammer drive *Journal of Mining Science* Vol 52 No 1 pp 135–145

[9] Chervov VV and Chervov AV 2014 Temperature of compressed air and exhaust in air hammer with elastic ring valve *J. Fundament. Appl. Min. Sci.* Vol 1 No 2 pp 192–198 (in Russian)

[10] Chervov VV et al RF Patent No 2105881 *Byull. Izobret.* 1998 No 6

IOP Conf. Series: Earth and Environmental Science **262** (2019) 012074 doi:10.1088/1755-1315/262/1/012074

Automated analysis of operating cycles of impact machines

VV Timonin*, EM Chernienkov and TV Timko

Chinakal Institute of Mining, Siberian Branch, Russian Academy of Sciences, Novosibirsk, Russia

E-mail: *timonin@misd.ru

Abstract. The results of research and analysis of operating cycles of hydraulic and pneumatic impact machines are described. The effectiveness of programming support in engineering of new machines and development of working models is substantiated. The analysis of a construction design is performed to enhance its energy efficiency.

1. Introduction

The costs of drilling can make 30% of the cumulative expenditures in drilling-and-blasting [1]. For this reason, drilling equipment is of particular concern.

Drilling is now mostly is performed by rotary–percussive method using down-the-hole air-driven hammers (DTH). From the international experience, the range of DTH use is from 110 to 250 mm, large diameter hole-making involves roller-bit drilling. In rocks with uniaxial compression strength more than 100 MPa, air percussion drilling is more efficient though the penetration rate lowers slightly as hole length increases as the impact impulse source is located directly at the bottomhole. Such percussion layout allows high-accuracy drilling path [2].

The downhole hydraulic hammer drills driven by fluid under pressure to 0.4 MPa ensure higher penetration rate in hard rocks as compared with DTH air hammers. One the other hand, hydraulic hammer drill can be used in low-jointed rocks; otherwise, the drilling column is highly likely to become stuck due to inefficient mud flush at high floe rate of drilling fluid [3].

The Theory, Design and Operation of Mine Drilling Equipment Workshop held by the Chinakal Institute of Mining in 2017 has shown that the domestic manufacturers of DTH hammers lack tools and procedures to determine energy parameters of hydraulic and air-driven pulse-generating systems. Downhole machines are designed with the present energy parameters without estimations of their values.

Hydraulic and air hammers with free piston allow varying drilling mode. There are three modes: pre-drilling (energy source pressure is lower than its operating pressure); normal drilling mode; bottomhole cleaning (higher consumption of energy source). The impulse action of impact machines is largely influenced by physical and mechanical properties of rocks, energy source supply pressure, etc. [4, 5]. Therefore, it is necessary to find exact energy parameters for different operating modes of downhole impact machines.

Foreign developers of downhole pneumatic percussion equipment increased impact energy by means of rising pressure of energy source (by now to 3.5 MPa). According to the foreign experience, the domestic DTH hammer designs should satisfy a few requirements such as: wide range of compressed air pressure (1–3 MPa); back valve to prevent sludging-up in drilling in water-cut rocks and using air-and-water mixture; no perforation holes in the walls of the drill housing to avoid early wear under friction of drill fines; air exhaust to the bottomhole; valvle-free air distribution system; high efficiency (0.85–1.1 m^3/kW·min); long life of piston (3/5 thou m—granite/basalt) and drill bits (0.6/1.2 thou m—granite/basalt) [6].

2. Development stages of downhole impact machine

At the present time, development of a downhole impact machine includes:

—setting required characteristics (hole diameter, useful life, energy source pressure, etc.);

—developing basic diagram (no valve; back valve);

—simulation modeling (air distribution system design with CAD software [7] to determine patterns of inlet and outlet edges, volumes of work chambers, length of power stroke, etc.);

—construction diagram;

—manufacturing (production effectiveness);

—benchmark testing (determination of actual energy characteristics for further debugging of the machine);

—industrial trial (in-situ operation).

Energy parameters of air distribution system are determined at the Institute of Mining using the procedure proposed in [8]. Figure 1 demonstrates the block diagram of a test bench. Using the control panel, the the blow time is recorded based on the signals from the impact sensor *1*, and the air pressure charts in the working chambers and supply line are recorded based on the readings from transducers *2* and *3*, respectively. Then, the analog signals from the transducers are sent through the amplifier to the analog-to-digital converter and loaded to a computer file.

As a rule, when the pressure charts are recorded, one working cycle is processed. In case of unsteady-state operation of the impacting mechanism, the most typical working cycle is selected, and the resultant force is plotted as function of the cycle time. The plot is then interpreted based on Sudnishnikov's theorem on "movement of mass during action of force" [9].

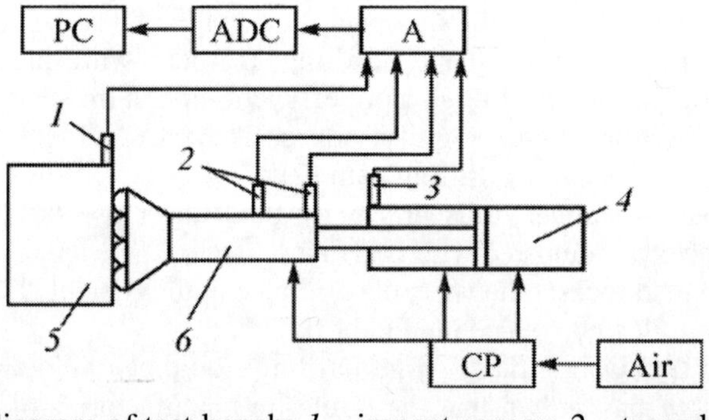

Figure 1. Block diagram of test bench: *1*—impact sensor; *2*—transducers of pressure in the chambers of power and idle strokes; *3*—transducer of air pressure in supply line; *4*—pneumatic actuator of feed; *5*—rock specimen; PC—personal computer; ADC—analog-to-digital converter; A—amplifier; CP—control panel.

Determination of areas and centers of gravity of the impulses takes much labor, especially in case of the impulses of complex geometrical form [8].

For the interpretation of the pressure charts, the Institute of Mining has developed a program in the object-oriented language C# (Figure 2). The program is supported by various versions of Windows OS. A user downloads huge data text files and displays them in graphical for. Windows of the program menu interact with each other. For the convenience, the analog window offers prompting messages. The program accelerates the process of interpretation of large bulks of data as against the grapho-analytical method. One more advantage consists is visual display of data and intuitive interface adaptable to a specific user. The program also enables addition of modules and windows, which widens its application field. Furthermore, the program searches and treats possible errors, e.g., in entry of input data of parameters. Each error and its cause are identified in an individual dialog box, which allows prompt error control.

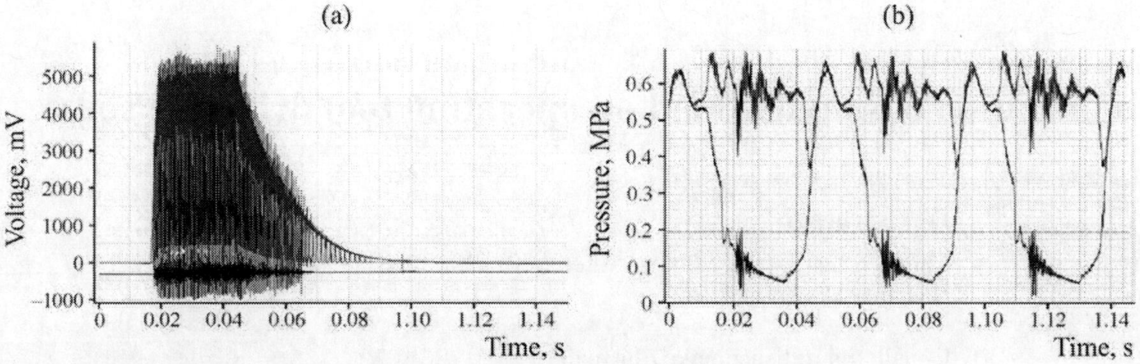

Figure 2. Illustration of automatic calculation of working cycles: (a) work diagram of the machine; (b) operating cycles after processing.

3. Conclusion

The computation program for working cycles of percussive machines computerizes determination of energy parameters of air- and hydraulic-driven percussive machines with a free moving piston. Sampling of a number of impact cycles improves accuracy of the energy parameters.

Acknowledgements

This study was carried out in the framework of the Basic Research Program, Project Registration No AAAA-A17-117122090003-2.

References

[1] Rshetnyak SP and Paladeeva NI 2012 Basic trends of development of technology for drilling blasting in open mining *Gornaya Tekhnika* Vol 2 No 1 pp 2–8
[2] Fox B 2011 *Atlas Copco Drilling Solutions LLC* Garland Texas USA
[3] Bruce A, Lyon R and Swartling S *The History of DTH Drilling and the Use of Water-Powered Hammers* Available at: http://ww.w.geosystemsbruce.com/v20/biblio/288-HistoryofDTHDrilling-Final.pdf (accessed at 30.05.2018)
[4] Oparin VN, Timonin VV and Karpov VN 2016 Quantitative estimate of rotary-percussion drilling efficiency in rocks *J Min Sci* Vol 52 No 6 pp 1100–1111
[5] Timonin VV, Alekseev SE, Karpov VN and Chernienkov EM 2018 Influence of DTH Hammer Impact Energy on Drilling-with-Casing System Performance *J Min Sci* Vol 54 No 1 pp 53–60
[6] Eremenko VA, Karpov VN, Filatov AP, Kotlyarov AA and Shahtorin IO 2014 Improvement of ore breakage toward pre-blasted rocks in rockburst-hazadous mining *Gornyi Zhurnal* No 1 pp 50–55
[7] Shakhtorin IO 2017 Debugging of impact machines using modern software *J Fundament Appl Min Sci* Vol 4 No 1 pp 72–76
[8] Esin NN 1965 *Research Methodology for Down-the-Hole Hammers* Novosibirsk: SO AN SSSR (in Russian)
[9] Sudnishnikov BV 1949 *Some Issues of the Theory of Impact Machines* Novosibirsk: GGI ZSF AN SSSR (in Russian)

IOP Conf. Series: Earth and Environmental Science **262** (2019) 012075 doi:10.1088/1755-1315/262/1/012075

Formation of impulse load on operating tool driven in soil

IV Tishchenko

Chinakal Institute of Mining, Siberian Branch, Russian Academy of Sciences, Novosibirsk, Russia

E-mail:*ighor.tishchienko70@mail.ru

Abstract. The author describes the experimental estimation procedure of impulse load generated in a rock-breaking rod overcoming resistance of soil. Oscillograms of power impulses in a pipe column are obtained in operation of generators with different energy efficiency and frequency. Penetration rate of a rod with selected geometry under vibration–percussion action is calculated. For comparison, the static force required for hammerless driving of rod in soil is found.

1. Introduction

At all stages of construction and underground mining connected with earthwork, a set of special technologies is used. It includes reinforcement of foundations under old and new structures, construction of cementation piles, bridge pillars trestle bridges and power lines, reinforcement of excavation and embankment slopes, construction of shields and screens, making channels for placement of closed design utility lines [1]. Percussion driving of various rod element in structures in elastoplastic subsurface layers of rock mass and soil have found wide application in actual practice [2–4].

Among diverse technological tools, a certain place belongs to impulse-generating machines with convert energy of compressed air to cyclic to-and-fro motion of piston. The latter, via some intermediate assemblies, interacts with the driven body by transmitting kinetic energy to it and performing useful mechanical work [5, 6].

The Institute of Mining, SB RAS, has been for a long time carrying out investigations aimed at improvement of this class percussive tools to meet incremental needs of construction industry [7–9]. It is impossible to improve efficiency of vibro-percussion technologies without comprehensive study into interaction of elements in the oscillating system of impulse generator–driven element–soil.

The next research cycle was focused on finding mechanisms of change in the characteristics of power pulse transmitted to a pipe column and in the displacement rate of the latter in elastoplastic soil under the action of different sources of cyclic impact in combination with addition static load.

2. Experimental setup and results

The body penetrated in soil was a steel pipe column *1* (Figure 1) composed of two sections with a threaded coupling. The outer diameters of these components were 33 and 40 mm, respectively. The pipe head 1.9 m long, interacting with soil, was equipped with a cone with a nose angle of 60°. At the opposite end of the second element, different model percussion units *2* were arranged. The coupling was universal reducing nozzle *3* with internal conical cavity with a nose angle of 4°to the pipe axis, which allowed reliable fixating of the impact source without auxiliary tension pullers. Additional static load was created by a set of loads *4* with an overall weight of 50 kg. The force transmission was implemented via an intermediate damping element *5* and a bolt *6* placed in the crosshole of the driven pipe.

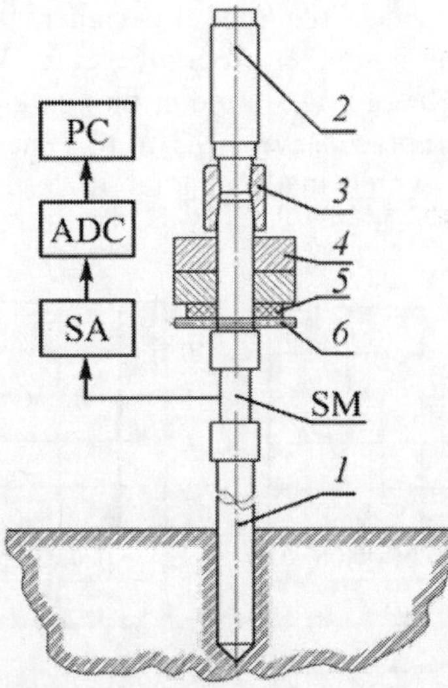

Figure 1. Experimental design: *1*—pipe column; *2*—percussion unit; *3*—reducing nozzle; *4*—additional load; *5*—damper; *6*—bolt; SM—strain meter; SA—signal amplifiers; ADC—AD converter; PC—portable computer.

The measurement instrumentation included a strain meter SM represented by a pipe with a length of 250 mm and diameter of 33 mm with the glued resistive strain sensors, a block of amplifiers of electric signal SA and an AD converter connected to a computer PC. Force calibration test of the recording channel was carried out using a hydraulic press equipped with a class 04 test pressure gage. Incorporation of the gage in the pipe column was done with screwed connections.

The shock pulses was generated by different models of air-driven hammers with pistons 0.5–4 kg in weight, with the common design of air distribution system with elastic valve in the back stroke chamber, typical of all machines of the class Typhoon [10, 11]. Specifications of the employed impact devices are compiled in Table 1.

Table 1. Specification of impact devices.

Parameter	Impact device model				
	T-0.5	T-1	T-1-2.4	T-2	T-4
Hammering part weight, kg	0.5	1	1	2	4
Working pressure, MPa	0.6	0.6	2.4	0.6	0.6
Blow energy, J	4.5	9	10	17	35
Blow frequency, Hz	23	13	41	6.4	9.0
Recoil force, kN	0.23	0.23	0.92	0.23	1.20
Dimensions, mm:					
Body diameter	40	40	40	40	90
Length	190	355	370	620	460

Soil was represented by sandy clay in natural occurrence, with decreased moisture content.
Experimentation involved some preparations:
—making of pilot holes 0.5 m deep from the soil surface to direct the pipe column vertically;
—preliminary penetration of the pipe head 1.9 m long;

—connection of the second pipe, assembly of the force sensor and vertical ruler 2 m long, with a scale interval of 1 mm, for the continuous monitoring of penetration process.

The pipe column driving was video recorded by camera ONY HDR-SR10E for step-frame processing of the resultant video using computer program Picture Motion Browser and for calculating average penetration rate per 0.1 m. Recording started at the mark of 2.3 m and ended at the depth of 3.5 m. The impact generators were replaced every 0.3 m of penetration. For reliability, each variant was repeated not less than 3 times, which made in total 15 tests and subsequent driving of 3 pipe columns.

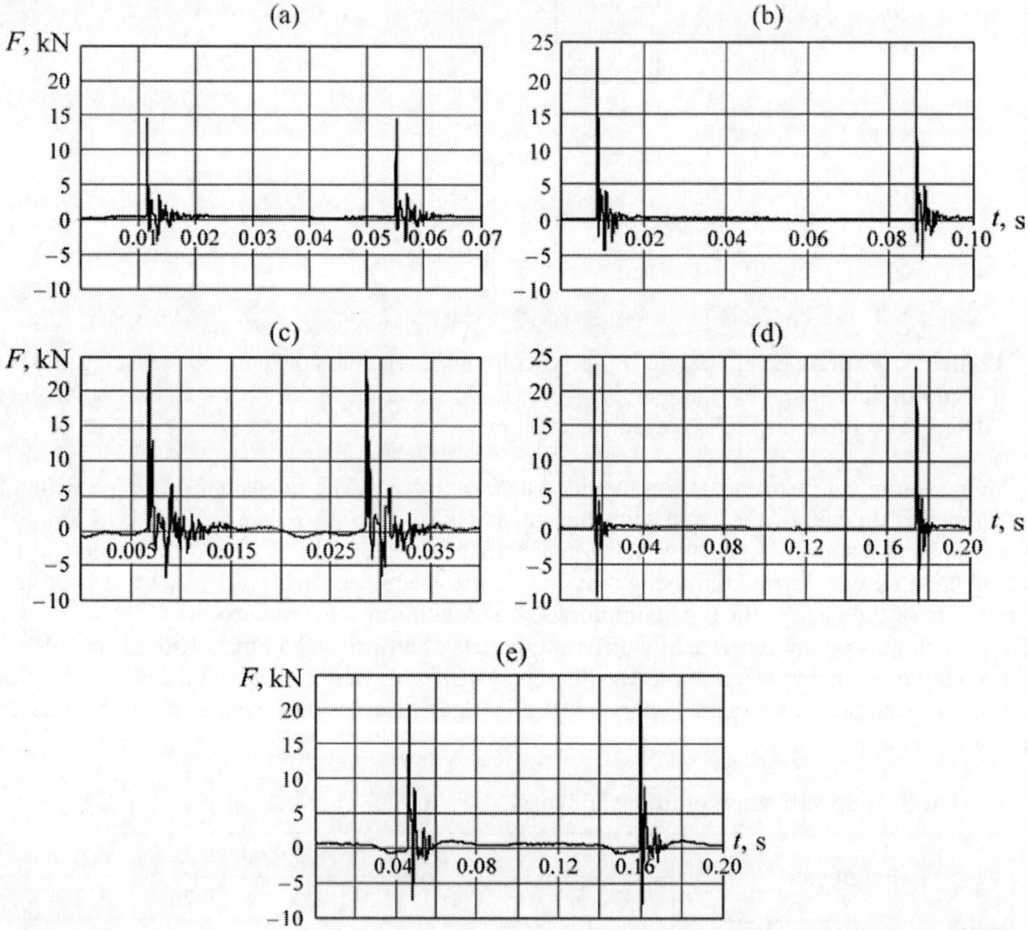

Figure 2. Amplitude and frequency of impact impulses in pipe column with different percussion sources: (a) Tayphoon-0.5; (b) Typhoon-1; (c) Typhoon-1-2.4; (d) Typhoon-2; (e) Typhoon-4.

Figure 2 shows the plots of power impulses generated in the drive pipe column by different sources of external cyclic percussion action in combination with constantly applied static force of gravity load. It is seen that the maximum amplitude of impact transferred to the driven pipe by the majority of the air hammers ranges as F_{max} = 20–25 kN (Figures 2b–2e). This parameter is independent of the dimension of the impact tools having considerable difference in the initial energy efficiency (table). An exception is the impact impulse generator represented by Typhoon-0.5 with the lowest weight of the piston, m_p = 0.5 kg, and the lowest blow energy E = 4.5 J. In the meanwhile, the amplitude of the force recorded in the pipe is never higher than F_{max} = 15 kN (Figure 2a). In the period between the neighbor blows, the soil–pipe–additional load–impact tool system is at rest. At the same

IOP Conf. Series: Earth and Environmental Science **262** (2019) 012075 doi:10.1088/1755-1315/262/1/012075

time, it is seen in Figures 2c–2e that the resulting force acquires negative values at some stages. This is reflective of insufficient weight of the added load G = 0.5 kN required to completely balance the reactive recoil F_r = 0.92 and 1.20 kN (table) in operation of Typhoon-1-2.4 and Typhoon-4, respectively, which acts in opposite direction to the pipe penetration. In the driven element, tensile stresses arise, which, together with elastic response of soil and under insufficient adhesion between the pipe surface and hole walls, can cause separation of the head nozzle from the hole bottom.

After the accomplishment of all tests, the video records were processed, the process of pipe displacement in soil was analyzed and the average penetration rates were calculated for each mode of pneumatic hammers. The results are presented in Figure 3.

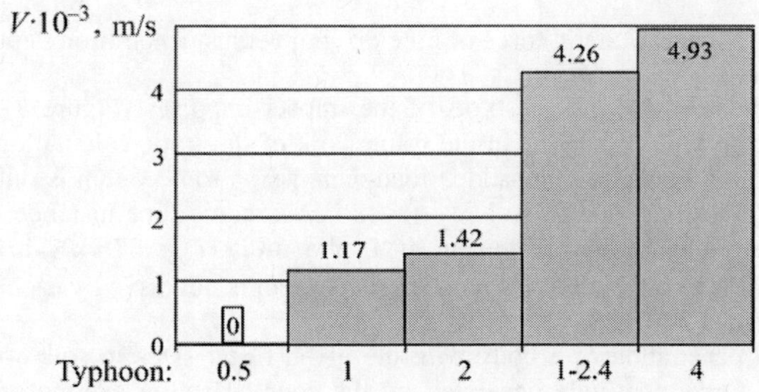

Figure 3. Average penetration rates of pipe under action of different impact sources.

According to these results, in the given conditions of the experiments, the start of the stable penetration of the pipe column at the rate $V = 1.17{\cdot}10^{-3}$ m/s is observed under action of Typhoon-1 hammer with the characteristics m_p = 1 kg and E = 9 J. Then, with more powerful hammers, the process of penetration accelerated considerably. It should be mentioned that the air hammers Typhoon-1-2.4 (m_p = 1 kg, E = 10 J, p = 2.4 MPa) and Typhoon-4 (m_p = 4 kg, E = 35 J, p = 0.6 MPa) have close results. For instance, despite considerably different dimension, weight of piston and blow energy, the penetration rates of the machines are $V = 4.26{\cdot}10^{-3}$ m/s and $V = 4.93{\cdot}10^{-3}$ m/s, respectively, which is the difference of not more than 15%. Evidently, for Typhoon-1-2.4, the lower energy component of the impact capacity is compensated by the higher blow frequency f = 41 Hz, which is achieved thanks to transition from the standard compressed air pressure p = 0.6 MPa to the higher value p = 2.4 MPa at the same pre-blow velocity of the piston.

Finally, the static force pipe driving in the same soil was determined in an additional experiment. The force generator was the hosting device, and the force was transferred to the pipe via a rope-and-block system. The force measurement system was the same set of elements as in the vibration–percussion tests (Figure 1). We determined resistance of soil to penetration of a compound rod composed of elements with diameters of 33 and 40 mm, with a conical nose. The results are depicted in Figure 4 as the plot of change in the driving force with the increasing depth of the pipe penetration. It is seen that in the main section of the path, from the mark of 2.4 m to the mark of 3.4 m, the resultant force of resistance ranges as: R = 16–20 kN. In the starting and ending sections, R = 14–15 kN.

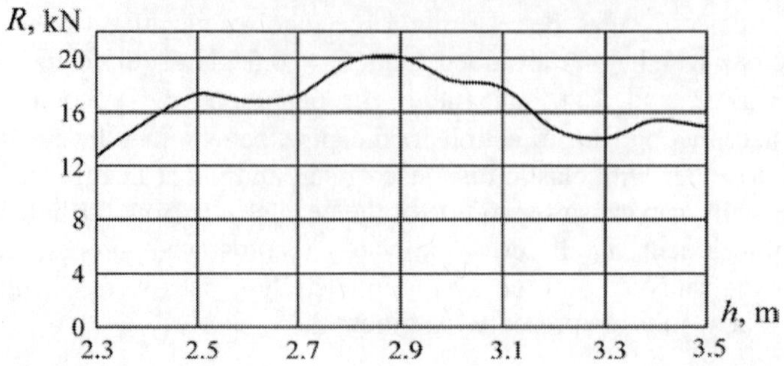

Figure 4. Static force of pipe driving versus penetration depth.

These data correlate well with the plots of the impact impulses (Figure 2) and and the average penetration rates of the pipe (Figure 3) in the framework of the static calculation model, according to which translational motion of the pipe–added load–percussion tool system is only implemented if the total frontal and lateral drags due to soil resistance is overcome. For instance, the amplitude of the impulse load transferred to the driven element by Typhoon-0.5 (F_{max} = 15 kN) is insufficient for plastic deformation of soil at the hole bottom; as a result, the system is not drive by each blow and the average penetration rate is zero ($V = 0$).

For the assured penetration of a body with the given linear sizes in soil with the total resistance R = 16–20 kN, the force amplitude generated by the coupled action exerted on the pipe by various impact generators and added load G = 0.5 kN is set at the level of F_{max} = 20–25 kN. In this case, the translational motion velocity of the pipe ranges as $V = (1.17\text{–}4.93)\cdot 10^{-3}$ m/s and depends on the energy E and frequency f of blows of different pneumatic hammers, and on the value of the additional static force G. With regard to physical and mechanical properties of soil, as well as to backward displacement of the pipe column under elastic response of soil and recoil of the percussion tool, the mentioned relation in the first approximation is given by [12]:

$$V = \left[\frac{E\eta}{R-G} - \frac{(Q+F_0)\lambda}{qSk+G}\right] f,$$

where η is the coefficient of the impact impulse transfer; Q is the elastic response of soil, N; F_0 is the recoil force, N; λ is the backward displacement of the pipe column, m; q is the normal lateral pressure of the hole walls, Pa; S is the total area of lateral contact between the driven element and soil, m^2; k is the metal–soil friction factor.

3. Conclusions

1. In the technologies of vertical driving of metal structures by percussion with application of additional static load exceeding recoil of pneumatic hammer and approximately making not more than 5–10% of the stable amplitude of the impact force: $G \geq F_0 \approx (0.05-0.1)F_{max}$ makes it possible to reduce backward displacement of the driven system under elastic response of soil and, as a consequence, to improve efficiency of the driving process. It is advisable to use gravity loads of required weight and place them in the driven pipe with an intermediate damping flexible suspension.
2. Comparison of the results obtained with Typhoon-1-2.4 and Typhoon-4 pneumatic hammers having the same pre-impact velocity of the piston (4.2–4.5 m/s) proves that, under condition that these machines ensure the ultimate energy of unit blow ($E \geq 9$ J), required to overcome resistance of soil, the solution to the problem of improvement of driving efficiency can be connected with the considerable expansion of the impact frequency range (41 Hz as against 9 Hz). Such approach allows reaching similar velocity characteristics ($V = 4.26\cdot 10^{-3}$ m/s and $V = 4.93\cdot 10^{-3}$ m/s, respectively) at the smaller weight and size of high-frequency percussion facilities with the piston weight m_p = 1 kg and E = 10 J as compared with m_p = 4 kg of the hammer with E = 35 J.

Acknowledgements
This study was carried out in the framework of the Fundamental Research Program, Project No AAAA-A17-117122090003-2.

References

[1] Smorodinov MI (Ed) 1974 *Handbook on Civil Works, Basements and Foundations* Moscow: Stroyizdat (in Russian)

[2] Kershenbaum NYa and Minaev VI 1985 *Horizontal and Vertical Percussive Drilling* Moscow: Nedra (in Russian)

[3] Rybakov AP 2005 *Trenchless Pipelaying Technologies, Theory and Practice* Moscow: PressByuro (in Russian)

[4] Kun G, Scheuble L and Schlick H 1993 *Trenchless Laying of Blind Pipelines* Moscow: Stroyizdat (in Russian)

[5] Kostylev AD, Grigorashchenko VA, Gileta VP et al 1987 *Pneumatic Drill Hammers in Construction* Novosibirsk: Nauka (in Russian)

[6] Smolyanitsky BN, Chervov VV, Trubitsyn VV et al 1997 New pneumatic impact machines for special civil construction *Mekhanizats. Stroit.* No 7 pp 12–17

[7] Smolyanitsky BN, Repin AA, Danilov BB et al 2013 *Enhanced-Efficiency and Improved-Life Impulse-Forming Machines for Long Borehole Drilling in Rocks* Novosibirsk: SO RAN (in Russian)

[8] Smolyanitsky BN, Tishchenko IV, Chervov VV, Gileta VP and Vanag YuV 2008 Sources for productivity gain in vibro-impact driving of steel elements in soil in special construction technologies *Journal of Mining Science* Vol 44 No 5 pp 490–496

[9] Smolyanitsky BN, Tishchenko IV and Chervov VV 2009 Improvement prospects for air hammers in building and construction works *Journal of Mining Science* Vol 45 No 4 pp 363–371

[10] Chervov VV, Smolyanitsky BN, Trubitsyn VV, Chervov VV and Tishchenko IV RF Patent No 2462575 Percussion Action Machine *Byull. Izobret.* 2012 No 27

[11] Tishchenko IV 2014 Variable percussion power pneumatic impulse generators *Izv. vuzov. Stroit.* No 6 pp 79–87

[12] Gurkov KS, Klimashko VV, Kostylev AD et al 1990 *Air-Driven Percussion Machines* Novosibirsk: IGD SO RAN (in Russian)

Adaptation of geotechnologies to mineral mining conditions in permafrost: Challenges

SM Tkach

Chersky Institute of Mining of the North, Siberian Branch, Russian Academy of Sciences, Yakutsk, Russia

E-mail: tkach@igds.ysn.ru

Abstract. The author addresses problems of efficient mineral mining in the zone of permafrost. It is illustrated that geotechnologies for the development of solid minerals in permafrost rocks should make use both of positive and negative factors of internal and external environment. The article reports about the research findings of the Chersky Institute of Mining of the North towards improvement and development of efficient and eco-friendly innovative technologies, methods and equipment for mineral mining and processing to be highly adaptable to the conditions of the permafrost zone.

1. Introduction

Mineral mining is the major industry promoting sustainable social and economic development of a country and the well-being of the nation. The mining industry in Russia faces problems governed by negative trends of the recent decades. One of the key problems is the outrun of increase in the cost of mineral mining as against the increase in the output [1]. A root cause of this situation is depletion of easily accessible and rich mineral deposits. For this reason, the average content of useful components in mineral deposits planned for or currently under development steadily declines, mining goes to deeper levels, processing circuits are fed with rebellious materials and, eventually, the cost of mining rise while productivity drops as scientific and technological advance falls behind the appreciation of mining operations.

Currently, in view of increasing depletion of mineral and hydrocarbon reserves in the middle of the country, the new mining projects steadily drift to poorly developed regions in the east and north-east. These regions feature vast territories, weak infrastructure and low density of population, remoteness from processing and consumption facilities, expensiveness of goods and services, deficiency of highly skilled personnel, etc. The natural environments (physiographic and geological) have dominant influence on economic efficiency in all sectors of industry, including mining. The problem is particularly acute in the zone of permafrost (which is 64% of the territory of Russia). The permafrost zone is distinguished for the subarctic climate (long and cold winters, short and hot summers), difficult ground conditions in rock mass of very poor quality and high moisture content (ice content), multiple cycles of freezing and thawing in the course of mining, poor knowledge on local geology and outdated economic-and-geological evaluation of mining efficiency. Located in these areas, proven reserves registered in the state balance sheet in the late 1900s (large deposits of solid minerals, unique in terms of quality and quantity) lack proper and responsible investors in the context of market economy, which impedes the industrial development in these territories of Russia.

2. Mineral mining in permafrost conditions

An illustration of the internal and external bars for the efficient development of mineral wealth in the country is the Republic of Sakha (Yakutia) the third of the immense territory of which lies in the Arctic zone. Yakutia is the largest region both in the Russian Far East and in entire Russia, covering the area over 3 million cubic kilometers. At the same time, the population of the region is 950 thousand people (1100 thousand people before 1990). By the end of the previous century, Yakutia possessed mineral reserves of unique variety, quality and quantity [2].

Started in the mid-1990s and continuing currently, the dynamic commercial development of mineral reserves in the north-east of Russia, including Yakutia, called for the proper scientific support of mining operations in the conditions of extremely severe nature and climate. Naturally, in 1970 Academician N.V. Chersky initiated an idea to found a department of mining within the existing Institute of Physicotechnical Problems of the North in the Yakutsk Division of the Siberian Branch of the Russian Academy of Sciences. Later on in 1980, on the Institute base, the first academic institution in the region—Institute of Mining of the North named after Chersky—was founded.

In the main research areas (integrated development of mineral reserves in conditions of permafrost; thermal physics and geomechanics of permafrost rocks and rock mass with regard to anthropogenic factors), the Institute's scientists have obtained a series of fundamental and applied results in geomechanics and mine thermophysics, open and underground mining technologies in permafrost zone, processing and deeper conversion of raw materials, geophysical studies of soil and rock mass, geostatistics and geoconomy of subsoil management in the North, etc. These achievements made a framework for the development of new mining machines and technologies, concepts and practical guidelines, and many of them were successively trialed or implemented at different times in mines in the North [3].

Upon transition to market economy in the 1990s, considerable portion of proven reserves was abandoned. Thus, it is of the current concern to refine the methods of investigation, appraisal and economic-geological re-evaluation of such mineral deposits to identify cost-effectively mineable objects. In the sphere of industrial geology and geostatistics, the Institute's scientists developed theory of grain size analysis for geo-materials, mathematical modeling of statistic and special distribution of minerals and basic texture types of ore, assaying of rocks and ore, as well as appraisal of places and ore bodies [4]. The models of quality management in mineral mining, processing and delivery are worked out. The concept of transition to graduated quality standards, with regard to geological conditions per block and natural laws of cluster and discrete structure of ore substance is proposed. A descriptive (verbal) and analytical model is constructed for a complex and dynamic georesouce–geotechnological transformation–product and effect system functioning. A new method is put forward and justified for assessing effect of ore dilution on mine performance by assuming dilution as a sump of the conventional geotechnical element and new structural and economic-geological constituents governed by standards of delineation and appraisal of mineral reserves. Application of this method to advance the earlier developed model show that the economic geopotential of solid mineral deposits, given the modern geotechnologies as well as loss and dilution standards, is under-used. This fact emphasized the demand for new and efficient regulations and technologies for the transformation of georeserves into products and new resources [1, 5].

The permafrost ground facilitates wide application of the ground-penetrating radar methods. The Institute has designed a dedicated geo-radar set with the appropriate software support for the geophysical research. This instrumentation enables sufficiently reliable appraisal of mineral deposits, as well as information on detail structure and cryogenic condition of permafrost rock mass. The method is trialed in the appraisal of Mayat deposit diamonds, as a result of which a paleochannel and geological dislocations have been revealed in the zones of high concentration of minerals as per the data of geological assaying (Almazy Anabara company) [6, 7].

The basic knowledge framework for solving mining problems in the conditions of the North is the developed and persistently modified computation methods for geomechanics and mine thermophysics backed by continuous monitoring of thermomechanical behavior of permafrost [8–11]. In particular, science-based technologies have been offered for low-energy control of thermal conditions in underground mines in the North For example, implementation of heat-shield shotcreting of underground excavations in the permafrost zone has appreciably increased stability of openings in mines as well as improved safety and operational comfort of personnel in mines of Aldanslyuda of NorNickel and ALROSA. The other projects include 2D and 3D mathematical models and software system for solving problems of heat exchange between rocks and backfills during slice mining of ore bodies, considering heat development due to binder hydration and phase transition of moisture in permafrost rocks; bundled software for monitoring and adjustment of thermomechanical behavior of rocks at the bottom of headframe foundation at vertical shafts in Internatsionalny, Mir, Udachny and Aikhal mines, ensuring long-term failure-free operation; procedure and software system for joint computation of ventilation and heating modes in the nonstationary mine network in the permafrost zone. The investigation results in the form of efficient methods of mine support, ventilation and heat regulation, etc. are included in project documentation for underground and placer mining of diamonds at ALROSA (pipes Internatsionalnaya, Nyurbinskaya and Botuobinskaya; placer Solur-Vostochnaya).

In the field of improvement of geotechnologies and adapting them to low-temperature conditions based on the outcome of long-term research into properties of frozen backfill, mechanisms of its placement and interaction with permafrost formations, instructional guidelines are made for evaluation of structural constraints and technology of frozen backfilling in different conditions of geomechanical and thermal behavior of permafrost rock mass. The technology has been successfully introduced at the Badran deposit of Zapadnaya Mining Company in the Republic of Sakha (Yakutia) [12, 13]. For open pit coal mining in the permafrost zone, based on the investigations of heat and moisture conditions in rock mass and in broken-rock disintegration, grain size analysis of broken rocks, dragline cycle time in the periods of different temperatures, the recommendations are made for evaluation and selection of efficient parameters for coal mining with direct internal dumping of broken overburden rocks prone to re-adfreezing [14]. A continuous flow process technology has been industrially introduced in mining deep-buried placer near the Kuranakh River (South Yakutia). Commercial use of a conveying&rotating system has been implemented in such climatic conditions for the first time [15].

In the field of geomechanics, jointly with Usolmash (Usolye-Sibirskoe), a brand-new technology and equipment of modular mobile ore processing plant has been full-scale trialed in gold ore assaying. The plant is composed of new-designed and patented machines (combined impact crusher DKD-300, impact centrifugal mill TsMVU-800, pneumatic separator POS-200). The dry method of disintegration and milling as well as pneumatic method of ore pretreatment allows three-times reduction in process water consumption and makes it possible to use the plant in cold seasons for ore processing and pre-treatment, which thereby prolongs the period of production in the conditions of long winter and short summer [16, 17].

This article presents a short list of R&D projects accomplished by the fellows of the Chersky Institute of Mining of the North, SB RAS.

3. Conclusions

In the context of emphasis Russia lays on advance socio-economic advancement in the under-developed areas of East and North-East of the country, chiefly by means of the immense mineral wealth management, it is necessary to aim maximum efforts of theory and practice in the field of mining toward solution of wide range of problems connected with modernization and innovative development of efficient and eco-friendly geotechnologies, methods and equipment to the utmost adaptable to the unique geographic and climatic conditions of actual operation.

References

[1] Tkach SM 2014 Regional features of the development of mining sciences and industry in the north-east of the country *J. Fundament Appl. Min. Sci.* Vol 1 pp. 314–320 (in Russian)

[2] Tkach SM 2006 *Methodological and Geotechnical Aspects to Improve Efficiency of Developing Ore and Alluvial Deposits of Yakutia* Yakutsk: Inst. Merzlotoved. SO RAN (in Russian)

IOP Conf. Series: Earth and Environmental Science **262** (2019) 012076 doi:10.1088/1755-1315/262/1/012076

[3] Tkach SM 2015 35 years of work of the Institute of Mining of the North named after N. V. Chersky of the Siberian Branch of the Russian Academy of Sciences in solving the problems of the development of the permafrost *Gornyi Zhurnal* No 4 pp 4–9
[4] Tkach SM, Batugin SA and Barakaeva ID 2016 Objective assessment of ore dilution—A promising way of enhancing mineral resource potential *Gornyi Zhurnal* No 1 pp 37–40
[5] Batugin SA amd Cherny ED 1998 Theoretical Framework for Assaying and Appraisla of Mineral Reserves Novosibirsk: Nauka (in Russian)
[6] Omel'yanenko AV and Fedorova LL 2006 *GPR Studies of Permafrost* Yakutsk: YaNTs SO RAN (in Russian)
[7] Sokolov KO, Prudetsky ND and Fedorova LL 2016 Mapping of paleoils watercourse at placer deposits of permafrost by georadiolocation method GIAB No 8 Special Issue 21 pp 514–522
[8] Galkin AF 2000 *Thermal Regime of Underground Structures in the North* Novosibirsk: Nauka (in Russian)
[9] Nikol'sky AM, Ordin AA, Kurilko AS, Klishin VI and Kulakov VN 2014 *Non-Pillar Method of Underground Mining of Placer Deposits in Yakutia* Novosibirsk: Nauka (in Russian)
[10] Kurilko AS, Khokholov YuA, Drozdov AV and Solov'ev DE 2017 Control of ground temperatures under headframes and collars of vertical shafts: A case study of the Udachny diamond mine (Yakutia) *Earth`s Cryosphere* Vol XXI No 5 pp 82–91
[11] Khokholov YuA and Kiselev VV 2016 Estimation of the efficiency of heat storage in the air conditioning systems of ore mines of the North *Fund. i Prikl. Issled. v Sovr. Mire* No 15-1 pp 80–83
[12] Neobutov GP and Petrov DN 2011 Increasing the efficiency of ore extraction using an ice and stone gobbing *Izv. Samar. Nauch. Tsentra RAN* Vol 13 No 1(5) pp 1274–1276
[13] Zubkov VP, Neobutov GP and Petrov DN 2017 Enhancing completeness and quality of underground gold ore extraction in the Republic of Sakha (Yakutia) *Gornyi Zhurnal* No 4 pp 53–57
[14] Panishev SV and Kaimonov MV 2017 Technical approach to prediction of dragline productiveness in blasted rock handling at open pit mines in permafrost zone *J. Min. Sci.* Vol 53 No 4 pp 702–707
[15] Ermakov SA and Burakov AM 2005 The flow technology of the development of the gold-ore deposit in Yakutia *Gornyi Zhurnal* No 12 pp 131–133
[16] Matveev AI, Grigor'ev AN, Filippov VE and Gladyshev AM 2003 Preliminary technological tests of new equipment of modular ore-dressing plant *Obog. Rud* No 5 pp 25–26
[17] Matveev AI, L'vov ES and Vinokurov VR 2016 New in ore preparation—Devices for crushing and grinding of multiple impact action *GIAB* No 8 Special Issue 21 pp 242–252

IOP Conf. Series: Earth and Environmental Science **262** (2019) 012077 doi:10.1088/1755-1315/262/1/012077

Comparison of mineral valuations by international codes of CRIRSCO and GKZ for Kyrgyzstan

BT Tolobekova* and DA Bespalov**

Institute of Geomechanics and Exploration of the Earth, National Academy of Sciences of the Kyrgyz Republic, Bishkek, Kyrgyzstan

E-mail: *burulkant@bk.ru, **divorses@mail.ru

Abstract. The authors compare valuations by GKS (RF State Commission of Reserves) and CRIRSKO for the transition of Kyrgyzstan to the international methods of mineral appraisal.

1. Introduction

Economic globalization, technological advance and expansion of business relations worldwide make numerous mining projects available for the international capital. All parties in this case are interested in standards for reporting of mineral resources and reserves as mining projects feature high risks.

The Committee for Mineral Reserves International Reporting Standards CRIRSCO has drastically contributed to the promotion of consistent reporting standards in the world. These standards are included in very similar codes published and accepted by the relevant professional agencies in Australia, Canada, South Africa, USA, Great Britain, Ireland, Russia and other countries.

The JORC code effective in Australia enjoys the highest success and popularity in the field of mining in the world thanks to simplicity, intelligibility and applicability to all kinds of minerals [1].

2. International approaches to the mineral valuation

Enhancing investment attractiveness and innovation potential in the mining sector needs international approaches to valuation of minerals. Productive yield of the mining industry of Kyrgyzstan in 2016 totaled 106 billion 86 million of Kyrgyzstan soms (approx 1 billion 600 million dollars). Private business invested to exploration 4.8 billion soms (approx 71 million dollars). The country spent 35.790 million soms (500 thou dollars) for replenishment of mineral reserves. The figure 1 demonstrates that the main product of the mining industry in Kyrgyzstan is gold, and its production rises.

The majority of the investors are from the developed countries and require transparent appraisals based on the international practice and the guarantorship secured by CRIRSCO. Currently CRIRSCO includes such countries and regions as (in brackets are the codes developed on the CRIRSCO standards) Australia and New Zealand (JORC), Canada (CIM), USA (SME), South Africa (SAMREC), Chile (IMEC), Europe (PERC), Russia (NAEN), Mongolia (MPIGM), Brazil (CBRR) and Kazakhstan (KAZRC).

Russian, Kazakhstan and Mongolia, the nearest neighbors of Kyrgyzstan, have already been engaged in the process of transition to the international reporting standards of mineral valuation, while Kyrgyzstan is merely making attempt to declare the transition. By now, Kyrgyzstan uses the mineral assessment system based on recommendations provided by the State Commission on Reserves (GKZ). It seems interesting to analyze the experience gained by GKZ in the transition to CRIRSCO standards in the CIS countries. For example, Russia has passed such stages as:

IOP Conf. Series: Earth and Environmental Science **262** (2019) 012077 doi:10.1088/1755-1315/262/1/012077

1990–2009—chaos in international reporting of hard minerals due to discordance on Russian and international classifications;

2006—CRIRSCO–GKZ RF working group set up for handling the problem;

2008—memorandum of understanding signed;

2009–2010—Guidelines on alignment of Russian minerals reporting standards and the CRIRSCO template prepared and approved by GKZ and CRIRSCO on Sep 28, 2010, Moscow;

2011—Russian CRIRSCO-aligned code developed for international market application.

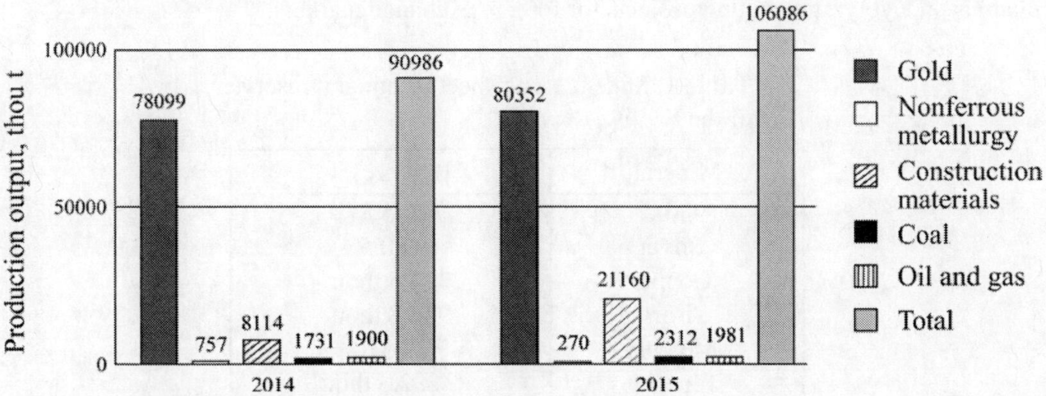

Figure 1. Mineral produced in Kyrgyzstan in 2014–2015 (as per 2016 Annual Report of the Board of the State Committee for Industry, Power Engineering and Subsoil Use in Kyrgyzstan. Available at: http://www.gkpen.kg/)

The problems arising in international reporting of mineral resources and reserves due to inconsistence of the international classification and that of GKZ included:

—diverse perceptions of harmonization between the Soviet and international categories;

—common misunderstanding of the GKZ system and the difference between the two systems;

—expensive removal of contradictions: doubling of operations with the concurrent assessment of mineral reserves / resources by GZK and international systems.

As a result of inconsistency and chaos, the estimates of mineral reserves by GKZ and international systems differ fantastically [2].

Introduction of the CRITSCO Code in Kazakhstan proceeded in a few stages. The working group involved representatives of large producers operating on the Kazakhstan market; Chair of the Committee Mike Armitage, Chair of the Board of Directors at SRK Consulting, was the Chair of the GKZ RF–CRIRSCO Committee; Co-Chair was A. A. Nadyrbaev, Deputy Chair of the Committee of Geology and Subsoil Use of the Ministry of Industry and New Technologies in the Republic of Kazakhstan.

On June 10, 2011 Kazakhstan Mining Club held the first meeting on Harmonization of Kazakhstan and International Reporting Standards; in 2014 the CRIRSCO–Kazakhstan code was prepared; in 2015 preparation of the KAZRC code was undertaken and the memorandum of understanding was signed between the Ministry for Investment and Development and CRIRSCO. On June 14, 2016 Kazakhstan was elected to be the tenth member of CRIRSCO.

Currently there is a great discrepancy between the estimates of GKZ and CRIRSCO. The mining companies that have accepted this reporting system make more than 80 % in the total capital of the mining industry. For this reason, Kazakhstan, in order to promote investment in the mining sector, should urgently pass to international mineral valuation systems accepted by the majority of countries–investors. The process of transition can take much time which can be shortened if there is a political will. Minerals enrolled in the state balance sheet and attractable for foreign investors are compiled in Table 1.

As seen in Table 1, Kyrgyzstan holds more than 565 t of gold reserves. This fact can attract foreign investors and initiate very profitable projects under adequate, clear and plain conditions provided by the CRIRSCO codes imposing minimum requirements on international stock exchangers and intended to guarantee that the published Report of geological exploration results, mineral resources and mineral reserves contain all information necessary for the investors to evaluate advantages/benefits and disadvantages/risks connected with a mineral mining project. The current GKZ code lays extra risks and expenditures on foreign investors; in the most of them make calculations on Kyrgyzstan mining projects for foreign exchange markets.

Table 1. State balance sheet of mineral reserves of the Kyrgyz Republic

Mineral	Reserves, t
Gold	565.8
Silver	480.5
Copper	314.5 thou
Iron	943.8 thou
Mercury	39 510
Fluorite	2 282.3 thou
Tin	186 761.4
Tungsten	117 233.2
Bismuth	5 082.6
Lead	30.0 thou
Antimony	263 968.0
Zinc	17.6 thou
Arsenic	65 200.0
Molybdenum	2 523.1
Rare earths	51.5 thou

Table 2. Mineral reserves evaluated by the CRIRSCO family codes

Mineral deposit	Code	Stock exchange
Taldybulak–Talass	JORC, (SAMREC)	ASX
Andash	JORC	ASX
Aktash	JORC	ASX
Zhamgyr	PERC	LME
Bozymchak	PERC	LME
Chaarat	PERC	LME
Kutessai II	JORC (CIM NI-43-101)	TSE
Kumtor	CIM NI-43-101	TSE
Shambesai	JORC	ASX
Dzherui	SME, PERC	LME
Kurandzhailyau	JORC	ASX
Dolpran	JORC	ASX
Tuyuk-Kargasha	JORC	ASX

Thus, 7 deposits are evaluated by the JORC code, and the largest deposit Kumtor is appraised the Canadian CIM NI-43-101 system. An advantage of the international codes of mineral valuation, among other things, is the rise in the value of a deposit by one third (Table 3) [3, 4].

Table 3. Advantages of the CRIRSCO code introduction

	Implementation period, months	Increment in value of deposit
Initial cost of mineral deposit		X
Strategy (definition)	1–2	X + 50 %
Scoping study	2–3	X×2
Estimated reserves C_1 by GKZ	**13–21**	X×3
Estimated reserves by the JORC or N1 43-101 codes	**3–4**	X×5
Pre-feasibility study	6–12	X×10
Validation	12–24	X×15
Feasibility study	6–12	X×20
Bankable feasibility study	12–18	X×50
Construction feasibility study	10–15	X×100

Many deposits have already been evaluated by the codes of the CRIRSCO family, which imposes a duty on the government to undertake early transition to the international codes of mineral valuation.

3. Conclusions

The JORC code is based on the market relations and reduction in corruption risks for investors.

At the present time, augmentation of investments in the mining sector of Kyrgyzstan is connected with the earliest transition to the international codes of mineral appraisal.

The nearest neighbors—Russia, Kazakhstan and Mongolia—have already implemented such transition and overcome some problems and contradictions, which offers advantages to Kyrgyzstan to avoid the difficulties by having learnt experience of these countries.

References

[1] Collection of international mining codes, JORC, VALMIN *Australian Guidelines for the Assessment and Classification of Coal Resources* Moscow: Gornaya Kniga 2015

[2] Yenshin N Comparison of the current reporting standard for mineral resources and reserves according to GKZ and CRIRSCO *KAZRC Standards* Available at: http://www.mining.kz/images/Miningclubkz/MCkz21/presentation/KAZRC_CRIRSCO.pdf

[3] Rogalsky AV 2017 *Issues of the Implementation of CRIRSCO Standards in the Kyrgyz Republic with the Assistance of the Kyrgyz Society of Nondenominational Experts* Bishkek Available at: http://mining.kz/images/Miningclubkz/ MCkz21/ presentation/KMA_Kyrgyz_Mining_Association.pdf

[4] Report of the Consulting Company Nblzoloto Moscow on Assessment of the Growth in the Cost of Deposits, Depending on the Degree of Preparation of the Technical and Economic Documentation

IOP Conf. Series: Earth and Environmental Science **262** (2019) 012078 doi:10.1088/1755-1315/262/1/012078

Influence of added aerodynamic resistance on stabilization of an air flow in mine air heater

AM Tsyba and BL Gerike*

Federal Research Center of Coal and Coal Chemistry, Siberian Branch, Russian Academy of Sciences, Kemerovo, Russia

E-mail: *gbl_42@mail.ru

Abstract. Investigation into the wind speed effect on operation of open fanless air heater in Gramoteinskaya mine, Yuzhkuzbassugol, enabled to find that at a wind speed of more than 10 m/s from the leeward side of the building the heated air is discharged into the atmosphere. An experiment was performed on a physical model in the air channel of a boundary shaft at Gramoteinskaya mine in terms of the similarity conditions: geometric, kinematic and dynamic ones. At the stable fan operation mode the airflow velocity in the channel was measured before the aerodynamic resistance and after it for 10 minutes at regular intervals. Additional aerodynamic resistance provides at average twofold reduction in the airflow speed.

1. Introduction

Nowadays Kuzbss mines: Abasheskaya, Alardinskaya, Gramoteinskaya, Esaul'skaya, etc. employ open fanless air heaters (AH) where water is used as a heat carrier. Investigation into the wind velocity effect on AN operation at Gramoteinskaya mine, Yuzhkuzbassugol Co., made it possible to establish that the heated air is forced to discharge into the atmosphere at wind velocity of more than 10 m/s from leeward side of a building [1]. By Kemerovo Hydrometeocenter weather data, windy weather with snowstorms is specific for about 116 days (62 days in average) in Kuzbass region, when wind velocity is more than 10 m/s [2]. This indicates that AH works approximately one third of heating period with deviations from the schedule in terms of wind velocity (3–4 m/s) [3]. To fulfill acting regulations requires to lower twofold an air flow velocity at the air heater inlet [4]. The air, supplied to mine after its pass through AH sections, should correspond to the acting mine ventilation regulations at the expense of mine air pressure drop.

The problem can be solved by stabilization of an air flow supplied through the air intake ports by control of an air flow amount and velocity in front of AH sections thanks to mounting of additional aerodynamic resistances. The semipermeable shield of design parameters and porosity (0.628) is selected as an additional aerodynamic resistance [5].

2. Numerical modeling and mine experiment results

The physical model was designed to hold the respective experiment in order to find whether the additional aerodynamic resistance of 0.628 porosity is capable to lower twofold the air flow velocity at AH inlet ports. In physical modeling the nature of the phenomenon is preserved, but the scale is changed [6]. In the study case the problem arises on similarity conditions for the semipermeable shield and its model. Two physical processes are considered similar, given that the compatible parameters of both processes are within rigorously defined relations [7]. These relations are three similarity types: geometric, kinematic, and dynamic ones.

The geometric similarity is fulfilled if dimensions of all similar elements of the model and nature differ at the same proportion. Thereto, the similar elements should be positioned at the same angles to the ram air velocity vector, viz., the below relations should be true [7]:

$$\frac{l_1}{l_2} = k_l,$$
$$\frac{S_1}{S_2} = \frac{l_1^2}{l_2^2} = k_l^2 = k_S, \tag{1}$$
$$\frac{V_1}{V_2} = \frac{l_1^3}{l_2^3} = k_l^3 = k_V,$$

where l is linear dimension, m; S is area or surface, m^2; V is volume, m^3.

To model the ram air flow from leeward side, the model should be mounted normally to the air flow.

The processes are kinematically similar, provided that geometric similarity (1) is fulfilled in them, if the below relations are true for all the pairs of time intervals during which the similar phenomena proceed [7]:

$$\frac{t_1}{t_2} = k_t,$$
$$\frac{v_1}{v_2} = \frac{l_1 t_2}{t_1 l_2} = \frac{k_l}{k_t} = k_v, \tag{2}$$
$$\frac{a_1}{a_2} = \frac{v_1 t_2}{t_1 v_2} = \frac{k_v}{k_t} = \frac{k_v^2}{k_l} = k_a,$$

where t is time, s; v is velocity (wind velocity V in our case) , m/s; a is acceleration, m/s^2. Time, wind velocity, and acceleration in natural conditions and in the model for all the pairs of time intervals are equal.

The processes are dynamically similar, if the kinematic similarity (2) is fulfilled and if relations for all the pairs of comparable forces are equal.

When considering dynamically similar processes it is explicit that the forces (gravity force, pressure, friction inertia, etc.) acting in them are different, and it is not simple to fulfill the dynamic similarity conditions for all the forces, in some cases it is impossible to do. In this connection the problem is solved approximately, considering the principal dominating forces, constituting the character of the hydraulic process [8].

When a body moves in ram air, the main acting forces are friction forces, so the dynamic similarity is fulfilled, if geometric (1) and kinematic (2) similarities are fulfilled, as well as Reynolds number values, calculated at comparable points of both flows, are equal.

Consider magnitudes in Reynolds number formula for a shield with square cells [9]:

$$\mathrm{Re} = \frac{Va}{\nu},$$

where $V = V_1/m$ is average air velocity in shield cells, m/s; V_1 is wind velocity in front of the shield, m/s; a is dimension of cell side, m (Figure 1); $m = a^2/b^2$ is porosity factor of semipermeable shield; b is step of a cell, m; ν is kinematic air viscosity, m^2/s. Then

$$\mathrm{Re} = \frac{V_1 a}{m\nu}. \tag{3}$$

IOP Conf. Series: Earth and Environmental Science **262** (2019) 012078 doi:10.1088/1755-1315/262/1/012078

Kinematic air viscosity and wind velocity under natural conditions and in the model are the same, so to fulfill equality of Reynolds number at comparable points of the flows implies that geometric parameters of model cells are fulfilled at 0.628 porosity.

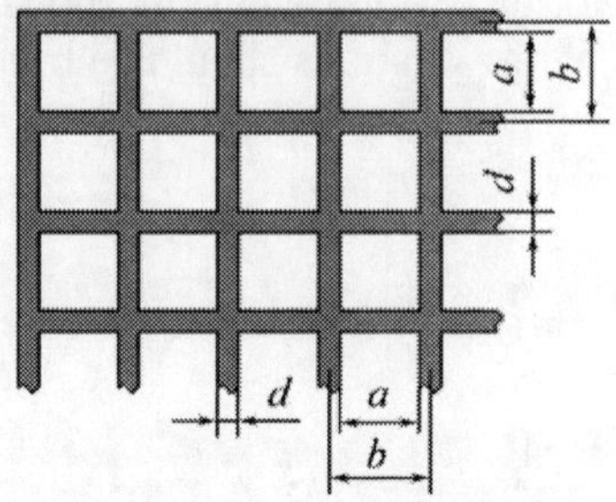

Figure 1. Semipermeable shield with square cells.

The model of a semipermeable shield is made of hot-rolled steel with circular cross-section under GOST 2590-88 V: rod diameter d = 8 mm; critical deviation + 0.3 – 0.5; rod cross-section area 0.5027 cm^2, mass of 1 m of profile 0.395 kg. width and height of the shield are 980 mm. the design size of a cell is a = 0.03 m. The final cell appeared rectangular in shape because of poor manufacturing precision; number of vertical rods was 27 u., horizontal rods were 28 u.

The actual porosity factor of the model was determined by the ratio of cell area to total shield area. The shield area was 0.9702 (0.98×0.99) m^2. Number of cells was 702 (26×27) u., average cell area was 8.35×10^{-4} m^2, total area of cells was 0.5862 ($8.35\times10^{-4}\times702$) m^2. Porosity was 0.6.

The experiment was held in an air channel of Gramoteinskaya boundary shaft. The air channel was made of metal shields for air supply to the mine with the help of fans. The cross-section of the air channel is 3.4 m^2 (1.52×2.0). The experimental device was mounted at straight section of 5 m in length in the air channel (Figure 2).

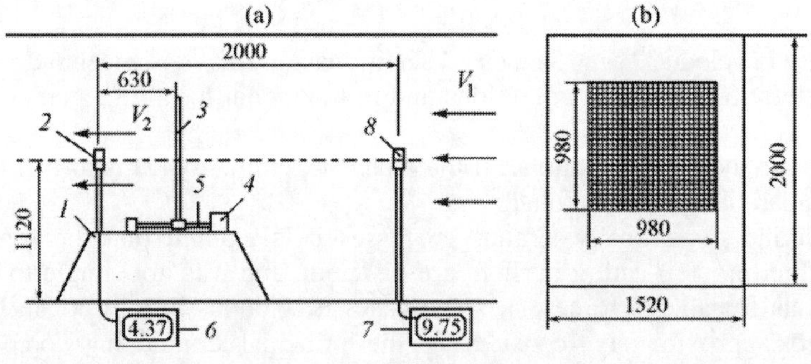

Figure 2. Scheme of the semipermeable shield in an air flow: (a) side view: (b) cross-section (*1*—table; *2*—air flow velocity sensor; *3*—semipermeable shield; *4*—drive motor for a semipermeable shield; *5*—screw; *6*, *7*—digital display of velocity meter; *8*—air flow velocity meter; *9*—air channel.

The velocity meter *8* (SDSV 01), mounted in the middle of the channel cross-section to measure an air flow velocity. Its readings are reported in digital display *7* beyond the channel. The table is mounted in depth of the channel where the analogue air flow velocity meter *2* is provided at the level, identical to the level of sensor *8*; the aerodynamic resistance, made as a metallic grate of 0.98×0.98 m in dimensions with porosity of 0.6. The readings of sensor *2* are at digital display *6* beyond the channel.

At the stabilized fan operation mode the air flow velocity was measured in the channel prior to aerodynamic resistance and post it, the measurement results recorded at digital display were read visually and recorded for 10 min at equal time intervals. By 92 air flow measurements the curves were plotted (Figure 3) and the readings were compared in Table. So, the actual value V_1 is 10.04 ± 0.99 m/s, V_2 is 5.17 ± 0.58 m/s.

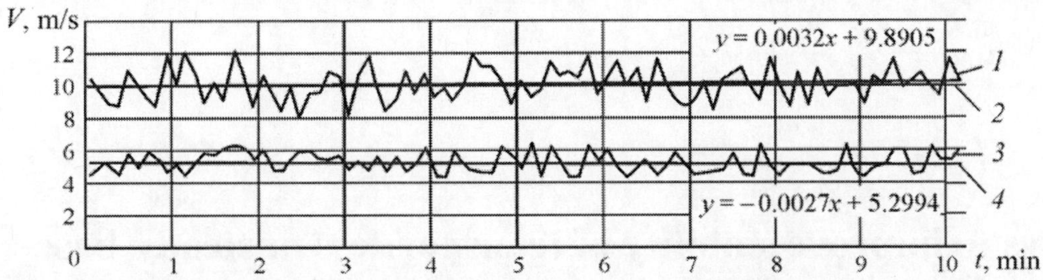

Figure 3. Variations in air flow velocity before and after supplementary aerodynamic resistance: *1*—air flow velocity, measured prior to aerodynamic resistance V_1; *2, 4*—trend line; *3*—air flow velocity, measured after aerodynamic resistance V_2.

Comparison of variations in airflow velocity prior to V_1 and post V_2 additional aerodynamic resistance

Parameter, m/s	min	max	Mean arithmetic, V, m/s	Mean square deviation σ	Cutoff value V_1
V_1	8.19	11.98	10.04	0.99	9.05÷1.03
V_2	4.25	6.3	5.17	0.58	4.59÷5.75

3. Conclusions

The body moving in air or ram air flow are subjected to friction force, so in physical modeling the dynamic similarity is fulfilled if the condition of geometric and kinematic similarities are fulfilled as well as Reynolds number values calculated for comparable points of both air flows under natural conditions and in the model are equal.

To gain equality of Reynolds number values at comparable points it is necessary that geometric parameters of model cells correspond to porosity of the semipermeable shield.

The scatter of experimental and in-situ measurements (about 10%) of air flow velocity is due to the fact that the air flow in the air channel was actuated by rotation of fan blades which failed to provide a uniform air flow motion, unlike air mass motion at an open site.

The factual porosity of the semipermeable shield is 0.60 and provides twofold reduction in the air flow velocity in average.

References

[1] Tsyba AM 2012 Mine heater installation with regulation of air currents in sections *Energy Security of Russia—New Approaches to the Coal Industry Development: Proc. XIV Int. Conf.* Kemerovo pp 18–21
[2] *Climate of the Kemerovo Region* URL: http://meteo-kuzbass.ru/pogoda/climat/january. URL: http://meteo-kuzbass.ru/pogoda/climat/january.]
[3] *Guide to the Coal Mine Ventilation Design* Makeevka Donbass 1989 (in Russian)
[4] Tsyba AM and Gehrike BL 2013 Substantiation of the design parameters of additional aerodynamic resistance of the mine heater installation *Vestn KuzGTU* No 3 (97) pp 27–30
[5] Tsyba AM and Gehrike BL 2012 Increase of the efficiency of the mineless fanless heater installation *Vestn. KuzGTU* No 4 (92) pp 14–16
[6] Alabuzhev PM, Geronimus VB, Minkevich LM and Shekhovtsov BA 1968 *Theory of Similarity and Dimensions Simulation* Moscow Higher School (in Russian)
[7] Abramovich GN 1969 *Applied Gas Dynamics* Moscow Nauka (in Russian)
[8] Geyer VG 1970 *Hydraulics and Hydraulic Drive* Moscow Nedra (in Russian)
[9] Altshul AD 1970 *Hydraulic Resistance* Moscow Nedra (in Russian)

IOP Conf. Series: Earth and Environmental Science **262** (2019) 012079 doi:10.1088/1755-1315/262/1/012079

Ways to improve stability of transportless unsteady-base dumps in terms of Mokhovsky open pit, Kuzbassrazrezugol Holding Co.

TA Tsymbalyuk* and NA Nemova**

Chinakal Institute of Mining, Siberian Branch, Russian Academy of Sciences, Novosibirsk, Russia

E-mail: *orsa@list.ru, **nemova-nataly@mail.ru

Abstract. Under consideration are issues of overburden stockpiling in mining with direct dumping. The dependence of resistance of the deposit rock in a breakage area versus a shear generated by disturbance of its monolithic status at the contact with rock refuse is established. Technological schemes of coal seam mining by a hydraulic excavator and overburden removal by draglines at Mokhovsky coal open mine, Kuzbassrazrezugol Holding Company, are proposed in the present paper.

1. Introduction

In Russia the direct dumping process is proved to be of actual value to remove overburden at coal open mines: Taldinsky, Krasnogorsky, Sibirginsky, Chernigovsky mines in Kuznetsky Coal Basin, Mugunsky in the Irkutsk Region, Irbeisky, Nazarovsky in the Kranoyarsk Krai and so on. The mining processes with overburden rehandling to internal dumps are employed in development of horizontal and flat mineral deposits. They are also efficient to mine coal-seam outcrops in development of inclined and steep deposits with filling of a worked out space with overburden. At open-pit coal mines the search for ways to expand application area of mining with direct dumping is an actual problem [1–3]. The aim of the present paper is to master the coal seam mining process and to improve the internal dump stability in terms of Mokhovsky coal pit, Kuzbassrazrezugol Co.

2. Features of the mining system

Parent rocks of coal bearing seams at Sartakinsky production area, Mokhovsky coal pit, involve fine- and very fine-grained sandstones of variable hardness, aleurolites and scarcely argillites. Sandstone used to be hard, dense of 4–6 hardness ratio in Protod'yakov's scale.

Varieties of course- and medium-size aleurolite are clay sandstones with physical-mechanical properties close to those of sandstones of 3–6 hardness ratio in Protod'yakov's scale. Fine and very fine aleurolites are close to argillites, stability is medium or weak at hardness ratio of 4–6. Argillites occur in a scarce amount and present homogeneous, viscous, hardly bound rocks, known for weak stability, prone to swelling, its hardness ratio is 3–4.

Stockpiling of internal direct dumps on the above rocks serving a weak base for overburden material deposition aggravates their stability.

The mining and geological conditions at Sartakinsky production area are complicated by the following factors:

—the floor of seam 6 is weak and represented with intensively jointed aleurolites and argillite beds;

—underlying jointed sandstones in the floor of seam 6 are confined to below-occurring feeding water-bearing horizon.

The height of parent rock bench is 25 m, width of a pass is 35 m, width of broken rock disintegration depends on a height of a bench to be blasted and amounts to 30–85 m, the maximum total height of a dump can reach 50–53 m. Thereupon, the re-excavation factor varies within 0.25–1.0, average discharge factor is 0.16.

At the production area the preliminary loosening was followed with removal of parent overburden rock by ESh-11/70 dragline with bucket capacity of 10 m^3; excavated overburden was discharged to fill a worked out space at the bottom floor of the lower coal seam 6; bedding angles of the seam vary from 7 to 13°, thickness along layer intersection is 5–6 m [2].

The extraction of the seam by longitudinal passes at the production area inevitably resulted in glides of internal dumps stockpiled on its floor. Moreover, a length of Severny production block in the mining zone did not exceed 400–500 m, but at the neighboring worked out section an internal dump was founded by using the transport mining system in the close vicinity to the production block. The constrained space made it difficult for transport devices to operate under the floor conditions in the production zone of coal seam 6 with increasing mining depth. 11 million cubic meters of overburden were re-excavated and removed from available stockpiles at the neighboring section in order to build an entry trench; thus, the industrial waste-to-ore ratio increased up to about 9 m^3/t, though its critical value being 8 m^3/t. Considering all the above factors served the grounds to cancel temporarily the coal seam mining in Sartakinsky block. The mining operations in future depended on solution of the following problems: to develop the process enabling to develop coal seam 6 without transport driving in the production zone along the floor and to improve the stability of a dump, founded on a weak base.

According to the analysis of mining operations the production zone of coal seam 6 becomes accessible for transport only through available rock piles. The mining scheme is shown in Figure 1.

To improve stability of dumps it was decided to arrange mining front at angle 25° to the coal seam strike. At such a pass the 7–13° incline angle to horizon in cross-section normal to direction of a full-fledged pass tends to flatten down to 3–8°. In fact, such extraction of a coal stream is equivalent to mining by discrete short blocks with stowing of worked-out space with overburden material.

In every pass the extraction of coal seam 6 starts with entry to the roof from an available dump of neighboring section where the transport mining system is practiced (height mark is +191 m). In extraction by diagonal passes the production seam is "cut" by two excavator passes into two triangle prisms (Figure 1a), the tip of the first prism is in the beginning of every pass at the point of entry at roof of the coal seam; the tip of the lower prism locates in the end of a pass nearby northern end of the production block [4–6].

After final coal is extracted in the pass nearby the eastern wall on the floor of the coal seam; a waste rock layer is piled, thickness of this layer is equal to thickness of seam 6 (up to mark +191 m) in order to provide an entry way to the roof of the seam in the follow-on pass and so on. Width of this pile is equal to width of a bench (35 m).

In view of the fact that the floor of coal seam 6 in the zone of direct dumping system is represented with jointed aleurolites with interlayers of frayed material, it is necessary to employ engineering preparation of the base designed for stockpiling of waste material at the attempt to eliminate the factors which are capable to induce an unstable state and to break stable parameters of dumps in the period of the coal pit exploitation. The engineering preparation implies that after coal seam extraction is finished, it is required to manufacture a loosened layer zone of triangle shape in cross-section of average depth up to 6 m in view to provide a greater stability of a would-be waste dump (Figure 1b) [7].

Loosening of rocks in the floor of the seam is performed by blasthole charges. Blastholes are drilled with drill rigs, capable to operate on an inclined base. It is possible to make a leveled site by using a ripper-dozer (Figure 2).

IOP Conf. Series: Earth and Environmental Science **262** (2019) 012079 doi:10.1088/1755-1315/262/1/012079

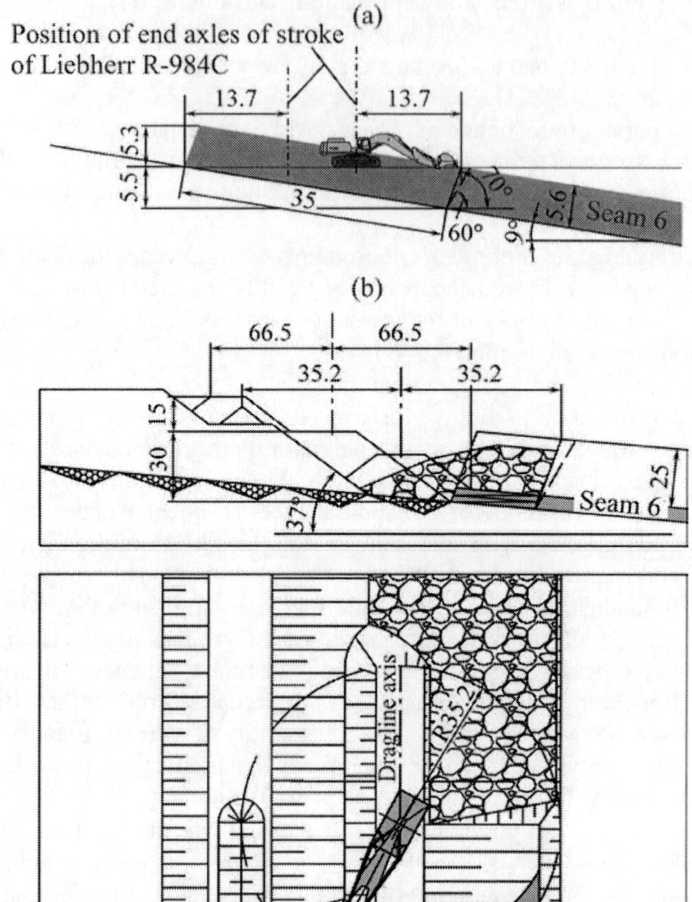

Figure 1. The process scheme to mine Severny production block: (a) coal seam 6; (b) overburden overlying seam 6 by ESh-11/70 dragline.

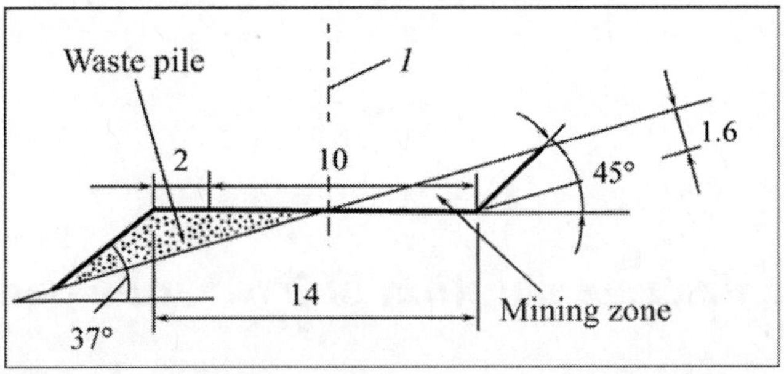

Figure 2. Plan of a mounting pad for drill devices: *1*—axis of a stroke of drill machine SBSh-250MNA-32 (MBR-160A-34); *2*—waste pile.

The small level of blasting in the course of loosening zone production exerts an insufficient effect on stability of available waste dumps.

As it is stated above, an incline of excavator pass in the full-scale front of mining operations amounts up to 8° and is directed towards a neighboring mining area; so the above loosened sites work as a drainage in the dump base, they provide water run-off and improve long-lasting stability of a dump [7–8].

3. Conclusions

It is established that engineering preparation of the base of a would-be waste dump provides 15–20% increase in resistance to shear in floor rocks at a loosened site thanks to breakage of its consolidation at contact with waste rocks, thereto the acting stress is lowered twofold approximately due to redistribution of stresses, the stability factor of the lower dump level grows from 1.3 up to 1.9, while the stability factor of the dump as a whole increases by 30%. To conclude, the opportunity appears to expand the application scope of mining with direct dumping.

The new-proposed technological scheme to mine a coal seam by a hydraulic excavator guarantees the wanted specifications dealing with disposition and operation of the mining and transport machinery.

References

[1] Rzhevsky VV 1985 *Open Mining Operations* Moscow: Nedra (in Russian)

[2] Khashin VN 2014 Technological ways to increase the stability of internal heaps on the sections of Kuzbass *Synopsis of Cand Tech Sci Thesis* Prokopievsk (in Russian)

[3] Smagin VP 1997 Experience and directions for improving non-transport technological schemes *Miner's Week Proceedings* Moscow: MGGU pp 161–164

[4] Gvozdkova TN 2006 Substantiation of the structure of nontransport schemes for the development of suites from three flat coal seams *Synopsis of Cand Tech Sci Thesis Kemerovo* (in Russian)

[5] Shpakov PS, Dolgonosov VN, Yunakov YuL and Shpakov MV 2015 Analytical method for calculating slope stability on a weak base of unlimited power *GIAB* pp 92–99

[6] Cheskidov VI, Bobylsky AS and Reznik AV 2017 Methodological basis for calculating the parameters of nontransport technological schemes for the open development of the formation of gently sloping coal seams *Journal of Mining Science* Vol 53 No 2 pp 95–102

[7] Cheskidov VI, Bobylsky AS 2017 Technological and ecological aspects of the overburden casting of overburden rocks in the Kuzbass sections *Journal of Mining Science* Vol 53 No 5 pp. 299–304

[8] Cheskidov VI and Norri VK 2016 The nontransport technology of overburden operations on the Kuzbass sections: state and prospects, *Journal of Mining Science* Vol 52 No 4 pp 725–731

IOP Conf. Series: Earth and Environmental Science **262** (2019) 012080 doi:10.1088/1755-1315/262/1/012080

Mechanism of discrete rotation of hydraulic perforator tool

M Uraimov[1]*, VE Eremyants[2] and SI Kvitko[1]**

[1]Institute of Machine Engineering, National Academy of Sciences, Bishkek, Kyrgyz Republic
[2]Kyrgyz–Russian Slavic University, Bishkek, Kyrgyz Republic

E-mail: *uraimov47@yandex.ru; **eremjants@inbox.ru

Abstract. The background for engineering a hydraulic rotary–percussive hydroperforator is discussed. The structural layouts of the drilling tool and hydroperforator rotary mechanism are presented. In the rotary mechanism, percussion and rotation assemblies are powered by the same source. The distinctive feature of the drilling method implemented in the developed perforator is shown.

1. Introduction

Hand-held dills are widely used in mining, construction and roading. By mode of functioning, these tools are grouped into hammer drill and rotary hammer drills [1]. In hammer drills, drill rod is rotated during forward or backward stroke of piston by helical pair and crickwork. In rotary hammer drill, drill rod is rotated independently from piston motion by a special device.

The existing drills are mainly driven hydraulically and need large compressors. Deficiency of such set is low efficiency, pollution with oil fog formed in return air exhaust and limited blow energy. According to [1], rotary hammer drills possess low torque, and the capacity spent to rotation of the tool is less than 15% of its percussion capacity.

Jackhammer drills meant for destruction of hard coatings also have drawbacks. For this reason, they are replaced recently by hydraulically driven hammers supplied by small hydro stations. The latter set of equipment has at the least three times as high efficiency as the set of a pneumatic hammer and a compressor station. Hydraulic hammers are lighter in weight, more convenient in operation and ecologically friendly.

It was mentioned in [2] that "...hydroperforators for rotary–percussion drilling machines, at the same size and weight as air-drills, transfer 2–3 times as much power to the drilling tool and increase drilling efficiency by 2–3 times." At the same time, due to no exhaust, hydroperforator produce noise 5–15 dB lower than air drills. Furthermore, as shown in [3], hydroperforators have more rational design of a striking piston, which, at the same blow velocity, creates smaller stress in drill rod as against pistons of air-drills. As a consequence, durability of drill rigs is increased.

Finally, air-drills are low-efficient in highland conditions. By the data from [5], energy and process performance of air-drills depends considerably on elevation of a working site. For example, in operation at the elevation of 4–4.5 thousand meters, performance of air-drills can decrease by 2.4–4.4 times, while the energy requirement of the process will grow by 1.3–2.4 times.

In this regard, it can be stated that one of the promising ways of improvement of hand drills is their transition to hydraulic energy. This trend is already implemented abroad. For instance, Atlas Copco (Sweden) has produced a hand-held hydraulic rotary hammer drill of small weight. Judging from literature sources, no hydraulic hammer drill yet exists. At least, it is unannounced. In this connection, the present research aims to develop design and concept flow chart of hydraulically driven hammer drill.

2. Rotation mechanism of hydraulic perforator tool

It is necessary that drilling tool rotates after blow, which in air-drills is executed by helical pair representing a screw mechanism with high angle of inclination and large lead of screw. These mechanisms are of modest reliability and, moreover, have the transfer torque constraints.

Application of air-drill design solutions in hydraulic hammer drills is not always possible. This is connected with low reliability of helical mechanism and, also, with difficult generation of high torque using the helical pair.

Thus, design of a hydraulic hammer drill needs both a hydraulic drive and a drill rod rotator more reliable than the helical mechanism. In the known design of hydroperforators, drilling tool is rotated using a special drive. At the same time, some designs of hydroperforators connect rotation of drilling tool with the percussion mechanism [5, 6].

One of the scenarios of the drilling rod rotation provides using a rotary hydraulic motor (Figure 1). It consists of a housing *1*, swivel nut *2* and a stator *3* arranged in the central opening of the housing *1*. The stator is immovable relative to the housing, and the nut can turn around its axis by a present angle. In the nut *2* a rotor *4* with rollers *6* is placed. The internal surface of the rotor is made as a hexagonal opening, where a drilling rod *5* is inserted and rotates together with the rotor. Along the outer diameter of the rotor, in parallel to its rotation axis, grooves are made to accommodate the rollers.

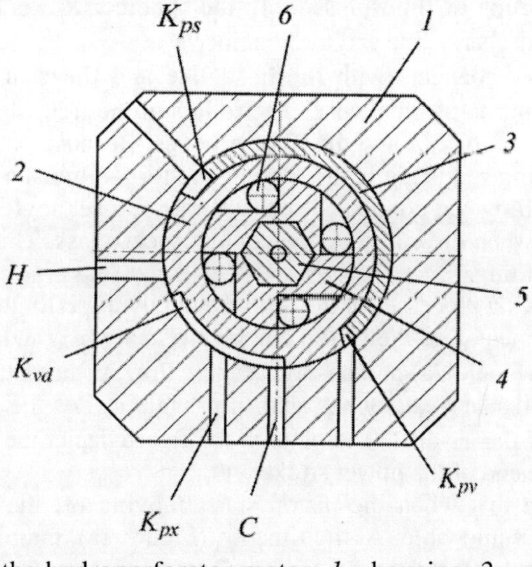

Figure 1. Layout of the hydroperforator motor: *1*—housing; *2*—rotation nut; *3*—stator; *4*—rotor; *5*—drilling rod; *6*—roller.

In this configuration of the rotor grooves, during clockwise rotation of the nut *2*, rollers roll to the wider part of the groove and transfer no torque from the nut to the rotor *4*. In the counterclowise rotation of the nut, the rollers are jammed between the nut and rotor in the narrow part of the groove, ad the nut turns counterclockwise with the rotor. In this fashion, the rotation nut with the rotor and rollers represents a free wheel unit which transfers torque in one direction, clockwise in this case.

Operation of a hydroperforator means synchronous functioning of the impact mechanism and rotary hydraulic motor. In between the surfaces of the rotation nut and the stator, the working chambers K_{PS} and K_{PV} are made (Figure 1). In order that the nut makes backward–rotational motion, the chambers connect in turn with the head and discharge lines. Switching of fluid flows in these chambers is ensured via switching channels connecting the chambers with the percussive mechanism.

This allows control over motion of the hydraulic motor rotation nut depending on position of the piston of the percussive mechanism.

Between the hydraulic motor housing and the rotation nut surface, there is the percussion mechanism control chamber *Kvd*. This chamber connects in turn with the head *H* and discharge *C* channels, in-between which there is a channel connecting the control chamber *Kvd* with the controlled chamber *Kpx* of the percussive mechanism.

It should be mentioned that in the air-drills, rotation of the drilling tool by the helical mechanism occurs when the drilling tool as a result of recoil from rock mass experiences no rotation resistance. The role of the rotation mechanism only reduces to repositioning of the drill bit from one place to another.

Rotation of the drilling tool by the rotary hydraulic motor happens at the moment of change in direction of the forces acting on the piston. In hydraulic percussive mechanisms, switching of the forces takes place twice within a single working cycle, at the end of the idle and power strokes [7]. Accordingly, the rotation nut of the hydraulic motor makes two turns within the working cycle.

The torque transfer from the rotation nut to the drilling tool only takes place on one direction dependent on the direction of the grooves of the free wheel unit. In the scenario in Figure 1, the torque transfer to the drilling tool is possible in clockwise rotation of the nut. This is ensured when the chamber *Kpv* connects with the head line and the chamber *Kps* connects with the discharge line.

Figure 1 shows position of the rotation nut 2 after completion of the drilling rod turn. The chamber *Kps* is connected with the head line, and the chamber *Kpv*—with the discharge line. For this reason, the nut under the action of fluid pressure in the chamber *Kps* is held in this position unit this chamber connects with the discharge line and the chamber *Kpv*—with the head line.

When the chamber *Kpv* connects with the head line and the chamber *Kps* connects with the discharge line, the rotation nut turns clockwise. The rollers of the free wheel unit are jammed between the internal surface of the rotation nut and the rotor grooves. Because of that, its further rotation goes with the rotor and the drilling rod. This is the way of the torque transfer from the rotation nut to the rotor. The rotor and the drilling rod connected with it turns at a set angle. In this case, the turn of the drilling rod can take place when the cutting edge contact rocks mass. Thus, rotation of the drill rod is accompanied by the rock cutting.

For comparison of working processes in the air-drill and hydroperforator, Figure 2 gives the plots of the velocity *U* and displacement *Y* of the piston, angular velocity ω_{rod} and turn angle φ_{rod} of the drilling rod for the hydroperforator (Figure 2a) and air-drill (Figure 2b). At the bottom of Figure 2, there are the plots of the change in the angular velocity ω_{nut} and turn angle φ_{nut} of the rotation nut of the hydraulic motor and the helical nut of the air-drill. The plots in the figure depict the variabt wgeb tgedrilling tool rotates after blow, i.e., at the end of the power stroke.

It is seen in Figure 2a that when the piston starts moving off the collision plant ($Y = 0$), the rotation nut 2 (Figure 1) is immovable. At the time t_{nut} (Figure 2a), the rotation signal is sent, and the nut turns in the time Δt_{nut} by the angle φ_{nut}. The piston is at the distance Y_c from the collision plane in this case. At the time t_1 a signal is sent to switch the direction of the force acting on the piston. Under the action of this force, the piston velocity drops to zero, then it displaces toward the drilling rod and hammers on its rear end; then the rotation signal is sent, and the drilling rod turns by the angle a поршень. Под действием этой силы скорость движения поршня-ударника снижается до нуля φ_{rod}. At the same time, the other side rotation signal is sent to the nut, and the nut turns together with the rod in the time Δt_{rod} by the angle φ_{nut}. When the rod rotation is completed, the signal is sent to switch the force acting on the piston and the piston moves toward the collision plane. The drilling rod remains immobile and is held in this passion by the free wheel unit. Then, the cycle is repeated.

In the air-drill (Figure 2), the idle stroke of the piston is accompanied by the rotation of the drilling rod until stop of the piston at the coordinate Y_0. In the power stroke of the piston, the helical screw with the drilling rod are held immobile by the clickwork. The helical pair screw, which is the piston stock, turns backward by the angle φ_{screw}.

The vital difference of the hydroperforator with the rotation mechanisms consists in the face that after blow on the drilling rod end, the piston remains immobile within the time t_{rod} as it is influence by the power stroke force which presses the piston to the drilling rod. As a result, during rotation of the drilling rod, rock cutting takes place.

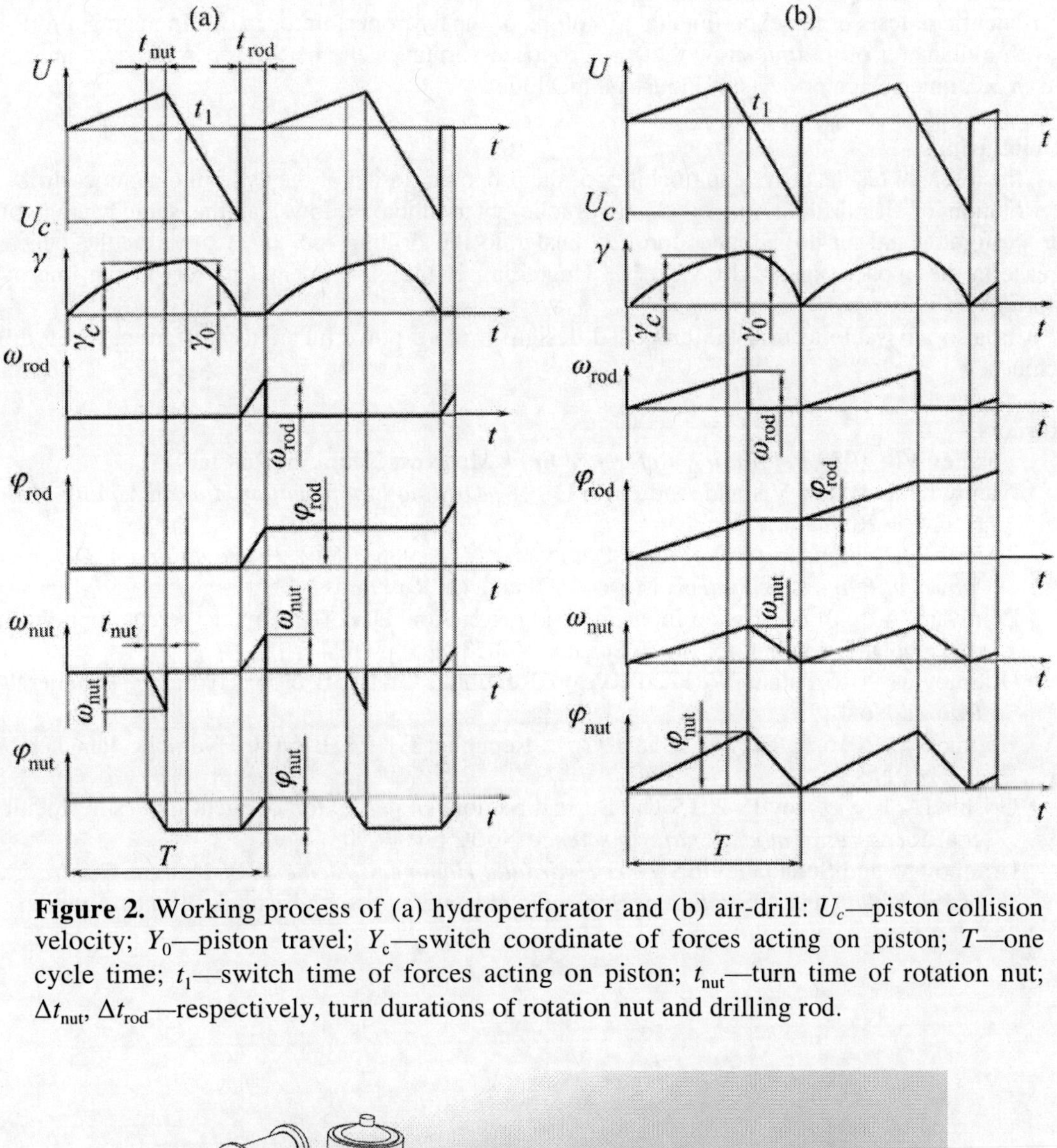

Figure 2. Working process of (a) hydroperforator and (b) air-drill: U_c—piston collision velocity; Y_0—piston travel; Y_c—switch coordinate of forces acting on piston; T—one cycle time; t_1—switch time of forces acting on piston; t_{nut}—turn time of rotation nut; Δt_{nut}, Δt_{rod}—respectively, turn durations of rotation nut and drilling rod.

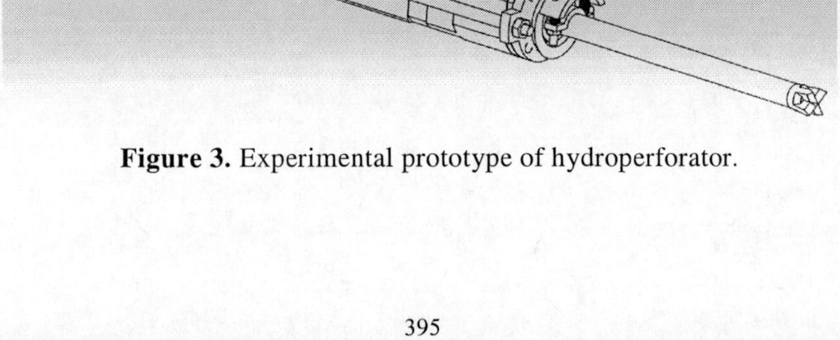

Figure 3. Experimental prototype of hydroperforator.

Based on the discussed scheme, the experimental prototype of the hydroperforator is developed (Figure 3). Its weight is 42 kg at the estimated impact capacity of 2.1 kW, torque of 108 N·m and the head line pressure of 10 MPa. The hydraulic motor capacity calculated based on the fluid pressure and flow rate is 1.49 kW, which is 70% of the impact capacity of the perforator. For comparison, the air-drill PT63 of the same weight develops the torque of 32.2 N·m and ensures percussive capacity of 1.8 kW.

Benchmark tests of the experimental prototype of the hydroperforator in drilling using a four-edge bit with a diameter of 43 mm show that the proposed variant of the perforator is efficient and can be used to continue design projects of industrial machines.

3. Conclusions

Thus, the research results prove exploitability of the proposed design of the hydraulic hammer drill.

Rotation of the drilling rod by the hydraulic rotary motor allows, at the same weight of the hydroperforator and air-drill, higher torque transfer to the drilling tool by 3 times at the concurrent increase in the impact capacity by 16–17%. Upgrading of the experimental prototype can improve its performance.

It is also advisable to trial the proposed design in more powerful perforator arranged on drilling machines.

References

[1] Vasiliev VM 1989 *Peforators: Reference Book* Moscow: Nedra (in Russian)

[2] Ivanov KI, Latyshev VA and Andreev VD 1987 *Drilling Equipment in Mineral Mining* Moscow: Nedra (in Russian)

[3] Alimov OD, Manzhosov VK and Eremyants VE 1985 *Blow. Propagation of Deformation Waves in Percussive Systems* Moscow: Nauka (in Russian)

[4] Eremyants VE 2017 Variation in energy and production data of pneumatic percussiuon machines in the uplands Journal of Mining Science Vol 53 No 4 pp 694–701

[5] Uraimov M 2015 Patent No 1726 Kyrgyz Republic: C1 E21B 6/06. Hydraulic Puncher *Byull. Izobret.* No 4

[6] Uraimov M 2016 Patent No 1958 Kyrgyz Republic: C1 E21B 6/06. Hydraulic Puncher *Byull. Izobret.* No 5

[7] Grishin IA, Knyazbaev JS 2015 The Basic directions of perfection of agent modes in flotation of coal *Successes of modern natural science* No 12 pp 87–90

[8] Uraimov M and Sultanaliev BS 2003 *Hydraulic Hammers. Elements of Design. Generalization of Production and Operation of Hydraulic Hammers Impulse* Bishkek: Ilim (in Russian)

IOP Conf. Series: Earth and Environmental Science **262** (2019) 012081 doi:10.1088/1755-1315/262/1/012081

Laboratory studies of geomechanical properties of deep-level rocks

OM Usol'tseva*, AA Eremenko, YuN Shaposhnik, PA Tsoi and VN Semenov

Chinakal Institute of Mining, Siberian Branch, Russian Academy of Sciences, Novosibirsk, Russia

E-mail: *usoltseva57@mail.ru

Abstract. The article describes the lab-scale tests of the deformation and strength characteristics of rocks extracted from great depths. Samples of high-grade ore, cupriferous ore, disseminated ore, hornfel and limestone were subjected to tension, uniaxial compression and triaxial compression, and the failure envelopes were plotted. The change in the value of cohesion determined from Mohr's envelopes in tension and uniaxial compression as well as in tension and triaxial compression is analyzed. The values of relative changes in cohesion and in uniaxial and triaxial compression strengths obey strong correlation.

1. Introduction

The underground construction and mining industries operating at great depths permanently deal with problems of geomechanics. In order to avoid or minimize unnecessary expenditures in mineral mining it is required to have a reliable prediction of rock mass behavior. Deep-level rock mass experiences complex stress state, which should taken into account in geotechnologies. The actual level of deep-level stress can exceed the uniaxial compression strength determined in lab-scale tests. Accordingly, a failure envelope plotted using the data of uniaxial compression and tension tests can offer inadequate characteristic of strength of deep-level rocks. For this reason, it is necessary to study behavior of rocks in different stress states, in particular, under triaxial compression. Many published studies are available on mechanical properties of rocks under triaxial compression and failure envelope construction, for example, [1–9].

This study is aimed to characterize strength properties of five rock types: high-grade ore, cupriferous ore, disseminated ore, hornfel and limestone sampled at a depth of 1600 m, subjected to different loading (tension, uniaxial compression and triaxial compression) and to construct their failure envelopes.

2. Laboratory test results

Core samples were loaded on Instron 8802 sevo-hydraulic testing machine at a loading program set by displacement of grips at a rate of 0.1 mm/s. In the course of loading, the continuous records of the change in the length and diameter of a sample, axial load and, additionally under the triaxial compression, lateral pressure were filed. Preparation of samples, experimentation and determination of strength characteristics followed the procedures under the state standards GOST 28985–91, 21153.2–84, 21153.3–85, 21153.8–88 and 21153.7–75. The uniaxial and triaxial compression tests were carried out on cylindrical samples 60 mm in length with a diameter of 30 mm; tensions tests—samples 30 mm long with a diameter of 30 mm. Using the 60/30 mm samples, the dynamic Young modulus and dynamic Poisson's ratio were determined. The images of the core samples are shown in Figures 1a–1e.

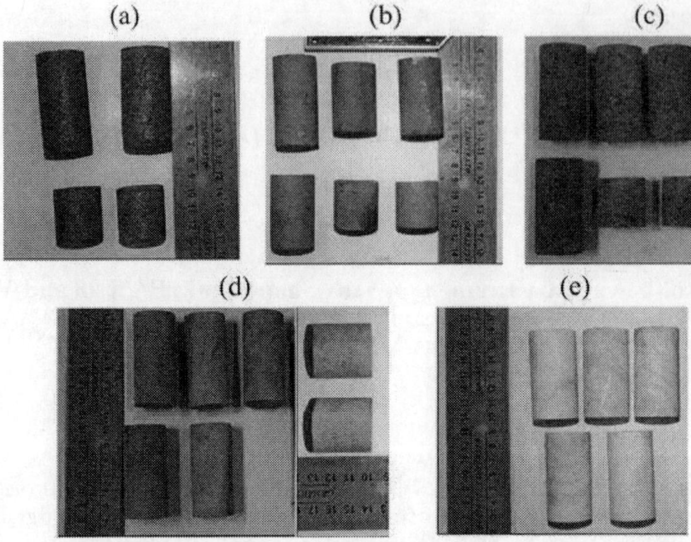

Figure 1. Core samples before testing: (a) high-grade ore; (b) cupriferous ore; (c) disseminated ore; (d) hornfel; (e) limestone.

The triaxial loading was carried out by the Karman scheme: simultaneous axial force and lateral pressure (hydrostatics) up to a certain value of the lateral pressure (5 MPa in these tests), then the axial force was increased until failure of the core samples. Deformation curves were plotted for each sample: axial stress–axial strain; axial stress–transverse strain. The strength characteristics of the core sample under uniaxial and triaxial compression, namely, limit strength, Young's modulus and Poisson's ratio, are compiled in Table 1 (averaged values).

Table 1. Strength test results for core samples: average values of limit strength σ^{lim}, Young's modulus E and Poisson's ratio ν.

No.	Rock type	Test	Limit strength σ^{lim}, MPa	Young's modulus E, GPa	Poisson's ratio ν
1	High-grade ore	Uniaxial compression	95	12.406	0.219
		Triaxial compression 5 MPa	160	18.264	0.159
		Tension	8	—	—
2	Cupriferous ore	Uniaxial compression	150	9.684	0.225
		Triaxial compression 5 MPa	230	15.944	0.168
		Tension	23	—	—
3	Disseminated ore	Uniaxial compression	90	13.169	0.208
		Triaxial compression 5 MPa	127	17.221	0.196
		Tension	17	—	—
4	Hornfel	Uniaxial compression	103	11.562	0.205
		Triaxial compression 5 MPa	157	17.979	0.173
		Tension	27	—	—
5	Limestone	Uniaxial compression	67	9.601	0.219
		Triaxial compression 5 MPa	121	14.977	0.175
		Tension	12	—	—

IOP Conf. Series: Earth and Environmental Science **262** (2019) 012081 doi:10.1088/1755-1315/262/1/012081

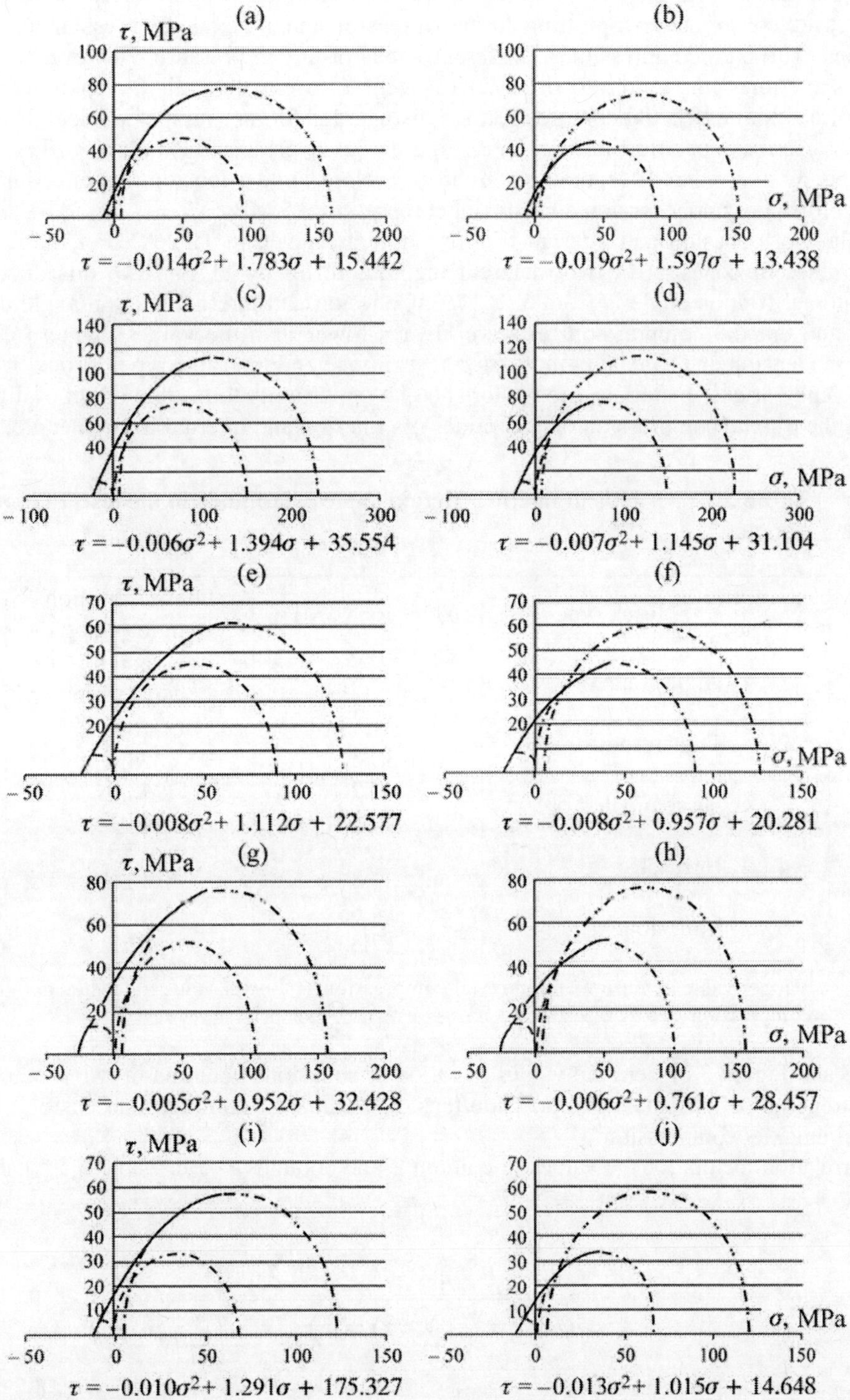

Figure 2. Resultant failure curves for: (a), (b) high-grade ore; (c), (d) cupriferous ore; (e), (f) disseminated ore; (g), (h) hornfel; (i), (j) limestone under tension and triaxial compression (at 5 MPa) in (a), (c), (e), (g), (i), under tension and uniaxial compression in (b), (d), (e), (h), (j). Dashed line—tension; dash-and-dot line—uniaxial compression; dash-and-two dots—triaxial compression; solid line—Mohr's envelope.

Failure curve is a Mohr's envelope in the normal and shear stress coordinates. According to [5, 6], failure curves are constructed by: plotting circles of tension and uniaxial compression and adding them with 3 circles of triaxial compression at different levels of lateral pressure. However, as experiments show, it is sometimes impossible to draw an easy curve to envelope all 5 semicircles. Eventually, either data of uniaxial or triaxial compression are disregarded, or an average envelope is drawn, which inaccurately describes specific limiting state. Figures 2a–2j present the failure curves plotted for 5 types of rocks by two methods separately: Mohr's envelopes under tension and uniaxial compression and Mohr's envelopes under tension and triaxial compression (5 MPa).

The values of cohesion and internal friction angle are given in Table 2. For the tested types of rocks, the values of cohesion in two different methods differ by 11 to 18%; this difference in the values of internal friction angle is from 5 to 17%. Cohesion and internal friction angle determined in the tension and unaxial compression tests are always lower than the values of these characteristics obtained in the tension and triaxial compression tests in one and the same type of rock. Unfortunately, Due to deficiency of core materials, the authors had no opportunity to test the effect of different lateral pressures in the triaxial compression on the values of cohesion and internal friction angle.

Table 2. Cohesion and internal friction angle determined in the tests of core samples.

No.	Rock type	Cohesion c, MPa	Internal friction angle φ, deg
1	High-grade ore	13.44* 15.44**	58.0* 60.75**
2	Cupriferous ore	31.11 35.55	48.9 54 4
3	Disseminated ore	20.28 22.58	43.8 48.1
4	Hornfel	28.46 32.43	37.3 43.6
5	Limestone	14.65 17.33	45.4 52.2

*Upper value in tension and uniaxial compression; **Lower value in tension and triaxial compression

Table 3 and Figure 3 present the values of ratios of cohesions obtained in two methods and ratios of limit strengths of triaxial compression/tension; uniaxial compression/tensionl and triaxial compression/uniaxial compression.

The correlation factors for the values in column 2 and columns 3–5 in Table 3 are calculated from the formula:

$$r = \frac{n\sum_1^n xy - (\sum_1^n x)(\sum_1^n y)}{\sqrt{(n\sum_1^n x^2 - (\sum_1^n x)^2)(n\sum_1^n y^2 - (\sum_1^n y)^2)}},$$

where x stands for $(c^{**} - c^{*}) / c^{*}$, y—for $\sigma_{\text{tcom}}^{\text{lim}} / \sigma_{\text{tens}}^{\text{в}}$; $\sigma_{\text{ucom}}^{\text{lim}} / \sigma_{\text{tens}}^{\text{lim}}$; $\sigma_{\text{tcom}}^{\text{lim}} / \sigma_{\text{ucom}}^{\text{lim}}$. The calculated correlation factors are given in Table 3.

IOP Conf. Series: Earth and Environmental Science **262** (2019) 012081 doi:10.1088/1755-1315/262/1/012081

Table 3. Ratios of cohesions determined by two methods, as well as ratios of limit strengths in the tests of triaxial compression/tension, uniaxial compression/tension and triaxial compression/uniaxial compression.

No	Rock	$(c^{**}-c^{*})/c^{*}$	$\sigma^{\lim}_{\text{tcom}}/\sigma^{\text{в}}_{\text{tens}}$	$\sigma^{\lim}_{\text{ucom}}/\sigma^{\lim}_{\text{tens}}$	$\sigma^{\lim}_{\text{tcom}}/\sigma^{\lim}_{\text{ucom}}$
1	High-grade ore	0.15	0.95	0.92	0.41
2	Cupriferous ore	0.14	0.90	0.85	0.35
3	Disseminated ore	0.11	0.87	0.81	0.29
4	Hornefel	0.14	0.83	0,74	0.34
5	Limestone	0.18	0.90	0.82	0.45
Correlation factor			0.379	0.165	0.954

c^{*}—cohesion from the results of tension and unaixal compression tests;
c^{**}—cohesion from the results of tension and triaxial compression tests

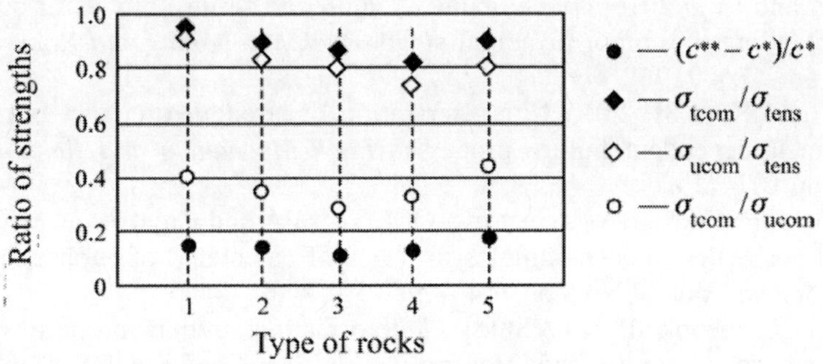

Figure 3. Relationship between ratios of strength characteristics (axis of ordinates) for different types of rocks: 1—high-grade ore; 2—cupriferous ore; 3—disseminated ore; 4—hornfel; 5—limestone.

It follows from Table 3 data that there is a strong correlation between the values of $(c^{**}-c^{*})/c^{*}$ and $\sigma^{\lim}_{\text{tcom}}/\sigma^{\lim}_{\text{ucom}}$ at the correlation factor 0.954.

Regarding the change in the internal friction angle, a satisfactory correlation was observed with neither value of rock properties.

3. Conclusions

Based on the experimental data on strength characteristics of rocks (high-grade ore, cupriferous ore, disseminated, hornfel and limestone) sampled at a depth of 1600 m, it has been concluded that the values of limit strengths and moduli of deformation under triaxial compression at the lateral pressure of 5 MPa exceed the values of these characteristics under uniaxial compression by 40–80 and 30–60%, respectively; the values of Poisson's ratio are lower by 10–40% in the latter case.

The difference in the values of cohesion obtained by two methods: Mohr's envelope under tension and uniaxial compression and Mohr's envelope under tension and triaxial compression (5 MPa) is from 11 to 18%; this difference in the values of internal friction angle—from 5 to 17%. Both cohesion and internal friction angle calculated by the results of the tension and uniaxial compression tests are always less than these values calculated from the tension and triaxial compression tests for one and the same type of rocks. The values of relative change in cohesion in two specified methods and uniaxial/triaxial compression strength ratios obey strong correlation dependence at CF of 0.954. The revealed law can be used in calculations of cohesion of deep-level rocks.

Acknowledgements

The studies were supported within the framework of the Fundamental Research Program, Projects Nos. AAAA-A17-117121140065-6 and AAAA-A17-117121140065-7. The tests were carried out using the facilities of the Shared Use Center for Geomechanical, Geophysical and Geodynamic Measurements under auspices of the Siberian Branch of the Russian Academy of Sciences.

References

[1] Zhang P, Mishra B and Heasley KA 2015 Experimental investigation on the influence of high pressure and high temperature on the mechanical properties of deep reservoir rocks *Rock Mechanics and Rock Engineering* Vol 48 Issue 6 pp 2197–2211

[2] Huang D, Gu D, Yang C, Huang R and Fu G 2016 Investigation on mechanical behaviors of sandstone with two preexisting flaws under triaxial compression *Rock Mechanics and Rock Engineering* Vol 49 Issue 2 pp 375–399

[3] Peng J, Rong G, Cai M, Wang X and Zhou C 2014 An empirical failure criterion for intact rocks *Rock Mechanics and Rock Engineering* Vol 47 Issue 2 pp 347–356

[4] Li X, Du K and Li D 2015 True triaxial strength and failure modes of cubic rock specimens with unloading the minor principal stress *Rock Mechanics and Rock Engineering* Vol 48 Issue 6 pp 2185–2196

[5] Kaiser PK and Kim BH 2015 Characterization of strength of intact brittle rock considering confinement-dependent failure processes *Rock Mechanics and Rock Engineering* Vol 48 Issue 1 pp 107–119

[6] Zhigalkin VM, Rychkov BA, Usol'tseva OM, Tsoi PA and Chynybaev MK 2011 Estimation of strength properties of rock samples in terms of calculated of Mohr's envelopes *Journal of Mining Science* Vol 47 No 6 pp 714–721

[7] Tsoi PA and Homenok IP 2017 Study of deformation-strength properties of sandstone under multistage triaxial compression *Journal of Physics: Conference Series* Vol 894 Art 012104 (7 p) DOI: 10.1088/1742-6596/894/1/012104.

[8] Jiang Q, Zhong S, Cui J, Feng XT and Song L 2016 Statistical characterization of the mechanical parameters of intact rock under triaxial compression: an experimental proof of the jinping marble *Rock Mechanics and Rock Engineering* Vol 49 Issue 12 pp 4631–4646

[9] Usol'tseva OM, Tsoi PA and Semenov VN 2016 Effect of the structure of layered rocks and geomaterials on strength and deformation characteristics under triaxial compression and tension *Interexpo GEO-Siberia-2016: 12th Int. Congr. & Expo. Subsoil Use. Mining. Trends and Technolgies in prospecting Exploration and Mining* Vol 4 pp 49–54 (in Russian)

Automatic system to control hydraulic drill operation

VB Vasiliev* **and AP Muslimov**

Institute of Machine Science, National Academy of Sciences, Bishkek, Kyrgyzstan

E-mail: *imash.nankr@gmail.com

Abstract. Under consideration is the operation principle of the automatic control system to manage the drilling modes in terms of variable rocks in view to increase the cutting tool durability, to lower drilling costs, and to improve the new-made hole quality. The modern requirements for drilling processes imply automation of the control processes as it is impossible to achieve the above goals with manual control of a drilling machine. Automation rests on the use of high-speed automatic systems, capable to provide automatic prompt changeover from one mode to another, along with routine adjustment of drilling modes.

1. Introduction

Drilling is a routine, but cost process in the mining industry. Moreover, it is actually known as a complicated method requiring high professional skills of workers and operators, responsible for quality and production cost of hole-making [1, 2]. Thereto, the production cost of hole making depends on a variety of other factors: drilling mode, factor of drilling machine utilization, its performance, durability of a tool. The drilling regimes: drill feed rate, drill rotation velocity, air hammer blow frequency, etc. are prime drilling parameters constituting operation life of the tool, and its capacity utilization factor [1, 2].

A drilling mode is computed and prescribed based on physical-and-mechanical properties of a rock mass to be drilled, but not always they match it because of a variability of rock properties especially in making long-length holes. The high-class skills of the drill operator imply just-in-time detection of variations in rock properties, proper adjustment of a drilling mode. The human factor does not always guarantee a success, durability of the tool can fail and lead to an accident; and in both cases production cost of hole making increases with a respective violence of hole quality.

The modern requirements for drilling velocity and hole-making grade became more rigorous and the conventional control of drill operation got really impracticable. The application of automatic system for control and adjustment of principal drilling parameters acquires a profound actuality [2].

Automation of drilling techniques as a many-years advance resulted in a good variety of available approaches and methods to adjust drilling modes. However the processor control systems with sophisticated tuning components often fail to meet present-day demands for speed performance. All the above provides the sound grounds to develop so called two-circuit automatic systems.

2. Automatic control system layout

The investigation into modern hydraulic drill machines with automatic control functions revealed that automation means for their hydraulic systems are only adjustable oil pumps to provide delivery control. The adjustable oil pumps have less operational life at higher cost as compared to uncontrolled pumps, so in a number of applications the adjustable pumps are economically unsound. Besides, delivery control in hydraulic systems often has a negative effect on a fluid flow rate in all the hydraulic lines, thus appreciably increasing an inertia capacity and transition time of the system [2, 3].

Throttle control eliminates the above cited drawbacks and provides speedy response of the automatic system [4, 5].

A novel automatic system to optimize drilling control is developed to drill expendable wells (Figure 1). The specific optimal tool feed, its rotation velocity, the impactor actuation moment are adjusted by three automatic fluid-flow-rate controllers *13*, *14*, *15*.

The operation principle of the automatic control system runs as follows: internal combustion engine *1* drives concurrent rotation of three oil pumps *2*, *3*, *4*. Pump *2* supplies a working fluid to hydraulic line of the percussion device *18*, where automatic fluid-flow-rate controller *13* is mounted to vary energy and impact frequency. Percussion device *18* is actuated automatically when rotation torque reaches a certain value at drill tool and is intended to break hard rock. Pressure reducing valve *11* is a component of regulator *13*, and is connected to its inlet and outlet lines and functions to eliminate pressure difference in a regulator and overflow of a fluid excess into a bin under throttling.

Pump *3* supplies the working fluid to hydraulic line of motor *16* rotating the drill tool. Distributer *8* provides reverse rotation of motor *16*, fluid-flow controller *14* varies its rotation velocity.

Pump *4* rotates motor *17* to feed the drill tool, includes distributor *9* and fluid-flow regulator *15*. Motor *17* generates velocity and drill feed force.

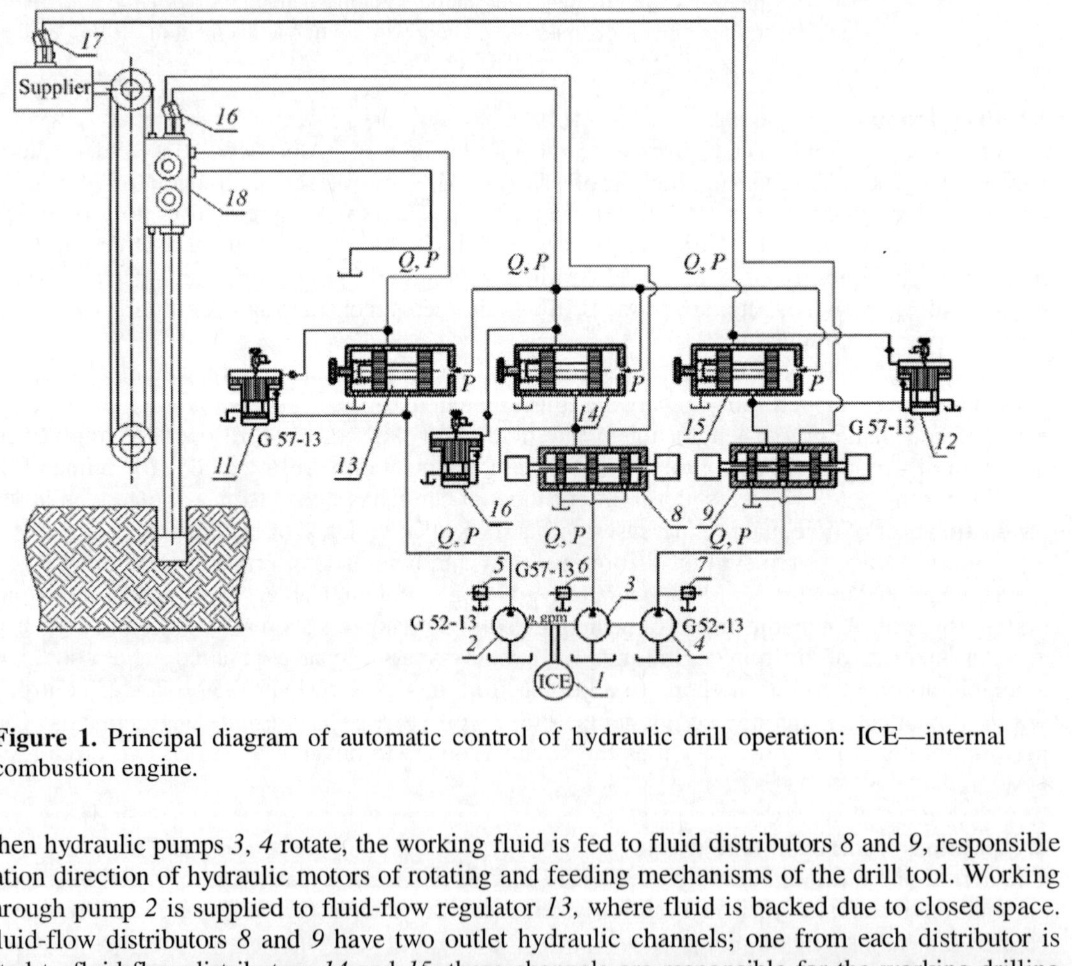

Figure 1. Principal diagram of automatic control of hydraulic drill operation: ICE—internal combustion engine.

When hydraulic pumps *3*, *4* rotate, the working fluid is fed to fluid distributors *8* and *9*, responsible for rotation direction of hydraulic motors of rotating and feeding mechanisms of the drill tool. Working fluid through pump *2* is supplied to fluid-flow regulator *13*, where fluid is backed due to closed space. Each fluid-flow distributors *8* and *9* have two outlet hydraulic channels; one from each distributor is connected to fluid-flow distributors *14* and *15*, these channels are responsible for the working drilling moments. Other two channels are responsible for secondary operations, such as pulling out of a drill string or demounting of a drill machine.

Control of sleeve valve of fluid-flow distributors *13*, *14* and *15* is performed by pump pressure mains, connected to air pressure line of motor *16* of drill-tool rotation device. Sleeve valves are tuned in such a way that if pressure in the line under their control starts to grow the sleeve lines move to the left side; thereto regulator *15* diminishes fluid flow rate, while regulators *13* and *14* increase it.

In the drilling process the pressure generated in the control lines characterizes the torque moment on the drill tool. This pressure moves sleeve valves *13*, *14* and *15* in the preset positions providing the design drill modes. Given that hardness of drilled rock varies, let, to a higher value, then the torque moment on the drill tool increases; this leads to increase in pressure in control main lines and position of sleeve valves of fluid-flow regulators changes. Hereupon, sleeve valves of regulators *14* and *15* start to move to the left, so the rate of drill feed lowers, but its rotation velocity grows. Sleeve valve of regulator *13* remains in the previous position, because it is tuned to move under pressure higher than that in sleeve valves of regulators *14* and *15*. Assuming that hardness of the rock to be drilled continues to vary to higher level and pressure in the control channels reaches such values that the sleeve valve of regulator *15* moves to close the fluid-supply channel, and the tool feed is cancelled. However the sleeve valve of regulator *14* maximally opens fluid-supply channel and the tool rotation will be maximum high; regulator *13* opens and actuates percussion mechanism to break hard rocks.

To conclude, the new-proposed automatic system has the range to adjust drilling modes of three executive members of the drill machine and after all provides transition from one kind of drilling to another.

3. Conclusions

1. Advantage of throttle adjustment of a drilling mode is substantiated.
2. The automatic system is developed to control the hydraulic drill operation modes.
3. The development of such automatic systems enables to improve drill durability, to lower an accident risk in drilling operation to higher the drilling performance and quality of new-made holes.

References

[1] Gafiyatullin RK, Ignatiev OV and et al 1978 *Modes and Automation of Rotational Drilling* Moscow: Nedra (in Russian)
[2] Medvedev IF 1986 Drilling Modes and Selection of Drilling Machines Moscow: Nedra (in Russian)
[3] Medvedev IF 1971 Determination of the cost of drilling blasting holes in underground stoping *Gornyia Zhurnal* No 11
[4] Besekersky VA and Popov EP 2003 *The Theory of Automatic Control Systems* Saint Petersburg: Professiya (in Russian)
[5] Goodwin GK, Grebe SF and Salgado ME 2004 *Designing of Control Systems* Moscow: Binom (in Russian)

Design of automatic fluid flow regulator to control hydraulic drill operation modes

VB Vasiliev

Machine Science Institute, National Academy of Sciences, Bishkek, Kyrgyzstan

E-mail: imash.nankr@gmail.com

Abstract. The author considers the design and operation principle of the automatic fluid flow regulator intended to control a fluid flow rate in a hydraulic system of a drilling rig with automatic adjustment of drilling modes in terms of variable properties of a rock to be drilled. The data on regulator tests and its output specification are reported.

1. Introduction

The main advantages of the hydraulic machinery are capacity for high-effort transmission and simplicity in operation and automation. The application scope of hydraulic means of automation tends to expand to realize the complete integrated automation in different industries [1, 2].

Apparatus, machine or a technical facilty where it is necessary to maintain or to update one or few parameters under a certain law is known as "controllable object". A set of devices applied to realize the automatic control of a process is termed a controller or a regulator.

In the mining industry automation of the drilling process acquires specific significance as it is one of the most complicated, widely used, but cost processes [3, 4]. Most drilling facilities especially to make underground holes operate on the hydraulic systems, where a fluid-flow rate is adjusted by contrable oil pumps, appreciably cutting down operation live of a drill rig [4]. It is the author's opinion, that the reason for such popularity of controlable pumps is simplicity in designing of an automatic system and the lack of fluid-flow regulators operating on the automatic control principle. The analysis of available fluid-flow regulator types gives the grounds to state that all the serial-produced divices function to control a fluid flow rate, disregarding variable external factors. When using this–type regulators the flowrate is adjusted manually or with application of auxilliary devices: gear wheels, rails, or hydraulic executive devices characterized with a slow transition, and a risk of lower precision of a new-set operation mode.

To regulate drilling modes implies availability of a mechanism capable to change automatically a fluid flow rate in view of variations in external factors. The problem is that in the world practice there are not so many such regulators, and, moreover, they are specified with a rather complicated control system.

2. Design of automatic liquid flow regulator and experiment results

Figure 1 demonstrates a scheme of an automatic fluid-flow regulator developed at Machine-Science Institute, National Academy of Sciences, Kyrgyzstan.

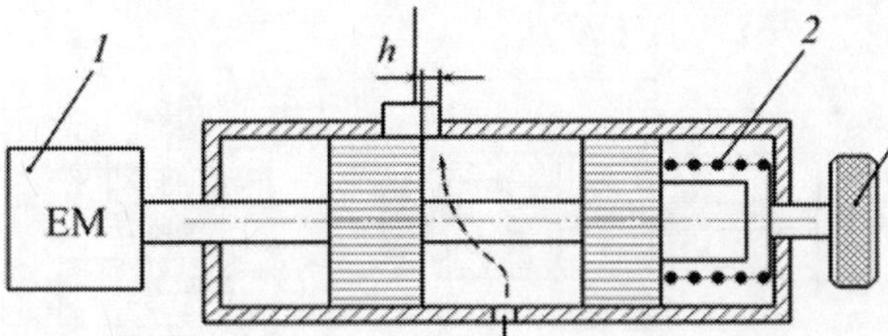

Figure 1. Scheme of an automatic fluid-flow regulator: EM—electromagnet.

Under action of electromagnet *1* sleeve valve overcomes spring force *2*, moves to the right and reduces gap *h* in a slot for the working fluid, thus cutting down a fluid flow rate. Under increase in input voltage of electromagnet coil the magnetic field force is growing and the sleeve valve changes its position proportionally to this change. Under reduction in magnetic field force the sleeve valve moves, the slot increases under action of spring *2*. Screw *3* is intended to set a wanted spring stiffness depending on a required pressure in the pressure mains (Figures 2 and 3).

Figure 2. General view of a fluid-flow regulator.

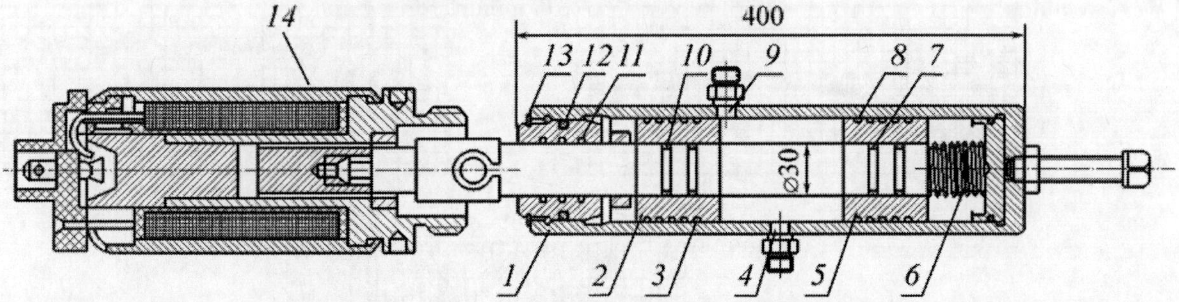

Figure 3. Diagram of electromagnetic fluid-flow regulator.

The regulator consists of case *1*, sleeve valves with rods *2*, *5* with cuts *3*, *8* in order to lower oil-leakage factor, spring *6*, cap with sealers *11*, *12*, *13*, electromagnet *14* (Figure 3). The sleeve valve axis has cuts *7*, *10* for ring-shaped rubber sealers, inlet and outlet holes *4* and *9*, respectively, for the working fluid flow. The sleeve valves are managed by electromagnet *14* (Figure 3) through changing the electromagnetic field force.

Reduction valve G-57-13 *2* is provided in the regulator design to eliminate oscillations of a sleeve valve and to obtain a straight-line output characteristic of the regulator, as well as to exclude pressure difference between inlet and outlet main lines (Figure 4).

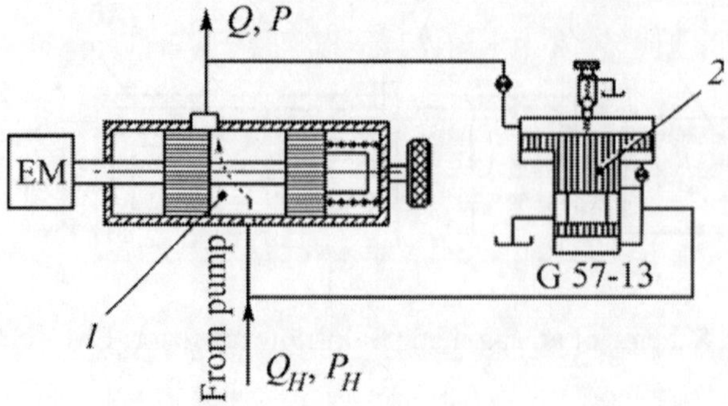

Figure 4. Scheme of reduction valve mounting on a fluid-flow regulator.

Thereto, condition $P_P - P$ =const is fulfilled disregarding load, then fluid flow rate Q is $Q = K_3$, where $K_3 = \mu\pi d\sqrt{2g/\gamma\Delta p}$ = const , $\Delta P = P_P - P$, где P_P—pump pressure; μ = 0.65—flowrate factor; d—diameter of sleeve valve; P—pressure in output main line of the regulator; Q_P—fluid flow rate through pump.

Experimental tests of the fluid-flow regulator were undertaken at a specially designed hydraulic stand. The oscillograms characterizing relation of hydrocylinder velocity versus a size of the working slot at the regulator were established using this stand (Figure 5).

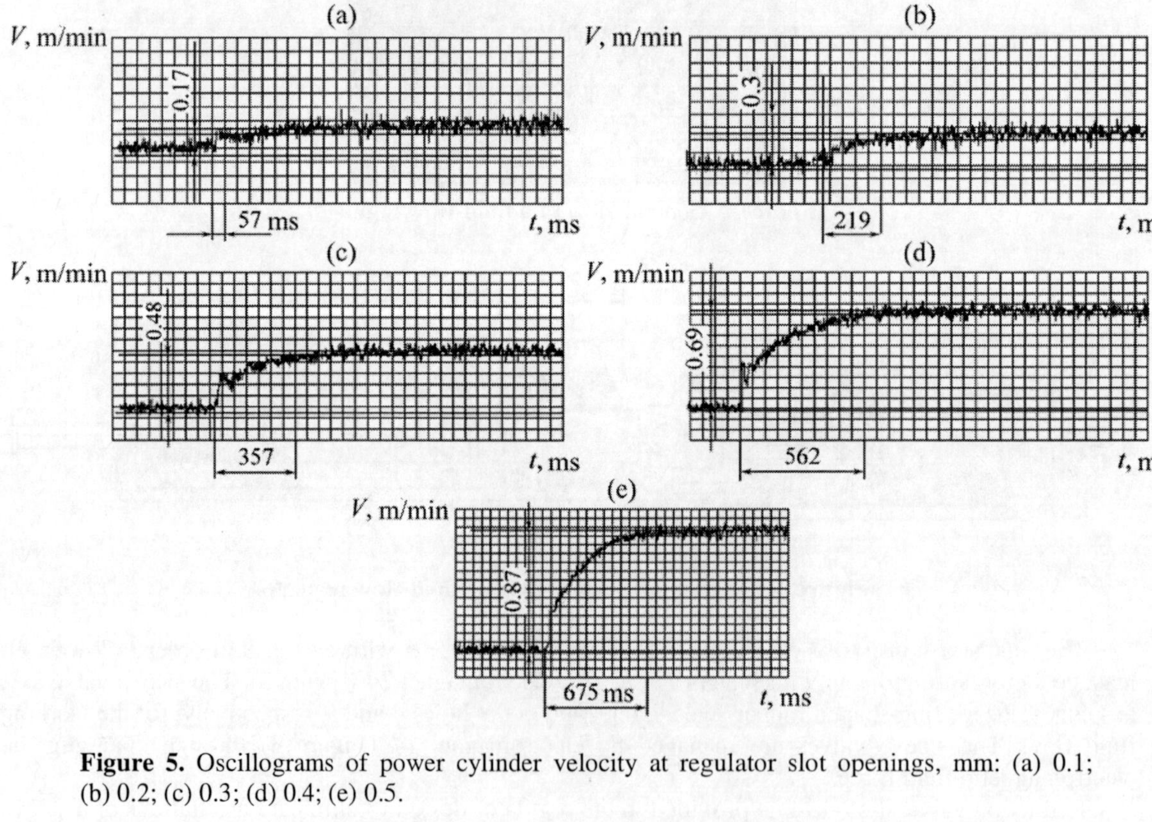

Figure 5. Oscillograms of power cylinder velocity at regulator slot openings, mm: (a) 0.1; (b) 0.2; (c) 0.3; (d) 0.4; (e) 0.5.

In Figure 6 theoretical and experimental relationships of variations in power cylinder velocity versus a size of regulator slot are plotted by the data reported in the table.

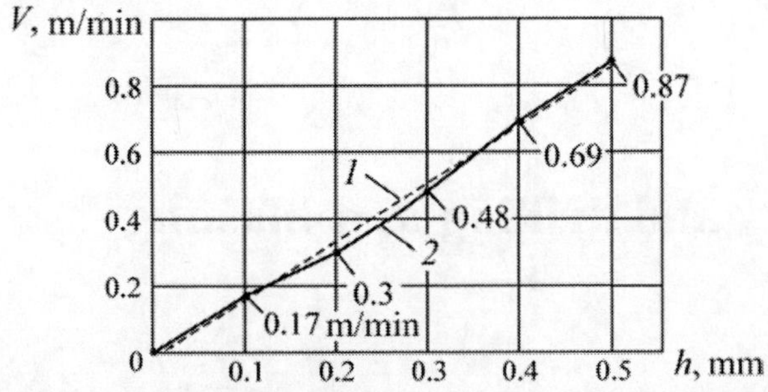

Figure 6. Relationship of cylinder rod velocity versus a size of regulator slot *h*: *1*—theoretical characteristic; *2*—experimental characteristic.

Output parameters of a fluid-flow regulator.

h, mm	0.1	0.2	0.3	0.4	0.5
V, m/min	0.17	0.3	0.48	0.69	0.87

3. Conclusions

1. Fluid-flow regulator with automatic control is developed.
2. The fluid-flow regulator is applicable in hydraulic systems where the automatic fluid-flow adjustment is required versus variable external factors.
3. The efficiency of the novel regulator is justified theoretically and experimentally. The output characteristic of the regulator is determined, and the transition time values are obtained for adjustment of velocity of a hydraulic cylinder rod as an example.

References

[1] Besekersky VA and Popov EP 2003 *The Theory of Automatic Control Systems* Saint Petersburg: Professiya (in Russian)
[2] Goodwin GK, Grebe SF and Salgado ME 2004 *Designing Control Systems* Moscow: Binom (in Russian)
[3] Gafiyatullin RK., Ignatiev OV et al 1978 *Modes and Automation of Rotational Drilling Process* Moscow: Nedra (in Russian)
[4] Medvedev IF 1986 *Drilling Modes And Selection of Drilling Machines* Moscow: Nedra (in Russian)

IOP Conf. Series: Earth and Environmental Science 262 (2019) 012084 doi:10.1088/1755-1315/262/1/012084

Pit wall stability and drilling-and-blasting

SN Zharikov

Institute of Mining, Ural Branch, Russian Academy of Sciences, Yekaterinburg, Russia

E-mail:*333vista@mail.ru

Abstract. The author describes discipline of wall control blasting in open pit mining. The introduction procedure of the special drilling-and-blasting technology at ultimate pit limits includes analysis of blast effects on pit wall rock mass, finding of consistent patterns of wave processes in rocks, determination of introduction between charges in perimeter blasting depending on rock mass strength, semicommercial testing of cutback methods, staging of process blasting in critical areas as well as identification of drilling-and-blasting efficiency criteria. Analysis of seismic wave propagation in rocks under dynamic impact and initiation of stress waves allows setting parameters of destructive effect of explosions in production blocks. At the same time, determination of seismic stability in the zone of destructive effect of blasts makes it possible to select the most sound and moderate drilling-and-blasting regimes and to maintain pit wall slope stability.

1. Introduction

Increasing angles of pit wall slope decreases amount of wall control blasting and improves economic efficiency of open pit mining. On the other hand, there is a heightened risk of pit wall failure as a result of geodynamic movements due to rock pressure and under short-terms shock loads of explosions. Estimation of pit wall slope stability at a certain angle is based on measurements of rock mass displacements in different directions within a long period of time [1–7], and short-term deformations causing local block movements are disregarded in justification of stable slopes. In this regard, inaccurate wall blasting can violate stability of pit walls and result in disastrous earth sliding [8–13].

2. Methods and results of research

Miners have become very interested recently in wall control blasting at the stage of transmission to ultimate pit limit. Evidently, even rise in cost of wall control blasting at higher quality and adequate control results in increased safety of mining, stability of pit walls and, consequently, in higher economic efficiency of open pit mining in whole. For example, elimination of earth slide is connected with cessation of mining or, sometimes, total arrest of performance with corresponding damage and, sometimes, death. For this reason, improvement of mining efficiency is not only an item of cost but weighty contribution to progression, which largely governs future efficient performance of a mine.

The Institute of Mining, UB RAS, is engaged in R&D on introduction of special drilling-and-blasting technology at ultimate pit limits. In order to abate dynamic impact of blasting on pit wall slope stability, seismic stability of rock mass is comprehensively investigated. The related procedures are described in [14–23]. Based on the research findings on seismic stability of rocks, a special drilling-and-blasting technology is developed for the ultimate pit limit stage. The shielding parameters are determined

IOP Conf. Series: Earth and Environmental Science **262** (2019) 012084 doi:10.1088/1755-1315/262/1/012084

for rock bands depending on characteristics explosives and diameters of the wall control blasting holes These characteristics are described in expressions (1)–(4), as well as in Figure 1 and Table 1.

The pressure on blast hole walls, P, MPa is:

$$P = \frac{r_{\text{charge}}^2}{2\, r_{\text{hole}}^2} P_d , \tag{1}$$

where r_{charge} and r_{hole} are the radii of explosive charge and blast hole, respectively.

The detonation wave pressure P_d, MPa:

$$P_d = \frac{\rho_{\text{expl}} D^2}{1+n} , \tag{2}$$

where ρ_{expl} is the explosive density, kg/m^3; D is the velocity of detonation, m/s; n is the index of explosion product expansion.

The fracturing radius, m:

$$R_{\text{frac}} = r_{\text{hole}} \sqrt[3]{(P / \sigma_{\text{dyn}})^2} , \tag{3}$$

where σ_{dyn} is the allowable dynamic ultimate strength of rocks, MPa.

The distance pressure in rocks after explosion of a cylindrical charge, P_R, MPa:

$$P_R = P \left(\frac{r_{\text{hole}}}{R} \right)^{1.5} \tag{4}$$

where R is the distance from the explosion, m.

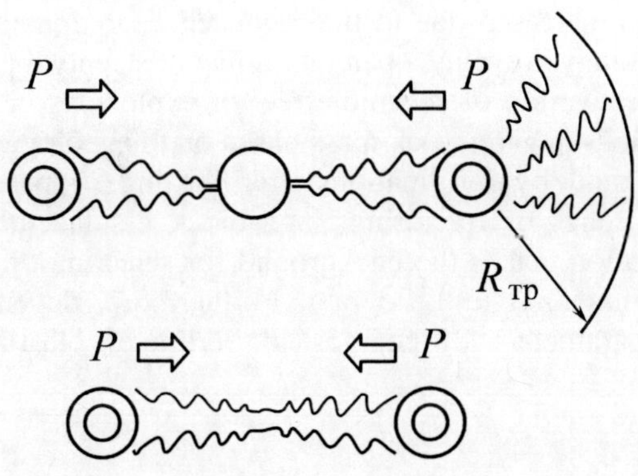

Figure 1. Analysis of pressures in determination of rational parameters of perimeter blasting.

The technological approaches are tested full scale in the framework of the program approved by management of a mine. The program describes implementation stages, lists necessary assets and dictates the control routine. After the program is implemented and the semicommercial tests are carried out, the results are analyzed, ad an in-house regulatory document on drilling-and-blasting at the ultimate pit limits is elaborated (production procedures). The expected results of this technology introduction include:

- Minimization of impact of drilling-and-blasting on pit wall stability;
- Increased safety of operation under high benches;
- Reduced cost of marketable products.

Table 1. Fragment of chart of spacing determination for blast holes in row using the table of values calculated from (1)–(4).

Rock	Tensile strength, MPa	Dynamic ultimate strength, MPa	Fracturing radius, m			
			Ammonite 6 ZHV			
			D_{hole}/d_{charge}			
			215/60	246/60	215/90	246/90
Carbon cataclasite	0.36	0.47	2.3	2.2	4.0	3.8
Carbon quart-carbonate rocks	0.56	0.73	1.7	1.7	3.0	2.8
Carbon metasiltstone	0.42	0.54	2.1	2.0	3.6	3.5
Metasomatite	0.53	0.69	1.8	1.7	3.1	3.0
Heavily damaged metasomatite	0.24	0.31	3.1	2.9	5.3	5.0
	Distance from charge, m		215/60	246/60	215/90	246/90
Perimeter blast pressure, MPa	1		1.66	1.56	3.75	3.50
	2		0.59	0.55	1.32	1.24
	3		0.32	0.30	0.72	0.67
	4		0.21	0.19	0.47	0.44
	5		0.15	0.14	0.33	0.31
	6		0.11	0.11	0.25	0.24
	7		0.09	0.08	0.20	0.19
	8		0.07	0.07	0.17	0.15

A key competitive advantage of this technology is its high efficiency when mining operations are very intensive. Safety is also increased due to the regulated operations at the ultimate pit limits. The application area of this technology is cycling open pit mining of steeply dipping mineral deposits.

The R&D involves investigation of seismic effect of explosions in order to correlate actual and calculated admissible vibrations in given rock mass based on the geological exploration documents, on the one hand. On the other hand, by deformations after blasting, properties of soil are approximately determined, and estimated zones of fracturing, inter-block displacements and residual strains are identified [24]. This information makes the background for selecting drilling-and-blasting patterns to be included in the semicommercial test program. Furthermore, depending on soil properties and explosive characteristics, parameters of charges in perimeter blasting meant for shielding are determined [25].

3. Conclusions

The introduction of the special drilling-and-blasting technology in wall control blasting is accompanied by investigation of wave processes in rocks, which governs the choice of the corresponding patterns. An important element of such investigations is refinement of properties of soil under short-term shock loading.

The technical approaches to seismic stability assessment of rocks include pre-calculation, comparison of the calculated data with the experimental measurement result, refinement of rock mass properties and identification of possible failure zones. Such scope of the research allows obtaining the broadest picture of wave propagation in specific rock mass and, consequently, enables rational control. For the first turn, this relates wall control blasting and subsequent process explosions.

Acknowledgements

The studies were supported in the framework of State Contract No 007-00293-18-00, Topics No 0405-2018-0015 and 0405-2018-0001, Project No 18-5-5-10 with some extra funds from commercial contracts.

References

[1] Sashurin AD, Balek AE, Panzhin AA and Usanov SV 2017 Innovative technology for diagnosis of geodynamic activity in geological media and safety assessment of subsoil use objects *Gornyi Zhurnal* No 12 pp 16–20 DOI: 10.17580/gzh.2017.12.03

[2] Panzhin AA and Panzhina NA 2012 Satellite geodesy-aided geodynamic monitoring in mineral mining in the Urals *Journal of Mining Science* Vol 48 No 6 pp 982–989

[3] Panzhin AA 2012 Space-time geodynamic monitoring at the objects of subsoil use *Gornyi Zhurnal* No 1 pp 39–43

[4] Sashurin AD, Bermukhambetov VA, Panzhin AA, Usanov SV and Bolikov VE 2017 The impact of modern geodynamic movements on the stability of quarry sides *Porbl. Nedropolz.* No 3 pp 38–43 DOI: 10.18454/2313-1586.2017.03.038 Available at: https://trud.igduran.ru/edition/14

[5] Haile A 2004 A reporting framework for geotechnical classification of mining projects *AusIMM Bulletin* No 5 pp 30–37

[6] Castendyk DN and Webster-Brown JG 2007 Sensitivity analysis in pit lake prediction, Martha Mine, New Zealand, 1: Relationship between turnover and input water density *Chemical Geology* Vol 244 pp 42–55

[7] Boehrer B and Schultze M 2006 On the relevance of meromixis in mine pit lakes *Proceedings of 7th International Conference on Acid Rock Drainage–ASMR*, Lexington, Kentuky pp 200–213

[8] Read J and Stacey P 2009 *Guidelines for Open Pit Slope Design* CRC Press, Balkema

[9] Hendrik N 2006 *Converted Wave Seismology for Coal Exploration* Recorder: Canadian Society of Exploration Geophysicists

[10] Deschamps R and Yankey G 2006 Limitations in the back-analysis of strength from failures *Journal of Geotechnical Engineering* Vol 132 No 4 pp 532–536

[11] Zoteev VG *Evaluation of Applicability of Guidelines for Pit Wall Design in Open Pit Mining in Russia* Available at: https://mining-media.ru/ru/article/ogr/

[12] Kahriman A 2004 Analysis of parameters of ground vibration produced from bench blasting at a limestone quarry *Soil Dynamics and Earthquake Engineering* Vol 24 No 11 pp 887–892 DOI: org/10.1016/j.soildyn. 2004.06.018

[13] Kumar R, Choudhury D and Bhargava K 2016 Determination of blast-induced ground vibration equations for rocks using mechanical and geological properties *Journal of Rock Mechanics and Geotechnical Engineering* Vol 8 No 3 pp 341–349 DOI: org/10.1016/j.jrmge.2015.10.009

[14] Bersenev GP, Senin LN and Menshikov PV 2007 Blasting in confined spaces *Vzryv Delo* No 97/54 pp 56–66

[15] Menshikov PV 2007 Impact of the shock wave on buildings and structures on the earth's surface *Vzryv. Delo* No 97/54 pp 77–82

[16] Bogatsky VF and Pergament VKh 1978 *Seismic Safety During Blasting Operations* Moscow: Nedra (in Russian)

[17] Bogatsky VF and Friedman AG 1982 *Protection of Engineering Structures and Environment from the Harmful Effects of Industrial Explosions* Moscow: Nedra (in Russian)

[18] Fadeev AB 1972 *Crushing and Seismic Effects of Explosions in Quarries* Moscow: Nedra (in Russian)

[19] *Seismically Safe Blasting Technology Procedure* Sverdlovsk: IGD MCHM SSSR, 1984

[20] Zharikov SN and Shemenev VG 2013 Impact of blasting on stress state of rock mass and geodynamic phenomena *Izv. vuzov. Gorn Zh.* No 3 pp 90–97

[21] Bondarenko IF, Zharikov SN, Zyryanov IV and Shemenev VG 2017 *Blasting in Kimberlite Open Pits of Yakutia* Yekaterinburg: IGD OrO RAN (in Russian)

[22] Zharikov SN, Shemenev VG and Kutuev VA 2017 Ways to clarify properties of rocks in drilling and blasting *Ust. Razv. Gorn. Territor.* Vol 9 No 1 pp 74–80. DOI:10.21177/1998-4502-2017-9-1-74-80

[23] Zharikov SN and Kutuev VA 2017 The study of the seismic effect of the explosion in the quarry for the introduction of a special technology of work on the formulation of the boards in the limit position *Deep structure, Geodynamics, Thermal Field of the Earth, Interpretation of Geophysical Fields: All-Rus. Conf. Proc.* pp 179–183 (in Russian)

[24] Knyazev DYu and Zharikov SN 2004 Seismic effect of explosions in underground mines *Vzryv. Delo* No 112/69 pp 251–261

[25] Zharikov SN and Kutuev VA Charge initiation schemes to ensure high-performance operation of the cycle link of the cycle-flow technology *Izv. Ural. Goz. Univer.* No 3 pp 76–79

IOP Conf. Series: Earth and Environmental Science **262** (2019) 012085 doi:10.1088/1755-1315/262/1/012085

Mining shovel energy efficiency and its relationship with the explosive fracture of rocks

SN Zharikov* and VA Kutuev

Institute of Mining, Ural Branch, Russian Academy of Sciences, Yekaterinburg, Russia

E-mail: *333vista@mail.ru

Abstract. Production cycles are correlated with the processes of drilling and blasting in rocks. It is shown that energy efficiencies of explosive fracture and shovel operation are interconnected with the face advance velocity. Modeling results on operation per one run of rope power shovels with different bucket capacities are presented. The found approximate relation for the determination of energy efficiency of excavation is based on the interconnection of explosive fracture of rocks, geometry of face and dynamics shoveling.

1. Introduction

Operation of open-cast mining shovels is extensively described in the scientific and technical literature. At the same time, it is yet difficult to determine mining shovel capacity for specific conditions of different open-cast mines based on the practical polices available [1, 2].

The production sequence of an open cast mine features a range of interrelated and interdependent work processes (process stages). Efficiency of every process stage is governed, first and foremost, by the cost/quality ratio of an output product and, for the second thing, by the compliance of the product with the requirements of the optimum technology of the product treatment at the next process stage. The most appropriate index and criterion of the entire production sequence efficiency is assumed an integral energy consumption per unit marketable product [3, 4]. In this fashion, the inaccurate definitions of productivity, energy efficiency or any other criterion introduces errors in the process control systems based on definite laws and interrelations. Practical experience shows that determination of criteria and adjustment of interrelations in the course of open-cast mine operation never lose topicality.

2. Numerical modeling and results

Regarding drill and blast operation, energy interconnections are quite certain [3, 5–7]. The relation between the specific energy characteristics of drilling and explosive fracture of rocks is refined in [7]. It is suggested to characterize energy of rotary drilling process using the index of specific energy per 1 cubic meter of drilling in rocks and energy of explosive fracture by the overall specific work of explosion or specific explosive energy a:

$$E = \frac{Nt}{q_{\text{rock}}}, \quad \text{J/m}^3, \tag{1}$$

where q_{rock}—rock yield per 1 m of blasthole, m^3; t—average time of 1 m of drilling, h; N—motor power consumption, kW.

$$A = q_{pr} Q_{sp} = \frac{Q_{sp} m_{ex}}{V}, \quad \text{J/m}^3, \tag{2}$$

where q_{pr}—project powder factor; Q_{sp}—specific heat of explosion per 1 kg of explosive, MJ/kg; m_{ex}—explosive weight, kg; V—volume of blasted rocks, m^3.

The specific explosive energy is a universe index as it both defines blastability of rocks, conforms with the theoretical description of blasting [8–13] and communicates with the energy inputs in the related processes such as drilling as the latter also spends energy for rock breakage. Setting such relations makes it possible to determine powder factors by energy consumptions in drilling.

Specific energies of drilling and explosive fracture of rocks have the same dimensionality:

$$A = \left[\frac{\text{J}}{\text{m}^3}\right] \leftrightarrow E = \left[\frac{\text{J}}{\text{m}^3}\right]. \tag{3}$$

This means that the specific energies of drilling and explosive fracture per 1 m^3 of rocks are directly proportional. The ratio of such energies for understanding the amount of work (in Joules) in blasting as against the amount of work in drilling per 1 m^3 in a specific type rock is determined experimentally. This ration is different in rocks of different hardness and jointing. The input energy is an indication of the rock hardness and jointing all together. For a certain rock, this expression is given by [7]:

$$\frac{A_i}{E_i} = \Pi_i, \tag{4}$$

where A_i—energy consumption of explosive fracture in i-th type rock; E_i—specific energy of drilling in i-th type rock; Π_i—proportionality factor.

Found in [7], the interrelation represented by expressions (3) and (4) has the specific physical sense as the dimensionality (J/m^3) is a unit of measurement of pressure (Pa). Accordingly, the interrelation between the specified energy characteristics of processes is consistent with the theoretical provisions on physics of rock disintegration. An attempt to set a similar relation with the efficiency of mining shovel reveals some mechanisms described below. In [14] it is show that the overall energy efficiency of mining shovel operation and explosive fracture of rocks lowers as the velocity of face advance is increasing; furthermore, it is supposed that the latter may be an interconnection criterion, including the other related processes. The comparison of the overall energy inputs in excavation on blasting in a model rock block yields similar results for different mining shovel machines (Figure 1).

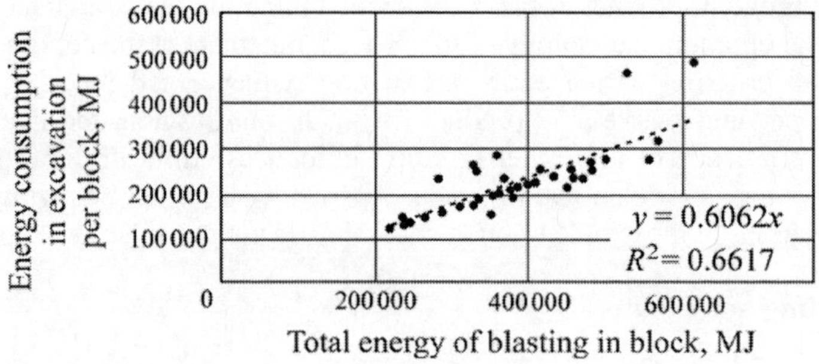

Figure 1. Total energy efficiency of excavation versus total energy of blasting in rock block for mining shovel EKG-10.

The further efforts undertaken to compare energy efficiencies (per 1 m^3) had no success and revealed no clear relation. Probably, this was connected with the geometrical parameters chosen for the modeling. In the meanwhile, it is still of importance that blast energy is spent for rock breakage (mostly uncontrollable) and mining shovel energy is taken by pressing, hoisting and turning the bucket. For this reason, physically, this mechanism is difficult to present on the flat. Nonetheless, the parameter of the face advance velocity is worthy of considering [15]:

$$v = \frac{Q}{S}, \tag{5}$$

where Q—mining shovel capacity; S—face area.

The length of an excavation block was assumed the same in the model experiments and, thus, withdrawn from the overall energy consumption. As a result, specific energies multiplied by face area were interrelated. The face area has the strongest geometrical sense and has influence on the process energy efficiency as it is related with the face advance velocity [14]. Figure 2 depicts the described relationship for mining shovel EKG-10.

All in all, operation of four models of shoveling machines was modeled: EKG-8I, EKG-10, EKG-12.5 and EKG-20s. The motor capacities were calculated in terms of forces of a shovel per one run according to [16]. After interpretation of the modeling results, the relation determined for mining shovels with bucket capacity of 8–20 m^3 provided that the face width is smaller or equal to the shoveling width is given by:

$$E_{\text{sp}}^{\text{ex}} \approx \frac{0.7AQ}{vS}, \tag{6}$$

where $E_{\text{sp}}^{\text{ex}}$—specific energy of excavation, J/m^3; A—specific energy of explosion, J/m^3; Q—mining shovel capacity, m^3/shift; S—face area, m^2; v—project velocity of face advance, m/shift.

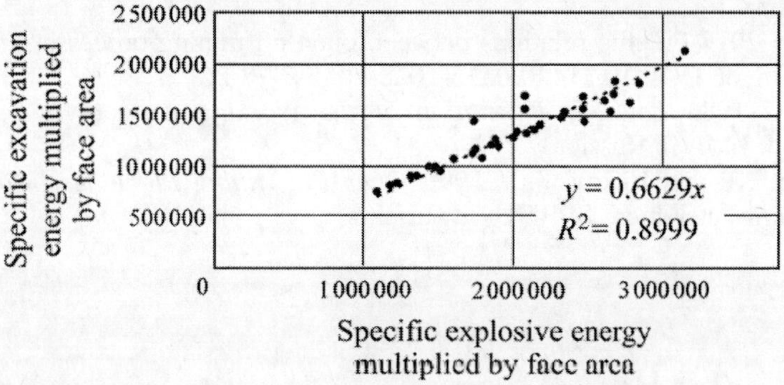

Figure 2. Relationship between the mining shovel energy multiplied by face area and the explosive energy multiplied by face area for mining shovel EKG-10.

3. Conclusions

Knowing specific explosive energy in a rock block and capacity of s mining shovel, with the found face area after blasting and by setting the wanted velocity of the face advance, it is possible to determine energy efficiency of shovel machine operation.

During the research, the approximate relation has been set between the energy efficiency of rope power shovels with bucket capacity of 8–20 m^3 and energy efficiency of blasting with regard to the shoveling machine capacity, face area and the required velocity of face advance.

Acknowledgements

The study was supported in the framework of State contract No. 007-00293-18-00 (Topics Nos. 0405-2018-0015 and 0405-2018-0001, Project No. 18-5-5-10), as well as some economic agreements.

References

[1] Zharikov SN 2017 About features of crawler excavator performance *Izv. Vuzov. Gorn. Zh.* No 1 pp 11–17

[2] Zharikov SN 2018 Improving excavator performance calculation *Zap. Gorn. Inst.* Vol 229 pp 56–61 DOI: 10.25515/PMI.2018.1.56

[3] Zharikov SN and Shemenev VG 2013 Evaluation procedure of energy intensity of technological processes in the drilling and blasting–crushing–grinding cycle in open-cast ore mining, *Gornyi Zhurnal* No 10 pp 83–85
[4] Tangaev IA 2002 The energetics of the processes and systems in open-pit mining and ore preparation Study Guide Bishkek–Moscow: Kyrgyz-Russian Slavic University (in Russian)
[5] Tangaev IA 1986 *The Energy Intensity of the Processes of Extraction and Processing of Minerals* Moscow: Nedra (in Russian)
[6] Tangaev IA 1978 *Drillability and Blastability of Rocks* Moscow: Nedra (in Russian)
[7] Zharikov SN 2011 The relationship of the specific energy characteristics of processes of rolling cutter drilling and explosive destruction of rocks *Dr Sci (Eng) Dissertation* Yekaterinburg 133 p (in Russian)
[8] Orlenko LP 2002 *Physics of Explosion* Moscow: Fizmatlit Vol 1 (in Russian)
[9] Cook MA 1974 *The Science of Industral Explosives* Ireco Chemicals
[10] Pokrovsky GI 1980 *Explosion* Moscow: Nedra (in Russian)
[11] Kornilkov MV 2008 *Fracture of Rocks by Explosion: Lectures*, Yekaterinburg: USMU (in Russian)
[12] Sadovsky MA 2004 *Selectals: Geophysics and Physics of Explosion* VV Adushkin (Ed) Moscow: Nauka (in Russian)
[13] Adushkin VV and Spivak AA 1993 *Geomechanics of Large-Scale Explosions* Moscow: Nedra (in Russian)
[14] Zharikov SN 2017 Finding relations between open pit mining processes *Izv. Vuzov. Gorn. Zh.* No 7 pp 78–86 DOI: 10.21440/0536-1028-2017-7-78-86
[15] Arsentiev AI 1986 *Laws of Formation of the Working Area in a Quarry: Study Guide* Leningrad: LMI (in Russian)
[16] Khmyznikov KP and Lykov YuV 1999 *Open-Pit Mining Machines. Mining Shovels: Study Guide* Saint-Petersburg: SPBGU (in Russian)

Problems and solutions in underground mining of non-ferrous and precious metal deposits in Yakutia

VP Zubkov and DN Petrov*

Institute of Mining of the North, Siberian Branch, Russian Academy of Sciences, Yakutsk, Russia

E-mail: *petrovdn74@mail.ru

Abstract. The article considers the experience gained in underground mining in Yakutia in terms of geological conditions of local ore bodies, as well as the mining methods used and equipment employed. The main factors of permafrost, having positive and negative effects on the safety and efficiency of mining are identified. The research data on deterioration of underground mining conditions during transition to the subpermafrost levels are given. The solutions proposed make it possible to improve the efficiency of underground mining of ore bodies with lower content of commercial minerals, without infrastructure available and in complex geological conditions, including subpermafrost reserves.

1. Introduction

The analysis of underground ore mining in Yakutia shows high variability in types and characteristics of ore bodies as well as in types and characteristics of enclosing rocks. For instance, Protodyakonov index of hardness in host rocks of non-ferrous and precious metals in Yakutia ranges from 3 to 22 [1, 2].

The deposits are mostly represented by ore bodies and lodes in difficult ground conditions—small expanded and necked areas, apophyses and branchings, faults and folds. The ore bodies are from 0.5 to 3.5 m thick and dip at an angle from 20 to 90°.

The local terrain is mostly mountainous which governs accessing the deposits by means of adits or inclined shafts. It often happens that access to ore is gained via exploration adits and other kind tunnels that are sometimes enlarged for the convenient operation of mining machines.

Ore bodies are usually prepared for mining. In the course of the development and face-entry drivage, the information on occurrence conditions and mineral contents is updated. The amount of work in this case is higher than in the standards, as well as the cost of the support and maintenance of underground openings increases [3].

2. Underground ore mining in Yakutia

Subject to geological and geotechnical considerations, Yakutia mines use mining systems with shrinkage, sublevel stoping, with long extraction panels, with frozen backfill, with delayed installation of support and backfilling (Table 1).

Ore is broken using portable jack hammers and drawn from blocks by scraper loading hoists. Drilling involves no dedusting, thus dust content of air in work areas reaches 600–1000 mg/m^3. In mining with long extraction panels, column pillars are left every 20–25 m for the roof support. After completion of stoping in a blocks, scrapers, brooms and a scarper excavator are used for cleaning-up.

Ore is hauled by self-propelling load–haul–dumpers and, sometimes, scarper excavators (in number of 5) to a distance of 100–200 m.

Table 1. Brief characterization of geology for some ore bodies in underground mines in Yakutia.

Mine	Morphology, dip angle and thickness of ore body	Hardness index of ore / enclosing rocks	Stability of ore / enclosing rocks	Mining system	Dilution of ore / enclosing rocks, %
Badran	Ore columns; 21–34°; 0.5–3.6 m	14–16 / from 3–4 to 14	Average stability / unstable	Room-and-pillar with frozen backfill; room-and-pillar with delayed support installation and dry rock fill	(0–3) / (17–21)
Duet	Echelon lodes; 0–45°; 0.2–3 m	12–18 / 10–16	Stable / average stability	Long extraction panels along the strike	(5–11) / (57–63)
Nezhdanins k	Ore bodies with veins; 55–80°; 0.2–12 m	15–16 / 15–16	Unstable / average stability	Stoping with shrinkage in blocks under level drifts	(8–12.5) / (17–19)
Sarylakh	Ore lodes; 60 0°; 0.2–5.2 m	9 / 7	Unstable / stable	Stoping with shrinkage in blocks	(4.6–6.5) / 30.4
Yursky	Small-folded lenticular quartz lodes; 0–80°; 0.2–2.5 m	16–22 / 17–18	Stable / stable	Long extraction panels along the strike	8–12 / (49–63)
Sentachan	Ore lodes; 50–80°; 0.1–6.9 m	8–9 / 7–8	Stable / stable	Stoping with shrinkage	(9–15) / (15–40)

Ore hoisting to the surface uses belt conveyors and electrical haulage, sometimes load–haul-dumpers or scarper excavators. Productivity of labor is under 1.0–1.3 m^3/man/shift.

Underground mining inflicts no damage to the ground surface, as a rule. The main contamination factors are surface infrastructure of underground mines, gangue stock piles and processing plants.

It is worthy of mentioning that the geological exploration data use mine planning and design show not every fine detail of geological occurrence, shape and structure of the ore bodies. As a consequence, it is typical that many mines produce ore with lower content of useful component due to the systematic overvaluation of ore amount and quality in geological appraisal of deposits [4].

Table 2. Key factors of permafrost to affect mining efficiency.

Factors of heat and moisture content in the permafrost ore bodies	
Facilitatory effect	Adversity
Higher load-bearing capacity and stability; No groundwater inflow in stopes; Very low gas content and gas permeability; Low-rate oxidation processes in mine are and surrounding rock mass under negative temperature	Low temperature in stopes worsens health conditions; Impossibility of using water for dust suppression; Congelation after warm air flow in stopes; Adfreezing of rocks after contact with water or warm air; Increased energy consumption by destruction of rocks; Very low electrical conductivity, which impedes equipment ground

All deposits occur in the blanket zone of permafrost. The temperature in rock mass varies from – 4–5 C° (Yursky mine) to – 10–12 C° (Badran mine). Permafrost has both positive and negative effect on different processes in mining, as well as on safety and efficiency of mines (Table 2).

Furthermore, as mining operations go to deeper levels, in the transition zone (between permafrost and positive temperature rock mass), enclosing rocks become instable, and increased gas content, water content and jointing is observed.

For example, in Badran mine, as soon as mining reached lower boundary of permafrost, stopes began suffering from instability, rock falls and local collapses became frequent, support systems were ruptured in development entries. The research accomplished by the Institute of Mining of the North showed that the strength of enclosing rocks mass was affected by higher temperature and moisture content (ultimate tensile strength reduced from 6 to 2 MPa) [5].

In stoping with shrinkage in Sarylakh and Sentanchan mines, high-rate roof falling was observed in adjacent rocks in the transition zone. In this connection, Sentanchan mine undertook studies aimed to revaluate parameters of stopes. According to the research findings, frozen rocks and ore have rather high strengths in compression and tension. However, under thawing, milonitized siltstone in the zones of schistocity becomes 5 times weaker while antimonite ore preserves its strength characteristics [6].

Instability of enclosing rock mass grows with depth and is a feature of all ore deposits in the zone of permafrost [7–9].

The geological deterioration in combination with the disagreement between the geological appraisal and factual amount and quality of ore reserves in the conditions of mining without infrastructure and permanent transportation routes considerably reduce cost efficiency of mining.

One of the ways of improving performance of ore mines is application of systems with cemented backfill that are very popular in the similar geological conditions in Alaska, Canada and Norway [10–12]. Unfortunately, the use of cemented backfill in Yakutia is complicated by the high cost of materials and total absence routine transport services.

The efficient underground mining of ore bodies in Yakutia requires scientifically sound and resource-saving technologies adaptable to the conditions of permafrost. The Institute of Mining of the North, after the integrated research of thermal and moisture conditions, as well as the systems of mining in permafrost, offers some guidelines towards prompt response to variation in geological situation and the desired completeness of ore extraction [13].

For example, if exposed frozen enclosing rocks are stable, it is efficient to use the system of stoping with shrinkage. At the same time, in subpermafrost mining, this system is inefficient as the enclosing rock mass is unstable, and secondary dilution and ore loss increase. In the latter conditions, for the best mine performance, it is suggested to use the system of stoping with shrinkage and a flexible canopy. Based on the findings of lab-sale research and data of full-scale testing in Bakyrchi mine in Kazakhstan and Nezhdaninsk mine in the Republic of Sakha (Yakutia), the Institute of Mining of the North has designed and manufactured a flexible synthetic canopy. The design of the canopy is protected by the author's certificate. The capacities of the put forward alternatives for mining in different geological conditions of ore bodies (dip angles 70–90º, thickness 2.5–5.5 m) were tested in the lab-scale physical modeling of the ore draw. The process solutions developed based on the test results enable ore loss to be decreased by 2 times and ore dilution to be reduced by 5 times as compared with the process of ore drawing without the canopy.

Badran ore mine used the room-and-pillar method of mining, which failed to ensure completeness and quality of gold extraction. The loss and dilution of gold reached 28–30 and 30–35%, respectively. Aiming to reduce gold ore loss and dilution in Badran mine, the Institute of Mining of the North has developed and introduced at a commercial level the system of room-and-pillar mining with the delayed frozen backfill. By now, mining with this system has produced more than 900 thousand tons of gold ore at the loss of 0–3% and dilution of 17–21%. After accomplishment of the integrated theoretical and experimental investigations, the recommendations on the design and implementation of frozen backfilling are made.

IOP Conf. Series: Earth and Environmental Science **262** (2019) 012086 doi:10.1088/1755-1315/262/1/012086

There some ore deposits in Yakutia, for instance Nezhdaninsk gold ore, Sarylakh and Sentachan antimony–gold ore, Upper Menkeche silver–lead–zinc ore, which allow using various systems of mining with caving owing to specific occurrence. The systems with caving feature high productivity and comparatively low cost, which make them profitable in mining ore bodies with low content of useful components given no essential expenditures connected with the ground control [14, 15]. Nonetheless, in order to apply this technology in the conditions of Yakutia, different constraints of underground mining should be analyzed, for instance, possible adfreezing of broken ore in stoping area.

For the assessment of the effect exerted by thermal and moisture conditions on mine performance in permafrost zone, it was studied how the time of holding broken ore prone to adfreezing in stoping area affects the ore loss in case of different moisture content of the ore. The broken ore loss due to adfreezing grows considerably as its moisture content rises from 0.5 to 1.0% at the time of holding the ore in the stoping area for longer than 1 day. Thus, at this stage of the studies, it is possible to handle the problem of ore adfreezing by reducing the time of its holding in stopes.

The determined relationship between the moisture content and loss of ore will be used in development of recommendations on draw technologies of ore prone to adfreezing in the conditions of negative temperatures in underground mine in the permafrost zone.

3. Conclusions

1. The current situation in underground ore mining in Yakutia is characterized with the deterioration of geological conditions, inconsistency of the geological appraisal and actual quality and amount of ore reserves, as well as with total absence of surface infrastructure and permanent transportation routes, which drastically degrades efficiency of conventional underground mining technologies.

2. The variation in the strength and stability of permafrost rock mass depending on temperature is the key cause of reduction in underground mine performance in the permafrost zone. Accordingly, it is required to develop appropriate programs for monitoring changes in strength, stability and temperature of rock mass as mining front goes to deeper levels.

3. The profitability of underground ore mines in Yakutia to be risen calls for a package of measures, including managerial and technical, based on the scientifically sound and resource-saving technologies with regard to specific occurrence conditions of ore bodies in the permafrost zone.

References

[1] Tkach SM 2006 *Methodology and Geotechnology for Imporving Effiiency of Mining Ore Deposits and Placers in Yakutia* Yakutsk: IMZ SO RAN (in Russian)

[2] Tapsiev AP and uskov VA 2014 Development of mining methods for the thawed zone conditions of the Nezhdaninskoe gold deposit *Geo-Sibiria & Expo–2014: Conference Proceedings* Novosibirsk (in Russian)

[3] Kursakin GA 2002 *Technology of Mining Gold Veins* Vladivostok (in Russian)

[4] Romanova ER and Batugina NS 2016 Mining industry in the econiomy of the Republic of Sakha (Yakutia): State-of-the-art, problems and challenges *Gornyi Zhurnal* No 9 pp 23–26

[5] Neobutov GP, Zubkov VP and Petrov DN 2014 Stability of rock exposures in undeground mines in permafrost *Nauchnoe Obozrenie* No 8 pp 941–944

[6] Petrov AN and Nikiforov LA 2011 The experience of undegrorund mining in Santachan gold–antimony deposit *GIAB* Special Issue 10 pp 185–191

[7] Klishin VI and Vlasov VN 2002 Problems of undeground mining of kimberlite deposits in permafrost zone *Permafrost Enginering: V Int. Symp. Proc.* Yakutsk Vol 2 pp 12–15

[8] Pavlov M, Semenov YuM, and Sosnovsky LI 2014 Design of stable pillars and rooms for undeground mining of inclined veins in Irokinda gold mine in the conditions of permafrost *GIAB* No 10 pp 21–27

[9] Iudin MM 2011 Effect of geo-cryology on underground ore mining in the North *GIAB* Special issue 10 pp 97–101

[10] Coil D et al *Gold Mining Methods* Available at: http://www.groundtruthtrekking.org/Issues/MetalsMining/GoldMiningMethods.html
[11] Scales M 2012 One of a kind. Lac des iles palladium mine expands *Canadian Mining Journal* February–March pp 22–24
[12] *The Björkdal Mine. Underground Mining. Horn International* 2015 Available at: http://hornonline.com/the-bjorkdal-mine
[13] Zubkov VP, Neobutov GP and Petrov DN 2017 Enhancing completeness and quality of underground gold ore extraction in the Republic of Sakha (Yakutia) *Gornyi Zhurnal* No 4 pp 53–56
[14] Eevertovsky VM 2014 Systems of undegrround mining with bulk caving at Irgiredmet *Zolotodobycha* No 193 Available at: http://zolotodb.ru / news/11164
[15] Savich IN and Mustafin VI 2015 Prospects and designs of block and sublevel caving *GIAB* Special Issue 1 pp 419–429

AUTHOR INDEX

Institute of Physics
Dirac House, Temple Back
Bristol BS1 6BE UK

ISSN: 1755-1307
ISBN 978-1-5108-8857-9

CURRAN ASSOCIATES INC.
proceedings.com